THE U.S. ATLAS OF NUCLEAR FALLOUT 1951-1970

VOLUME I. TOTAL FALLOUT

ABRIDGED GENERAL READER EDITION

THE U.S. ATLAS OF NUCLEAR FALLOUT

VOLUME I: TOTAL FALLOUT

1951-1970

ABRIDGED GENERAL READER EDITION

RICHARD L. MILLER

LEGIS BOOKS
Div of LEGIS CORP

THE WOODLANDS, TEXAS

LEGIS BOOKS
Division of LEGIS CORP.
P.O. Box 7888
The Woodlands, Texas 77387-7888

Printed in the United States of America

Printing number

1 2 3 4 5 6 7 8 9 10

Library of Congress Cataloging-in-Publication Data

Miller, Richard L. (Richard Lee)
The U.S. Atlas of Nuclear Fallout: Volume I, Total Fallout Abridged General Reader Edition
1. Nuclear weapons–Testing.
2. Radioactive fallout–United States.
3. United States–History.
4. United States–Geography.
5. Science Reference.

ISBN 978-1-881043-13-3 (Softcover Edition)

This edition is dedicated to the memory of my mother, Ruth E. Miller

THE U.S. ATLAS OF NUCLEAR FALLOUT 1951-1970

VOLUMES IN THE SERIES

Volume I: Total Fallout (Abridged General Reader Edition)	ISBN 978-1-881043-13-3
Volume II: County Comparisons	ISBN 978-1-881043-29-4
Volume III: Eastern U.S.	ISBN 978-1-881043-28-7
Volume IV: Western U.S.	ISBN 978-1-881043-27-0
Volume V: Calculations	ISBN 978-1-881043-31-7

REVISION LIST R 2 .0

This marks the first revision of the draft of *The U.S. Atlas of Nuclear Fallout 1951-1962 Volume I: Total Fallout Abridged General Reader Edition.* Revision 1.0 was published in the second quarter, 2002. Revision 2.0 was made available in the second quarter 2003. Changes to Revision 1.0 include:

1. CDC WONDER data incorrectly stated the dates as 1979-1996 for many cancer rates. The actual dates are 1979-1995. The designation was changed throughout the book to reflect this.
2. Some NCI Atlas values were listed incorrectly as 1970-1995. The correction shows the dates as 1970-1994.
3. Revision 2.0 notes that fallout was deposited on North America, not just "America."
4. p. 9: "2 percent of fallout" was clarified to read "2 percent fallout activity."
5. p. 10: Re: Americium-241 remaining after 50 years: the percentage of 99.49% was corrected to 92.294%. Credit to Karl Johanson, as well as for the suggestions to change "element" to the more precise term "isotope" throughout the book.
6. p. 14. "determine" changed to "estimate."
7. p. 15: "a nuclear device is essentially a ball of fissionable radioactive material surrounded by explosive." Karl Johanson correctly pointed out that not all weapons share this configuration. He also noted that not all nuclear weapons are made of (weapons grade) plutonium. Noted and corrected.
8. p. 15 "Neutron capture" changed to "neutron activation."
9. p.17. It was incorrectly stated that there are 39 million microns in one inch. The corrected value is 25,400 microns/inch.
10. p. 17. The diameter of the particle in the illustration is approximately 1 mm, not 0.5 mm.
11. p. 18. X-rays are not emitted by the nucleus, rather they are emitted from the electron cloud surrounding the atom. The correction appears in this revision.
12. p. 23.The designation "Mademooiselle" Marie Curie was incorrect on two levels. First, Ms Curie was married (thus was "Madame"), and secondly, the Curie was also named after her husband Pierre Curie. Noted in this revision.
13. p. 24. Noted that a square meter is equal to 10.76 square feet or equivalent ***to the area*** of a square 3 ft 3 inches on a side.
14. p. 24: "section" changed to "second."
15. p. 29. Kiloton (in terms of nuclear weapon strength) refers to an equivalent energy release from a thousand tons of ***TNT*** rather than dynamite.
16. p. 36. Typo correction: PB7: "not" changed to "no."
17. p. 47. Clarification re GM and GSD. "The greaer the geometric standard deviation, the greater is the uncertainty surrounding the estimated geometric mean (GM) and thus the greater the spread between the minimum and maximum values," changed to "The GSD indicates how tightly clustered the data are. The greater the geometric standard deviation, the greater is the uncertainty surrounding the estimated geometric mean (GM) and thus the greater the spread of the data." Credit Jim Hoerner for pointing out that the minimum and maximum values are secondary to the data spread.
18. p. 57. The map in the first edition incorrectly displayed the trajectory for Ranger Baker. The map in this revision displays the correct trajectory for shot Baker-2.
19. p. 576. "Americium-241is" changed to "Americium-241 is".
20. p. 596: re Sr-90 "half-life of approximately 28 years" changed to "half-life of approximately 29 years."
21. p.598: "Uranium 237 is a fission product of U-238" changed to "Uranium 237 is a product of Neptunium-237.
22. p. 598: It was pointed out that Beta radiation is in fact comprised of electrons. However the Nuclear Data Base at the Brookhaven National Laboratory includes separate categories for Betas, electrons and Augers—all of which are electrons.
23. p. 616: 95" is equivalent to 19 in 20 rather than 1 in 20. Corrected. Credit to Jim Hoerner.

24. pp 161-7: Typos: Capitalization correction, removal of extra period.
25. p. 619: Table I2-2 was changed to Table 11-2.
26. p. 620. "precipitation highly correlated with rainfall" changed to "precipitation highly correlated with fallout." Credit to Jim Hoerner for catching that one.
27. p. 634: "would become part of local watershed" changed to "could become part of local watershed."
28. pp. 658-670. Missing table header for R and maximum T-score corrected.
29. p. 656: IN115m changed to In115m.
30. p. 662: Tb61 changed to TB1 61.
31. p. 731: Typo: extra period removed.
32. Included Volume V: Calculations in the list of volumes in the series.
33. Full justification for all paragraphs.
34. Be7_ changed to Be7_*, W181_ changed to W181*, Zr95_ changed to Zr95_* etc. in tables designating values based on 1 Year postshot.
35. Table B-7: POSSON changed to POISSON.
36. Corrected 96 to 95 for WONDER data in Index.
37. All spacing: Before); After 8 pt. Line spacing: Exactly 14 pt.

REVISION LIST R 3 .0

1. Added fallout values for Hardtack II and the Underground series including Baneberry.
2. The Fallout-by-Series table has been dropped. The book length was limited to approximately 700 pages, so something had to go. Given the other information available in the book, this appeared to be redundant, and thus was cut.
3. New data for Ranger Baker-2 and Tumbler-snapper Easy, Fox and How. Deep-seated errors during the sorting process using Microsoft Excel allowed minor errors to slip into the Ranger Baker-2 data. Other mistakes occurred allowed underreporting of deposition rates for Tumbler-Snapper Easy and Fox. An error in the spreadsheet that calculated total fallout values from I-131 data resulted in a significant overestimate for shot How day 2. This resulted in a day 2 value that was incorrect by an order of magnitude. All of these errors have been addressed and corrected in this edition.
4. The cancer section has been dropped. The cancer-fallout information will be published in its own volume.

CONTENTS

ACKNOWLEDGEMENTS

This book would have been impossible without the dedication and commitment of the scientists of the National Cancer Institute, who produced the original I-131 deposition data. Theirs was the most difficult task of all, and they have my deepest appreciation and respect.

I would also like to thank the following software companies for donating or providing discounted software to support my research: Palisades Corporation (@Risk and Risk Optimizer); Attar Software (Xpert Rule); Decisioneering (Crystal Ball Predictor); Spotfire (Spotfire Pro); MapInfo (MapInfo Pro); Golden Software (Didger); Cytel Software (EGRET); Salford Systems (CART and MARS); and SPSS (Systat). And, of course, a tip of the hat to Microsoft, maker of the only software that should be classified as heavy equipment: Excel 97. Additionally, I would like to express thanks to the makers of Spreadsheet Assistant™ for Excel 97 and 2000. Without it, the work would have taken four times as long to complete.

Thanks to Lynn Anspaugh of the University of Utah for his help and suggestions regarding the Hicks Tables, and to Owen Hoffman of SENES, Oak Ridge, Tennessee, for his assistance, suggestions and patience regarding geometric means, geometric standard deviations, arithmetic means of log normal distributions, and Monte Carlo simulations.

It should be noted that Cliff Honicker performed the original investigations on rainouts in both South Dakota and Minnesota, and also kindly provided me with the AEC reports of investigations. The device specs were provided by the indefatigable nuke researcher Chuck Hansen.

Special thanks to Leah Stratmann, for her excellent work transcribing the 1983-4 NCI Cancer Atlas for the 513 Midwestern counties, and to the fine technical writers Roy Constible and Mary Constible for the many hours spent transcribing the Hicks Tables to Excel. Thirty radionukes and not a single mistake. Also a tip of the hat to my friends Brian Thorne, Walter Hirsch and Wally Cummins for their unflagging encouragement and optimism.

Additionally, I would like to thank the many librarians who spent time with the advance reviews and made suggestions for enhanced readability. They include: Patty Gross, Council Valley (ID) Library, Wendy Bossie, Caribou (ME) Public Library, and Symma Zerkow, Houston (TX) Public Library.

I would also like to thank sci-fi writer extraordinaire Karl Johanson and nuclear maven Jim (Jaime) Hoerner for their excellent copyedit work on what I thought was the final draft. Great job, gentlemen.

Finally, I would like to thank my mother, Ruth E. Miller and my wife, Kim and daughter Trace for their continued interest and encouragement and support throughout the long days and nights working on this project. This book belongs to them.

PREFACE
TO THE
GENERAL READER EDITION

WHY YOU SHOULD READ THIS BOOK

The era of aboveground nuclear testing at the Nevada Test Site began on January 27, 1951 with the detonation of Ranger ABLE. Starting in 1951 and continuing until 1962, the federal government detonated well over a hundred nuclear weapons in aboveground tests. It is well known that radioactive fallout fell on Nevada, Utah, Arizona, Colorado and Idaho. What is not generally known, however, is that the prevailing winds carried massive quantities of radioactive fallout to all of the remaining states, even those as far away as Maine and Florida. No county in any of the continental United States escaped fallout from the Nevada Test Site.

What did the fallout contain? How harmful were the radioactive elements in the fallout? What are the long-term health effects of exposure to radioactive fallout? How can an individual citizen know if he or she has been exposed to fallout—and what is the significance of the exposure?

Until recently, there have been few if any answers to these questions. Now, however, new data has been made available that sheds light on these questions. In fact, most of the above questions have now been answered.

This book uses recently published government data to show which counties in the United States had the heaviest dose of fallout and when that fallout came down. I hope the research in this book is a starting point for further research into what may potentially be a significant health risk.

I also hope the average reader will be able to use this data to learn where he or she was at the time the fallout came down.

SOME HISTORY

For over ten years significant quantities of radioactive materials from nuclear testing were deposited across North America, making their way into the food chain and exposing several generations of Americans to radiation. While the Atomic Energy Commission was aware that a thunderstorm could "wash" nearly all the radioactive material from a nuclear cloud and send it to earth within an hour, there was no effort to evaluate hot spots and rainouts downwind of the Test Site. Additionally, there were relatively few attempts to evaluate potential links between nuclear fallout and disease rates.

In 1981 Harry G. Hicks published tables that allowed researchers to theoretically determine deposition rates of over a hundred fallout components. Two years later, Congress authorized the National Cancer Institute to conduct a study to determine if thyroid cancer in the population could have been caused by radioactive iodine in fallout.

The NCI was given only *one year* to complete the study. It took them 14 years.

THE 1983-1997 NATIONAL CANCER INSTITUTE I-131 STUDY

On October 4, 1997 the National Cancer Institute announced that the radioactive iodine report was finally ready.

Data for each nuclear test, by fallout day was published for each and every county in the United States. Some counties in Utah, Nevada, Arizona and California were even divided into sub-counties based on geographical differences.

The report in its entirety ran to over 120,000 pages. Perhaps for that reason, the authors of the study decided against publishing the results in book form. Rather, the information was posted on the Internet on the National Cancer Institute Website under the heading "What's New."

Some of the pages had a rushed quality to them: some map titles were written in pencil and the cloud trajectories appeared more rudimentary than the originals from 1953 and earlier. County lists included some places, such as Washabaugh County, SD, that no longer existed—or, in the case of Columbus County GA, *never* existed (the NCI undoubtedly meant Muscogee County, which is the county surrounding the city of Columbus.)

Worse, for anyone trying to analyze the data, the alphabetical order for counties seemed to vary with the shot. In some tables Arizona came before Arkansas; in others the order was reversed. For many of the tests, the NCI tracked days of fallout beginning with day one (the detonation date) to day ten. But for one notoriously fallout-intensive shot, Buster-Jangle Easy, the fallout tracking began on day *three*. Others, like the subsurface test Buster-Jangle Uncle, were apparently only tracked for one day.

Several nuclear tests were combined, even though the events surrounding the detonation and subsequent cloud paths didn't really warrant it. One important nuclear test, shot Stokes in 1957, was listed as having produced *no* fallout, even though internal Atomic Energy Commission memorandums strongly suggested otherwise.

There were problems in deciphering the fallout deposition rates. The data included obscure and poorly documented statistical terms (more about that later), was difficult to download and nearly impossible to evaluate once downloaded.

From a larger perspective, the study had an enormous shortcoming: it included data for only *one* component of fallout, a form of radioactive iodine, I-131. This short-lived component of fallout accounted for only ***two percent*** of the total fallout activity deposited across the country.

Still, it was an extraordinary piece of work. In addition to county deposition of radioactive iodine (I-131), the authors included information about population exposure based on milk consumption. A reader (or more appropriately, an Internet surfer) had only to type in her birth date, county data and milk consumption information to determine who many rads her thyroid received from fallout. Unfortunately, the most important piece of information was left out: the National Cancer Institute never discussed whether exposures to radioactive materials in fallout could cause cancer.

WHY FALLOUT IN THE 1950S MATTERS IN THE 21ST CENTURY

Because of fallout from the Nevada Test Site, radioactive elements called radionuclides were deposited in every county of the United States. Many of these radionuclides became incorporated into living cells. It is known that atoms of strontium-90 and ruthenium-103 made their way to bones and teeth. Other radionuclides such as iron-59, copper-64 and rubidium-88 may have been incorporated into important enzymes.

While all of these radionuclides emitted radiation of one form or another, many, such as arsenic, beryllium, cadmium, curium, cobalt, rhodium, tin, uranium and yttrium are either suspected or recognized carcinogens.

Others, such as indium, lead, manganese, molybdenum, selenium, tellurium, and tungsten are suspected teratogens, capable of causing birth defects.

All of these elements, in their various radioactive forms (radioisotopes), were significant components of fallout.

Moreover, measurable quantities of many radionuclides can remain in the environment in significant quantities *fifty years* after detonation of the device that produced them. These radionuclides include manganese-54, cobalt-60, americium-241, curium-242, strontium-90, yttrium-90, cesium-137, barium-137m, promethium-147 and europium-155. In other words, radionuclides deposited in 1957 will still be with us in 2007. As a somewhat extreme example, consider the toxic fallout component americium-241: The amount of this isotope remaining after 50 years from just one shot—Teapot Apple II in 1955—is fully 92.294 per cent of the amount produced *at the time of detonation.*

This influx of radionuclides into the environment from the Nevada Test Site continued for over 12 years. While global fallout, due in part to dilution effects, painted the environment in broad-brush strokes, Nevada Test Site fallout was anything but uniform. Some counties received thousands of times the fallout as other counties. In addition, there is strong evidence that rainstorms were instrumental in concentrating the fallout in extremely small areas *within* county boundaries.

Despite this, there has *never* been a scientific study for the entire United States evaluating cancer rates or birth defects against deposition of total fallout. Moreover, there has *never* been an epidemiological investigation of a cancer cluster outside of Utah and Nevada, which has included nuclear fallout as a potential factor.

Did fallout in the 1950's and 1960's cause cancer? Given the extremely long latency period associated with many cancers, perhaps a better question is: did fallout in the 1950's and 1960's cause cancers in the 1990's?

After 50 years, there have been no government, university, or even foundation-sponsored studies to answer this basic question.

Hopefully this series of books will be a step toward that answer.

INTRODUCTION

In January 1983, the 97th Congress directed the National Institute of Health to evaluate the extent and effects of the fallout component iodine-131 (I-131) fallout resulting from the Nevada nuclear tests conducted in the 1950s and early 1960s.[1 2]

The task was finally completed and published in 1997. The report consisted of I-131 fallout deposition rates from nearly 60 aboveground and subsurface nuclear tests—for each of 3098 counties. The finished product included data for up to 20 fallout days subsequent to the shot date. All together, the report was equivalent to more than 120,000 pages of data.

Officials at the National Cancer Institute decided that publication of such a mass of data would be feasible only over the Internet. Interested citizens could download data for their own home counties and learn just how much I-131 was deposited from each test, and when. Not only that, they could enter their birth date and estimated milk consumption and read an estimate of the radiation exposure to their thyroid.

There were some problems, however. While some areas had a dense network of detection sites—there were 65 in New York state shortly after the Ranger detonations—the fixed radioactivity detection stations numbered only 97. Thus, there was a measure of uncertainty regarding fallout amounts when the county was a significant distance from the detection site. So, instead of a single value for fallout, say, 50 units of fallout per square foot (or in this case, square meter), the data came out more like this: "the value was somewhere between 5 units of fallout per square foot and 100 units of fallout per square foot."

One way of presenting this information would have been using minimums and maximums: For example, the above information might have read:

Minimum: 5 units of fallout per square foot.

Maximum: 100 units of fallout per square foot.

The explanation would have sounded something like this: "Because back during the 1950s the fallout was collected at only one place—the weather station—in your county, we don't really know exactly how much fallout came down in **your**

county. But we're 95 percent sure that the number was somewhere between 5 units per square foot and 100 units per square foot."

The authors didn't do that. Instead, the information presented over the Internet was given in terms of ***GM*** and ***GSD*** with no explanation of what ***GM*** and ***GSD*** actually meant. In practical terms, the information was limited to those individuals with (1) access to the Internet, (2) an ability to download and translate the data, (3) an education or experience that included graduate-level statistics, and (4) access to a spreadsheet and an ability to perform the complicated mathematics necessary to make sense of the information.

The GM actually meant ***Geometric Mean*** and the GSD referred to ***Geometric Standard Deviation.*** These were statistical terms that are rarely encountered even in college-level statistics courses. Essentially, the geometric mean was a measure of the average fallout deposition, while the GSD was a measure of the uncertainty surrounding that average. For a more detailed discussion of these terms, the reader may skip to the technical section at the end of this Introduction. However, a detailed knowledge of these terms is not a requirement for an understanding of the maps and tables included in this book.

I-131 AND TOTAL FALLOUT

The published NCI data included values for only a single radioactive component of fallout—one of the radioactive isotopes of the element iodine. This particular radioactive element (called a ***radioisotope*** or ***radionuclide***), makes up only two percent of the total fallout.

> **Fact:** Fallout from nuclear tests is made up of 150-200 radioactive isotopes called ***radionuclides.*** Some of these radionuclides include radioactive versions of common elements such as iron, copper, cobalt and sodium. One variety of radioactive iodine, I-131 typically makes up **two percent** of the total fallout.

Since radioactive iodine 131 makes up two percent of the total, one might think it possible to calculate total fallout simply by multiplying the I-131 amounts by 50. Actually, it is not as easy as that.

For one thing, each atomic test produced its own mix of fallout. One atomic test might produce very low levels of the radioactive isotope americium-241, while another test might produce much greater amounts.

Worse, radioactive versions of elements—*radioisotopes* or *radionuclides*-- have a unique characteristic that sets them apart from ordinary non-radioactive isotopes: they eventually change, over time, to *other* isotopes. The radioactive isotope of beryllium can change to an isotope of *lithium*. The radioactive isotope, lanthanum-144 changes or decays to an isotope of another element, cerium-144, which in turn decays to yet a different element, praseodymium-144. Each of these elements has a particular chemical and physical characteristic.

FACT: The number following the isotope symbol (I-***131***, for example) refers to the number of protons in the isotope. This number is unique for each radioactive isotope. All atoms of iodine-131 (I-131) for example, decay at a different rate and produce *different* amounts of radiation than do all atoms of iodine-***130*** (I-130).

FACT: Each nuclear test is unique in the **type** and **amount** of radioactive isotopes found in its fallout.

What was needed was some way to determine what ***kinds*** of radioactive isotopes were produced for *each particular atomic test*. Not only that, it was essential to know how much of *each* radioactive isotope was produced in *each* of the aboveground atomic tests.

Interestingly, that work had actually been done—and published in 1981— by a Livermore physicist named Harry G. Hicks. Through experimentation and long hours on a mainframe computer, Dr. Hicks calculated ratios for more than 125 radioactive isotopes produced by *every* aboveground test ever detonated at the Nevada Test Site—and even a few for the underground shots.

The work came to be called the Hicks Tables and, importantly, I-131 was among the radionuclides listed. With a knowledge of the amounts of radioiodine-131 levels produced by each atomic test—and a little mathematics—it was possible to finally estimate how much fallout each atomic test actually deposited on each of 3098 U.S. counties.

NAMING CONVENTIONS

In the 1950s they were called "atomic tests". In the 1960s and later, they were referred to as "nuclear tests." Some in the weapons community simply referred to them as "shots" or "events." All terms are synonymous with a detonation of a device that produces radioactive fallout. In most of the tables, this detonation is referred to simply as **shot**, followed by the name of the test or "event."

It is important to remember that all of the nuclear tests associated with this book refer to detonations that occurred at or very near the Nevada Nuclear Test Site in Nye County, Nevada. Pacific tests are not included, nor is the one that took place in a salt mine outside of Hattiesburg, Mississippi.

Each series of nuclear tests (or atomic tests, or shots—again, they all refer to the same thing) was given a name. The first series of tests in Nevada was termed ***Ranger***. The first nuclear test in the ranger series was named ***Able.*** It was detonated on January 27, 1951. The second test in the Ranger series was ***Baker***. Those familiar with military code designations for alphabet (Able, Baker, Charlie, Dog, Echo, Fox, Golf, etc) would probably have expected the third shot in the Ranger series to be named "Charlie." Instead, the third test in the Ranger series was named ***Baker-2.***

FACT: There were 6 major series of aboveground nuclear tests at the Nevada Test Site. They were:

1. Ranger	Jan 27-Feb 6 1951
2. Buster-Jangle	Oct 22-Nov 29 1951
3. Tumbler-Snapper	Apr 1-Jun 5, 1952
4. Upshot-Knothole	Mar 17-Jun 4 1953
5. Teapot	Feb 18-May 15, 1955
6. Plumbbob	May 28-Oct 7, 1957

The final nuclear test covered by the book is shot Sedan, which was technically an underground test detonated on July 6, 1962.

Through the first, second and third series, the detonation names generally followed the A, B, C, D—pattern. The eighth test in the Tumbler-Snapper series in 1952, shot ***How***, had a name that began with H—the eighth letter of the alphabet.

By 1953 the nuclear test names had become more personalized: **Annie, Nancy, Dixie** and **Simon** were all names of nuclear tests in this series. The 1955 nuclear test series included shot names such as **Turk, Hornet, Wasp, Bee** and **Ess. Shot Ess** apparently was so-named because it was detonated some sixty feet below the *surface* of the desert.

The final official aboveground series included nuclear tests named for scientists important in the history of physics. There were tests named **Boltzmann, Doppler, Kepler, Stokes, Fizeau,** and of course, **Newton.**

And some shots were named for both scientists and geological formations. While shot **Wheeler** could have easily been named for physicist John Wheeler, there also happens to be a mountain in New Mexico named Wheeler Peak. And, as luck or fate would have it, the nuclear cloud from this same test drifted over Wheeler, Texas.

In this book, the nuclear tests are generally referred to first by the series, then by the specific name in **all caps**, as in Upshot-Knothole HARRY. Abbreviated, the shot may be listed with the series name abbreviated (**R** =**R**anger, **BJ** = **B**uster-**J**angle; **TS** = **T**umbler **S**napper; **UK** = **U**pshot-**K**nothole; **TP** = **T**ea**p**ot, **PB** = **P**lumb**b**ob) followed by the sequence number of the shot in that series.

For example, shot HARRY was the ninth listed by the National Cancer Institute's I-131 survey in the Upshot-Knothole Series. Thus, this nuclear test can also be designated simply as ***UK9***. In some volumes in this series, I have included an additional bit of information—the day the fallout occurred relative to the day of the actual detonation.

In this case, the symbol **UK9-3** would refer to the *third* day of fallout resulting from the *ninth* shot in the **U**pshot-**K**nothole series, as listed by the National Cancer Institute.

Where space permits, shots are listed by the series name or abbreviation, followed by the shot name. Importantly some nuclear tests were *not* included in the NCI study. Others, such as Teapot BEE and Teapot ESS were combined into one. Also, some tests, such as Ranger-FOX (R-3) and Plumbbob STOKES (PB-7), which were associated with no fallout (according to the 1997 NCI study), were excluded from some lists and maps. A complete list of abbreviations used in the book is found at the end of this section.

MORE ON SHOT DAYS

The term "shot day" refers to a particular day after detonation. For example, TS1-1 refers to the *day of detonation* of the ***first*** shot in the Tumbler-Snapper series. That test was formally named Tumbler-Snapper ABLE. In this case, the day that TS-1 refers to is April 1, 1952. The term **TS1-2** refers to one day *after* detonation of Tumbler-Snapper 1—or April 2, 1952. TS1-3 refers to the second day after the detonation. Another way to look at it: the 2 in TS1-2 refers to the number of days the fallout has been in the environment, including the day of detonation. In this case, since TS1 refers to shot Tumbler-Snapper ABLE, and since that detonation occurred on April 1, 1952 (see above), TS1-3 refers to April 3, 1952.

TS3-5

5 refers to the number of days *after* detonation.

3 refers to the *third* shot in the *Tumbler-Snapper* series.

TS refers to the shot series, *Tumbler-Snapper* The different shot series include: Ranger (**R**), Buster Jangle (**BJ**), Tumbler-Snapper (**TS**), Upshot-Knothole (**UK**), Teapot (**TP**) and Plumbbob (**PB**).

WHAT *IS* NUCLEAR FALLOUT?

A nuclear device generally consists of a ball of fissionable material surrounded by explosives. This fissionable material is usually some form of weapons-grade plutonium or enriched uranium. Sometimes, another radioactive material, tritium, is added to the mix to make the blast more powerful. Other elements in nuclear devices might include americium-241 as part of the neutron source and beryllium to bounce neutrons coming from the plutonium *back* to the plutonium. Additionally, there are elements such as steel, carbon and copper that make up the structural components of the device. When the plutonium sphere fissions, the plutonium or uranium atoms may fracture into many different kinds of radioactive isotopes, from beryllium-7 to terbium-161. In all, the resultant mix may 200 different radioisotopes. Other structures that happened to be near the device can become radioactive through a process called neutron activation.

These structures could include the shot cabin or "shot cab", the copper electrical cables, the sodium and magnesium in the sand beneath the fireball, and of course, the iron shot tower—which weighs about a ton per vertical foot.

Most of these things will be vaporized by the fireball, only to re-form as tiny spheres ranging in size from a micron to maybe several thousand microns. There are 25,400 microns in an inch.

FACT: Fallout particles that are carried by the wind from the site of a nuclear detonation may be as large as several thousand microns, to as small as 1 micron in diameter. By way of comparison:

1. One inch is equal to 25,400 microns.
2. The droplets making up a summer cumulous cloud have an average diameter of 11.2 microns.

(From: Observed Microstructure of Warm Clouds. http://rams.atmos.colostate.edu/AT620/chapter6.)

Interestingly, most of the fallout resulting from a high altitude burst are sphere-shaped, while fallout from tower or low air bursts can consist of both spherical and irregularly-shaped particles. Those detonated *underground* produce fallout particles that are generally irregular in shape.[3]

Fallout particles large enough to *have* color can be colorless, gold-yellow, orange, red, brown, green or black. Some may be magnetic.[4]

Close-up, a fallout particle may look like a glossy, brown marble, or a jagged, irregular piece of rock.

FIG 3.0 FALLOUT FROM AN NTS TOWER SHOT

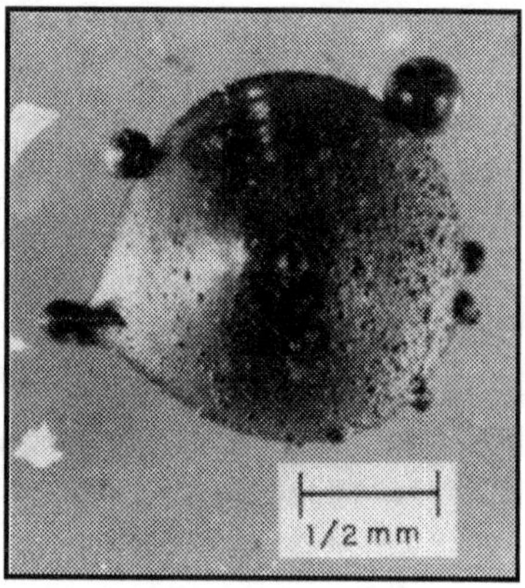

Figure 3.0 shows a photo of a 1 mm (about four-hundredths of an inch in diameter) sphere of fallout taken from one of the tower shots. Note the smaller spheres sticking to its surface.

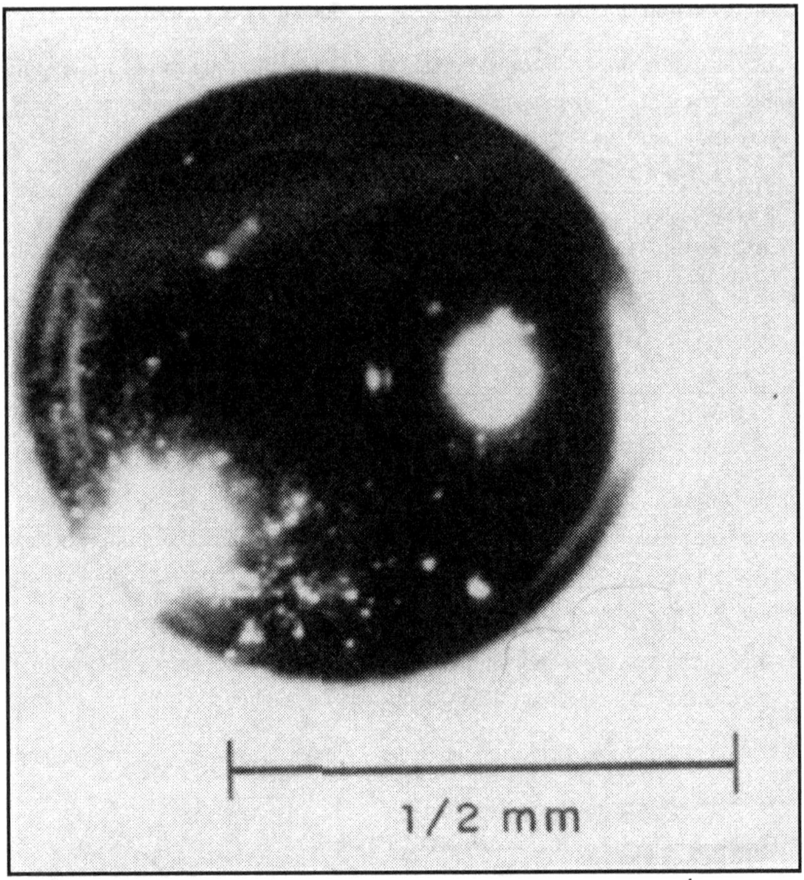

FALLOUT PARTICLE FROM AN NTS TOWER SHOT (2).[1]

Figure 4 is of a different kind of fallout particle with a brilliant, glossy surface. It was also the result of a tower shot.

WHY FALLOUT IS RADIOACTIVE

Fallout is radioactive because the elements that make up the fallout emit radiation. This radiation may be in the form of alpha particles, beta particles, low-speed electrons, gamma rays or x-rays. Alpha particles consist of a packet of two protons and two neutrons. Beta particles are essentially electrons.

X-rays are emitted from the electron cloud while the more energetic gamma rays come from the nucleus.[6]

> **FACT:** The frequency associated with normal household current is 60 cycles per second. That is, it reverses sixty times a second. The frequency of a **gamma ray** or **X-ray** can be as high as 1 x 10^{19} cycles per second—or 167,000,000,000,000 times that of household current.

Each time radiation—either alpha, beta, gamma, low-energy electrons or x-rays are emitted by a radioactive atom, that is a sign that the atom has changed from one isotope to another. For example, a radioactive atom of strontium may emit radiation and, in the process, become another isotope, radioactive yttrium. The yttrium atom might then emit radiation and become yet another element, non-radioactive (stable) zirconium.

It's important to remember that each radioactive element or radionuclide (also called radioisotope) has its own particular radiation emission rate. In the above example, radioactive strontium-90 has a half-life of 28.78 **years**, while yttrium-90—the isotope that strontium-90 changes to has a half-life of 64 hours—as it changes to zirconium-90 which is stable, and doesn't decay or emit radiation.

Another thing: each particular radioactive isotope emits its own unique type and energy of radiation. The radioactive element americium-241, for example, emits an alpha particle, strontium-90 emits only a low-energy beta particle, while cobalt-60 produces beta particles, gamma rays *and* X-rays.

Since each nuclear test produces a unique blend of radioactive elements, trying to calculate the total radiation from such a changing mixture of elements is not particularly easy.

> **THE FOLLOWING TECHNICAL SECTION MAY BE SKIPPED WITHOUT LOSS OF CONTINUITY.**

ATOMIC MASS UNITS

In the above example we used the names of elements followed by numbers to represent the various radioactive isotopes. In physics and chemistry, each element is represented by a specific symbol. Strontium is represented by Sr, yttrium by Y, cobalt is **Co**, iron is **Fe**, nickel is **Ni**, americium is Am, and so on.

You may remember that each atom is made up primarily of three objects—neutrons, protons and electrons. The standard atom is often visualized as a sphere, with the neutrons and protons packed together at the center of the sphere—at the core—and with the electrons forming the outer layers. The configuration of the electrons determines the chemical characteristics of the atom—that is, how it combines with other atoms to form molecules.

The configuration of the inner core of neutrons and protons, the ***nucleus***, determines how the atom acts physically. That is, whether it will emit radiation and change to another isotope of the same element or an isotope of a different element.

Scientists have long known that the number of protons determines the type of element. The element carbon has six protons in its nucleus, and therefore is assigned a specific ***atomic number*** on the chart of the elements: 6. Nitrogen has seven protons and is thus assigned the number 7 on the chart of the elements. The more protons an atom has, the heavier it is. Iron has 54 protons and lead has 82. Plutonium, an extremely heavy atom, has an atomic number of 94.

As it turns out, protons carry a positive charge, neutrons carry no charge, and electrons carry a negative charge. **In order to balance out the charges in an atom, the number of electrons should equal the number of protons**. Since the configuration of electrons determine the chemical characteristics of the atom, and since the number of protons actually determines the number of electrons that the atom has available, *it is really the number of protons that determines how an atom will act chemically.*

Scientists have long known that in the single, isolated atom the number of electrons equals the number of protons. And while the number of electrons might change from molecule to molecule, the number of protons remained fixed. This was why the ancient alchemists had such poor luck in "transmuting" lead to gold. They would have to somehow remove three protons from each atom of lead (82 protons per atom) to get atoms of gold (79 protons per atom.)

Scientists examining atoms originally surmised that the number of neutrons—the subatomic objects with no charge—equaled the number of protons. So, if an atom had, say 6 protons (which would make it the element carbon), then it also probably had 6 neutrons.

But early in the 20th century, scientists discovered, that while the number of protons per atom was fixed, **the number of neutrons might vary from atom to atom, even those with the same number of protons.** Since a neutron has no charge, a change in the number of neutrons wouldn't affect the number of electrons and thus the chemical characteristics of the atom. But the fact remained that some atoms of a given element were heavier than others, due to the extra neutrons. Atoms of the same element--having the same number of protons-- but differing numbers of neutrons are called ***isotopes*** of each other.

To differentiate between the atoms of an element, scientists decided to create the term **atomic mass unit.** The concept was simple: assume that the protons and neutrons **each weigh one atomic mass unit** (designated a.m.u.) Thus, if the sum of neutrons and protons in an atom of carbon, equaled, say, 14, then the isotope was assigned the name *carbon-14.*

As another example, the element cobalt has an atomic number of 27—which means simply that it has 27 protons in its nucleus. **There can be no other element with 27 protons in its nucleus. If there are 27 protons in the atom then it must be an atom of cobalt.** Now, if someone were to count the number of neutrons in an atom of cobalt, they would likely find 32 of them—on average. For reasons that go beyond the scope of this book, an atom where the number of neutrons is ***significantly greater or less than*** the number of protons tends to be unstable. As it turns out the cobalt atom is most stable when the mass number equals 59—that is, when there are 32 neutrons combined with 27 protons.

But there are also cobalt atoms with *more* than 32 neutrons. Some of the cobalt produced as a result of a nuclear detonation might have **33** neutrons—making it relatively ***unstable*** (compared with cobalt having only 32 neutrons. The unstable isotope of cobalt will tend to change to another more stable isotope —and in its "path to stability," emits radiation.

In our unstable cobalt example, we sum the atomic mass units for the neutrons and protons and find the mass equals **60**. In other words, when the atom of cobalt has 27 protons and 33 neutrons, then scientists define it as cobalt-60. Using symbols, the term would be written simply **Co-60** or occasionally, more in keeping with the older conventions in physics and chemistry, Co^{60}. A somewhat more modern method, one used in the 1997 National Cancer Institute report, for example, is ^{60}**Co**.

NOTE: The use of Atomic Mass Units *is simply a way for scientists to differentiate between atoms of the same element that have differing numbers of neutrons.*

For example, iodine-130 and iodine-131 both have the **same** number of protons (which is why they are both iodine.) However, iodine-131 has ***one more*** neutron than iodine-130, and as a result behaves differently with regard to the type of radiation emitted.

The number following the symbol refers to the sum of the neutrons and protons in atomic mass units. The important thing to remember is that the radioisotope we are talking about is defined by the symbol (i.e. **Co** for *cobalt*) plus the mass unit number (for example **60**—which means there are 27 protons and 33 neutrons in the nucleus). Taken together, **Co + 60 = Co^{60}** (or ^{60}Co.)

Atoms with unstable configurations of neutrons and protons eventually change to other, more stable isotopes. This process is called ***decay***. It's not the same as what we normally speak of when we talk about decay of organic materials. But it does mean that the atom is losing energy and is transforming itself from a *less* stable configuration to one that is *more* stable. Eventually, in fact, the atom will reach a point where the neutron-proton ratio is at its most stable configuration. It is at this point that the isotope is no longer radioactive.

NOTE: Given two atoms, if the symbols are the same, then each of the two atoms share the same chemical properties. That is, they each react with other chemicals in the same way.

Both **iodine-130** and **iodine-131** are iodine. *But because they have different numbers of neutrons* **iodine-130** and **iodine-131** are represent two ***different isotopes*** of iodine. *And* they act differently with regard to the radiation they emit.

Expressed another way, If two radioisotopes each have the same number of protons but differing numbers of neutrons, then their chemical properties of these two radioisotopes will be the similar, but their radiological properties will be different.

CURIES AND NANOCURIES

A radioactive element (***radionuclide, radioisotope***) that decays rapidly and thus emits more bursts of radiation per second is said to have a ***higher activity*** than a substance that decays slowly, emitting fewer bursts of radiation per second.

Scientists have invented a number of units to compare activities of radionuclides. One such unit cited in the 1997 National Cancer Institute Fallout study—and the one used in this book—is the Curie. The Curie is named after Pierre and Marie Curie, the French scientists who made numerous important discoveries about radiation. A radioactive substance is considered to be associated with Curie of activity if the radioactive atoms in the substances are decaying at a rate of 37,000,000,000 atoms per second. For those readers familiar with scientific notation, that's 3.7×10^{10} atoms per second (3.7 followed by ten zeroes is equal to 37 followed by nine zeros.)

FACT: **Scientific notation** is occasionally used in this book. Scientific notation is a shorthand way of writing very large or very small numbers. Suppose we wanted to write 37,123,456,789 in scientific notation.

1. First we round the 37,123,456,789 to whatever place we would like. Here, we will round down to simply **37,000,000,000.**
2. Now, we express the first two numbers as a value between 1 and 9.9, here **3.7.**
3. Finally, we divide 37,000,000,000 by 3.7 and get 10,000,000,000. In other words, 3.7 x 10,000,000,000 = 37,000,000,000.
4. Count the zeroes: 10.
5. Now we can write the second part of the number, that tells us by how many factors of ten we must multiply the base number (here, 3.7) to make it equal to the original number (37,000,000,000).
6. The answer is 10, or 3.7×10^{10}. The same value can be written as 3.7E+10.
7. If the value is 3.7×10^{-10} or 3.7E-10, this means it is very small. In decimals, the same value is equal to **0.00000000037.**

FACT: Another term of activity similar to a Curie is the **becquerel** (bq), named after French scientist Henri Becquerel. One bq is equivalent to **one disintegration or nuclear transformation per second**.

It must be noted that one Curie—about 37,000,000,000 disintegrations per second—represents significant activity. Anyone standing unprotected next to a vial of unshielded radioactive material producing one Curie of activity would likely be exposed to a lethal amount of radiation.

While the aboveground nuclear tests in the 1950s released fallout into the atmosphere equal to *millions* of Curies of activity, the fallout was dispersed so that by the time the tiny spheres of fallout reached the ground the overall activity per square foot was on the order of a ***billionth*** of a Curie. When something is a billionth of something else, scientists describe it using the term ***nano.*** Thus, a billionth of a Curie—equal to 37 bursts of radiation per second—is called a **nanoCurie.** The abbreviation for nano Curie is **nCi.**

Incidentally, one millionth of a Curie is a micro-Curie, abbreviated **μCi**. A **microCurie** is equal to a thousand nano-Curies or 37,000 bursts of radiation per second. On individual fallout days most counties in the United States received fallout in the **nanoCurie** amounts. Summed over the years from 1951 to 1962, the amounts are generally in the **microCurie** range. Thus, for some tables, you will see fallout given in nCi/sq meter and for other tables you will see fallout given in μCi/sq meter.

Again, the important thing to remember is that ***dispersion of a 1000-ft diameter nuclear cloud over thousands of cubic miles of air will result in a dilution of activity.*** As a consequence, the radioactivity of the fallout, when it finally fell to earth, was measured not in Curies, but in millionths and billionths of a Curie: microCuries (μCi/sq meter) and nanoCuries per square meter (nCi/sq meter)

CURIES, MICROCURIES, NANOCURIES AND BECQUERELS

In 1981 the term Curie was officially superceded by the term becquerel. However, the NCI data was reported in nanoCuries. To avoid confusion, the US Fallout Atlas series uses the Curie term rather than becquerel.

TO RECAP:
The fallout values used for the maps in **Sections 1-7** and for the Table in Section 9 are given in *nanoCuries.* A nanoCurie represents an activity level equal to one thousandth of a microCurie. **NanoCuries are represented in this book by the symbol nCi, and represent an activity equal to 37 disintegrations per second—which is equal to 37 becquerels.**

One becquerel (**bq**, sometimes seen as **Beq**) is equal to one (1) disintegration or transformation per second. Thirty-seven bq is equal to one nanoCurie---or 37 disintegrations (and concurrent emissions of radiation) per second.

In **Section 8**, the fallout values for counties with the highest levels of fallout are expressed in terms of **microCuries.** You will see the term microCurie represented by the symbol **µCi.** ***One microCurie is defined as radiological decay equal to 37,000 disintegrations per second.*** (37,000,000,000 divided by 1,000,000).

OTHER RECENT RADIOLOGICAL TERMS: GRAYS AND SIEVERTS

Other more recent radiological terms not used in this book include **Grays** and **Sieverts**. A **Gray (Gy)** represents an **absorbed dose** of 1 joule per kilogram. A **Sievert (Sv)** represents an ***equivalent*** **absorbed dose** in joules per kilogram of an organ or tissue. Mathematical weighting factors account for the type of radiation and the susceptibility of the organ absorbing the radiation.

SQUARE METERS AND SQUARE YARDS

The activity discussed earlier is given in terms of square meters. A square meter, abbreviated either ***sq meter*** or ***m²*** simply refers to an area equivalent to a square that is 39.37 inches on a side. In other words, a square meter is slightly bigger in area (about 1.2 times) than a square yard.

So, when you see the fallout level given as 10 nCi/sq meter (10 nanoCuries per square meter) you will know that this means there were 10 x 37 = 370 bursts of radiation per second from a plot of land about 1.2 times as large as a square yard.

FACT: A square meter is slightly larger in area than a square yard, and a little over ten times as large in area as a square foot.
1 square meter = 1.196 square yards
1 square meter = 10.76 square feet

The 1997 National Cancer Institute study gave fallout levels in terms of nanocuries per square meter (nCi/sq meter). Again, a square meter is equal to 10.76 square feet, equivalent in area to a square that is 3 ft. 3 inches on a side. Thus, if you read that fallout from a particular shot produced an average of 1 nCi/sq meter on a given county on a particular day, you may reasonably conclude that for each square plot of land 3.3 ft on a side, the fallout produced 37 **units** of radiation.

Why just units of activity? Because, while we know how fast the various mix of elements is changing (the activity level expressed in billionths of a Curie) we still don't know what particular radioactive isotopes are in that mix. Not yet, anyway. And since we don't know what radioisotopes are in the fallout mix, we are not able to say what forms of radiation are produced. (Thanks to some mathematics, and a closer look at some important research by government scientists we will eventually have that information, but that is for Volume II: Radionuclides.)

NOTE: Before one can determine what kinds of radiation are being emitted from nuclear fallout, one must **first** determine which radioactive isotopes are found *in* the fallout.

The subject of total fallout forms the basis for the first volume. Subsequent volumes characterize the fallout patterns by radionuclide and compare additional fallout values for each county.

Hopefully, it will provide a new perspective on an important part of our history and will stimulate further research in the area of nuclear fallout.

ORGANIZATION OF BOOK

Volume I of the **U.S. ATLAS OF NUCLEAR FALLOUT** is divided into the following main sections:

- Sections 1-8 include maps of total fallout organized by series and by specific nuclear test, arranged in chronological order.
- Section 9 includes tabulated data for counties with the highest average fallout, specifically, the arithmetic mean of the log normal distribution.
- Section 10 includes a brief discussion of some of the radioactive elements that are found in nuclear fallout—the radionuclides, also called radio**isotopes**.

All information regarding Total Fallout is based upon analysis of the 1997 National Cancer Institute I-131 study.

SECTIONS 1-8: MAPS AND CITY LISTS

Parts 1 through 8 include city lists and maps organized by nuclear test ("shot"). The information includes these components:

U.S. Weather Service trajectory map, 2-D maps of total fallout, 3-D maps of relative average fallout; shot characteristics, and occasionally, photos of the nuclear detonation.

CITY LIST WITH DETONATION DATA

When a nuclear test occurs, radioactive debris, consisting of shot tower remnants, volatilized dirt and bomb casing material ascend into the air. Debris is then carried (generally) east with the wind. The debris will usually follow different paths depending upon the wind direction and speed. For that reason, there are often several cloud trajectories associated with each nuclear test. Because of dispersion effects, as the nuclear clouds travel further from the test site they ***spread out***. After many nuclear tests, the Atomic Energy Commission, in conjunction with the U.S. Department of Defense, conducted flights to follow and determine the centerline positions of the nuclear clouds as they crossed the continent. The results are the maps included in this section.

Using the official Department of Defense maps in conjunction with modern highway maps, I attempted to determine which cities and towns were in the centerline path of the nuclear clouds. More than 10,000 sites were eventually identified. That list is included and typically begins the subsection for each nuclear test. In addition to a list of cities and towns in the path of the cloud is information regarding the shot date, time, area, sponsor (either Los Alamos or Livermore), yield in kilotons (equivalent energy release in terms of ***thousands of tons of dynamite)*** radiation level at ground zero, height of the burst and the cloud top height.

U.S. WEATHER SERVICE TRAJECTORY MAPS

Following the City List is a copy of the Atomic Energy Commission-U.S. Weather Service Trajectory Map of the cloud trajectory. Some, such as those associated with the 1951 Ranger series, are very basic and display few details.

2-D MAPS

Two-dimensional maps are presented showing areas of fallout by county. The information is organized two different ways: first by **equal intervals** of data and then by **equal numbers** of counties. Both types of maps—the equal interval maps and the equal number maps— represent the ***same total fallout values. The only difference is how the data is organized.***

Equal-interval maps were constructed by first dividing the fallout **values** into categories defined by equal fallout ranges (intervals), and then shading counties by fallout range.

Equal-number maps were constructed by first dividing the counties into groups containing equal numbers of counties, and then shading the counties based upon relative amounts of fallout among the groups.

Maps organized by equal number of counties typically show greater fallout coverage, while the maps organized by equal intervals show only the counties with the *highest* fallout amounts.

3-D MAPS: RELATIVE AVERAGE TOTAL FALLOUT

Here, three-dimensional maps of the United States represent the relative total fallout for each NCI-evaluated nuclear test. County borders are projected into the third dimension in proportion to their deposition of fallout *relative to the entire United States* and is based on the **equal interval map** (see above). Thus, only the counties with the highest relative amount of fallout will ascend any distance above the "surface" of the map. Some shots may include several 3-D maps with close-ups of hot spots or fallout paths.

CHARACTERISTICS

This section lists characteristics of the particular nuclear test, including size and weight of the nuclear device, associated radionuclides produced and fallout patterns. **The CHARACTERISTICS section includes technical material that can be skipped without loss of continuity.**

SECTION 9: COUNTIES OF HIGHEST AVERAGE FALLOUT

This section ranks the counties with highest average calculated levels of fallout for a particular nuclear test. Activity is given in microCuries/square meter of fallout deposition. A microCurie is equal to one thousand nanoCuries and a nanoCurie refers to any radioactivity that results in 37 decays (releases of radiation) per second.

Counties are ranked for fallout, with number 1 representing the highest estimated fallout levels to 3093 representing the lowest estimated fallout levels. Several counties near the Nevada Test Site were divided into sections labeled 1, 2, 3 etc. These represent different geological differences within counties that may have affected close-in fallout.

SECTION 10: RADIONUCLIDES

This section includes a brief description of some of the major radioactive elements found in fallout. Maps are included to show areas of significant deposition.

ADDITIONAL

INTRODUCTORY NOTES

RELATIVE FALLOUT VALUES

A primary purpose of this book is to show ***relative*** fallout levels for each county from each nuclear test. To do that, arithmetic means were calculated from the geometric means and geometric standard deviations given in the NCI tables. These arithmetic means were then used as central values to evaluate relative fallout levels. Where tables or maps indicate "**Average**," the term specifically refers to the *arithmetic mean of the log-normal distribution.*

NUCLEAR TESTS AND TEST SERIES ABBREVIATIONS

Nuclear tests and test series were assigned names by those in charge of the detonations. Some of these names, like ***Buster-Jangle, Tumbler-Snapper and Upshot-Knothole*** were quite long and involved. To save space in the tables, these names were abbreviated to ***BJ***, ***TS*** and ***UK***. To ensure consistency, shot names are spelled out, but represented in all caps, such as shot CLIMAX in the UK (Upshot-Knothole) series.

The following list includes the nuclear tests that are included in this volume. The first column lists the abbreviation of the **shot** (nuclear test) or **series** (group of shots.) The second column gives the ***full name*** of the shot or series. The third column lists the date of the shot or series, and the fourth column tells whether that shot or series was included in the National Cancer Institute's 1997 radioactive iodine study. Not all tests were included in the study. Some smaller detonations, such as **BJ Able** were considered too small to be included in the analysis.

MAJOR NUCLEAR TESTS
Conducted at the Nevada Test Site reported to have produced significant fallout, 1951-1970

ABBRE-VIATION	NUCLEAR TEST OR SERIES	DATES	INCL IN 1997 NCI STUDY?	Kt
R	Ranger Series	27 Jan-6 Feb 1951	Yes	---
BJ	BusterJangle Series	22 Oct-29 Nov, 1951	Yes	---
TS	Tumbler-Snapper Series	15 Apr-5 Jun, 1952	Yes	---
UpK	Upshot-Knothole Series	17 Mar-4 Jun, 1953	Yes	---
TP	Teapot Series	18 Feb-15 May, 1953	Yes	---
PB	Plumbbob Series	28 May-7 Oct, 1957	Yes	---
HT	Hardtack II Series	12 Sep-30 Oct, 1958	Yes	---
UE	Underground Series	Approx. 1961-Dec 1970	To 1970	---
R Able	Ranger Able	27 Jan 1951	No	1
R1	Ranger Baker	28 Jan 1951	Yes	8
R Easy	Ranger Easy	1 Feb 1951	No	1
R2	Ranger Baker-2	2 Feb 1951	Yes	8
BJ Able	Buster Able	22 Oct 1951	No	Low
BJ1	Buster Baker	28 Oct 1951	Yes	3.5
BJ2	Buster Charlie	30 Oct 1951	Yes	14
BJ3	Buster Dog	1 Nov 1951	Yes	21
BJ4	Buster Easy	5 Nov 1951	Yes	31
BJ5	Jangle Sugar	19 Nov 1951	Yes	1.2
BJ6	Jangle Uncle	29 Nov 1951	Yes	1.2
TS1	TS Able	1 Apr 1952	Yes	1
TS2	TS Baker	15 Apr 1952	Yes	1
TS3	TS Charlie	22 Apr 1952	Yes	1
TS4	TS Dog	1 May 1952	Yes	19
TS5	TS Easy	7 May 1952	Yes	12
TS6	TS Fox	25 May 1952	Yes	11
TS7	TS George	1 Jun 1952	Yes	15
TS8	TS How	5 Jun 1952	Yes	14
UpK1	UpK Annie	17 Mar 1953	Yes	16
UpK2	UpK Nancy	24 Mar 1953	Yes	24
UpK3	UpK Ruth	31 Mar 1953	Yes	0.2
UpK4	UpK Dixie	6 Apr 1953	Yes	11
UpK5	UpK Ray	11 Apr 1953	Yes	0.2

MAJOR NUCLEAR TESTS Conducted at the Nevada Test Site reported to have produced significant fallout, 1951-1970				
UpK6	UpK Badger	18 Apr 1953	Yes	23
UpK7	UpK Simon	25 Apr 1953	Yes	43
UpK8	UpK Encore	8 May 1953	Yes	27
UpK9	UpK Harry	19 May 1953	Yes	32
UpK10	UpK Grable	25 May 1953	Yes	15
UpK11	UpK Climax	4 Jun 1953	Yes	61
TP1	TP Wasp	18 Feb 1955	Yes	1
TP2	TP Moth	22 Feb 1955	Yes	2
TP3	TP Tesla	1 Mar 1955	Yes	7
TP4	TP Turk	7 Mar 1955	Yes	43
TP5	TP Hornet	12 Mar 1955	Yes	4
TP6-A	TP Bee	22 Mar 1955	NCI combined fallout from shots TP Bee and TP Ess	8
TP6-B	TP Ess	23 Mar 1966	Yes	1
TP7	TP Apple-1	29 Mar 1955	Yes	14
Wasp Prime	TP Wasp Prime	29 Mar 1955	No	3
TP HA	High Altitude test	6 Apr 1955	No	3
TP8	TP Post	9 Apr 1955	Yes	2
TP9	TP Met	19 Apr 1955	Yes	22
TP10	TP Apple-2	5 May 1955	Yes	29
TP11	TP Zucchini	15 May 1955	Yes	28
PB1A	PB Boltzmann	28 May 1957	Combined with PB Franklin	12
PB1B	PB Franklin	2 Jun 1957	Combined with PB Boltzmann	0.14
PB2	PB Wilson	18 Jun 1957	Yes	10
PB3	PB Priscilla	24 Jun 1957	Yes	37
PB4	PB Hood	5 Jul 1957	Yes	74
PB5	PB Diablo	15 Jul 1957	Yes	17
PB John	PB John	19 Jul 1957	No	2

MAJOR NUCLEAR TESTS
Conducted at the Nevada Test Site reported to have produced significant fallout, 1951-1970

PB6A	PB Kepler	24 Jul 1957	Combined with PB Owens	10
PB6B	PB Owens	25 Jul 1957	Combined with PB Kepler	9.7
PB7	PB Stokes	7 Aug 1957	Yes	19
PB8	PB Shasta	18 Aug 1957	Yes	17
PB9	PB Doppler	23 Aug 1957	Yes	11
PB10	PB Franklin Prime	30 Aug 1957	Yes	4.7
PB11	PB Smoky	31 Aug 1957	Yes	44
PB12	PB Galileo	2 Sep 1957	Yes	11
PB 13A	PB Wheeler	6 Sep 1957	Combined with PB Coulomb B and PB LaPlace	0.19
PB13B	PB Coulomb B	6 Sep 1957	Combined with PB Wheeler and PB LaPlace	0.3
PB13C	PB LaPlace	8 Sep 1957	Combined with PB Wheeler and PB Coulomb B	1
PB14	PB Fizeau	14 Sep 1957	Yes	11
PB15	PB Newton	16 Sep 1957	Yes	12
PB 16	PB Whitney	23 Sep 1957	Yes	19
PB17	PB Charleston	28 Sep 1957	Yes	12
PB18	PB Morgan	7 Oct 1957	Yes	8
HT1	HT Eddy	18 Sep 1958	Yes	0.08
HT2	HT Hidalgo	5 Oct 1958	Yes	0.07
HT3	HT Quay	10 Oct 1958	Yes	0.07
HT4	HT Lea	13 Oct 1958	Yes	1.4
HT5	HT Vesta	17 Oct 1958	Yes	0.02
HT6	HT Rio Arriba	18 Oct 1958	Yes	0.09
HT7	HT Socorro	22 Oct 1958	Yes	6.0
HT8	HT Wrangall	22 Oct 1958	Yes	0.11
HT9	HT Sanford	26 Oct 1958	Yes	4.9
UE1	UE Antler Figures given in this series for test UE1 refer to I-131 values	15 Sep 1961	Yes	2.6

MAJOR NUCLEAR TESTS Conducted at the Nevada Test Site reported to have produced significant fallout, 1951-1970

UE2	UE Danny Boy	5 Mar 1961	Yes	0.43
UE3	UE Platte Figures given in this series for test UE3 refer to I-131 values	14 Apr 1962	Yes	1.85
UE4	UE Eel Figures given in this series for test UE4 refer to I-131 values	19 May 1962	Yes	4.5
UE5	UE Des Moines	13 Jun 1962	Yes	2.9
S (UE6)	Sedan	6 Jul 1962	Yes	110
UE7	UE Johnie Boy	11 Jul 1962	Yes	0.5
UE8	UE Small Boy	14 Jul 1962	Yes	Low
UE9	UE Bandicoot Figures given in this series for test UE9 refer to I-131 values	19 Oct 1962	Yes	12.5
UE10	UE Pike Figures given in this series for test UE10 refer to I-131 values	13 Mar 1964	Yes	<20
UE11	UE Sulky	18 Dec 1964	Yes	0.09
UE12	UE Palanquin	14 Apr 1965	Yes	4.3
UE13	UE Pin Stripe	25 Apr 1966	Yes	<20
UE14	UE Cabriolet	26 Jan 1968	Yes	2.3
UE15	UE Buggy	12 Mar 1968	Yes	5.4
UE 16	UE Schooner	8 Dec 1968	Yes	30
UE 17	Baneberry Fallout from this test was the result of a venting accident.	18 Dec 1970	Yes	10

EXTRA COUNTIES

In the 1997 National Cancer Institute I-131 report fourteen counties near the Nevada Test Site were subdivided based upon geography and different fallout patterns. These counties are shown in the following illustrations. Table A2. 1 of the NCI report defines the parameters of each of the separate subdivisions. The following illustrations were taken from that report: *Estimated Exposures and Thyroid Doses Received by the American People from Iodine-I-131 in Fallout Following Nevada Atmospheric Nuclear Bomb Tests.* Appendices to the Report from the National Cancer Institute. U.S. Department of Health and Human Services: Oct 1997.) In this book, the subdivided counties are differentiated using numerals, as, for example, AZ, Coconino1, Coconino2, Coconino3, etc. The California subdivided counties are on the left, the Arizona subdivided counties are on the right.

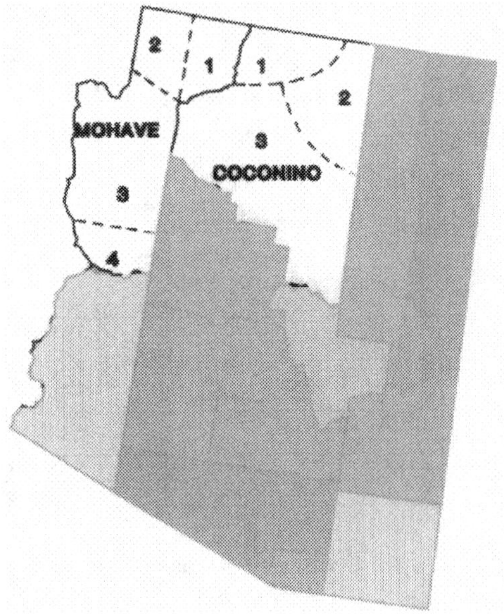

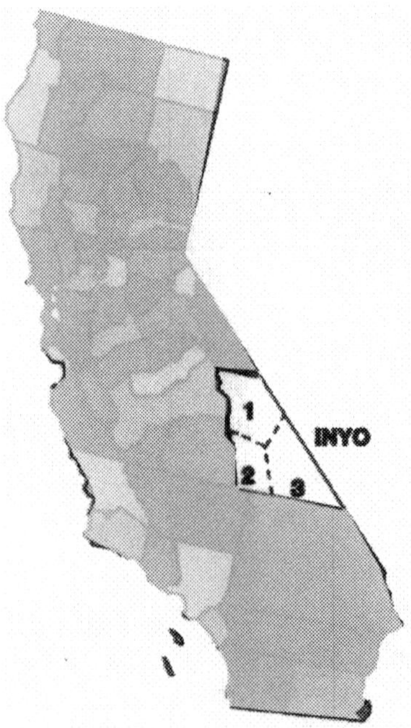

NEVADA COUNTIES SUBDIVIDED

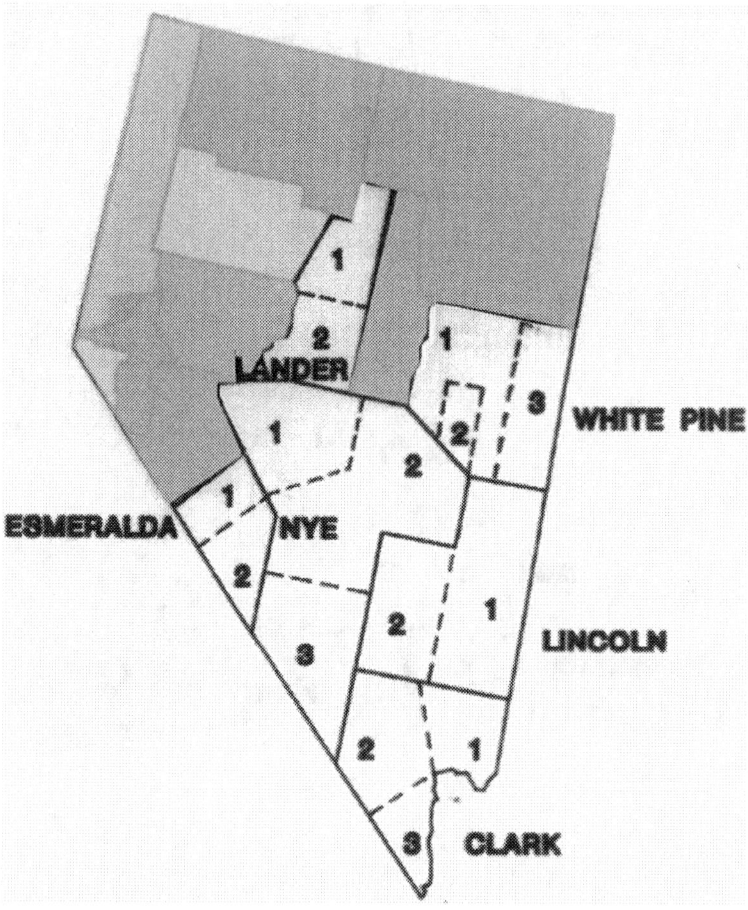

UTAH COUNTIES SUBDIVIDED

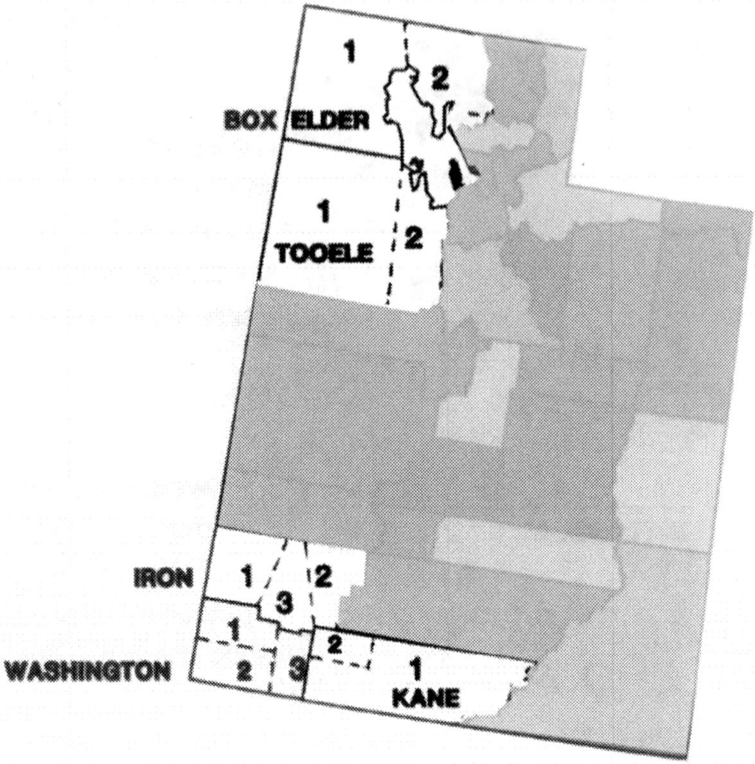

PRELIMINARY TECHNICAL DISCUSSION

(This section discusses the statistics behind the values. Readers not interested in the statistics behind fallout determination can skip this section without loss of continuity.)

POTENTIAL RAINOUTS

One of the main concerns of the Atomic Energy Commission during the aboveground test series was that a nuclear cloud might come into contact with a storm system and brings high levels of radioactive fallout to earth with the rain. Termed variously rainouts (when the nuclear cloud is *within* the rain or snow cloud) or washouts (when the nuclear cloud is *below* the rain cloud), the potential for high levels of radiation exposure during these events was quite real. An examination of the NCI data suggests that rainouts occurred with some regularity.

While there have been no government-sponsored epidemiological studies regarding communities exposed to the potentially intense fallout during rainouts, the Department of Defense was aware of the problem. A 1977 DOD-Energy Department book, *The Effects of Nuclear Weapons*, concluded that a rainfall of only 1.1 hours could—theoretically—scavenge 99 percent of a nuclear cloud: Included on the next page is a table published in 1977 detailing information regarding rainouts and washouts.

Absence of associated precipitation does not ensure low fallout. A significant fallout event in Troy, New York in 1953 was probably due to dry downbursts preceding a thunderstorm. Generally speaking however, rainfall can significantly increase the amount of fallout deposited. Subsequent volumes in this series will discuss rainouts in greater detail.

Table 9.74a

ESTIMATED RAINFALL DURATION FOR RAINOUT

Percent of Cloud Scavenged	Duration of Rainfall (hours)
25	0.07
50	0.16
75	0.32
90	0.53
99	1.1

Table 9.74b

ESTIMATED RAINFALL DURATION FOR WASHOUT

Percent of Cloud Scavenged	Duration of Rainfall (hours)		
	Light	Moderate	Heavy
25	8	1.6	0.8
50	19	3.8	1.9
75	38	7.7	3.6
90	64	13	6.4
99	128	26	13

TABLE 9.74 FROM THE 1977 DEPARTMENT OF DEFENSE BOOK : THE EFFECT OF NUCLEAR WEAPONS[8]

GEOMETRIC MEAN AND GEOMETRIC STANDARD DEVIATION

On October 4, 1997, the NCI published the long-awaited county-by-county radioiodine-131 report. Due to the volume of the material—over a hundred thousand pages—the values were published over the Internet. To obtain information about deposition levels for any given county, interested individuals with access to the Internet had only to download a data set. It came as something of a surprise that the each county deposition data set consisted of not one number but two, defined only as the ***GM*** and ***GSD***. As it turned out, the GM (the first value) referred to the *Geometric Mean* and the GSD (the second value) to the *Geometric Standard Deviation.*

In practical terms, the GM, or Geometric Mean referred to the average estimated fallout while the GSD or Geometric Standard Deviation referred to the uncertainty around that average. In other words, the GM represented the average value—say, 50 units per square foot—while the GSD represented the size of the range. The GSD was a mathematical expression that told the statistician how wide the range was on either side of the average. It answered the mathematical question: "plus or minus *what*?"

Here's another way to look at it: Suppose you wanted to know how much it rained over your house on a particular day. So you call the local weatherman and he tells you something like this:

> "Well, the rainfall at the weather station was 2 inches, but since you live a long way from the weather station, it's *likely* that you didn't get exactly the same amount. It's likely that you got some rain, and it might have been 2 inches, but more *likely* it was somewhere between, say a half an inch and six inches."

He might even go on to tell you that he was 95 percent sure that your house received somewhere between half an inch of rain and six inches of rain. That is, the chances were only 5 percent—one chance out of twenty—that the rainfall you received was ***less*** than half an inch or ***more*** than six inches.

But suppose you ask what's average? ***That's*** when things get complicated. Your weatherman might respond by asking *which* average do you want?

It seems that there are more than one kind of ***average.***

The *average* we all think of refers to the mathematical average that is found when the units are distributed normally—like in a classroom of 100 children with of differing heights. One or two will be exceptionally tall, one or two will be exceptionally short. The greatest number of kids will have heights that are exactly in between the tallest and shortest. If you plotted the children height versus the number of children with that height, the shape of the line would look like a bell (the bell-shaped curve). Mathematically, such a group would be called normally distributed, and the average would likely be at the exact midpoint of the bell.

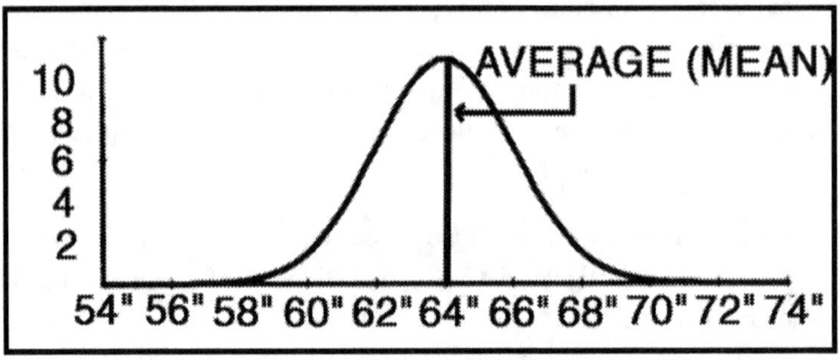

Fig.1: EXAMPLE OF A NORMAL DISTRIBUTION

But, what if all the 100 children were nearly the same size—four or five feet tall or so—except for five or six extraterrestrial children who were ***fifty*** feet tall? Then, the curve wouldn't be bell-shaped at all. It would look as though someone pushed the bell all the way to the ***left***, and the average with it.

Unfortunately, that's the way rain droplets—and nuclear fallout particles—are often distributed. Most are very small, but there are also a few larger particles many times the size of the small ones.

Worse, it is also how rainfall—and nuclear fallout particles deposition is distributed. Most parts of the country received small amounts; but some parts of the U.S. received many, many times that.

One way to evaluate such a distribution is to "compress" one set of units using logarithms and leave the other set of units as is. The result, called a ***lognormal*** distribution is simply one in which units on one axis (here, the y or vertical axis) of the chart are given in logarithms of the values while the units on the other axis are given in equal increments (such as 1,2,3. . .etc).

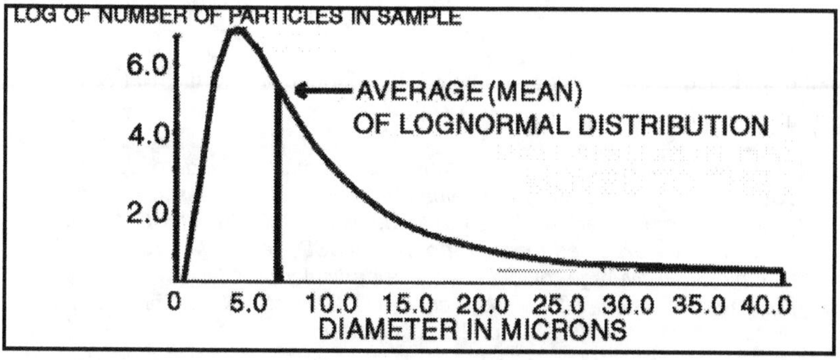

FIG 2. EXAMPLE OF LOGNORMAL DISTRIBUTION: THEORETICAL PARTICLE SIZE DISTRIBUTION FROM A SAMPLE OF FALLOUT (0-40 MICRONS)

Figure 2 shows how a scientist might graph the particle number vs. size ratios from a sample of fallout. Here, the smaller particles predominate.

Put another way, calculating the average deposition of fallout over an area is *not* the same as calculating the average height of a hundred children in a classroom. In a classroom of a hundred children normally distributed—and no extraterrestrial kids included— one can get a good idea of the average simply by finding the midpoint between the tallest and the shortest child—subtracting the minimum height from the maximum height and dividing by two.

But when the distribution is different, as in rainfall or nuclear fallout, the methods of calculating the average must also be different. The values found in this book represent a particular kind of average called the ***arithmetic mean of the lognormal distribution.***

The formula for determining the arithmetic mean of the lognormal distribution from the GM and GSDs found in the NCI study is given here:

$$\mathbf{AM = GM(exp((lnGSD)^2(0.5))}$$

Where: AM = Arithmetic Mean of the Lognormal Distribution;
GM = Geometric mean of the lognormal distribution
GSD = Geometric standard deviation of the lognormal distribution
exp = exponent e = 2.7183
ln = natural logarithm

Total fallout values in this volume represent the arithmetic mean of the lognormal distribution of fallout. They were calculated using the above formula.

In the abridged book version of their I-131 fallout study, the National Cancer Institute had this to say about GM and GSD:

> *"The uncertainties expressed in terms of geometric standard deviations, GSD, implying that 67% of the values in the distribution associated with a best estimate, GM, are expected to lie between GM/GSD and GM x GSD, while 97% are expected to range from GM/(GSD)^2 and GM x (GSD)^2. For example, if an average thyroid dose to a particular population group from a given test is listed with a best estimate, GM, of 0.4 rad and with an associated uncertainty, GSD, of 2.5, this means*
>
> *(a) That there is a 50% probability that the true value of the average thyroid dose is greater than 0.4 rad, and conversely, that there is a 50% probability that the average thyroid dose is lower than 0.4 rad; and,*
>
> *(b) That the distribution of the expected values is such that there is a 67% probability that the true value of the average thyroid dose likes between:*
>
> *GM/GSD = 0.4/2.5 = 0.16 rad; and*
>
> *GM x GSD = 0.4 x 2.5 = 1 rad.*
>
> *And that there is a 97% probability that the true value of the average thyroid dose likes between: GM/(GSD)^2 = 0.4/6.25 = 0.06 rad, and GM x (GSD)^2 = 0.4 x 6.25 = 2.5 rad."*
>
> *The estimates provided in the Annexes and in the Subannexes for the average doses to the various population groups show that the associated GSDs range, in general between 2 and 10, the lowest GSDs being usually related to populations living in the vicinity in the NTS in areas for which County Data Base or Town Data Base data were available. The highest GSDs are associated with the dose estimates for which the depositions of I 131 were assessed with the meteorological approach."*[9]

The important thing to remember is, while the geometric mean (GM) represents the average value, the geometric standard deviation (GSD) is the primary measure of uncertainty—the range of possible values around the geometric mean. The GSD indicates how tightly clustered the data are. The greater the geometric standard deviation, the greater is the uncertainty surrounding the estimated geometric mean (GM) and thus the greater the spread of data.[10]

Thus, just because a county is listed as having received a relatively high average value of total fallout does not necessarily mean that the county was in a hot zone. the county in question may have been part of an area where the uncertainty regarding the fallout deposition level was high. In the NCI study, counties located near the Nevada Test Site typically were associated with smaller geometric standard deviations than were counties located farther from the Test Site.

This book does not report fallout values in terms of geometric mean and geometric standard deviation. Rather, the values derived from the GMs and GSDs of I-131 deposition rates are reported as a particular kind of average—the arithmetic mean of the lognormal distribution.

Those interested in a more detailed discussion of the mathematical procedures associated with this work are encouraged to read the technical sections at the end of the book. Those seeking further information are directed to the original 2001 edition of the ***U.S. Atlas of Nuclear Fallout Volume I: Total Fallout*** and ***Volume V: Calculations***, published in 2002.

NOTES

[1] Eisler, Peter and Steve Sternberg, "Fallout: did It Harm? Study shows contaminants fell far from Nevada test site," USA Today. 26 Jul 1997.

[2] Section 7(a) of Public Law 97-414 directs the Secretary of Health and Human Services to "(1) conduct scientific research and prepare analysis necessary to develop valid and credible assessments of the risks of thyroid cancer that are associated with thyroid doses of Iodine 131; (2) conduct scientific research and prepare analysis necessary to develop valid and credible methods to estimate the thyroid dose of Iodine 131 that are received by individuals from nuclear bomb fallout; and (3)conduct scientific research and prepare analysis necessary to develop valid and credible assessments of the exposure to Iodine 131 that the American people received from the Nevada atmospheric nuclear bomb tests."

[3] Crocker, G.R., J.D. O'Connor and E.C. Freiling. "Physical and Radiochemical Properties of Fallout Particles," USNRDL-TR-899 15 June, 1965.

[4] Ibid.

[5] Glasstone, Samuel and Philip J. Dolan. The Effects of Nuclear Weapons. United States Department of Defense and the Energy Research and Development Administration. 1977. p. 411.

[6] Columbia Encyclopedia Sixth Edition. Columbia University Press. 2000.

[7] Estimated Exposures and Thyroid Doses Received by the American People from Iodine-131 in Fallout Following Nevada Atmospheric Nuclear Bomb Tests. U.S. Department of Health and Human Services, National Institutes of Health, National Cancer Institute. NIH Pub No. 97-4264. October 1997. Appendix A. p. A2.9.

[8] Glasstone. Op cit.

[9] NIH Pub No. 97-4264 Op cit. p. 10.5.

[10] Nuclear technical specialist Jim Horner suggested the wording regarding GM and GSD. Personal communication, Oct. 2002.

SECTION 1

RANGER SERIES

Jan 27-Feb 6, 1951

SHOT RANGER ABLE

(Not Included in the NCI Study)

RANGER: ABLE*

Detonation Date: 27 Jan 1951 • *Detonation Time:* 5:45 A.M. • *Area:* Frenchman's Flat • *Sponsor:* Los Alamos • *Yield:* 1 kt • *Radiation Level at Ground Zero:* 0.5 R/hr • *Height of Burst:* 1,060 ft • *Cloud Top Height:* 17,000 ft

10,000-Foot Trajectory: NEVADA: Lincoln County • UTAH: Orem, Provo, Vernal • COLORADO: Pueblo • KANSAS: (Crossed the border at Greeley, left Kansas near Wallace) • MISSOURI: St. Joseph, Cameron, Chillicothe, Brookfield, Atlanta, Bethel, West Quincy • ILLINOIS: Quincy, Jacksonville, Springfield • INDIANA: Lafayette, Kokomo, Marion, Berne • OHIO: Springfield, Dayton, Columbus, Canton • PENNSYLVANIA: Beaver Falls, Kittanning, Williamsport, Wilkes-Barre • NEW YORK: Port Jervis, Middletown, Poughkeepsie • CONNECTICUT: New Haven, New London • RHODE ISLAND: Newport • MASSACHUSETTS: Fall River.

30,000-Foot Trajectory: NEVADA: Caliente • UTAH: Cedar City, Panguitch, Moab • COLORADO: Crested Butte, Buena Vista, Colorado Springs • KANSAS: (The clouds crossed trajectories over Castle Rock at the border of Gove and Trego counties in Kansas), Topeka, Lawrence • MISSOURI: Kansas City, Harrisonville, Warrensburg, Sedalia, Columbia, St. Louis • ILLINOIS: Alton, Vandalia, Robinson • INDIANA: Bloomington, Columbus, Greensburg • OHIO: Hamilton, Athens, Marietta • PENNSYLVANIA: Uniontown, Lancaster, Reading • NEW JERSEY: New Brunswick.

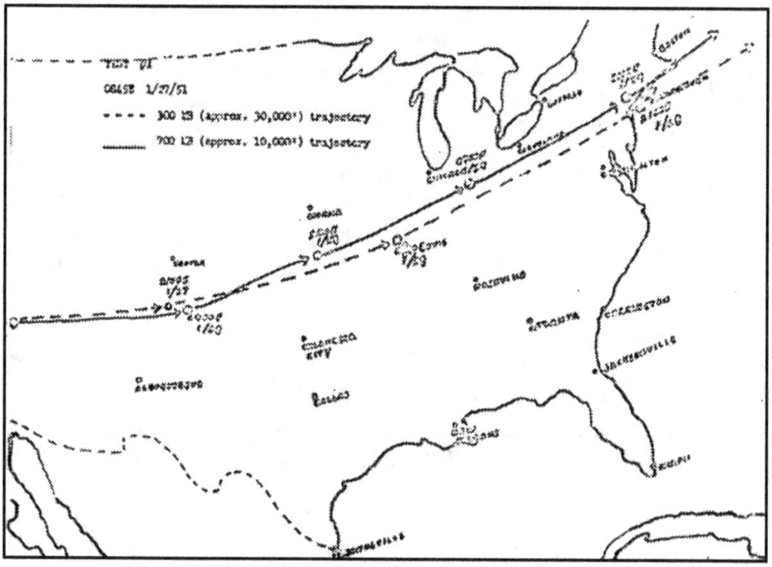

TRAJECTORY: RANGER ABLE

SHOT 1: RANGER BAKER-1 (R1)

Ranger: Baker-1

Detonation Date: 28 Jan 1951 • *Detonation Time:* 5:52 A.M. • *Area:* Frenchman's Flat • *Sponsor:* Los Alamos • *Yield:* 8 kt • *Radiation Level at Ground Zero:* 15 R/hr • *Height of Burst:* 1,080 ft • *Cloud Top Height:* 35,000 ft

10,000-Foot Trajectory: COLORADO: Boulder • MISSOURI: St. Joseph, Clark • ILLINOIS: Alton, Terre Haute • OHIO: Montgomery, New Philadelphia • PENNSYLVANIA: Kittanning • NEW YORK: Middletown • CONNECTICUT: New Haven • MAINE: Nantucket.

30,000-Foot Trajectory: UTAH: St. George • COLORADO: Fort Carson, Arapahoe (This trajectory crossed over into Kansas above 4,039-foot Mount Sunflower, the highest point in Kansas) • KANSAS: Palmer • NEBRASKA: Falls City • MISSOURI: Coatsville • IOWA: Burlington • INDIANA: South Bend • MICHIGAN: Gibralter • NEW YORK: Auburn, Glens Falls • NEW HAMPSHIRE: Portsmouth.

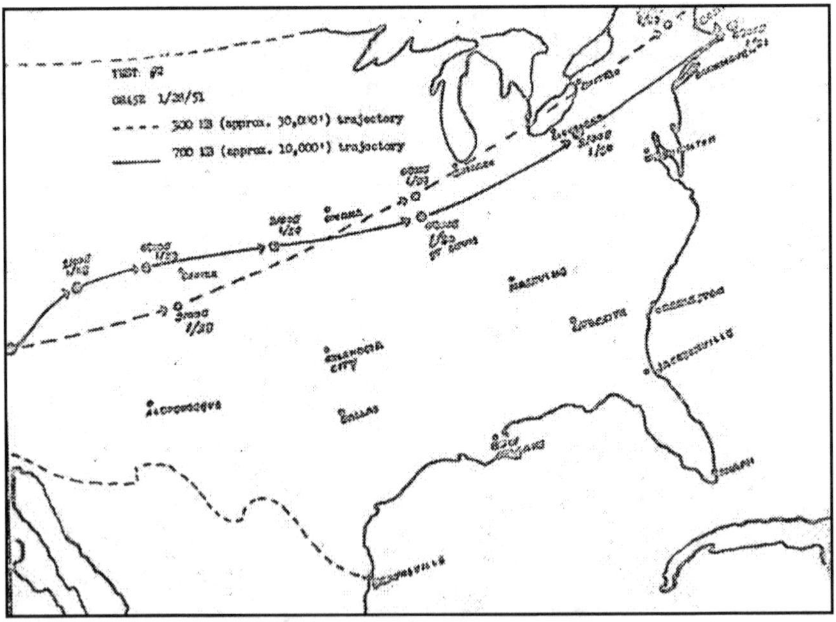

TRAJECTORY: RANGER BAKER

TOTAL FALLOUT: RANGER BAKER-1

Equal Intervals

nCi/square meter

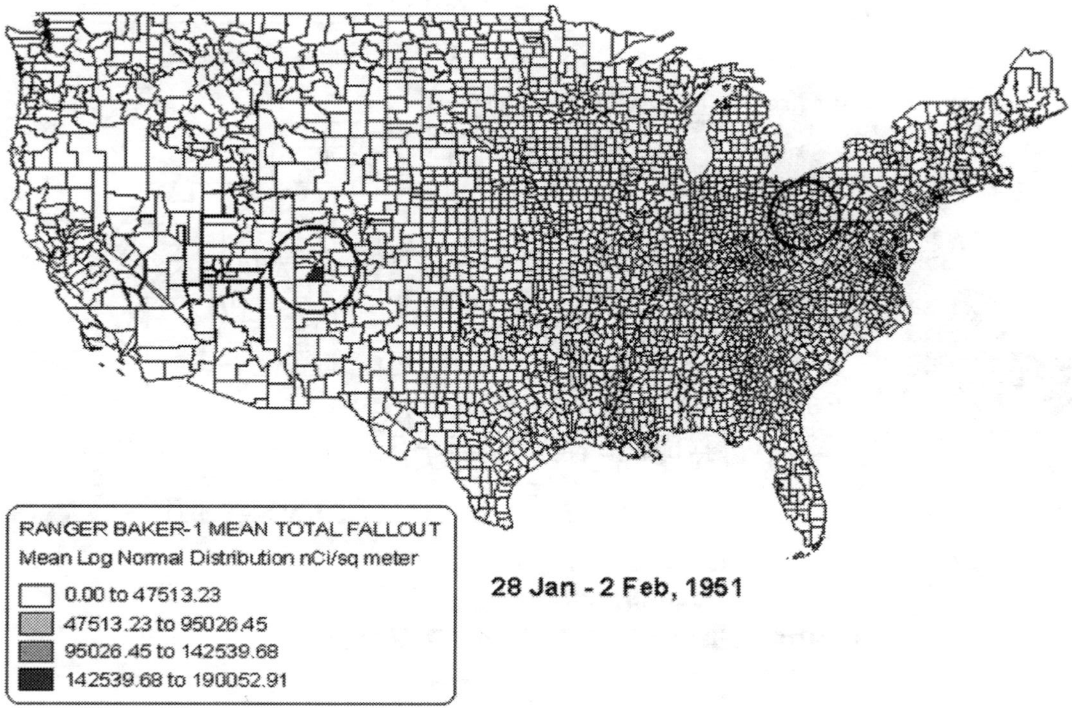

28 Jan - 2 Feb, 1951

TOTAL FALLOUT: RANGER BAKER-1

Equal Number

nCi/square meter

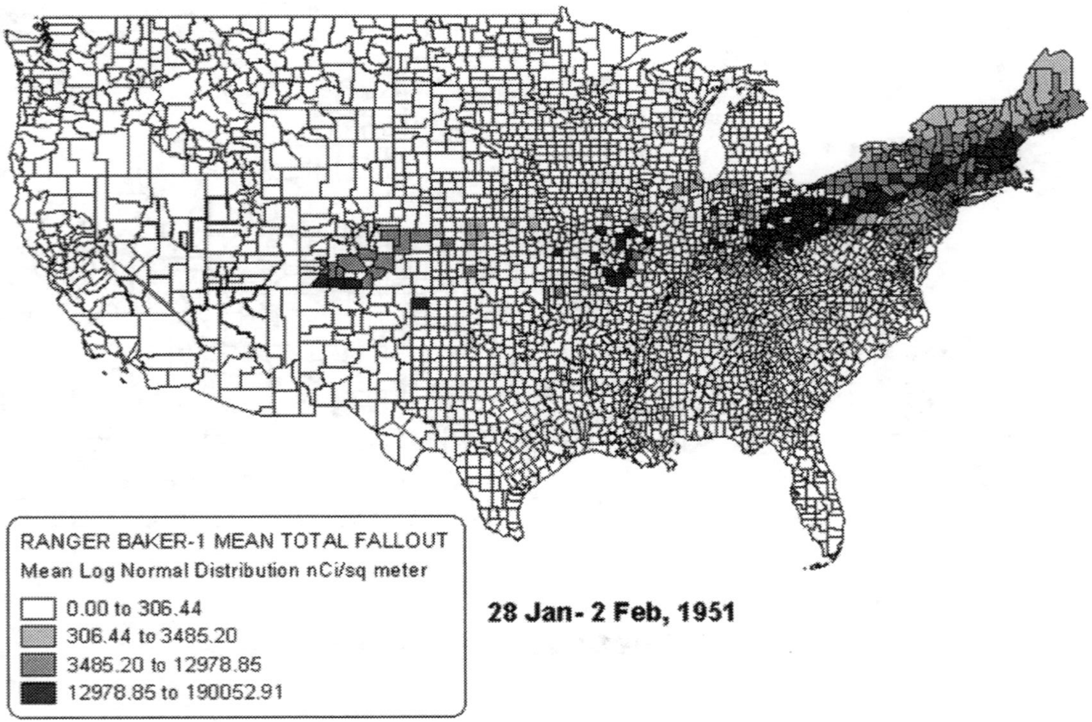

28 Jan - 2 Feb, 1951

RELATIVE TOTAL FALLOUT: RANGER BAKER

JAN 28, 1951 8 KT

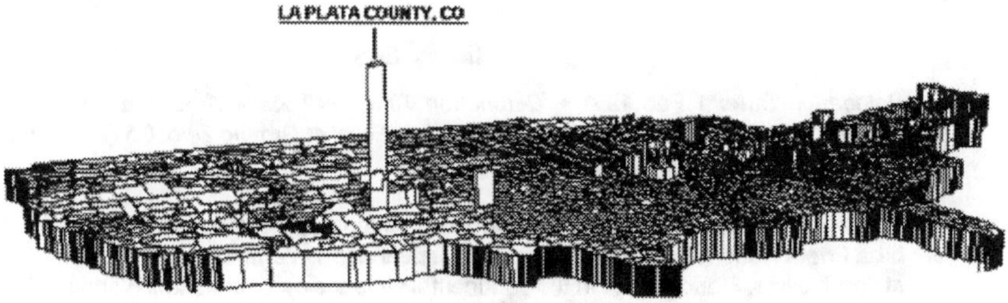

FALLOUT PATH: RANGER BAKER JAN 28-FEB 2, 1951

CHARACTERISTICS: RANGER BAKER-1

The test device weighed 10,800 lbs and resembled the “Fat Man” device dropped on Nagasaki. Ranger-BAKER-1 ranks fourth among the nuclear tests in the percentage of Np239 contained in the fallout. The deposition pattern of Ranger BAKER-1 was most similar to that of shot Buster-Jangle CHARLIE (R value: 0.32, T value: 19.07).

SHOT RANGER EASY
(Not Included in the NCI Study)

RANGER: EASY

Detonation Date: 1 Feb 1951 • *Detonation Time:* 5:47 A.M. • *Area:* Frenchman's Flat • *Sponsor:* Los Alamos • *Yield:* 1 kt • *Radiation Level at Ground Zero:* 0.5 R/hr • *Height of Burst: 1,080 ft* • *Cloud Top Height: 12,500 ft*

10,000-Foot Trajectory: Crossed San Bernadino County, California, into Mexico over the Cabeza Prieta National Wildlife Refuge. The nuclear debris came back into the United States at Fort Meyers, Florida, and left the continental airspace over Sebastian, Florida.

30,000-Foot Trajectory: TEXAS: El Paso, Austin, Beaumont • LOUISIANA: Baton Rouge • ALABAMA: Montgomery • NORTH CAROLINA: Durham • VIRGINIA: Nassawadox.

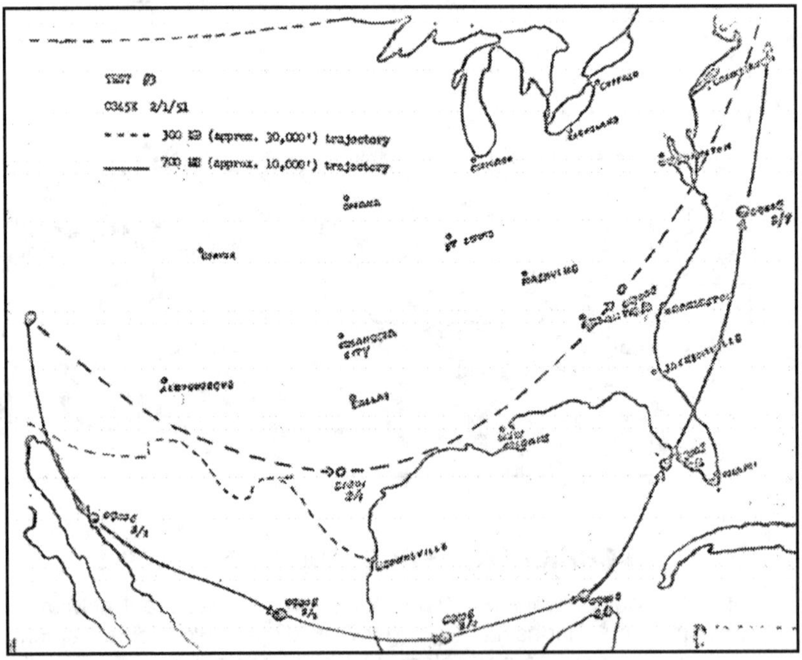

TRAJECTORY: RANGER EASY

SHOT 2: RANGER BAKER-2 (R2)

RANGER: BAKER-2

Detonation Date: 2 Feb 1951 • *Detonation Time:* 5:49 A.M. • *Area:* Frenchman's Flat • *Sponsor:* Los Alamos • *Yield:* 8 kt • *Radiation Level at Ground Zero:* 15 R/hr • *Height of Burst:* 1,100 ft • *Cloud Top Height:* 28,000 ft

10,000-Foot Trajectory: The nuclear debris crossed north of Las Vegas and into Arizona, passing north of Phoenix and leaving the state at Paradise. It entered Mexico airspace over the Alamo Hueco Mountains in New Mexico, then crossed back into the United States over Ruidosa, Texas. From there is passed out over Big Bend National Park, then to Carrizo Springs, Texas, and out over the Gulf at Aransas Pass, Texas. It entered Florida airspace at St. Petersburg and left near Melbourne.

30,000-Foot Trajectory: The nuclear debris at 30,000 feet passed over St. George, Utah, crossed the state and entered New Mexico airspace north of Crystal. From there it moved south of Los Alamos, then to Clovis, New Mexico, where it crossed into Texas near Bovina. The cloud passed over Petersburg and Waco and left the coast near Winnie, just north of Houston. Later it crossed the Florida peninsula, passing over Miami.

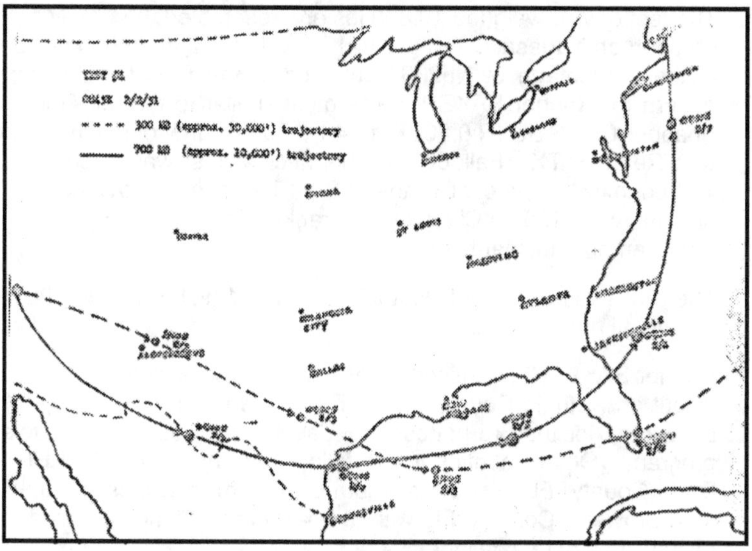

TRAJECTORY: RANGER BAKER-2

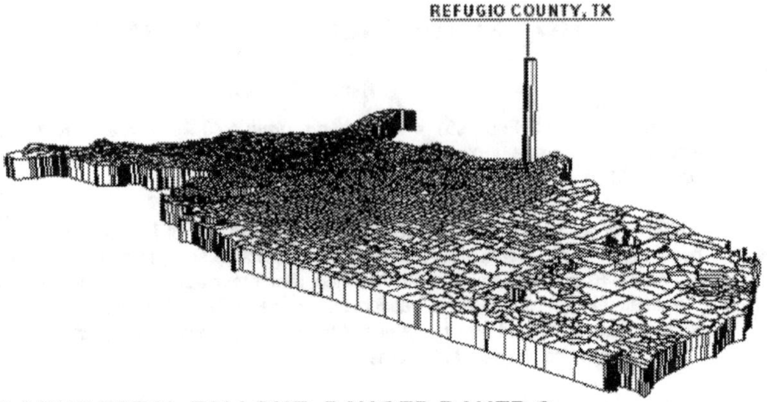

RELATIVE TOTAL FALLOUT: RANGER BAKER-2

CHARACTERISTICS: RANGER BAKER-2

The test device weighed 10,800 lbs and resembled the "Fat Man" device dropped on Nagasaki. Ranger BAKER-2 ranks fifth among nuclear tests in the percentage of Np239 contained in the fallout. The deposition pattern for Ranger BAKER-2 was most similar to that of shot Tumbler-Snapper CHARLIE (R:0.20, T: 11.49). The County of highest deposition was Refugio, TX. Fallout from this nuclear test was deposited across 182 counties (6 percent of the U.S.). The average fallout per County amount was 1302 nCi/sq meter, ranking this shot 46th for average fallout among nuclear tests.

The day 1 ratio of total fallout to I-131 was 139.1:1, ranking 9th highest among all nuclear tests.

Ranger BAKER-2 deposited an estimated maximum of 1.1 milliCuries of fallout on Refugio County, TX on Feb 3, 1951. The Refugio, TX fallout accounted for the 3rd hottest shot day in the history of the nuclear test program. Other counties in the fallout path of Ranger BAKER-2 were Dade County, FL with a maximum of 3.76 microCuries of total fallout and Broward County, FL with a maximum estimated value of 1.44 microCuries of total fallout.

TOTAL FALLOUT: RANGER BAKER-2

Equal Intervals
nCi/square meter

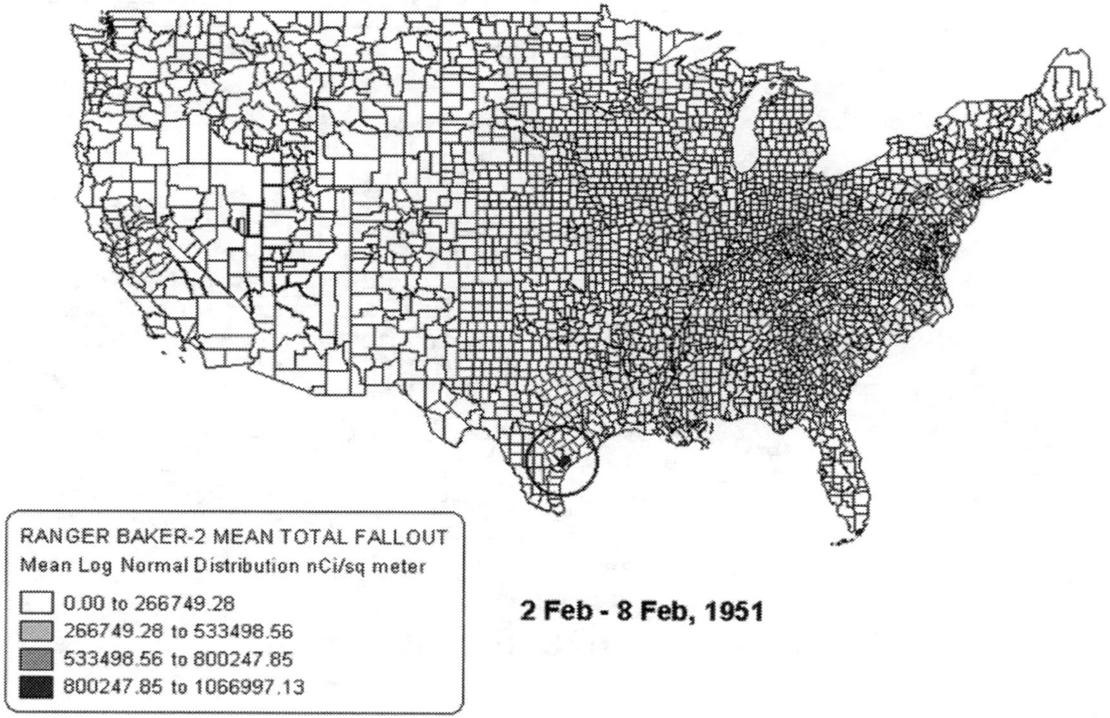

2 Feb - 8 Feb, 1951

TOTAL FALLOUT: RANGER BAKER-2

Equal Number
nCi/square meter

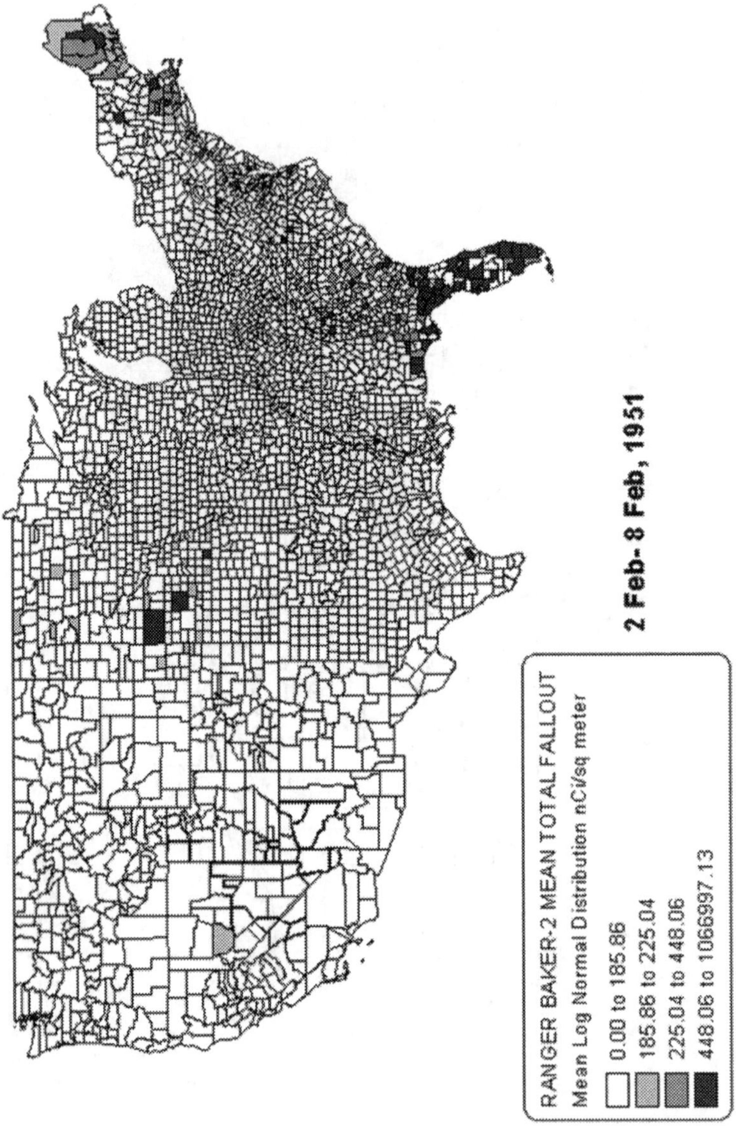

2 Feb- 8 Feb, 1951

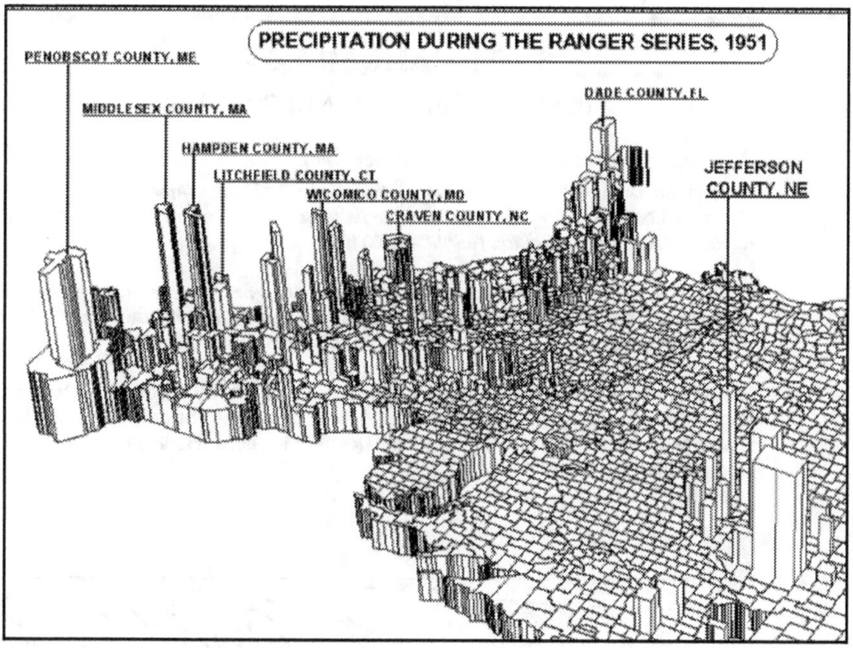
PRECIPITATION DURING THE RANGER SERIES, 1951
PENOBSCOT COUNTY, ME
MIDDLESEX COUNTY, MA
HAMPDEN COUNTY, MA
LITCHFIELD COUNTY, CT
WICOMICO COUNTY, MD
CRAVEN COUNTY, NC
DADE COUNTY, FL
JEFFERSON COUNTY, NE

SHOT RANGER FOX

(Not Included in the NCI Study)

RANGER: FOX

Detonation Date: 6 Feb 1951 • *Detonation Time:* 5:47 A.M. • *Area:* Frenchman's Flat • *Sponsor:* Los Alamos • *Yield:* 22 kt • *Radiation Level at Ground Zero:* 15 R/hr • *Height of Burst:* 1,435 ft • *Cloud Top Height:* 43,000 ft

10,000-Foot Trajectory: This portion of the nuclear cloud passed over Mohave County, Lake Mead, and Prescott, and left United States airspace east of Pedgregosa, Arizona. It crossed back into the U.S. at Laredo, Texas, and left the state at Corpus Christi.

30,000-Foot Trajectory: This trajectory moved north of the 10,000-foot trajectory, passing over Flagstaff, Arizona. After briefly crossing Mexico airspace, the debris crossed the Texas cities of El Paso, Roma, McAllen, Pharr, Donna, Mercedes, Harlingen, San Benito, and Brownsville.

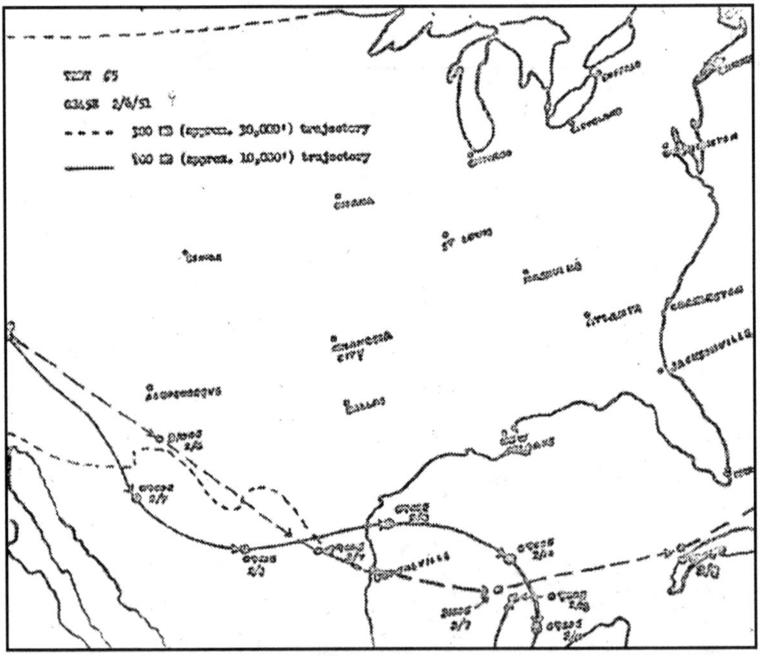

TRAJECTORY: RANGER FOX

SECTION 2

BUSTER-JANGLE SERIES

Oct 22-Nov29, 1951

SHOT BUSTER ABLE

(Not Included in the NCI Study)

BUSTER: ABLE

Detonation Date: 22 Oct 1951 • *Detonation Time:* 6:00 A.M. • *Area:* 7-5 • *Sponsor:* Los Alamos • *Yield:* Less than 0.1 kt • *Radiation Level at Ground Zero:* 300 counts per minute alpha radiation • *Height of Burst:* 100 ft • *Cloud Top Height:* 8,000 ft

8,000-Foot Trajectory: The cloud of debris passed north of Las Vegas and traveled on to Flagstaff, Arizona, circling down past the Colcord Mountains and Black Canyon, passing near Prescott then back over Flagstaff. From there, the debris traveled north, crossing into Utah over the Rainbow Plateau. The cloud followed Glen Canyon then drifted north over Roan Cliffs and East Tavaputs Plateau and into Colorado at Dinosaur. The cloud crossed into Wyoming at Power Wash and moved on to Rawlins. It left the state near Redbird and crossed into South Dakota near Parker Peak.

The cloud crossed the Badlands, then moved northeast, leaving South Dakota near Corona and entering Minnesota at Clinton. It then crossed over to Kingsdale, Minnesota, and into Wisconsin at Moose Junction. The BUSTER: ABLE cloud reentered Minnesota at Washburn and finally left United States airspace over Lake Superior.

CHARACTERISTICS: BUSTER ABLE

A first attempt to detonate the Buster-ABLE device met with failure. Due in part to faulty circuitry, nothing happened (a similar event would occur years later during the Plumbbob DIABLO event). When the device finally detonated, there was only a minimum yield. Official sources reported that the nuclear yield was “less than 0.1 kiloton” but unofficial sources suggest the yield was less than a pound.

The tower was damaged but still standing after the test.

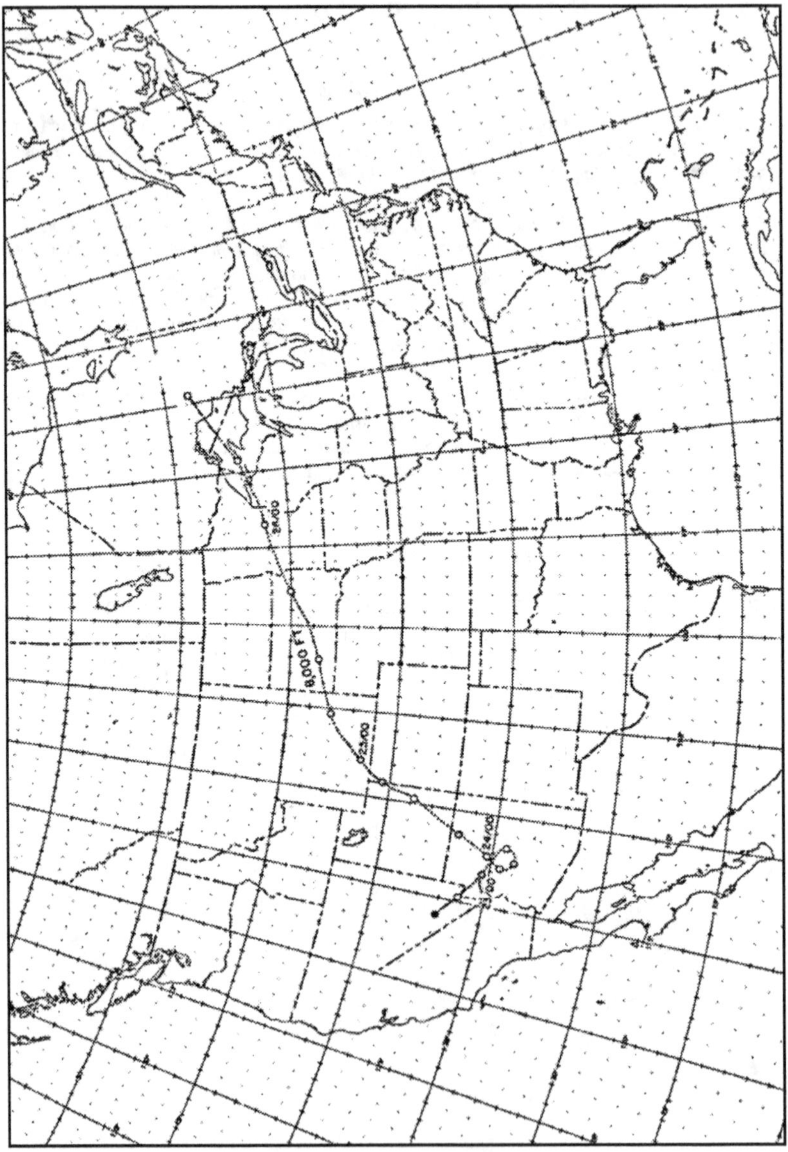

TRAJECTORY: BUSTER ABLE

SHOT 1: BUSTER BAKER (BJ1)

BUSTER: BAKER

Detonation Date: 28 Oct 1951 • *Detonation Time:* 7:20 A.M. • *Area:* 7-3 • *Sponsor:* Los Alamos • *Yield:* 3.5 kt • *Radiation Level at Ground Zero:* 6 R/hr • *Height of Burst:* 1,118 ft • *Cloud Top Height:* 31,700 ft

All trajectories traveled west over California in a band extending between Mount Whitney-Cape San Martin and Stove Pipe Wells-Guadalupe. The entire cloud crossed over San Luis Obispo County in California, circled, then moved east over the rest of the United States. One component of the 8,000-foot trajectory split off from the rest of the cloud and circled back over the Los Angeles area. The debris was tracked along four trajectories: the 8,000-, 10,000-, 24,000-, and 30,000-foot levels. Interestingly, three of the clouds—the 18,000-, 24,000-, and 30,000-foot levels—all crossed over the same point: Concordia, Kansas.

10,000-Foot Trajectory: This level moved north of Phoenix, Arizona, over Montrose, Colorado, then southeast toward Amarillo, Texas. It followed the Red River and crossed over the towns of Eldorado, Oklahoma, Wichita Falls and Denison, Texas (Dwight D. Eisenhower's boyhood home). It crossed over Hugo, Oklahoma, Ashtown, Texas, the state of Arkansas, and entered Mississippi near Tupelo. Other cities and areas under the 10,000-foot trajectory included: TENNESSEE: Chattanooga, Knoxville • VIRGINIA: Roanoke, Alexandria • Washington, D.C. • DELAWARE: New Castle • PENNSYLVANIA: Philadelphia • NEW YORK: New York • RHODE ISLAND: Providence • MASSACHUSETTS: Lexington, Boston • MAINE: Washington County.

24,000-Foot Trajectory: CALIFORNIA: Barrett • NEVADA: Laughlin, Lake Mead • UTAH: Hilldale, Cisco • COLORADO: Grand Junction, Denver, Hale • KANSAS: Goodland, Dresden, Blair, Concordia • MISSOURI: St. Joseph, Chillicothe, Brookfield, Macon, Hannibal • ILLINOIS: Jacksonville, Champaign, Danville • INDIANA: Logansport • OHIO: Celina, Greensburg, Youngstown • PENNSYLVANIA: Sharon, Tioga City (The 8,000- and 24,000-foot trajectories crossed over Elkland) • NEW YORK: Elmira, Glens Falls • VERMONT: Pawlet, West Hartford • NEW HAMPSHIRE: Hanover, Kearsarge • MAINE: (Near the White Mountain National Forest) Waterville, Bangor, Princeton.

30,000-Foot Trajectory: This followed the debris at the 24,000-foot level, the two crossing paths at Concordia, Kansas. The debris at the 30,000-foot level then crossed over Leavenworth, Kansas, and entered Missouri over Liberty. Traveling east, the cloud passed over Terre Haute, Indiana; Cincinnati, Ohio; and left the U.S. over Atlantic City, New Jersey.

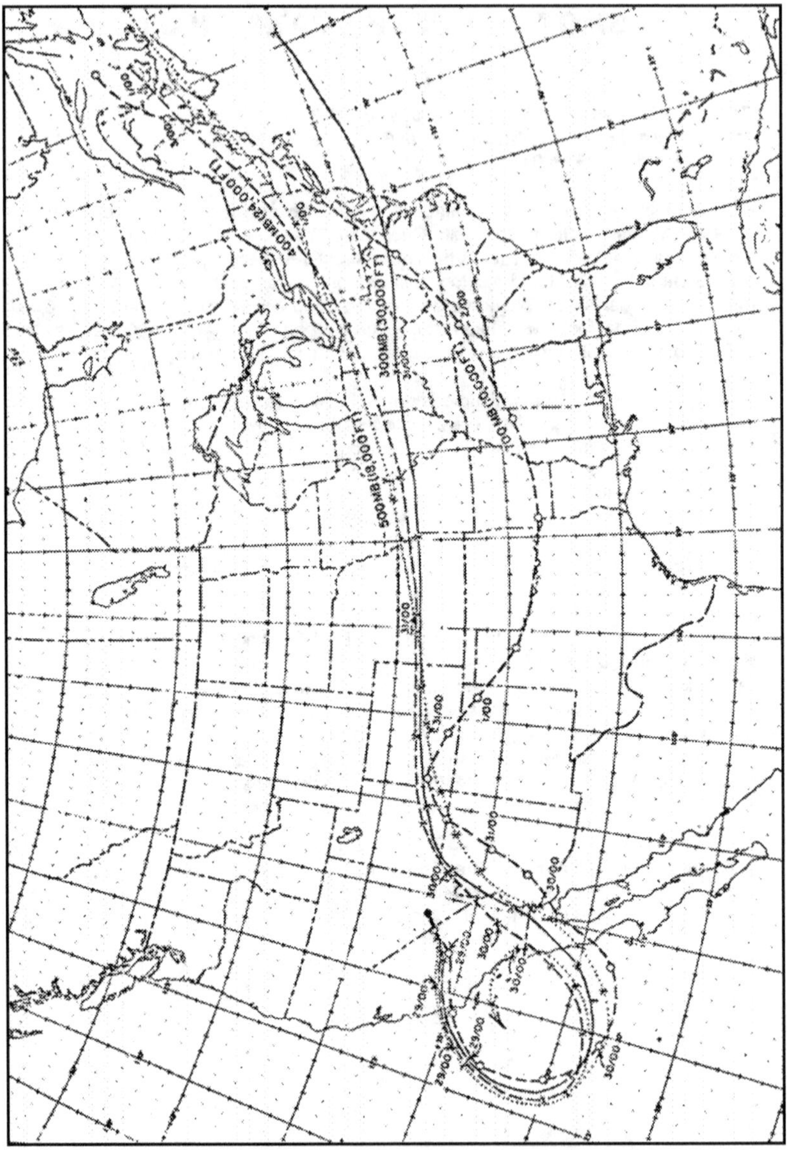

TRAJECTORY: BUSTER BAKER (BJ1)

TOTAL FALLOUT: BUSTER BAKER

Equal Numbers

nCi/square meter

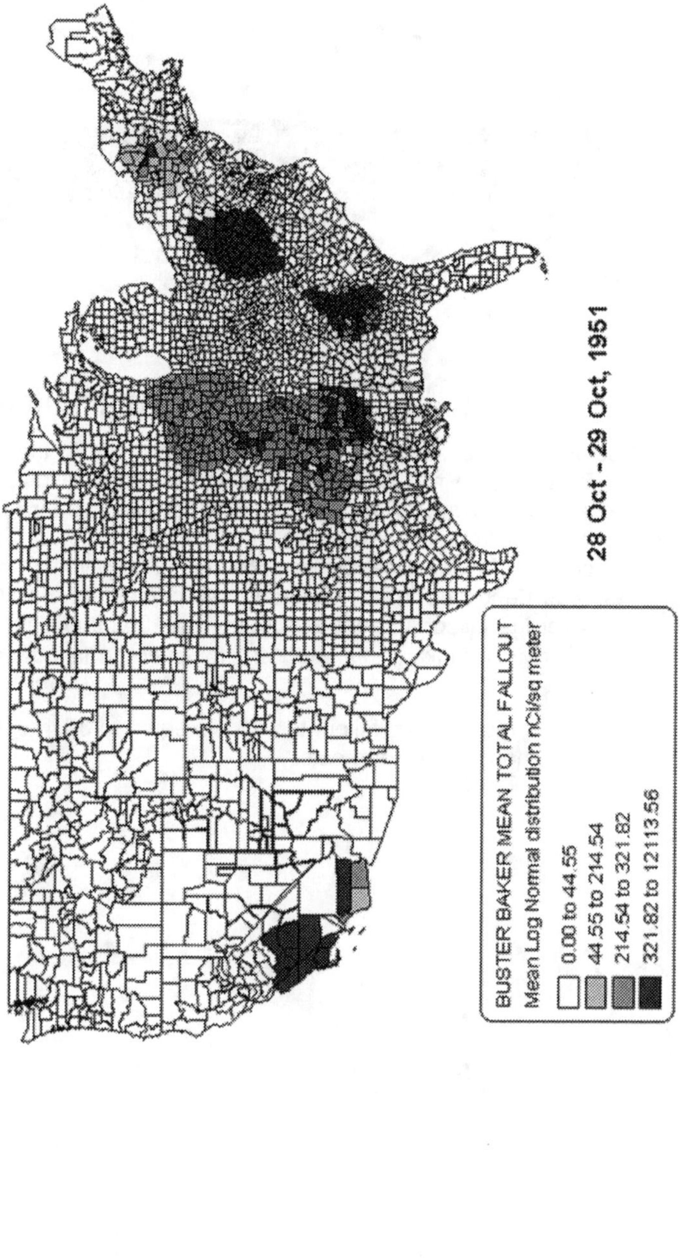

28 Oct - 29 Oct, 1951

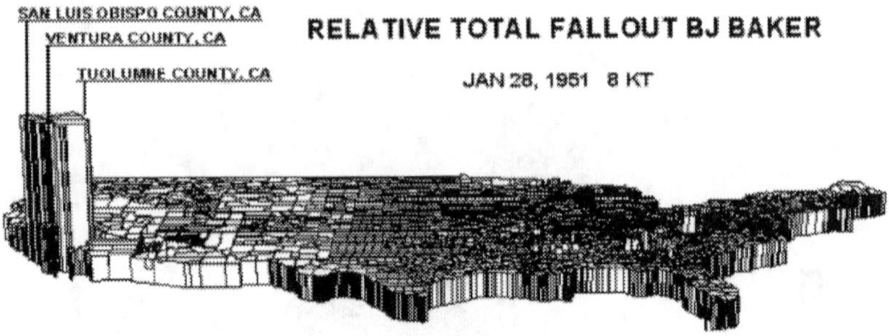

CHARACTERISTICS: BUSTER BAKER

Buster Baker ranks first among the nuclear tests in the percentage of Co-50, Eu-155, Pm-147 and Tc-99m contained in fallout, and 3rd in the percentage of Am-241 in fallout. The deposition pattern for this shot is most similar to that of Buster-Jangle CHARLIE (R:0.30, T: 17.64). The US county of highest deposition: Fresno, CA. Fallout from this nuclear test was deposited in 527 counties (17 percent of the U.S. counties.) The average fallout amount per county was 302.9 nCi/sq meter (0.3029 µCi/sq meter), ranking this shot 52nd for average fallout among nuclear tests.

The day 1 ratio of total fallout to I-131 is equal to 87.4 to 1, ranking 47th highest among all nuclear tests.

SHOT 2: BUSTER CHARLIE (BJ2)

BUSTER: CHARLIE

Detonation Date: 30 Oct 1951 • *Detonation Time:* 7:00 A.M. • *Area:* 7–3 • *Sponsor:* Los Alamos • *Yield:* 14 kt • *Radiation Level at Ground Zero:* 5 R/hr • *Height of Burst: 1,132 ft* • *Cloud Top Height:* 41,000 ft

10,000-Foot Trajectory: This part of the atomic cloud moved east, then south over Lake Mead. It crossed Phoenix, Arizona, then moved south into Mexico. It curved back to reenter the U.S. at Texas near the town of Lajitas. It then met up with the 18,000-foot trajectory over Houston, and crossed it again near Mobile, Alabama. From there, the radioactive cloud moved northeast over Columbus, Georgia and Charlotte and Greensboro, North Carolina. North of Richmond, Virginia, it began to curve east and crossed the East Coast over Atlantic City, New Jersey, but followed the coastline up to New York City. It then moved north crossing Danbury, Connecticut, and Pittsfield, Massachusetts, then followed the Vermont–New Hampshire border north to Canada. The cloud then followed the U.S.–Canada border until it left United States airspace.

18,000-Foot Trajectory: This trajectory crossed Death Valley, California, north of Bakersfield, continued west and left the coast over Point Conception, California. Once over the Pacific, however, it turned back east, crossing Mexico and then reentered United States airspace over Texas's Big Bend area. It crossed over San Antonio, Houston, and Beaumont–Port Arthur, Texas, and then entered Louisiana west of Lake Charles. The cloud crossed Louisiana and Mississippi and entered Alabama just north of Mobile. The debris continued on over Georgia and South Carolina, leaving the coast near Charleston.

24,000-Foot Trajectory: This trajectory crossed Fresno, California, and then, along with the rest of the atomic cloud, moved out over the Pacific. It then doubled back and entered Texas near Esperanza. It tracked just south of Lubbock, crossed Wichita Falls and Denison, Texas, then entered Arkansas airspace. Other cities it crossed included Little Rock, Arkansas; Memphis, Tennessee; Gilbert, West Virginia; Washington, D.C.; and Dover, Delaware.

30,000-Foot Trajectory: This trajectory moved west north of Los Angeles, crossed the coastline, then reentered the United States at Cowlic, Arizona. Later it crossed over Albuquerque and Santa Fe, New Mexico, then moved north, eventually crossing the 40,000-foot trajectory at Sterling, Nebraska. Other cities 30,000 feet below the atomic cloud: Indianola and Iowa City, Iowa; Rockford and Waukegan, Illinois; Lansing and Port Huron, Michigan; Toronto; Montreal; Presque Isle, Maine.

40,000-Foot Trajectory: This trajectory moved northeast over the northern tier of Colorado counties, crossing the cities of Fort Collins and Sterling. It then moved across the *second* southern tier of Nebraska counties, crossing the countryside just south of Lincoln. It followed the Missouri-Iowa border, then moved north over Burlington, Iowa, and Peoria, Illinois. Other cities and areas 40,000 feet under the atomic cloud: Fort Wayne, Indiana; Toledo, Ohio; the southern tier of New York counties from Jamestown to Kingston; and Worcester, Boston, and Lexington, Massachusetts.

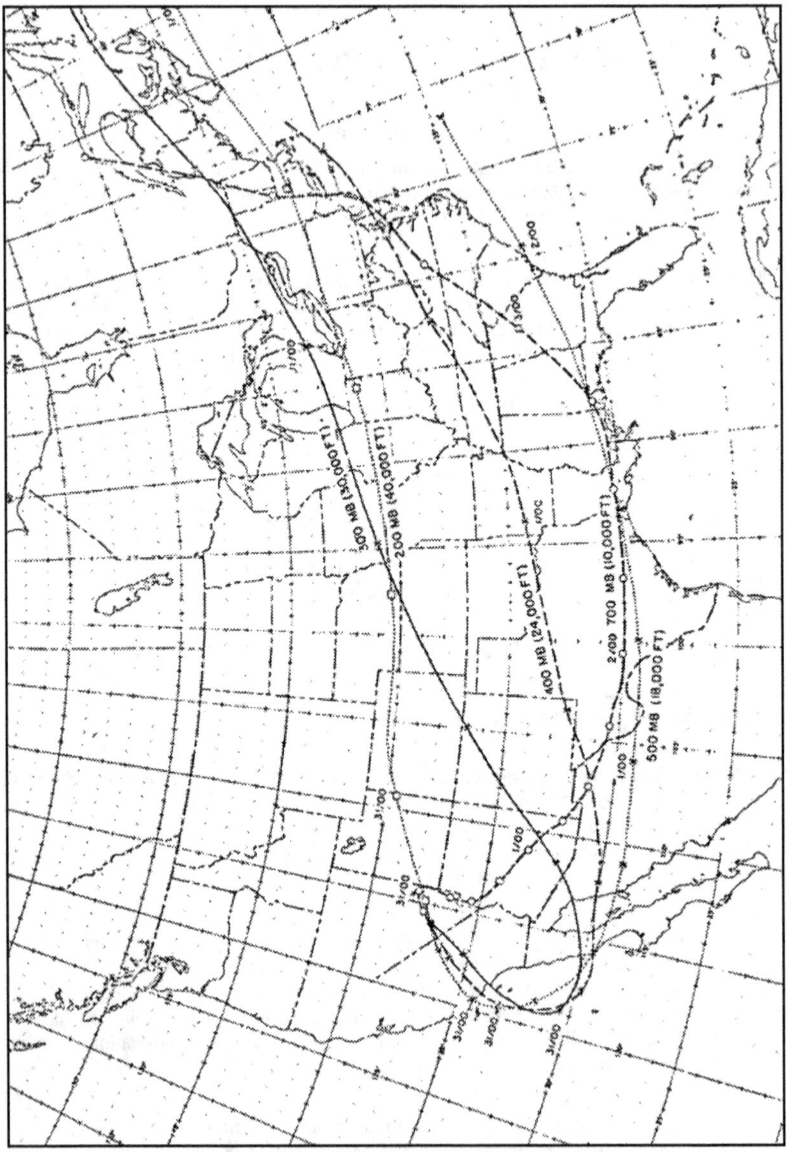

TRAJECTORY: BUSTER CHARLIE (BJ2)

CHARACTERISTICS: BUSTER CHARLIE

Like the devices in the Ranger series, the Buster CHARLIE test device resembled the "Fat Man" bomb detonated over Nagasaki. The deposition pattern for Buster CHARLIE is most similar to that of Buster DOG (R: 0.43; T: 26.85).

Fallout from this test was deposited across 95 percent of the U.S. or, 2937 counties. The average fallout amount per county was 12,266 nCi/sq meter, ranking this shot 8th for average fallout among the nuclear tests.

The day 1 ratio of total fallout to I-131 is 139.7 to 1, the 8th highest among all nuclear tests. Buster CHARLIE ranks 9th in deposition of Pr-144 and 21st in deposition of Te-127m across the U.S.

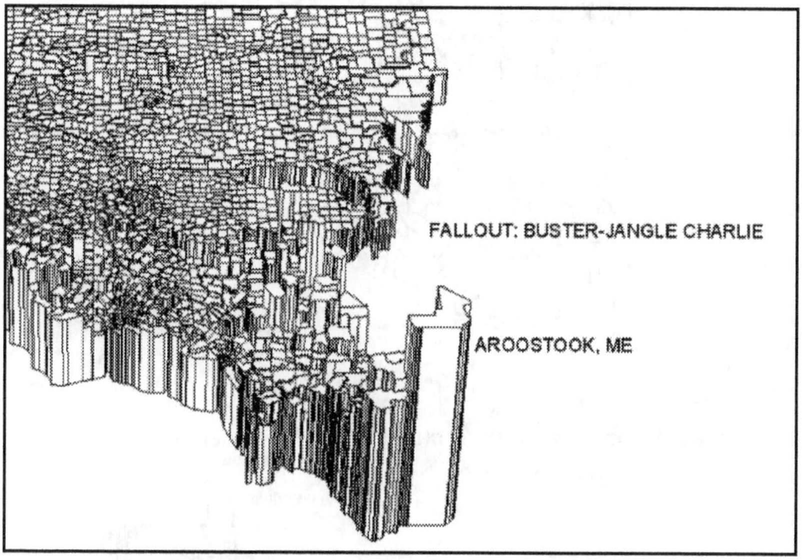

EAST COAST FALLOUT: BUSTER CHARLIE

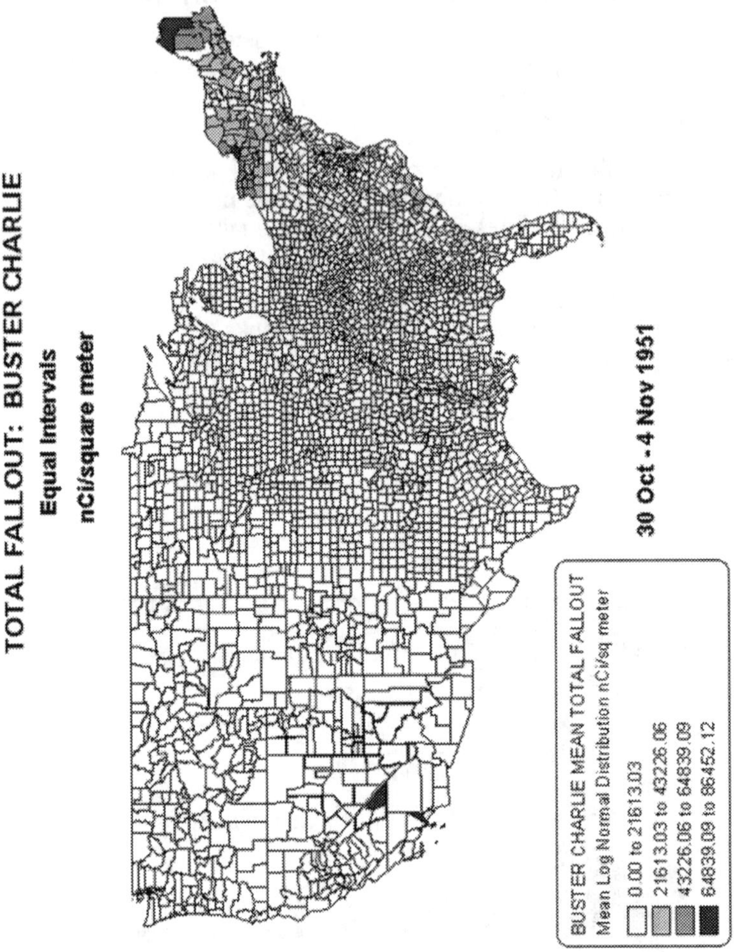
TOTAL FALLOUT: BUSTER CHARLIE
Equal Intervals
nCi/square meter
30 Oct - 4 Nov 1951
BUSTER CHARLIE MEAN TOTAL FALLOUT
Mean Log Normal Distribution nCi/sq meter
0.00 to 21613.03
21613.03 to 43226.06
43226.06 to 64839.09
64839.09 to 86452.12

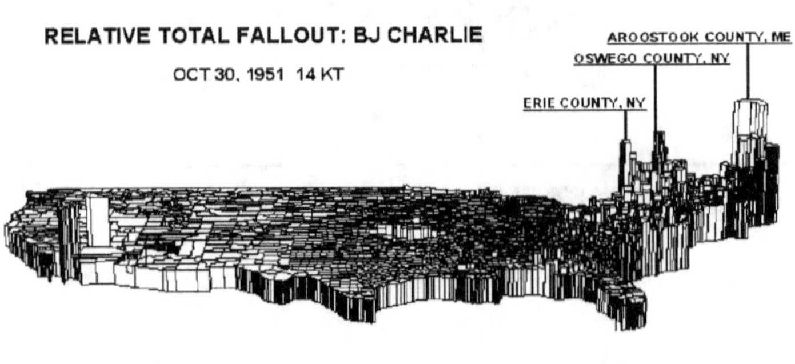
RELATIVE TOTAL FALLOUT: BJ CHARLIE
OCT 30, 1951 14 KT
AROOSTOOK COUNTY, ME
OSWEGO COUNTY, NY
ERIE COUNTY, NY

SHOT 3: BUSTER DOG (BJ3)

Buster: Dog

Detonation Date: 1 Nov 1951 • *Detonation Time:* 7:30 A.M. • *Area:* 7–3 • *Sponsor:* Los Alamos • *Yield:* 21 kt • *Radiation Level at Ground Zero:* 20 R/hr • *Height of Burst:* 1,416 ft • *Cloud Top Height:* 46,000 ft

The 18,000- 24,000- 30,000- and 40,000-foot levels all moved east in a wide arc, crossing over: Las Vegas, Nevada; Phoenix, Arizona; El Paso, San Angelo, Austin, Waco, Texas; and Shreveport, Louisiana. The band tightened over the north part of Madison County, Mississippi, at Camden (home of the Casey Jones State Museum), then spread in a wide arc over the entire Northeast, from a line from Knoxville, Tennessee–Ottawa, Canada, to a line from Asheville, North Carolina–Atlantic City, New Jersey.

The track of the atomic clouds passed over: NEVADA: Las Vegas, Henderson, Boulder City • ARIZONA: Dolan Springs, Hackberry, Dewey, Humboldt, Tonto Basin, San Carlos, Clifton • NEW MEXICO: Silver City, Tyrone, Bayard, Las Cruces, University Park, Mesquite • TEXAS: (Guadalupe Peak), Dell City, Pecos, Barstow, Royalty, Grandfalls, Big Lake, Sherwood, Vick, Eden, Millersview, Placid, Elm Grove, Bee House, Pearl, McGregor, Waco, Mexia, Red Lake, Palestine, Neches, Turney, New Salem, Gary, Carthage • LOUISIANA: Keatchie, Gloster, Castor, Lucky, Liberty Hill, Quitman, Vernon, Richwood, Raysville, Holly Ridge, Warden, Monticello • MISSISSIPPI: Onward, Valley, Yazoo City, Pickens, Goodman, Kosciusko, High Point, Sessums, Artesia, Columbus • ALABAMA: Fernbank, Belk, Corona, Jasper, Empire, Hayden, Oneonta, Rosa, Hendricks, Altoona, Mountainboro, Collinsville, Ringgold • GEORGIA: Summerville, Echota, Calhoun, Resaca, Ellijay, Cherrylog, Morganton • NORTH CAROLINA: Hayesville, Tusquitee, Kyle, Bryson City, Cherokee, Waterville • TENNESSEE: Johnson City, Elizabethton, Hunter, Carter, Shady Valley • VIRGINIA: Sugar Grove, Wytheville, Pulaski, Radford, McCoy, Newport, Blacksburg, Abbot, New Castle, Barbours Creek, Eagle Rock, Lexington, Greenville, Crimora, Lydia, Haywood, Boston, Rixeyville, New Baltimore, Gainesville, Arlington • Washington, D.C. • MARYLAND: Bethesda, Spencersville, Columbia, Baltimore, Sweet Air • PENNSYLVANIA: Oxford, West Chester, King of Prussia, Norristown, Philadelphia, Warminister, Newtown • NEW JERSEY: Trenton, Hopewell, Princeton, Piscataway, Wesfield, Cranford, Newark, Paterson, Hackensack, Paramus • NEW YORK: Millwood, Katona, Salem Center, White Plains • CONNECTICUT: Wilton, Naugatuck, Meridan, Middletown, Portland, Marlborough, Willimantic, Putnam, Harrisville • RHODE ISLAND: Woonsocket • MASSACHUSETTS: Millville, Bellingham, Holliston, Framingham, Needham, Wellesley, Newton, Brookline, Boston, Cambridge, Somerville, Malden, Lexington, Saugus, Lynn, Wakefield, Peabody, Salem, Gloucester, Pigeon Cove.

The 10,000-foot level moved south, crossing Mexico, then the southern tier of Texas counties: Zapata, Starr, Hidalgo, and Cameron.

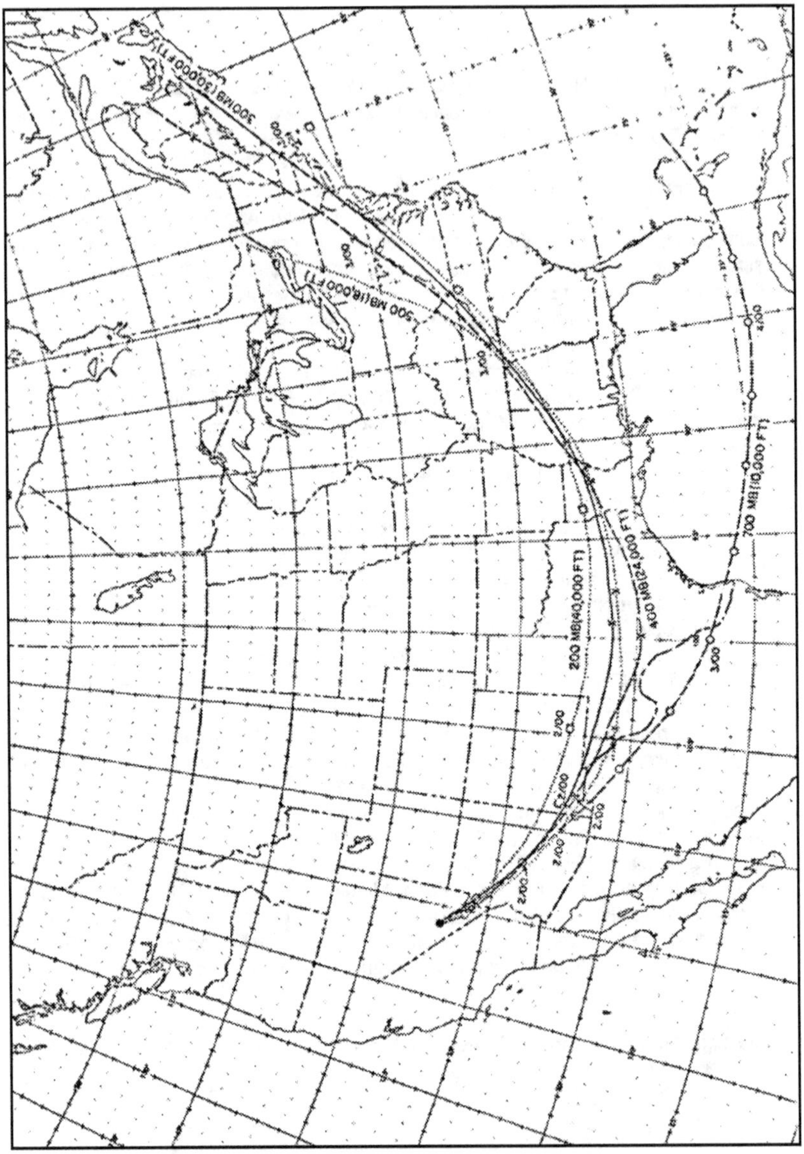

TRAJECTORY: BUSTER DOG (BJ3)

CHARACTERISTICS: BUSTER DOG

Among the nuclear tests, Buster DOG is 6TH for percentage of Co-58 and Co-60 in fallout. The deposition pattern for Buster DOG is most similar to that of Buster EASY. The counties of highest deposition include the New York counties of Chautauqua, Erie, Genesee, Niagara and Orleans. Estimated maximums for all: 210 μCi/sq meter of fallout on the same day, Nov 6, 1951.

Fallout from this nuclear test was deposited across 2163 counties (41 percent of the U.S.). The average fallout amount was 3.8 μCi/sq meter, ranking this shot 38th for average fallout among the nuclear tests.

The day one ratio of total fallout to I-131 for Buster DOG is 146.3 to 1, making it the 6th highest total fallout to I-131 ratio among all nuclear tests.

RELATIVE AVERAGE TOTAL FALLOUT BJ DOG

NOV 1, 1951 21 KT

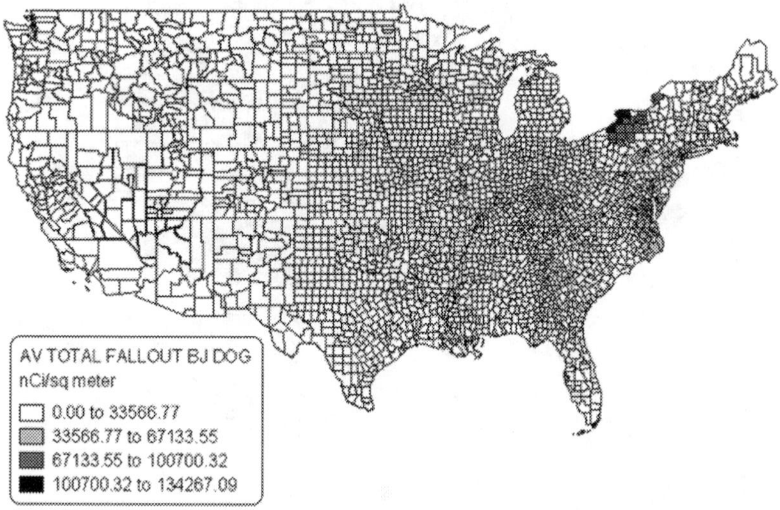

EASTERN US FALLOUT FROM SHOT BJ DOG

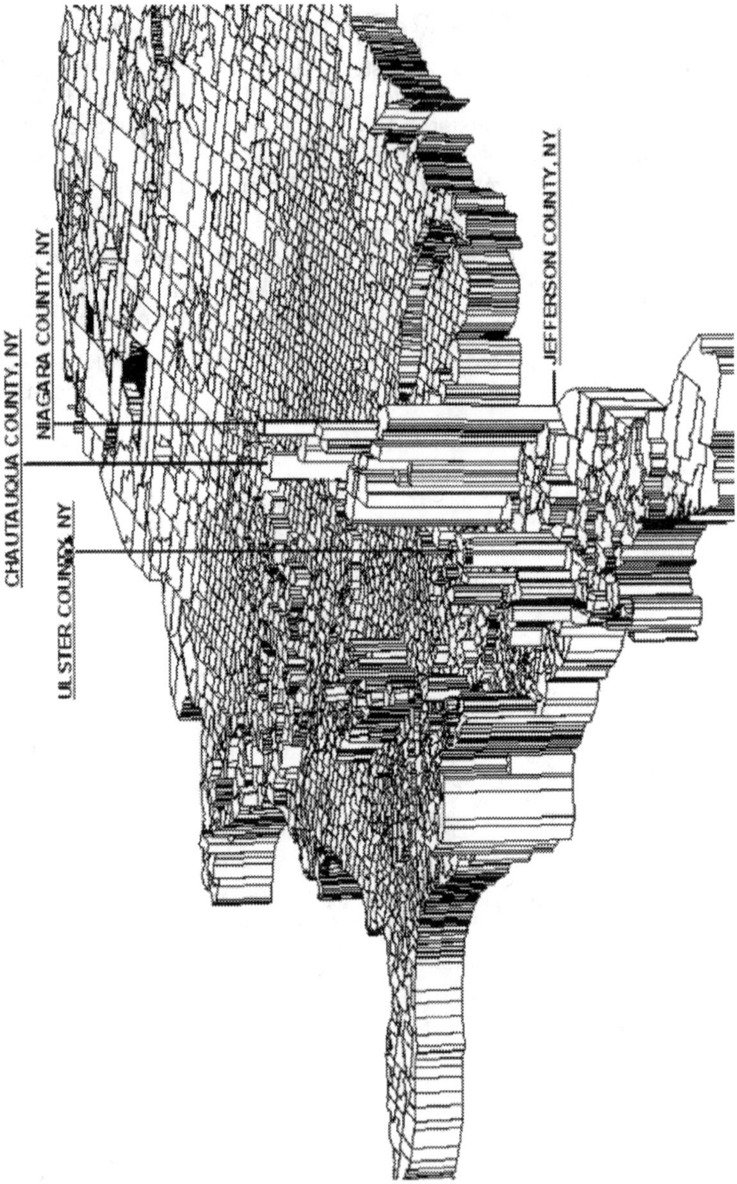

SHOT 4: BUSTER EASY (BJ4)

BUSTER: EASY

Detonation Date: 5 Nov 1951 • *Detonation Time:* 8:30 A.M. • *Area:* 7 • *Sponsor:* Los Alamos • *Yield:* 31 kt • *Radiation Level at Ground Zero:* 20 R/hr • *Height of Burst:* 1,314 ft • *Cloud Top Height:* 50,000 ft

The 24,000-, 30,000-, 40,000-, and 46,000-foot trajectories all crossed over Las Vegas, then moved southwest in a wide arc across the United States and Gulf. Among cities under the high-altitude nuclear clouds: Phoenix and Tucson, Arizona; Juarez, Mexico–El Paso, Laredo, San Antonio, Corpus Christi, Texas; Jacksonville, Florida.

10,000-Foot Trajectory: This part of the nuclear cloud moved southwest to Santa Barbara, Lompoc, and Santa Maria, California. It then followed the coastline to San Francisco where it crossed the 18,000-foot trajectory. It then moved back over Marysville, California, and into Nevada over the Smoke Creek Desert, where it continued Northeast. Over Dunphy, Nevada, it began to turn back to the south. It was last tracked over northern Milliard County in Utah.

18,000-Foot Trajectory: This part of the nuclear cloud followed essentially the same path as the 10,000-foot level, but after crossing that trajectory over San Francisco, it moved northeast, passing over Nevada, then Idaho. The cloud moved over the Teton Range into Wyoming, then drifted over Jackson and Thermopolis, finally leaving the state north of Kirtley. The cloud at 18,000-feet then moved across the northern tier of Nebraska counties before entering Iowa at Sioux City. Once in Iowa, it crossed over Storm Lake, Fort Dodge, Iowa Falls, and Cedar Falls. Soon after, it entered Illinois airspace at Elizabeth. From there it passed over Rockford, Woodstock, and Waukegan, then crossed into Michigan. Other cities 18,000 feet under the nuclear cloud included: MICHIGAN: Kalamazoo • OHIO: Toledo, Cleveland, and Youngstown • PENNSYLVANIA: Philadelphia • NEW JERSEY: Atlantic City.

RELATIVE TOTAL FALLOUT BJ EASY

NOV 5, 1951 31 KT

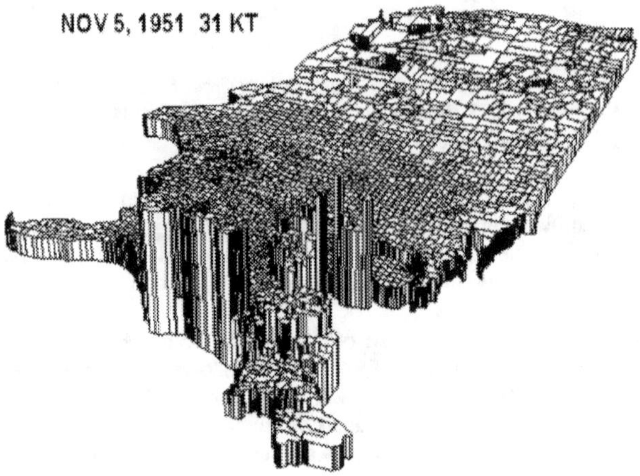

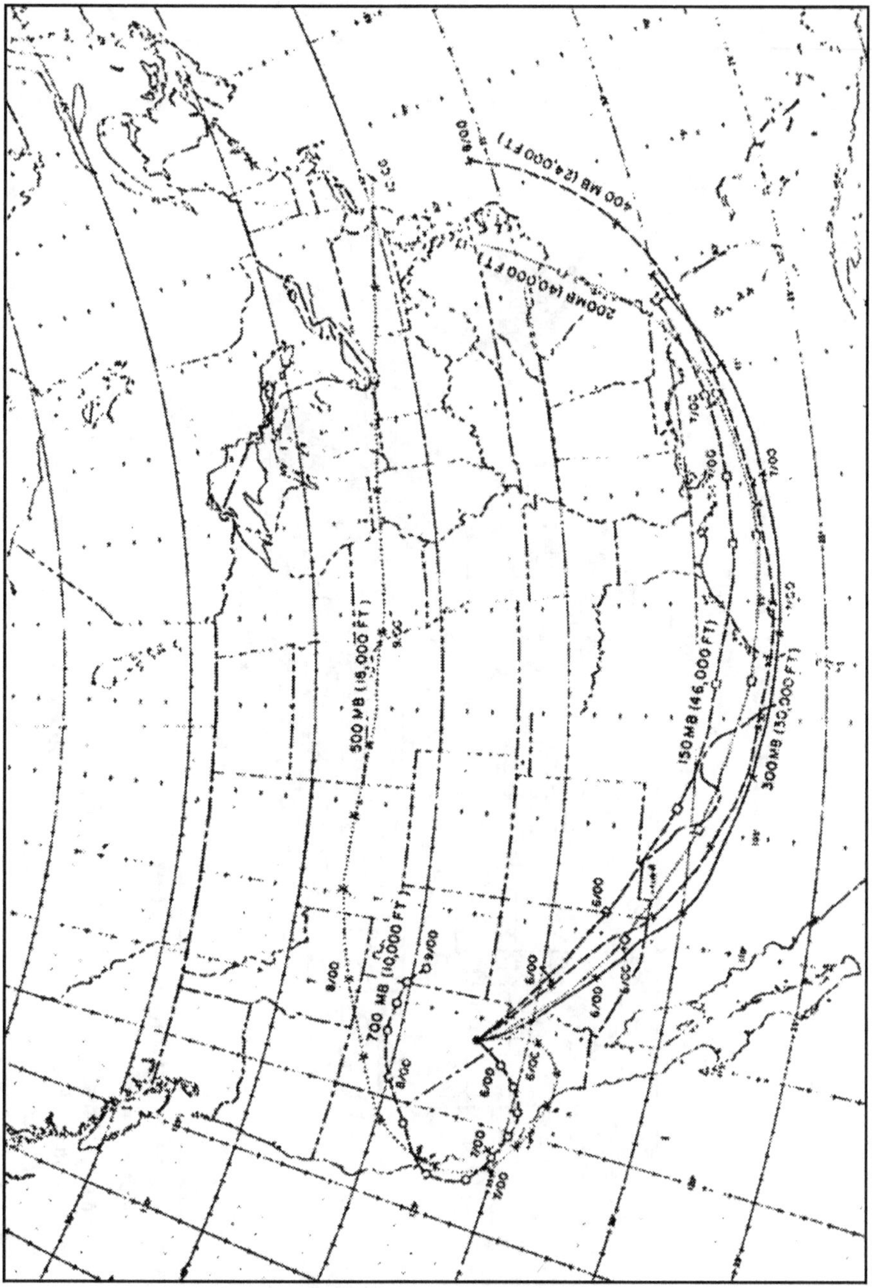

TRAJECTORY: BUSTER EASY (BJ4)

TOTAL FALLOUT: BUSTER EASY

Equal Intervals

nCi/square meter

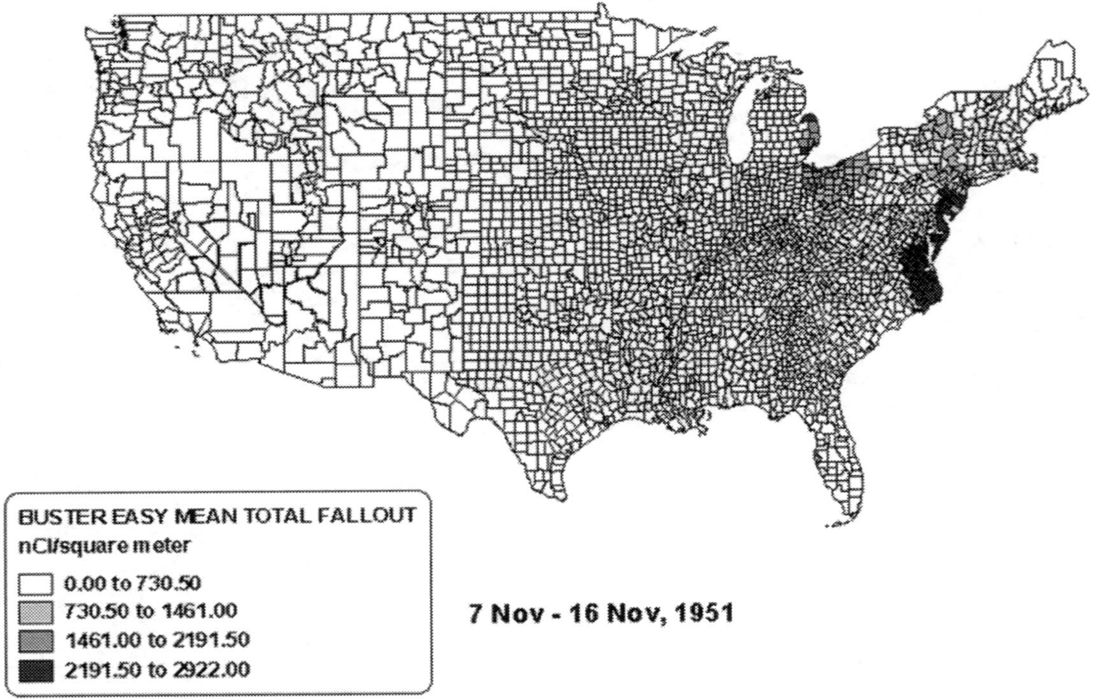

7 Nov - 16 Nov, 1951

TOTAL FALLOUT: BUSTER EASY

Equal Number

nCi/square meter

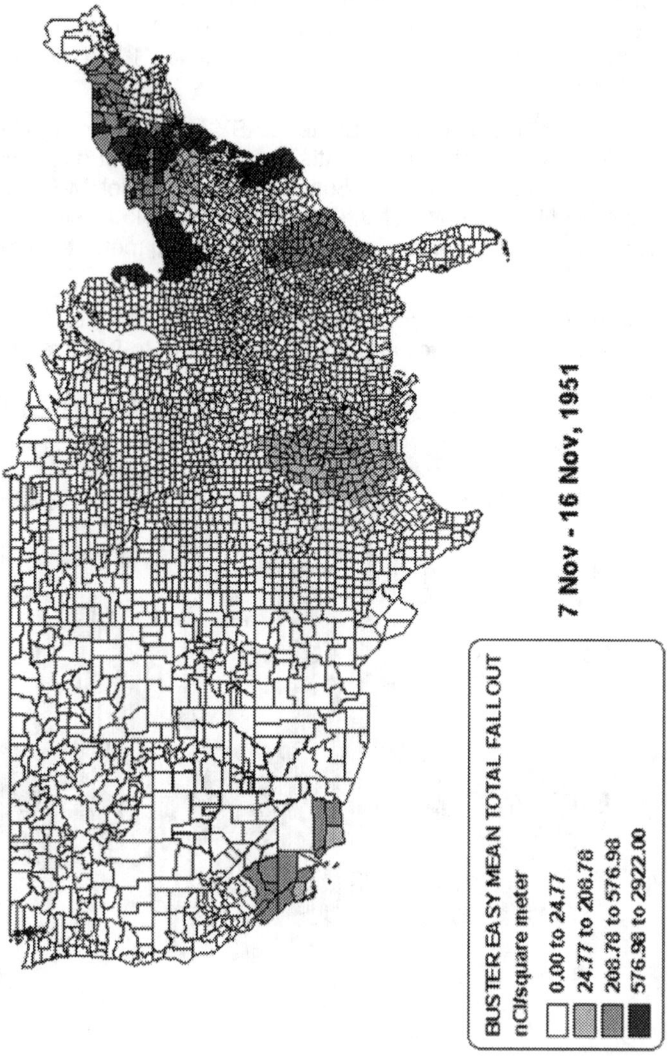

7 Nov - 16 Nov, 1951

CHARACTERISTICS: BUSTER EASY

Buster EASY ranks 2nd among the nuclear tests in the percentage of Co-58 in fallout and 6th for percentage of Tc-99m in fallout. Among the aboveground nuclear tests this shot ranks 1st in deposition of the radionuclide Nb-95m.

The deposition pattern for Buster EASY is most similar to Buster-DOG (R: 0.51; T: 33.07). The counties of highest fallout deposition included: Kent DE, New Castle DE, Sussex DE, Worcester MA, Burlington NH, Camden NJ, Cape May NJ, Cumberland NJ, Gloucester NJ and Mercer NJ. Estimated maximum activity: 24.35 μCi/sq meter total fallout on Nov 10, 1951.

The day one ratio of total fallout to I-131 was 121 to 1, ranking it 17th highest among all nuclear tests for total fallout-to-I-131 ratio.

RELATIVE TOTAL FALLOUT BJ EASY

NOV 5, 1951 31 KT

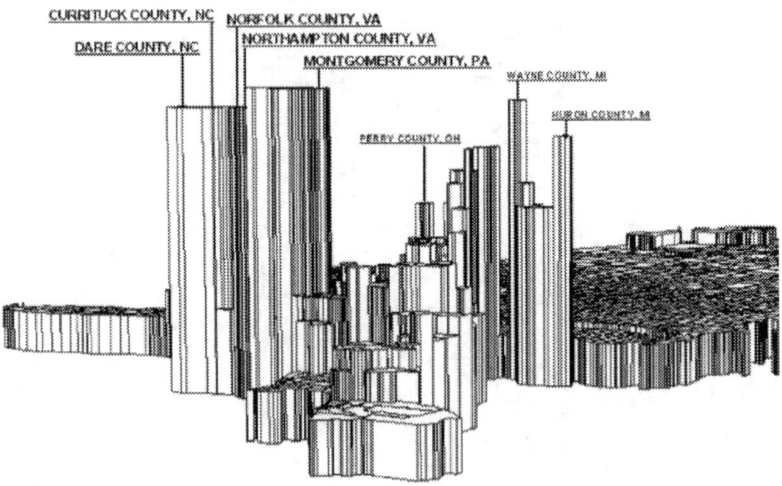

RELATIVE TOTAL FALLOUT BJ EASY

NOV 5, 1951 31 KT

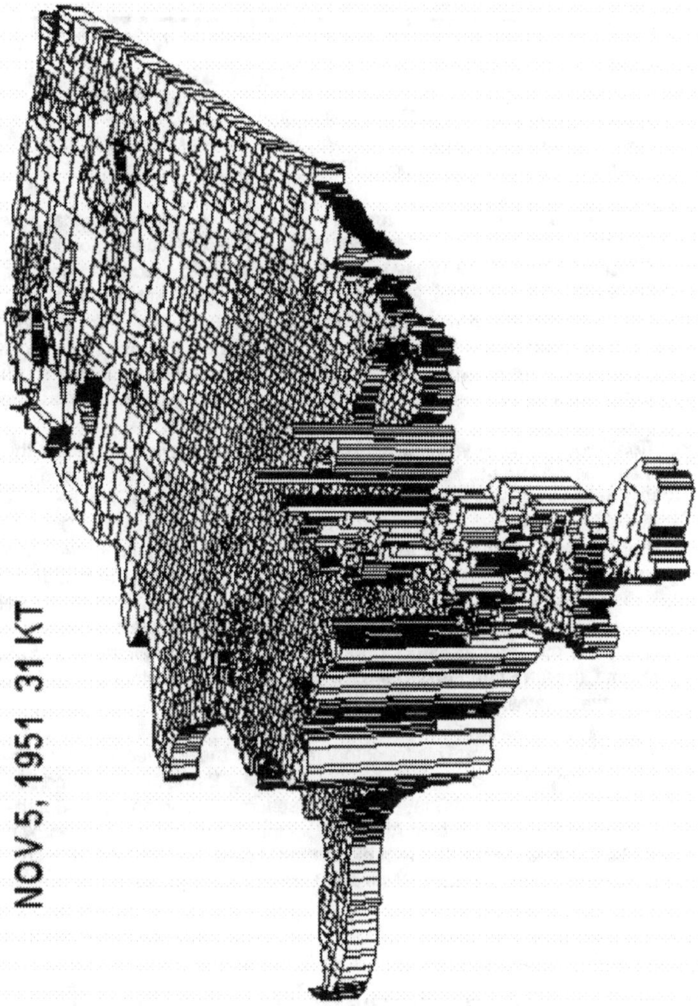

SHOT 5: JANGLE SUGAR (BJ5)

JANGLE: SUGAR

Detonation Date: 19 Nov 1951 • *Detonation Time:* 9:00 A.M. • *Area:* 9 • *Sponsor:* Department of Defense • *Yield:* 1.2 kt • *Radiation Level at Ground Zero:* 7,500 R/hr • *Height of Burst:* 3.5 ft • *Cloud Top Height:* 15,000 ft

Low-level debris from Sugar crossed north over Nye County, Nevada, passed along the border of Eureka and White Pine counties, and left Nevada through Elko County. It then crossed over the Duck Valley Indian Reservation and proceeded on toward Caldwell, Rayette, and Weiser, Indiana. From there, it curved west toward Huntington, Oregon, where it followed the track of now-U.S. 84 North across Baker, La Grande, and Pendleton. It then moved into the airspace of Washington State, where it passed over Richland, curled north around the Hanford Atomic Energy Works, and was then deflected south and west by the Saddle Mountains. The lower level was last detected over Yakima, Washington.

10,000-Foot Trajectory: Sugar's 10,000-foot section moved north over the Great Salt Lake, then crossed into the southeastern corner of Idaho at Malad City. It left the state near Irwin and entered Wyoming over the Teton Pass near Jackson. The cloud passed over Cody, Powell, and Lovell, then crossed into Montana airspace over the Crow Indian Reservation. From there it drifted over the villages of Bighorn, Sanders, Angela, and Union before it turned southeast toward Glendive. This section of the atomic cloud finally left the state at Wibaux. From there, the debris at 10,000 feet altitude entered North Dakota at Beach. It crossed Bullion Butte, New England, Elgin, Bismarck, Steele, Jamestown, Valley City, and finally left the state at Fargo. Upon entering Minnesota airspace, it crossed Georgetown, Ogema, (Itasca State Park), Deep River, Hibbing, and Virginia before leaving the state, and the United States, at Grand Portage, Minnesota.

14,000-Foot Trajectory: This trajectory moved northeast over Ogden, Utah and entered Wyoming airspace at Sage. It then crossed over to La Barge and the Wind River Indian Reservation. The cloud entered South Dakota airspace at Spearfish. It left South Dakota at Herreid and entered North Dakota at Ashley. From there, it moved over Jamestown and left the state at Grand Forks. The 14,000-foot cloud then moved into Minnesota airspace, crossing Thief Lake. It finally left the U.S. at Warroad, Minnesota, near Lake of the Woods.

CHARACTERISTICS: JANGLE SUGAR

Jangle SUGAR ranks 1st among the nuclear tests in the percentage of Nb-95m contained in fallout and 2nd for percentages of Sr-90, Y-90 and Y-91. The deposition pattern resulting from this test is most similar to that of Plumbbob SHASTA (R:0.37; T: 22.45). The county receiving the highest deposition of total fallout from Jangle SUGAR was Clearwater, ID. Other counties receiving high levels of fallout were Valley, ID and Boise, ID. Fallout from Jangle SUGAR was deposited across 615 counties (20% of the U.S.).

RELATIVE AVERAGE TOTAL FALLOUT: BJ SUGAR

NOV 19, 1951 1.2 KT

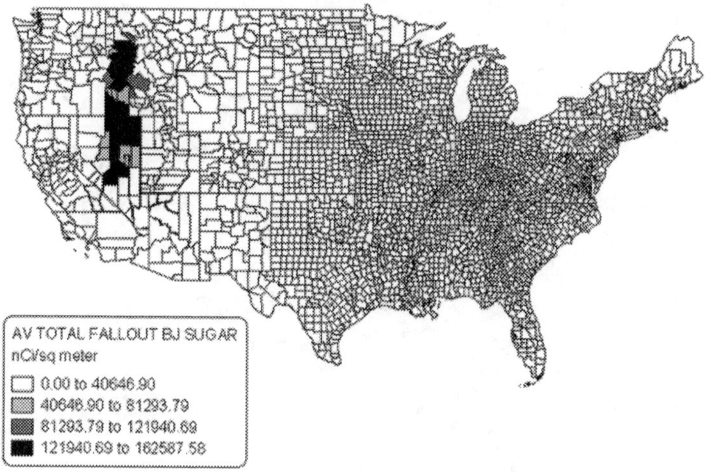

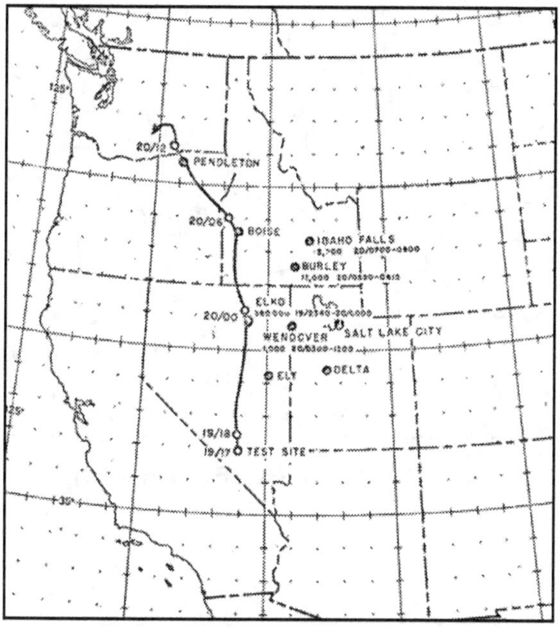

LOW-LEVEL DEBRIS: JANGLE SUGAR (BJ5)

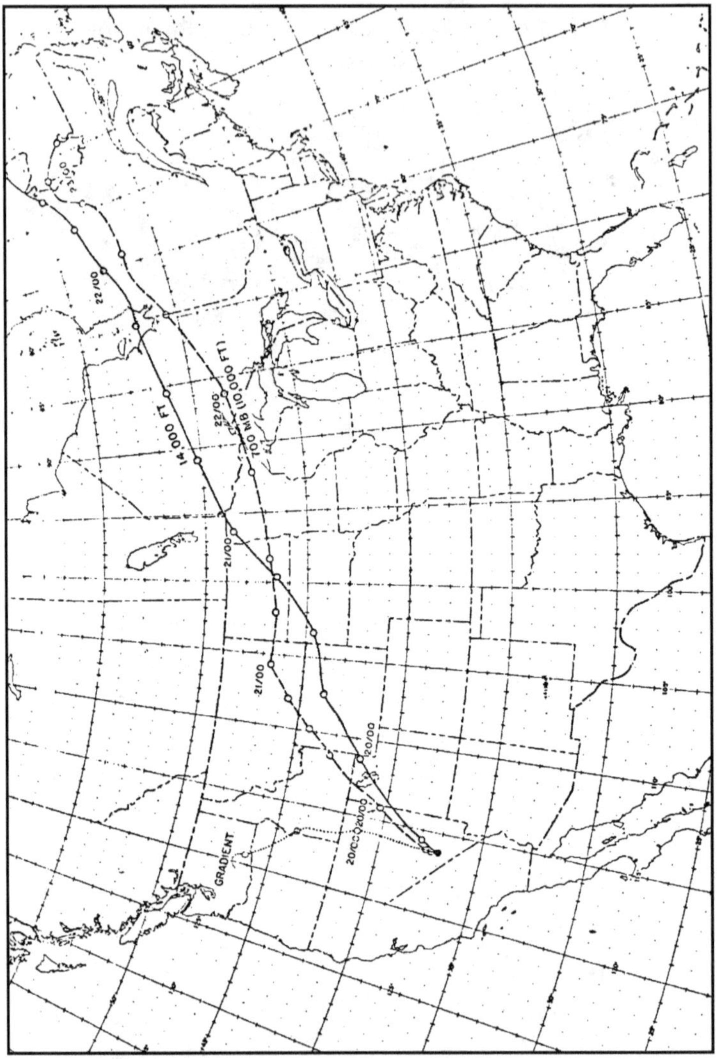

TR

AJECTORY: JANGLE SUGAR (BJ5)

SHOT 6: JANGLE UNCLE (BJ6)

JANGLE: UNCLE

Detonation Date: 29 Nov 1951 • *Detonation Time:* 12:00 Noon • *Area:* 10 • *Sponsor:* Los Alamos, Department of Defense • *Yield:* 1.2 kt • *Radiation Level at Ground Zero:* 7,500 R/hr • *Height of Burst: – 17 ft; underground burst: filled shaft in Nevada soil* • *Cloud Top Height:* 11,500 ft

The low, surface-level portion of the Uncle cloud traveled due north over Nye County, Nevada, across the Diamond Mountains and over Huntington, Jiggs, the Te-Moak Indian Reservation, and Elko. It split up into two clouds over North Fork, Nevada. Shortly thereafter, tracking of this portion of the Uncle cloud was terminated.

10,000-Foot Trajectory: This part of the cloud crossed Nye County and moved toward the northeast over Railroad Valley and Lund. It crossed into Utah at Trout Creek, then passed over Salt Lake City. It left the state near the town of Upton. After entering Wyoming airspace, it crossed the towns of Mills, Church Buttes, Farson, Rawlins, Como, Medicine Bow, Garrett, Dwyer, and Guernsey. Other cities under the section of the nuclear cloud at 10,000-foot altitude: NEBRASKA: Agate, Crawford, Chadron, Whiteclay • SOUTH DAKOTA: (Slim Butte Mountain), Oglala, Potato Creek, Norris, White River, Kimball, Mitchell, Sioux Falls • MINNESOTA: Luverne, Wilmot, Windom, Odin, Truman, Matawan, Blooming Prairie, Rochester, Winona • WISCONSIN: Ettrick, Black River Falls, Pray, Arpin, Neopit, Beaver, Wausaukee • MICHIGAN: Banat, Carney, Steuben, Seney, Paradise • MAINE: (Kelly Brook Mountain), (Rocky Mountain), St. John, Fort Kent, Notre Dame.

CHARACTERISTICS: JANGLE UNCLE

Jangle UNCLE ranks 2nd among the nuclear tests for percentage of U-237 in fallout. The deposition pattern from this test is most similar to that of Tumbler-Snapper BAKER (TS2) (R:0.22; T: 12.31). The county of highest deposition was White Pine, NV where the estimated maximum (95% confidence interval) on Nov 29, 1951 was 184.08 μCi/sq meter. Other high-deposition counties were Lincoln(2) NV with an estimated maximum activity of 139 μCi/sq meter and Nye(2) NV with an estimated maximum activity of 112.98 μCi/sq meter. The day one ratio of total fallout to I-131 for this shot is moderate at 98.6 to 1.

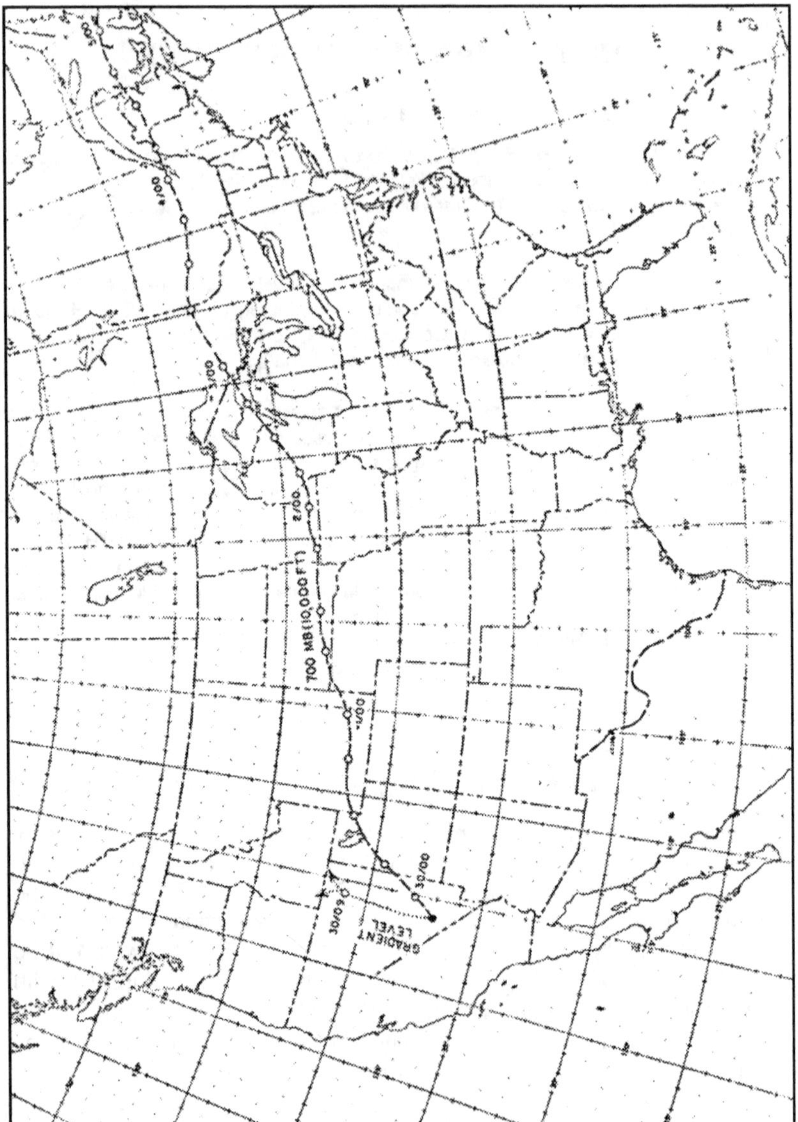

TRAJECTORY: JANGLE UNCLE (BJ6)

RELATIVE AVERAGE TOTAL FALLOUT: BJ UNCLE

NOV 29, 1951 1.2 KT

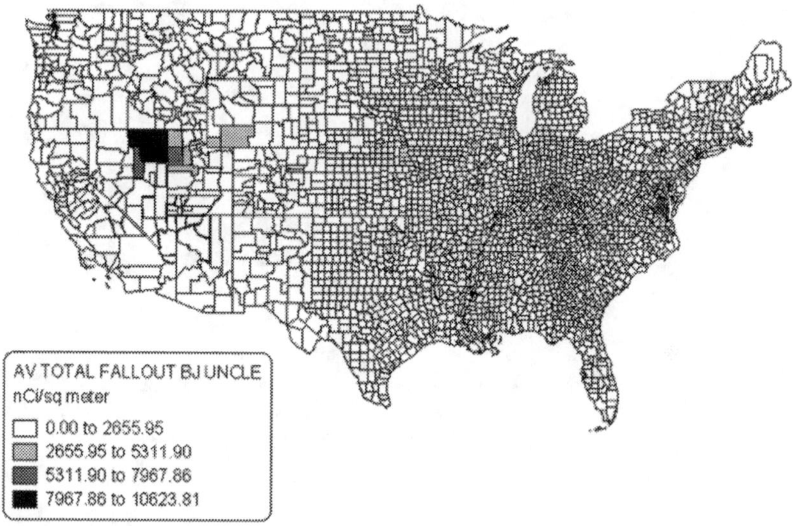

SECTION 3

TUMBLER-SNAPPER SERIES

Apr 1-Jun 5, 1952

SHOT 1: TUMBLER-SNAPPER ABLE (TS-1)

TUMBLER-SNAPPER: ABLE

Detonation Date: 1 Apr 1952 • *Detonation Time:* 9:00 A.M. • *Area:* Frenchman's Flat • *Sponsor:* Los Alamos, Department of Defense • *Yield:* 1 kt • *Radiation Level at Ground Zero:* 5 R/hr • *Height of Burst:* 793 ft • *Cloud Top Height:* 16,200 ft

10,000-Foot Trajectory: NEVADA: Alamo, Hiko, Pioche • UTAH: Wah Wah Springs, Oak City, Fountain Green, Milburn, Soldier Summit, Roosevelt, Bridgeport • WYOMING: Rawlins, Wheatland, Hawk Springs • NEBRASKA: Kimball • COLORADO: Sterling, Yuma, Kirk, Stratton, Cheyenne Wells • KANSAS: Syracuse, Ulysses, Arkalon, Hayne, Liberal • OKLAHOMA: Forgan, Beaver, Slapout, Harmon, Taloga, Geary, Seminole, McAlester, Page • ARKANSAS: Mena, Sulphur Springs, Hot Springs, Pine Bluff, Crumrod • MISSISSIPPI: Hillhouse, Batesville, Bonneville, Doskie • TENNESSEE: Lowryville, Topsy, Mount Pleasant, Columbia, Nashville, Portland • KENTUCKY: Glasgow, Campbellsville, Harrodsburg, Stamping Ground • OHIO: Georgetown, Hillsboro, London, Port Clinton • MICHIGAN: Detroit, Pontiac, Flint, Saginaw, Bay City, Clare, Big Rapids, Muskegon • INDIANA: Portage, Francisville, Sitka, Young America, Kokomo, Anderson, Rushville, New Trenton • OHIO: Cincinnati, Bethel, Georgetown, Portsmouth, Buckhorn, Gallipolis • WEST VIRGINIA: Ashton, Left Hand, Sutton, Harper, Martinsburg • MARYLAND: Baltimore • DELAWARE: Wilmington • PENNSYLVANIA: Philadelphia • NEW JERSEY: Trenton, Twin Rivers, Englishtown, Leonardo • NEW YORK: Lindenhurst, Mastic Beach, Southhampton • MASSACHUSETTS: Martha's Vinyard, Hyannis, Barnstable, E. Orleans.

16,000-Foot Trajectory: NEVADA: Rox, Carp • UTAH: St. George, Leeds, Mount Carmel Junction, Mexican Hat, Navajo Indian Reservation • COLORADO: Cortez, Durango,. Summitville, Alamosa, Fort Garland, Trinidad • OKLAHOMA: Black Mesa Summit (highest point in Oklahoma) near Kenton, Castaneda, Keyes, Eva, Guyman, Gray, Catesby, Woodward, Seiling, Longdale, Bison, Stillwater, Tulsa, Tahlequa, Stilwell • ARKANSAS: Mountainburg, Mountaintop, Clarksville, Russellville, Beebe, Brinkley, Marianna • MISSISSIPPI: Prichard, Senatobia, Abbeville, Tupelo, Tremont • ALABAMA: Bexar, Hamilton, Jasper, Birmingham, Bemiston, Woodland • GEORGIA: Texas, Hogansville, Topeka Junction, Macon, Irwinton, Wrightsville, Rocky Ford, Springfield • SOUTH CAROLINA: Ridgeland, Burton, Beaufort, Edisto Island.

CHARACTERISTICS: TS ABLE

As with most of the shots in the previous series, the device was based on the Nagasaki Mk4 bomb, which weighed in at 10,500 lbs. The shot Tumbler-Snapper ABLE ranks 2nd among the nuclear tests for percentage of Np-239 contained in fallout. The deposition pattern from this shot was most similar to that of Ranger BAKER-1 (R: 0.22; T: 12.24). The counties of highest deposition included the Arkansas counties of Franklin, Johnson, Montgomery, Newton, Perry, Pope and Yell; the Missouri counties of Barry and Christian; and the Kansas county of Crawford. Estimated maximum activity: 254 μCi/sq meter total fallout on April 3,1952. Other counties receiving high fallout with an estimated maximum fallout activity of 241 μCi/sq meter each were the Nebraska counties of Banner, Dawes and Sioux.

The day one ratio of total fallout to I-131 for Tumbler-Snapper ABLE is 206.7 to 1, which is the highest ratio among all aboveground nuclear tests, including SEDAN. Fallout from Tumbler-Snapper ABLE was deposited in 2089 counties (68 percent of the U.S.).

RELATIVE AVERAGE TOTAL FALLOUT: TS ABLE

APR 1, 1952 1 KT

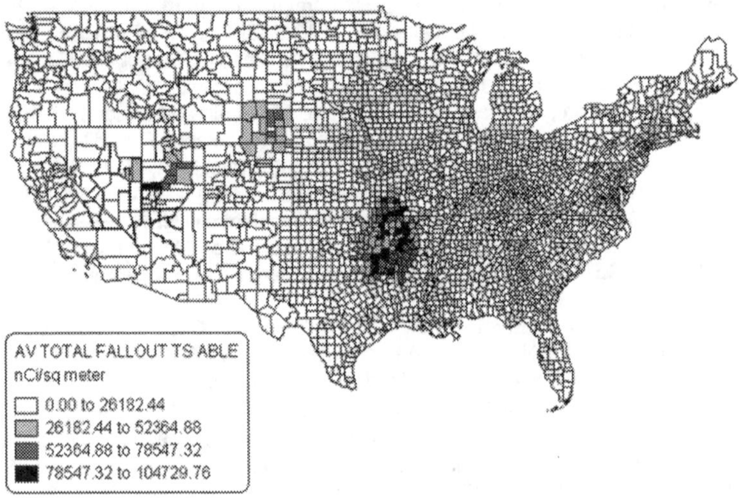

RELATIVE TOTAL FALLOUT: TS ABLE

APR 1, 1952 1 KT

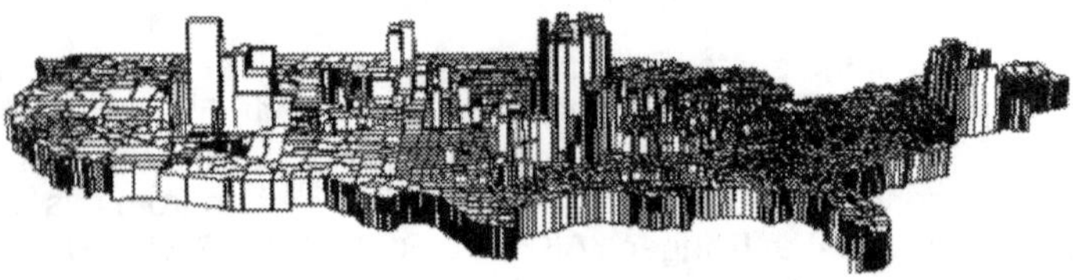

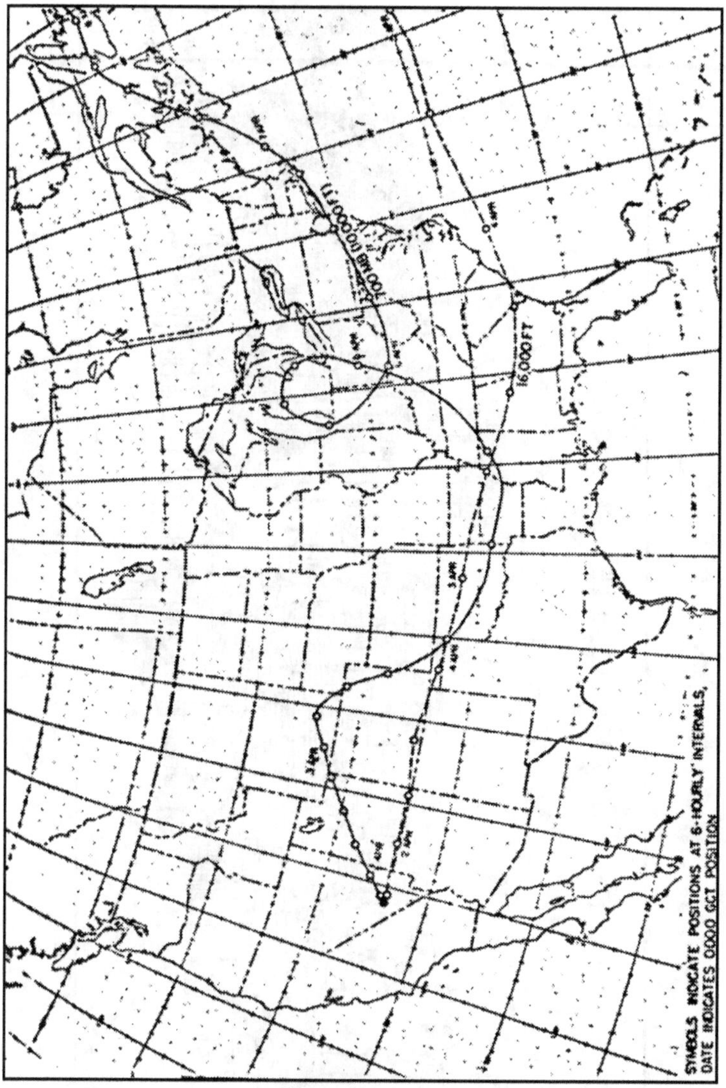

TRAJECTORY: TUMBLER-SNAPPER ABLE (TS1)

FALLOUT PATH: TS ABLE APR 1 1952

SHOT 2: TUMBLER-SNAPPER BAKER (TS2)

TUMBLER-SNAPPER: BAKER

Detonation Date: 15 Apr 1952 • *Detonation Time:* 9:30 A.M. • *Area:* 7–3 • *Sponsor:* Los Alamos, Department of Defense • *Yield:* 1 kt • *Radiation Level at Ground Zero:* 4.5 R/hr • *Height of Burst:* 1,109 ft • *Cloud Top Height:* 15,700 ft

10,000-Foot Trajectory: NEVADA: Goodsprings, Sandy • CALIFORNIA: Cima, Ludlow, San Bernadino, Redlands, Ontario, Pomona, Whittier, Buena Park, Long Beach, Redondo Beach, Inglewood • ARIZONA: San Luis, Somerton, Wellton, Growler, Quartzite, Bouse, Signal, Yucca, Kingman, Temple Bar • NEVADA: Overton, Glendale, Galt, Pioche • UTAH: Garrison, Provo, American Fork, Murray, Kerns, Salt Lake City, Syracuse • NEVADA: Shafter, Currie, Cherry Creek, Warm Springs, Alamo, Las Vegas • CALIFORNIA: Goffs, Vidal, Blythe, Palo Verde.

16,000-Foot Trajectory: NEVADA: Las Vegas • ARIZONA: Hualpi Indian Reservation, (Grand Canyon National Monument) • UTAH: Rockville, Springdale, Long Valley Junction, Hatch, Lund, Zane, Beryl, Hamlin Valley • NEVADA: Ursine, Pioche, Adaven, Warm Springs, Tonopah, Arlemont • CALIFORNIA: Benton Station, Benton, (Yosemite National Park), Bridgeport • NEVADA: Sweetwater, Babbitt, Hawthorne, Thorne, (Quartz Mountain), Austin, Winnemucca, Paradise Valley, Rebel Creek • IDAHO: Triangle, Silver City, Murphy, Nampa, Boise, Horsehoe Bend, Crouch, Warm Lake, Yellow Pine • MONTANA: Hamilton, Bearmouth, Helmville, Blackleaf, Conrad, Shelby, Sunburst, Sweetgrass.

RELATIVE AVERAGE TOTAL FALLOUT: TS BAKER

APR 15, 1952 1 KT

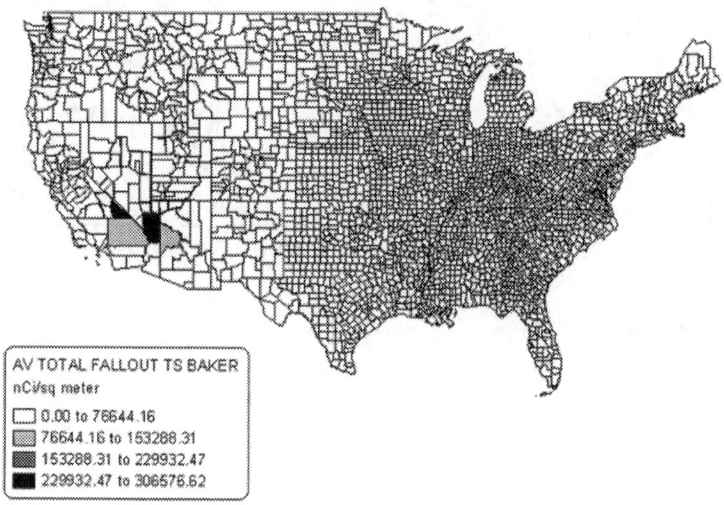

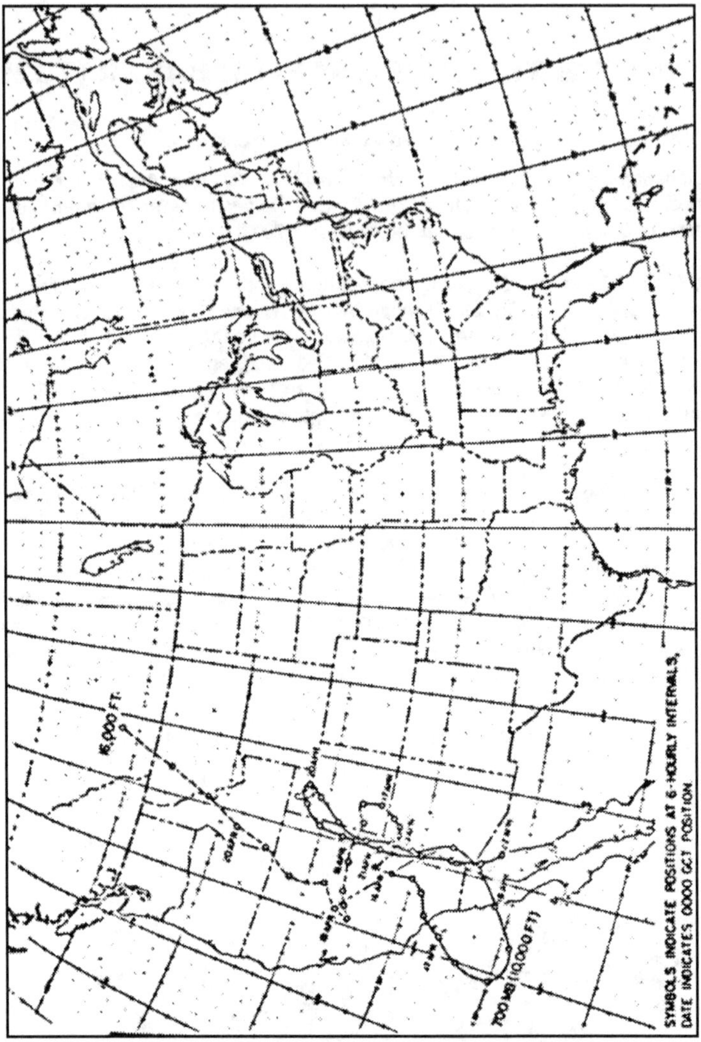

TRAJECTORY: TUMBLER-SNAPPER BAKER (TS2)

TOTAL FALLOUT: TUMBLER-SNAPPER BAKER

Equal Number

nCi/square meter

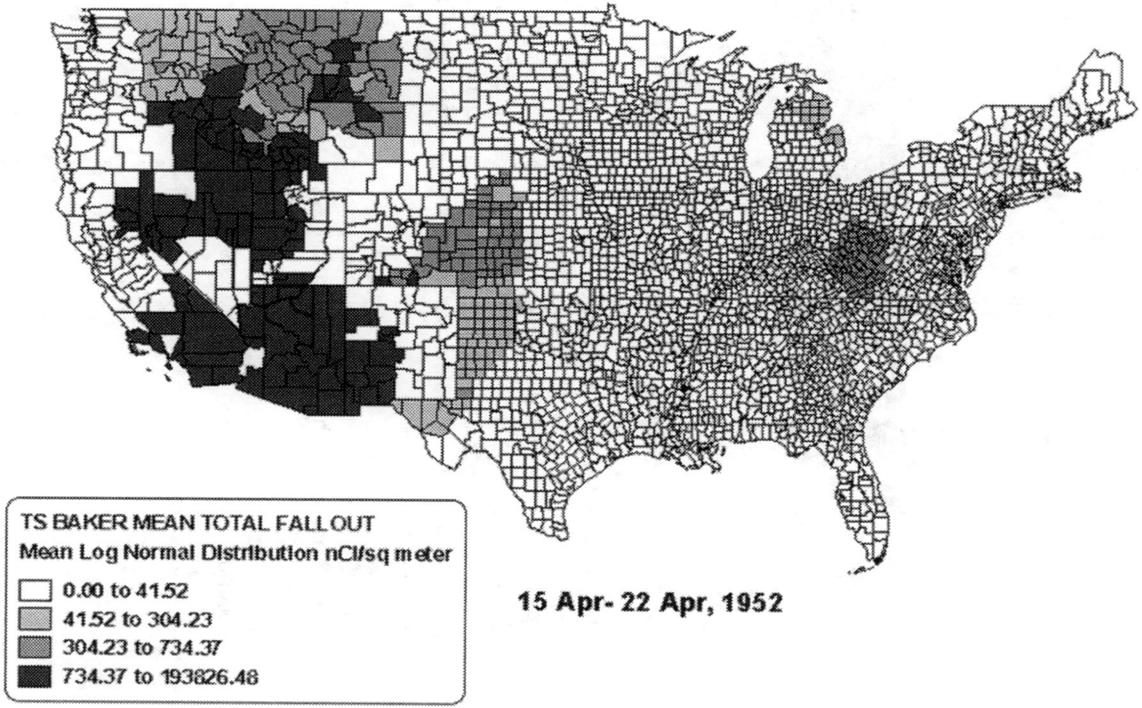

RELATIVE TOTAL FALLOUT: TS BAKER

APR 15, 1952 1 KT

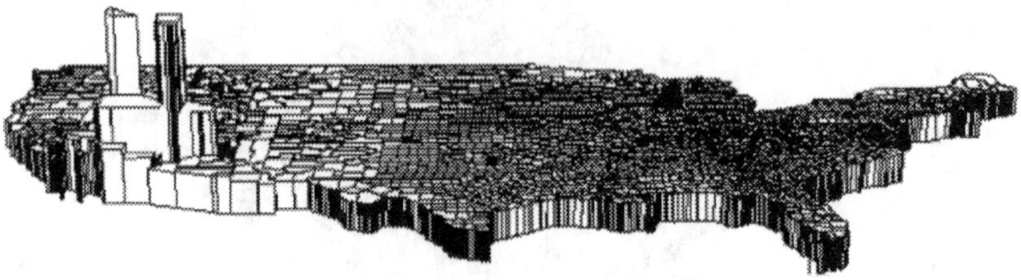

CHARACTERISTICS: TS BAKER

The Tumbler-Snapper BAKER device was based on the Nagasaki bomb design (Mk4) and weighed approximately 10,800 lbs. Tumbler-Snapper BAKER ranks 3rd for fallout percentage of Np-239 and U-239 and 4th for U-240. The deposition pattern for Tumbler-Snapper BAKER is most similar to Teapot CLIMAX (TP11). The county of highest deposition was Inyo, CA with an estimated maximum activity of 569.46 μCi/sq meter total fallout.

Fallout from this nuclear test was deposited on 406 counties, comprising only 13 percent of the total U.S. counties.

The ratio of total fallout to I-131 for Tumbler-Snapper BAKER ranks second highest among nuclear tests at 201.5 to 1.

SHOT 3: TUMBLER-SNAPPER CHARLIE (TS3)

TUMBLER-SNAPPER: CHARLIE

Detonation Date: 22 Apr 1952 • *Detonation Time:* 9:30 A.M. • *Area:* 7-3 • *Sponsor:* Los Alamos, Department of Defense • *Yield:* 31 kt • *Radiation Level at Ground Zero:* 0.15 R/hr • *Height of Burst:* 3,447 ft • *Cloud Top Height:* 42,000 ft

10,000-Foot Trajectory: NEVADA: Pahrump • CALIFORNIA: Shoshone, Tecopa, Fort Irwin, Pioneer Point, Little Lake, Kings, (Canyon National Park), June Lake, (Mono Lake), Coleville, Markleeville • NEVADA: Gardnerville, Carson City, Crystal Bay • CALIFORNIA: Truckee, Portola, Susanville, Bieber, White Horse, Tule Lake • OREGON: Malin, Beatty, Silver Lake, Millican, (Stevenson Mountain), Mayville, Condon, Arlington • WASHINGTON: Whitcomb, Horse Heaven Hills, Prosser, Ephrata, Soap Lake, Mansfield, Disautel, Havillah, Chesaw • MICHIGAN: Sault Ste. Marie, Barbeau, Hessel, Freedon, Boyne Fals, Lansing, Battle Creek, Kinderhook • INDIANA: Orland, Fort Wayne, Muncie, Connersville, Batesville, Madison • KENTUCKY: Carrollton, Frankfort, Sunnybrook • TENNESSEE: Chanute, Boatland, Crossville, Chattanooga • GEORGIA: Dalton, Carrollton, LaGrange • ALABAMA: Lafayette, Midway, Dothan • FLORIDA: Campbellton, Marianna, Apalachicola, Tampa, Arcadia, Homestead, Florida City, Key Largo.

18,000-Foot Trajectory: CALIFORNIA: Beacon Station, San Bernadino, Anaheim, Santa Ana, Buena Park, Whittier, San Mateo, San Francisco, Santa Rosa, Ukia, Willits, Blue Lake, Gasquet • OREGON: Remote, Swisshome, Tidewater, Eddyville, Willamina, Blaine, Westport • WASHINGTON: Grays River, Doty, Olympia, Bremerton, Burlington, Mount Baker • MINNESOTA: Laurel, Silverdale, Aurora, Castle Danger • WISCONSIN: Ironwood, Pulaski, Green Bay, Two Creeks • INDIANA: Gary, Hammond • ILLINOIS: Kankakee, Rantoul, Centralia, Carbondale • MISSOURI: Cape Girardeau, Belmont • KENTUCKY: Hickman • TENNESSEE: Nankipoo, Collierville, Rossville • MISSISSIPPI: Byhalia, Eupora, Meridian • ALABAMA: Bogueloosa, St. Stephens, Perdido • FLORIDA: Pensacola, Fort Meyers, Hialeah, Miami.

24,000-Foot Trajectory: This part of the cloud followed the California-Nevada, California-Arizona border, then up through San Diego. The other cities are: CALIFORNIA: El Centro, Victorville, Fresno, Merced, Modesto, Sacramento, Burney, Dorris • OREGON: Klamath Falls, Belknap Springs, Sandy • WASHINGTON: (Mount St. Helens), Puyallup, Auburn, Kent, Bellevue, Everett, Sumas • MINNESOTA: Grand Marais • MICHIGAN: Beacon Hill, Houghton, Negaunee, Garden, Traverse City, Battle Creek • INDIANA: Kendallville, Marion, Indianapolis, Lamar • KENTUCKY: Owensboro, Hopkinsville • TENNESSEE: Clarksville, Fort Campbell, Wayland Springs • ALABAMA: Florence, Birmingham, Montgomery, Dothan • FLORIDA: Chattahoochee, Tallahassee, Williston, Daytona Beach.

30,000-Foot Trajectory: NEVADA: Las Vegas • ARIZONA: Signal, Phoenix, Casa Grande, S. Tucson, Bisbee, Douglas • TEXAS: Eagle Flat, Rankin, Barnhart, San Marcos, Cuero, Victoria, Port O'Connor • FLORIDA: Crystal River, Ocala, Jacksonville • NORTH CAROLINA: Wilmington, Jacksonville, New Bern, Elizabeth City.

40,000-Foot Trajectory: ARIZONA: Grand Canyon, Flagstaff • NEW MEXICO: Zuni Indian Reservation, Grants, Los Alamos, Amistad • TEXAS: Hartley, Dumas, Borger, Canadian • OKLAHOMA: Regdon, Oklahoma City, Muskogee • ARKANSAS: Fayetteville, Mountain Home • MISSOURI: Pontiac, Mountain View, Ellington, Minimum, St. Marys • ILLINOIS: Rockwood, Mount Vernon, Bridgeport • INDIANA: Vincinnes, Indianapolis, Anderson, Muncie, Portland • OHIO: Lima, Lorain • NEW YORK: Buffalo, Rochester, Mexico, Camden, Port Henry • VERMONT: Montpelier, Barre • NEW HAMPSHIRE: Lisbon • MAINE: Augusta, Bar Harbor.

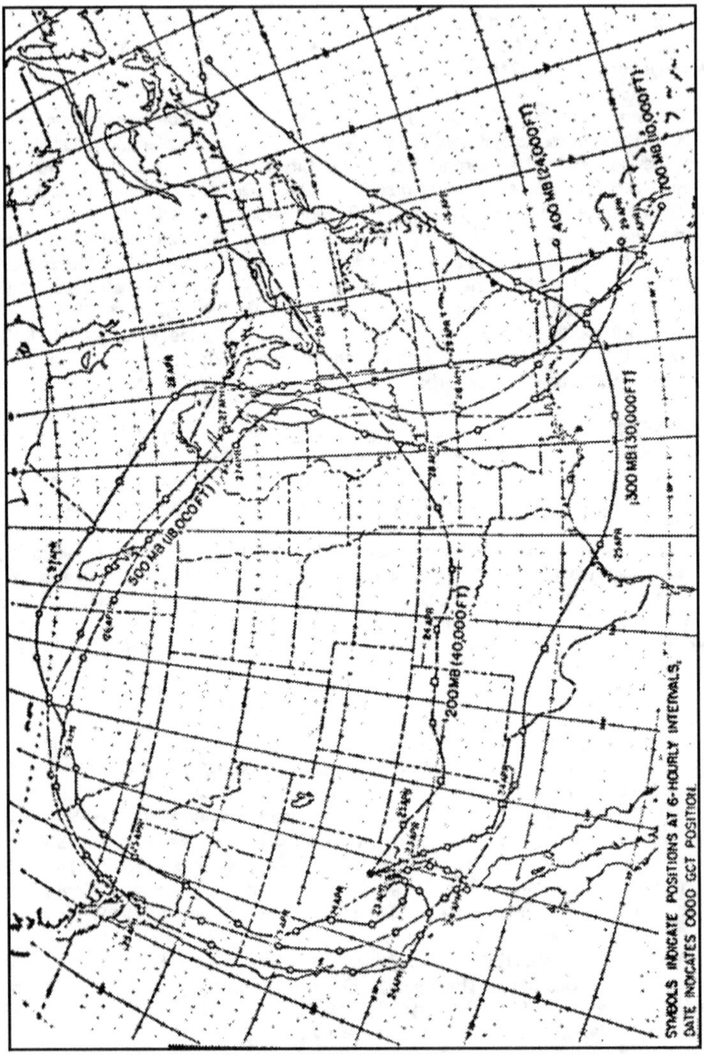

TRAJECTORY: TUMBLER-SNAPPER CHARLIE (TS3)

TOTAL FALLOUT: TUMBLER-SNAPPER CHARLIE

Equal Intervals

nCi/square meter

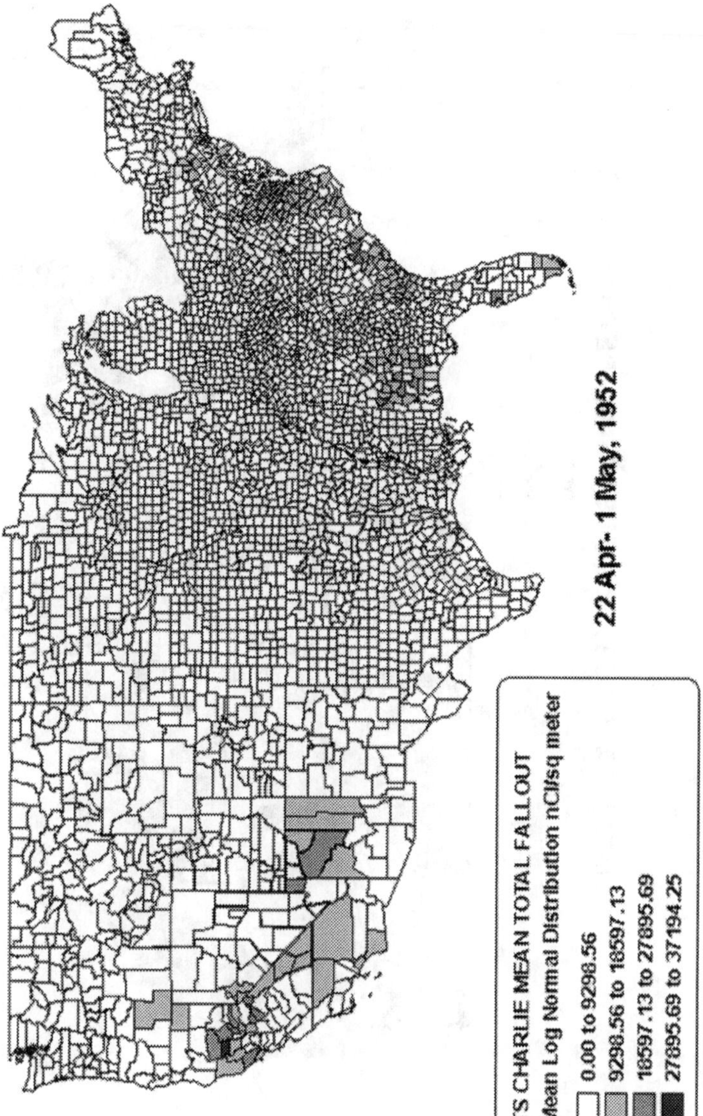

22 Apr- 1 May, 1952

TOTAL FALLOUT: TUMBLER-SNAPPER CHARLIE

Equal Number

nCi/square meter

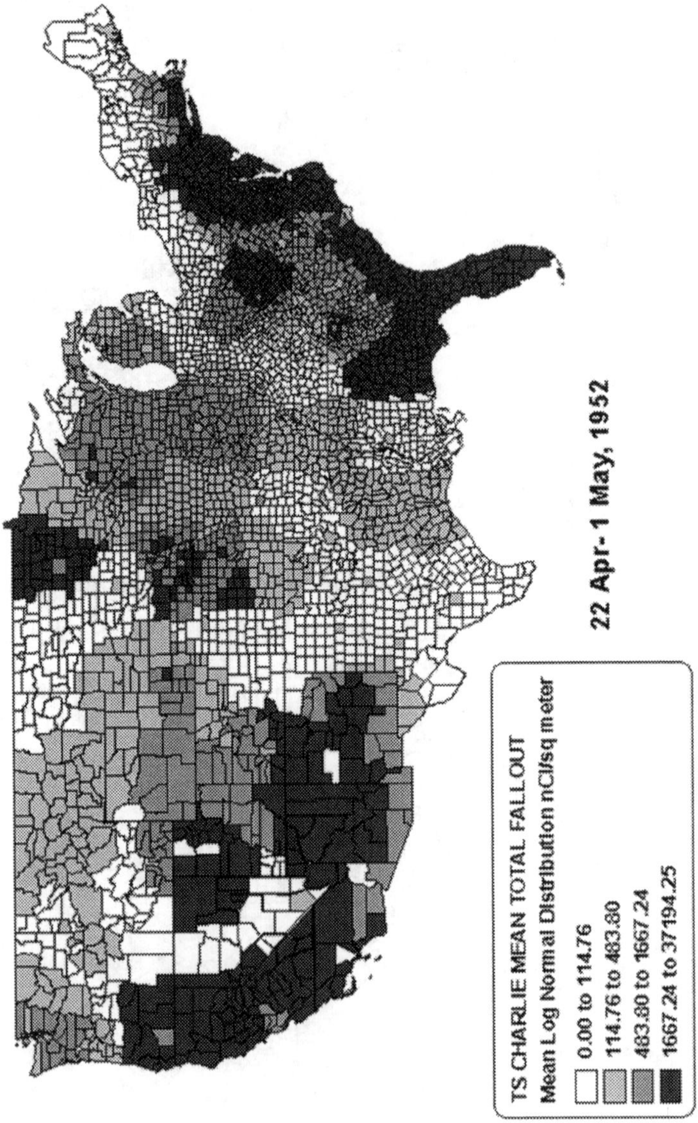

22 Apr- 1 May, 1952

FALLOUT IN EASTERN US FROM: TS CHARLIE

APR 22, 1952 31 KT

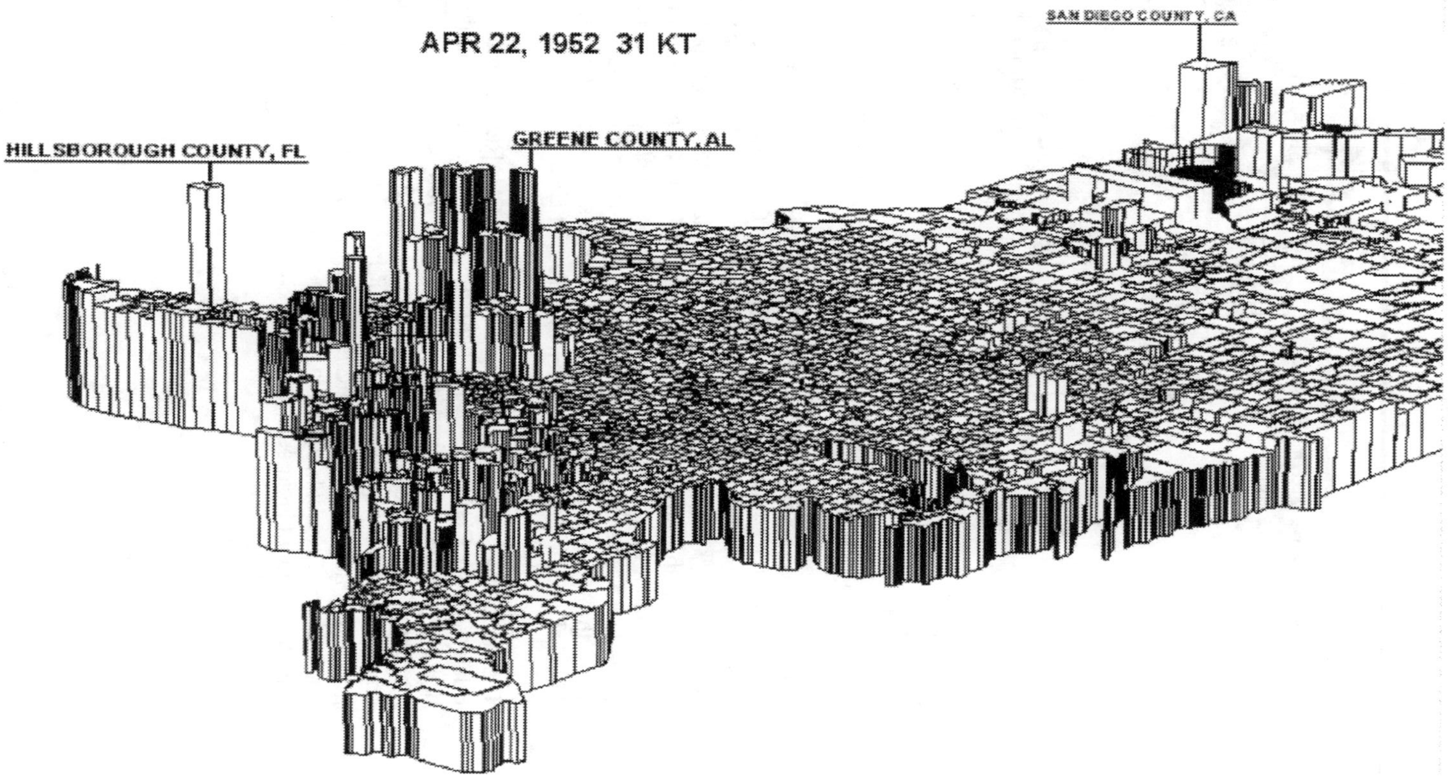

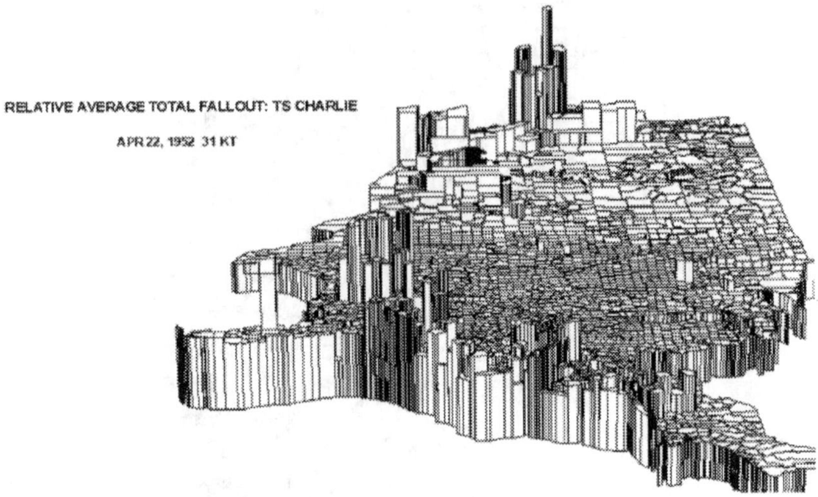

CHARACTERISTICS: TS CHARLIE

The device was based on the Mk4 design, similar to the Nagasaki bomb, and weighed 10,440 lbs. Tumbler-Snapper CHARLIE is unique in that it ranks 1st among the nuclear tests, for percentage of Co-57 and Pr-144 in fallout and 2nd for percentage of U-240 in fallout. In addition, this test ranks 1st in the amount of the radionuclide Pr-144 deposited across the U.S. Fallout from this test was deposited on 3062 counties, (99 percent of the U.S.). The day one total fallout-to-I-131 for this nuclear test is 153.3 to 1, the 4th highest among all nuclear tests (after Tumbler-Snapper ABLE, Tumbler-Snapper BAKER, and Teapot ESS).

The deposition pattern for Tumbler-Snapper CHARLIE is most similar to Buster-Jangle CHARLIE (BJ2) (R:0.26; T: 15.19). The county with the highest amount of fallout from this test was Glenn CA: Estimated maximum activity: 225.29 μCi/sq meter of fallout on April 25, 1952. Other counties receiving high levels of fallout from this nuclear test (estimated maximum activity: 163.28 μCi/sq meter each) included the Alabama counties of Conecuh, Geneva, Greene, Lowndes and Wilcox; and the Florida county of Holmes.

SHOT 4: TUMBLER-SNAPPER DOG (TS4)

TUMBLER-SNAPPER: DOG

Detonation Date: 1 May 1952 • *Detonation Time: 8:30 A.M.* • *Area:* 7, Target 3 • *Sponsor:* Los Alamos • *Yield:* 19 kt • *Radiation Level at Ground Zero:* 10 R/hr • *Height of Burst:* 1,040 ft • *Cloud Top Height:* 44,000 ft

18,000-Foot Trajectory: UTAH: St. George • ARIZONA: Flagstaff, Prescott, Truxton, Crozier, Valentine • UTAH: St. George, (Zion National Park), Cedar City, Parowan, Thompsonville, Castle Dale, Vernal • COLORADO: Dinosaur, Sunbeam • WYOMING: Hanna, Medicine Bow, Casper, Moorcroft, (Devil's Tower National Monument), Hulett • MONTANA: Alzada, Albion, Capitol • SOUTH DAKOTA: Camp Hook, Ladner, Ludlow • NORTH DAKOTA: Bowman, New England, Antelope, Washburn, McClusky, Devil's Lake, Park River, Cavalier, Pembina • PENNSYLVANIA: Erie, Oil City, DuBois, Clearfield, Carlisle, York, Lancaster • MARYLAND: Taneytown, Eldersburg, Parksville, Pikesville, Baltimore, Edgemore, Centreville, Henderson • DELAWARE: Hartley, Dover, Kilts Hummock • NEW JERSEY: Villas, Rio Grande, Wildwood.

CHARACTERISTICS: TS DOG

The deposition pattern for Tumbler-Snapper DOG is most similar to shot Teapot TESLA (TP3) (R:0.41; T:25.10). The county of highest deposition was Iron, UT with an estimated maximum activity of 40.34 µCi/sq meter on May 1, 1952. Other counties receiving high fallout from this nuclear test were the Indiana counties of Carroll, Clinton and Tippecanoe, each receiving an estimated maximum activity level of 22.66 µCi/sq meter on May 5, 1952.

The day one ratio of total fallout to I-131 for this test is 124.2 to 1, ranking it 12th among all aboveground nuclear tests.

FALLOUT PATH: TS DOG

MAY1, 1952 19 KT

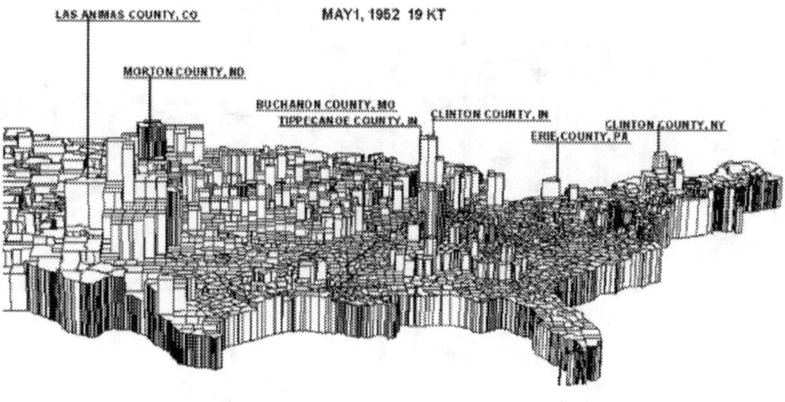

RELATIVE AVERAGE TOTAL FALLOUT: TS DOG

MAY1, 1952 19 KT

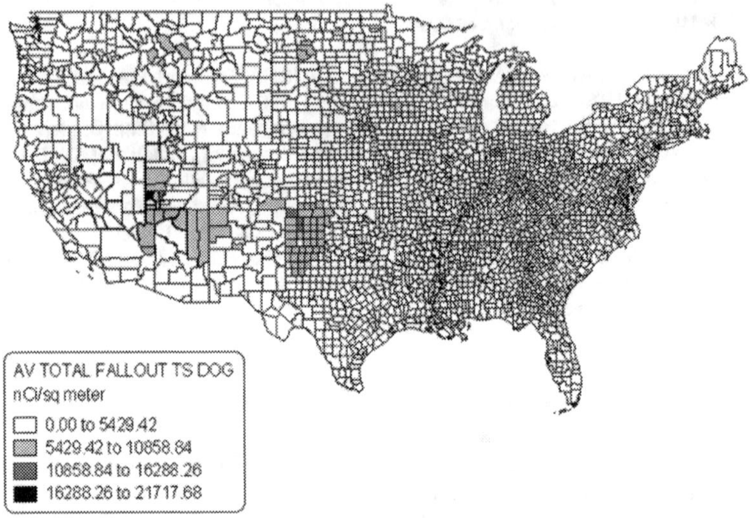

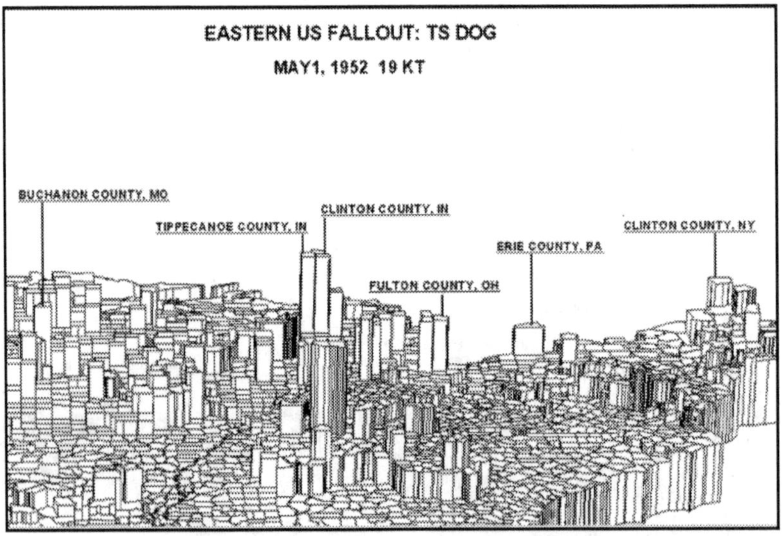

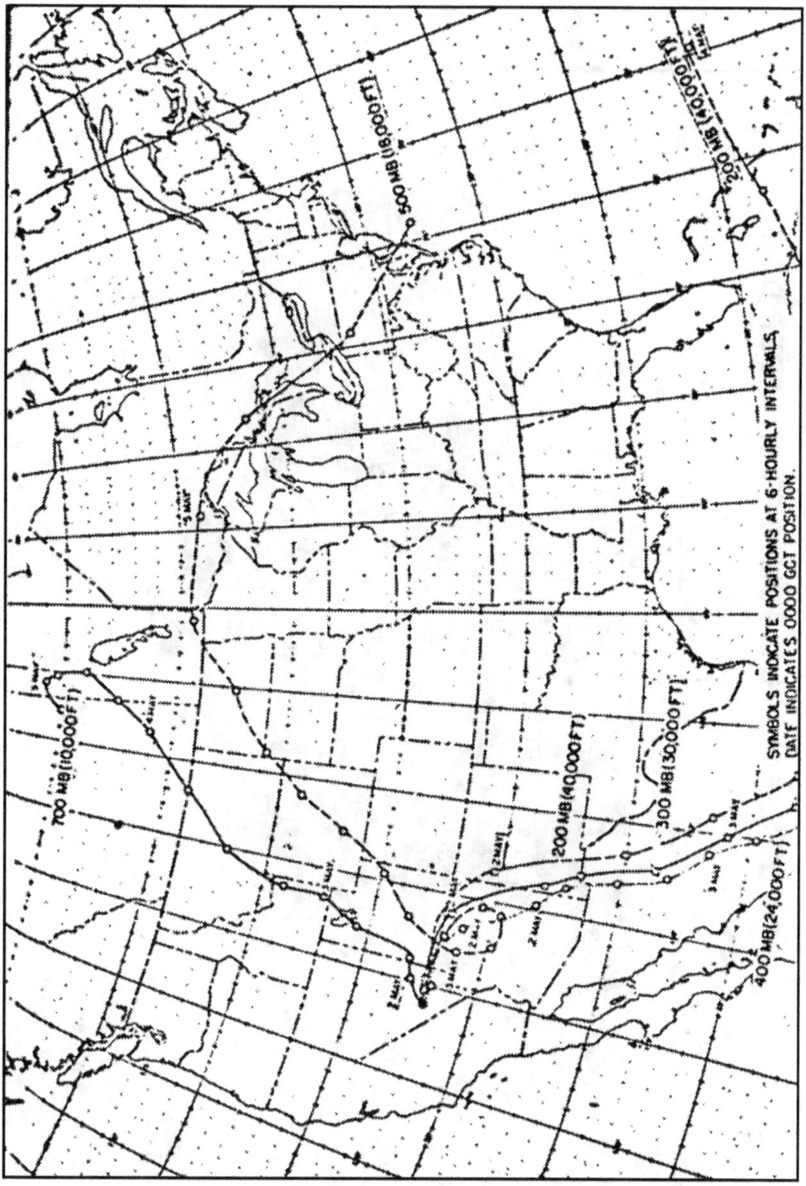

TRAJECTORY: TUMBLER-SNAPPER DOG (TS4)

TOTAL FALLOUT: TUMBLER-SNAPPER DOG

Equal Number

nCi/square meter

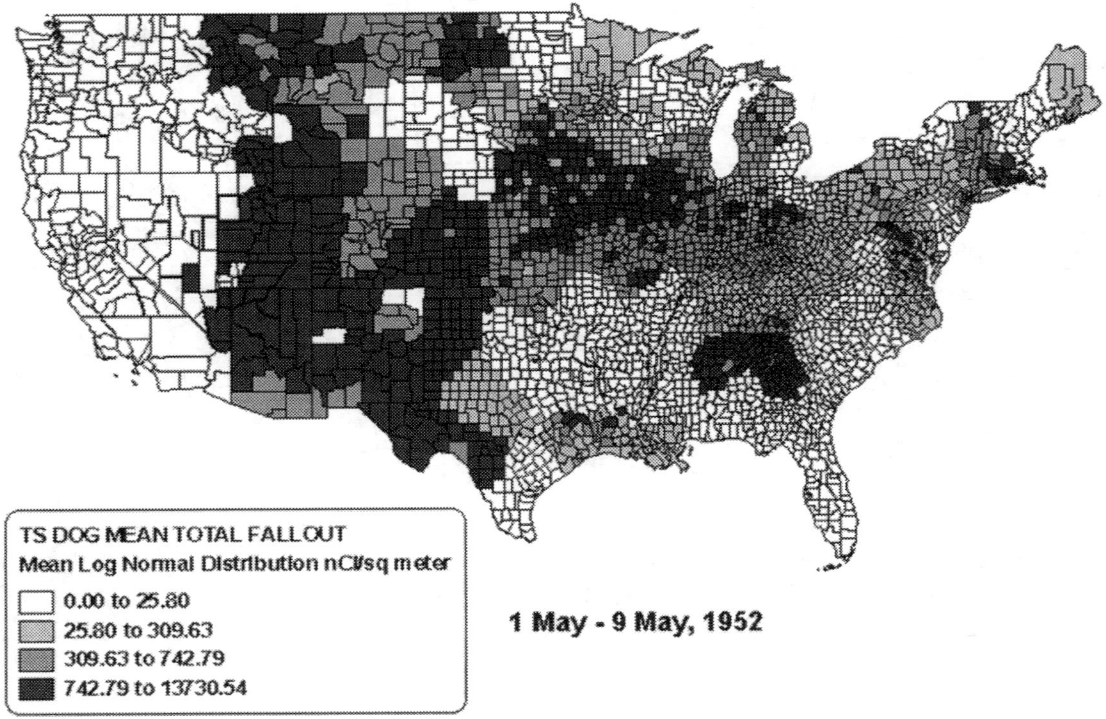

1 May - 9 May, 1952

SHOT 5: TUMBLER-SNAPPER EASY (TS5)

TUMBLER-SNAPPER: EASY

Detonation Date: 7 May 1952 • *Detonation Time:* 4:15 A.M. • *Area:* T-1 • *Sponsor:* Los Alamos • *Yield:* 12 kt • *Radiation Level at Ground Zero:* 500 R/hr • *Height of Burst:* 300 ft • *Cloud Top Height:* 34,000 ft

10,000-Foot Trajectory: NEVADA: (Black Rock Summit), Hobson, Cave Creek, Metropolis, Contact • IDAHO: Amsterdam, Twin Falls, Jerome, Shoshone, Richfield, Arco, Mud Lake, St. Anthony • WYOMING: Bechler, (Yellowstone National Park: Lone Star Geyser, Shoshone Lake, Lewis Lake, West Thumb, Yellowstone Lake, Sylvan Pass, East Entrance, Pahaska, Powell, Elk Basin, Deaver, Frannie, Parkman, Sheridan, Rocky Point, Colony • SOUTH DAKOTA: Arpan, Newell, Creighton, Cottonwood, Long Valley, Tuthill • NEBRASKA: Nenzel, Kennedy, Brownlee, Thedford, Broken Bow, Miller, Kearney, Red Cloud • KANSAS: Webber, Concordia, Manhattan, Burlingame, Prescott • MISSOURI: Hume, Nevada, Hope, Manes, Rector, (Taum Sauk Mountain), Ironton, Lixville, New Wells • ILLINOIS: Aldridge, Carbondale, Carriers Mills, Ridgeway • INDIANA: Mount Vernon, Evansville, Boonville, Lama, Alton, Corydon, New Albany • KENTUCKY: Louisville, New Castle, Corinth, Mount Olivet, Maysville • OHIO: Ripley, Waverly, McArthur, Nelsonville, McConnelsville, Caldwell, Bellaire • WEST VIRGINIA: Benwood, Bethlehem, Wheeling, Valley Grove • PENNSYLVANIA: Washington, Pittsburgh, Punxsutawney, DuBois, Johnsonburg, Betula, Port Allegany, Myrtle • NEW YORK: Ceres, Bolivar, Friendship, Wiscoy, Genesco, Avon, West Rush, Henrietta, Rochester, Rondequoit, Sea Breeze.

CHARACTERISTICS: TS EASY

The device used for this test was small, only 22 inches in length and weighing 625 lbs. The yield was equivalent to 12 kilotons of TNT. The deposition pattern for Tumbler-Snapper EASY is most similar to shot SEDAN (R:0.37; T: 21.92). The county of highest deposition was Tooele(1), UT with an estimated maximum activity of 878.70 μCi/sq meter on May 7, 1952, followed by Tooele(2), UT with a estimated maximum activity of 418.7 μCi/sq meter. Other counties with high fallout from this nuclear test were Box Elder(2) UT (estimated maximum activity 417 μCi/sq meter); Rich, UT (estimated maximum activity 417 μCi/sq meter); Davis, UT (estimated maximum activity 375.5 μCi/sq meter); and Utah, UT (estimated maximum activity 335.0 μCi/sq meter), all on May 7, 1952. Fallout from this test was deposited across 2937 counties, or 95 percent of the United States.

The day one ratio of total fallout to I-131 for Tumbler-Snapper EASY is relatively low at 82 to 1.

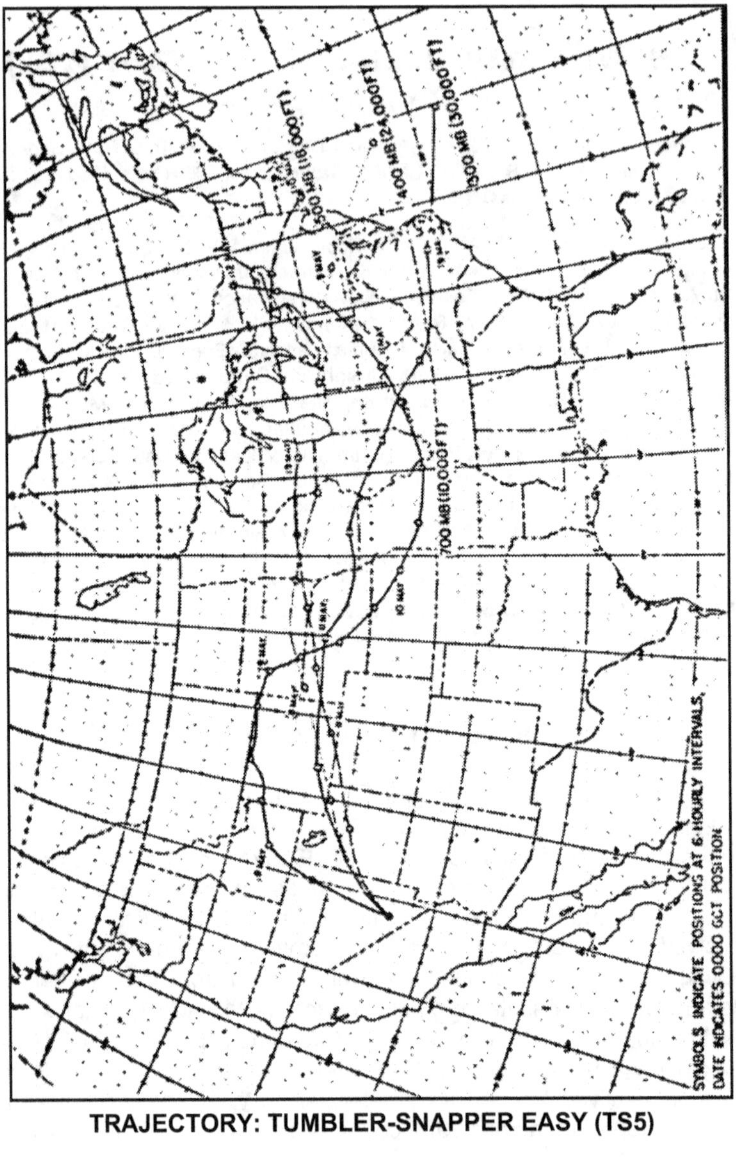

TRAJECTORY: TUMBLER-SNAPPER EASY (TS5)

RELATIVE AVERAGE TOTAL FALLOUT: TS EASY

MAY 7, 1952 12 KT

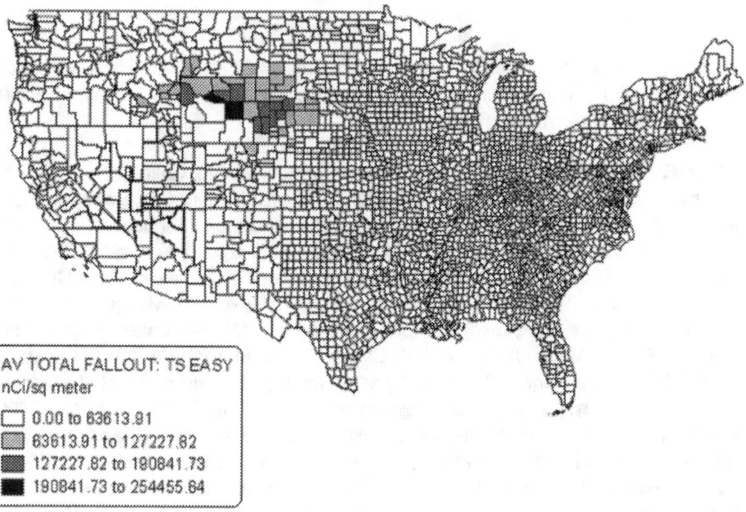

RELATIVE TOTAL FALLOUT: TS EASY

MAY 7, 1952 12 KT

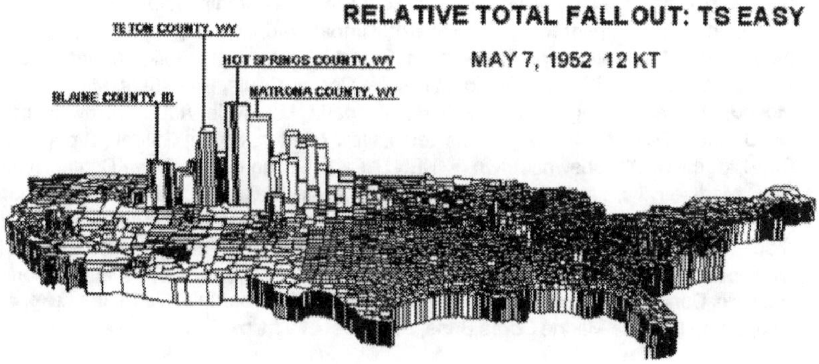

SHOT 6: TUMBLER-SNAPPER FOX (TS6)

TUMBLER-SNAPPER: FOX

Detonation Date: 25 May 1952 • *Detonation Time:* 4:00 A.M. • *Area:* 4 • *Sponsor:* Los Alamos • *Yield:* 11 kt • *Radiation Level at Ground Zero:* 2,000 R/hr • *Height of Burst:* 300 ft • *Cloud Top Height:* 41,000 ft

30,000-Foot Trajectory: NEVADA: Hilco, Ash Springs, Alamo, Caliente • UTAH: Minersville, Bicknell, Nanksville, Moab • COLORADO: Gateway, Orchard City, Bowie, Somerset, Oliver, Marble, Avon, Mount Powell, Heeney, Granby, Berthoud, Johnstown, Millikan, Greeley, Barnesville, Cornish, New Raymer • NEBRASKA: Chappell, Lomoyne, Tryon, Neligh, Hadar, Hoskins, Norfolk, Pender, Bancroft, Decatur • IOWA: Onawa, Dow City, Buck Grove, Manilla, Bagley, Perry, Gardiner, Moran, Elkhart, Baxter, Newburg, Belle Plaine, Cedar Rapids, Springville, Anamosa, Canton, Bellevue • ILLINOIS: Loran, Ridott, Rockford, Machesney Park • WISCONSIN: Beloit, Clinton, Delavan, Waterford, Wind Lake, Milwaukee • MICHIGAN: Muskegon, Ravenna, Sparta, Belding, Matherton, Maple Rapids, Eureka, Elsie, Oakley, Burt, Vassar, Caro, Gagetown, Bad Axe, Filion • NEW YORK: Madrid, Unionville, Potsdam, Parishville, Clear Lake, Saranac Lake, Lake Placid, Keene Valley, Moriah • VERMONT: Chimney Point, Cornwall, Salisbury, Sherburne Center, Windsor, Claremont, Deering, Nashua • MASSACHUSETTS: Lowell, Billerica, Woburn, Lexington, Medford, Arlington, Malden, Somerville, Cambridge, Boston, Hull, North Cohasset, Humarock, Kent Park, Rexhame, Ocean Bluff, Brant Rock, Green Harbor, Denis, South Yarmouth.

40,000-Foot Trajectory: NEVADA: Alamo, Ash Springs, Hiko, Caliente, Panaca, Carp, Elgin • UTAH: Milford, Cove Fort, Fillmore, Richfield, Mount Pleasant, Moroni, Price, East Carbon, Ouray, Bonanza • COLORADO: Massadonia, Blue Mountain, Elk Springs, Maybell, Sunbeam, Clark • WYOMING: Baggs, Dixon, Saven, Mountain Home, Woods Landing, Buford, Cheyenne, Carpenter • NEBRASKA: Kimball, Sidney, Lodgepole, Chappel, Ogallala, North Platte, Maxwell, Brady, Pleasanton, Grand Island, Aurora, York, Milford, Lincoln, Denton, Syracuse • MISSOURI: Tarkio, Maryville, Coffey, Chula, Brookfield, Marceline, College Mound, Duncan's Bridge, Holliday, Paris, Vandalia, Louisville, Auburn, New Hope • ILLINOIS: Alton, Edwardsille, Aviston, Bartelso, New Minden, Spring Garden, Broughton, Texas City, Ridgeway, Old Shawneetown • KENTUCKY: Spring Grove, Grove Center, Wheatcroft, Clay, Madisonville, Nortonville, Apex, Kirkmansville, Fearsville, Allegre, Claumour, Whippoorwill, Olmstead, Adairville • TENNESSEE: Orlinda, Portland, Bethpage, Hartsville, Carthage, Buffalo Valley, Bon Air, Roddy, Loudon, Glendale, Maryville, Townsend, Elkmont, Gatlinburg • NORTH CAROLINA: Asheville, Azalea, Vale, Reepville, Davidson, Concord, Albemarle, Candor, Manchester, Eastover, Clinton, Lyman, Fountain, Catherine Lake, Jacksonville, Camp Lejeune Marine Corps Base, Bear Inlet of Onslow Bay.

Note: Both the 30,000-foot and the 40,000-foot trajectory crossed over Chappel, Nebraska.

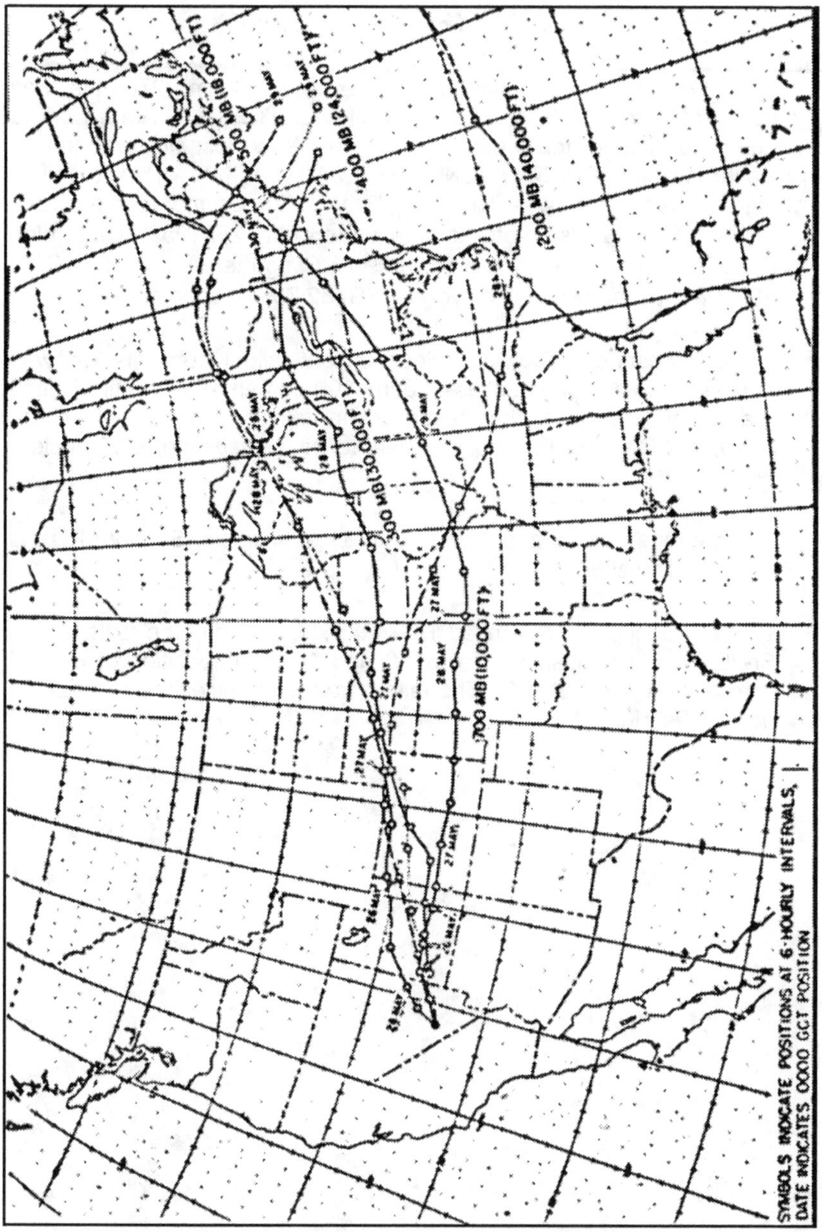

TRAJECTORY: TUMBLER-SNAPPER FOX (TS6)

CHARACTERISTICS: TS FOX

The FOX device, 40 inches in diameter and weighing 2700 lbs, was detonated atop a 300-foot tower at Nevada Test Site Area 4. The deposition pattern for Tumbler-Snapper FOX is most similar to that of Upshot-Knothole HARRY (UK9) (R: 0.29; T: 17.00). The county of highest deposition was Lincoln, UT with an estimated maximum. activity of 521 μCi/sq meter total fallout on May 25, 1952. Other high fallout counties included Iron UT (509.7 μCi/sq meter), Hinsdale, CO (99.7 μCi/sq meter) and Cheyenne, KS (52.73 μCi/sq meter,)

Tumbler-Snapper FOX was unique in that fallout from this test was deposited over 3086 counties—nearly 100 percent of the entire United States land area. In that respect, the fallout from shot Tumbler-Snapper FOX was the most dispersed of any aboveground nuclear test.

The day one ratio of total fallout to I-131 for Tumbler-Snapper FOX is relatively low at 96 to 1.

Tumbler-Snapper FOX ranked 12th in the deposition of the radionuclide Sr-90, 13th in deposition of Ru-106 and 14th in deposition of Cs-137 across the United States. The fallout from this nuclear test contained *no* Co-60 or Mn-54.

TOTAL FALLOUT: TUMBLER-SNAPPER FOX

Equal Number

nCi/square meter

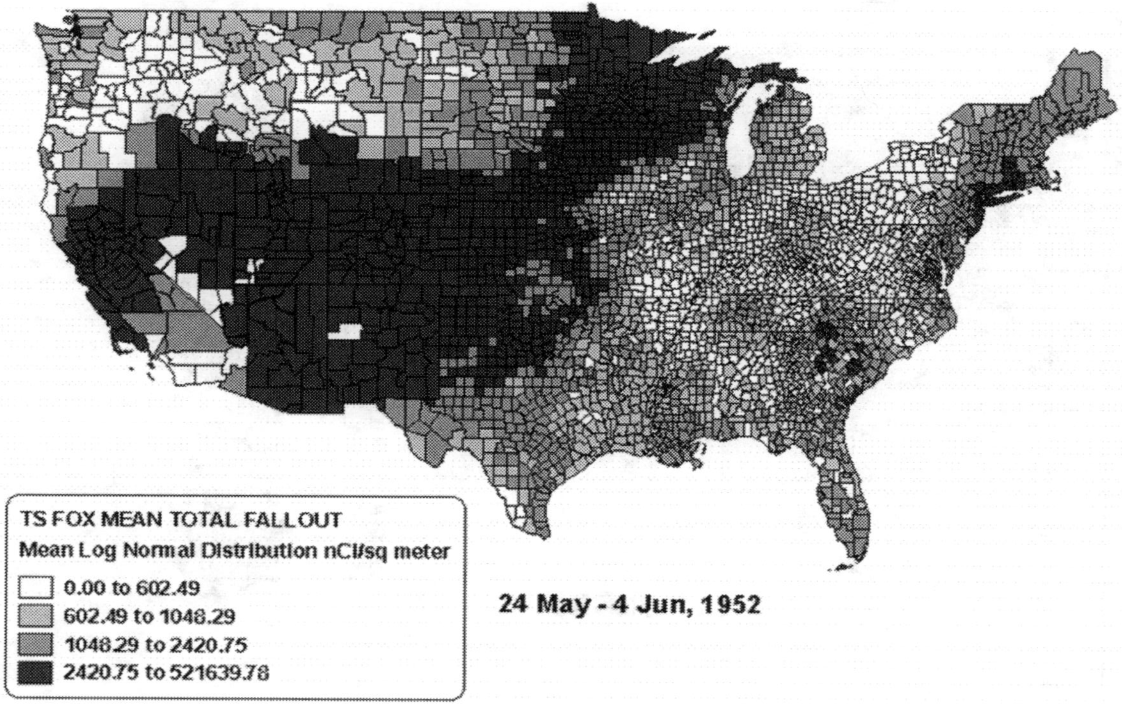

24 May - 4 Jun, 1952

RELATIVE AVERAGE TOTAL FALLOUT: TS FOX

MAY 25, 1952 11 KT

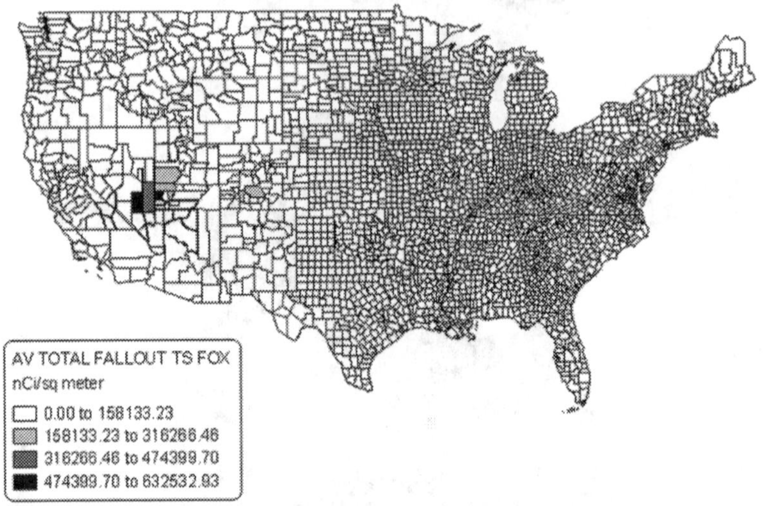

RELATIVE TOTAL FALLOUT: TS FOX

MAY 25, 1952 11 KT

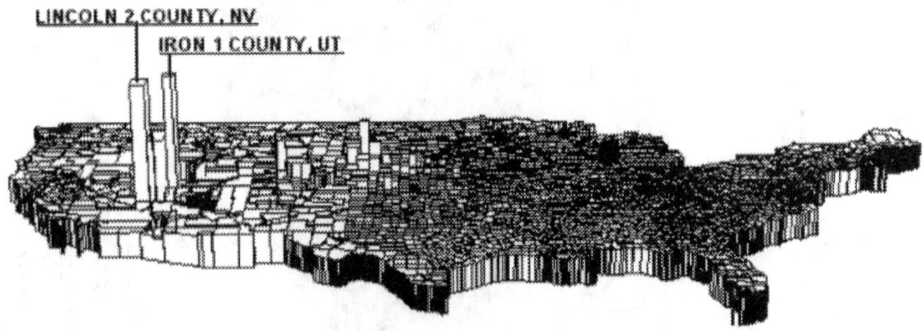

SHOT 7: TUMBLER-SNAPPER GEORGE (TS7)

TUMBLER-SNAPPER: GEORGE

Detonation Date: 1 Jun 1952 • *Detonation Time:* 3:55 AM • *Area:* 3 • *Sponsor:* Los Alamos • *Yield:* 15 kt • *Radiation Level at Ground Zero:* 1,000 R/hr • *Height of Burst:* 300 ft • *Cloud Top Height:* 37,000 ft

35,000-Foot Trajectory: NEVADA: Caselton, Montello • UTAH: Artesian City, Carey, Antelope River, Darlington, Winsper, Humphrey • WYOMING: (Yellowstone National Park, Northern Campgrounds) • MONTANA: Dean, Joliet, Silesia, Castle Rock, Brandenburg, Ekalaka, Mill Iron • SOUTH DAKOTA: Ladner, Ludlow, Lodgepole, Glad Valley, Isabel, Ridgeview, Agar, Miller, Iroquois, Oldham, Elkton • MINNESOTA: Ruthton, Aroca, Dundee, Lakefield, Dunnell • IOWA: Buffalo Center, Forest City, Hanford, Marble Rock, Horton, Sumner, Osborne • WISCONSIN: Cornelia, Seymour Corners, Cadiz Springs • ILLINOIS: Durand, Belvidere, Chicago • INDIANA: Coburg, Hamlet, Argos, Luther, Roanoke, Zanesville, Ossian, Curryville, Salem • OHIO: Celina, New Brennen, Lena, Thackeray, Springfield, Brookwalter, Bloomingburg, Austin, Alma, Beaver, Bloom, Pedro Forrestdale • WEST VIRGINIA: Shoals, Eloise, Holden, Christian, Lincoln, Woolsey, Rock, Ingleside • VIRGINIA: Point Pleasant, Shorts Creek, Fancy Gap, Canal • NORTH CAROLINA: Mount Airy, Pinnacle, Cedar Lodge, Cid, Ophir, Norman • SOUTH CAROLINA: Clio, Dillon, Mulins, Galivants Ferry, Toddville, Burgess, Brookgreen.

CHARACTERISTICS: TS GEORGE

The device was similar in size to the one used in the earlier FOX test: 2700 lbs and 40 inches in diameter. As with the FOX test, TS GEORGE was detonated from atop a 300-foot tower. The deposition pattern for Tumbler-Snapper GEORGE is most similar to that of the 1957 shot Plumbbob MORGAN (PB18) (R: 0.50; T: 31.78). The counties of highest fallout deposition included Knox, MO (estimated maximum activity: 1382 μCi/sq meter), Lewis, MO (estimated maximum activity: 1290 μCi/sq meter), Niobrara, WY (estimated maximum activity: 1199 μCi/sq meter), Appanoose, IA (estimated maximum activity: 1114 μCi/sq meter), and Scotland, MO (estimated maximum activity: 1111 μCi/sq meter)—all on June 3, 1952, and most associated with rain.

Tumbler-Snapper GEORGE deposited fallout on 2937 counties— 95 percent of the total counties in the United States, ranking it 3rd among the above ground tests for widespread fallout. Interestingly, this shot ranks 1st for deposition of the radionuclide Cs-137 (followed by Upshot-Knothole SIMON). It also ranks 1st in deposition of the radionuclides Nb-95 (again followed by Upshot-Knothole SIMON at 2nd place and Upshot-Knothole HARRY in 3rd place for this radionuclide.) From a radionuclide composition standpoint, Tumbler-Snapper GEORGE somewhat resembles the later shots UK SIMON and UK HARRY. Tumbler-Snapper GEORGE is also 2nd in the deposition of the radionuclide Nb-95m (with 1st place going to Buster-Jangle EASY and 3rd place held by Upshot-Knothole SIMON.)

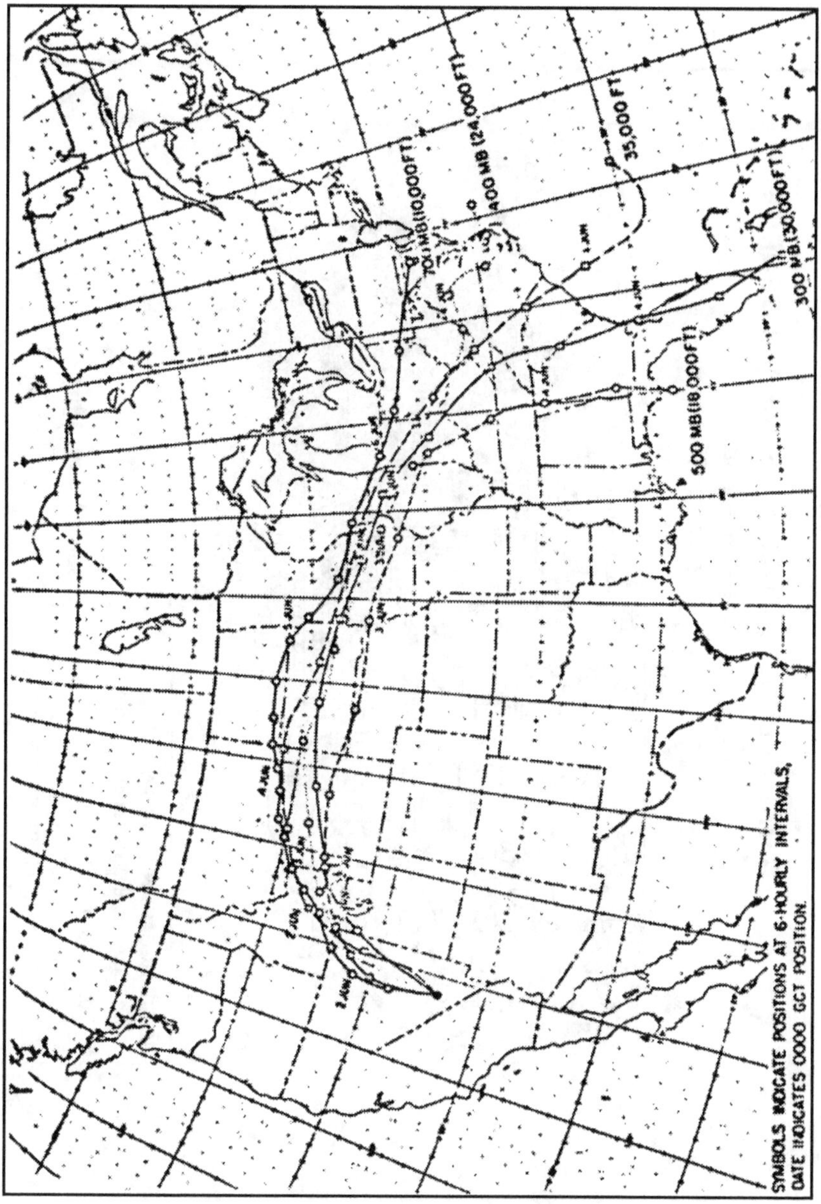

TRAJECTORY: TUMBLER-SNAPPER GEORGE (TS7)

TOTAL FALLOUT: TUMBLER-SNAPPER GEORGE

Equal Intervals
nCi/square meter

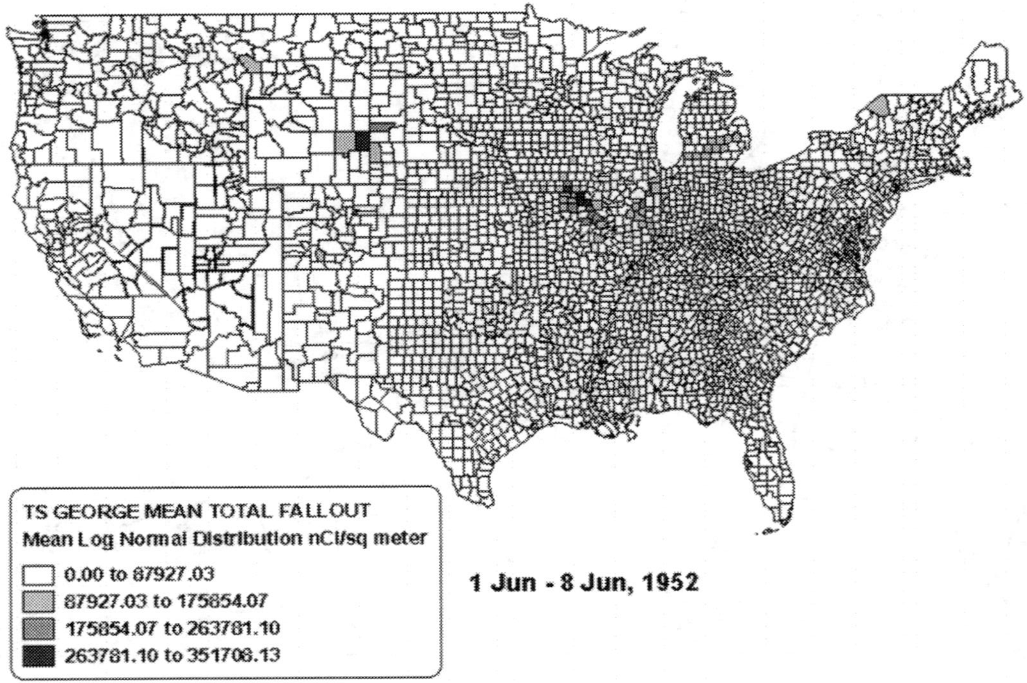

1 Jun - 8 Jun, 1952

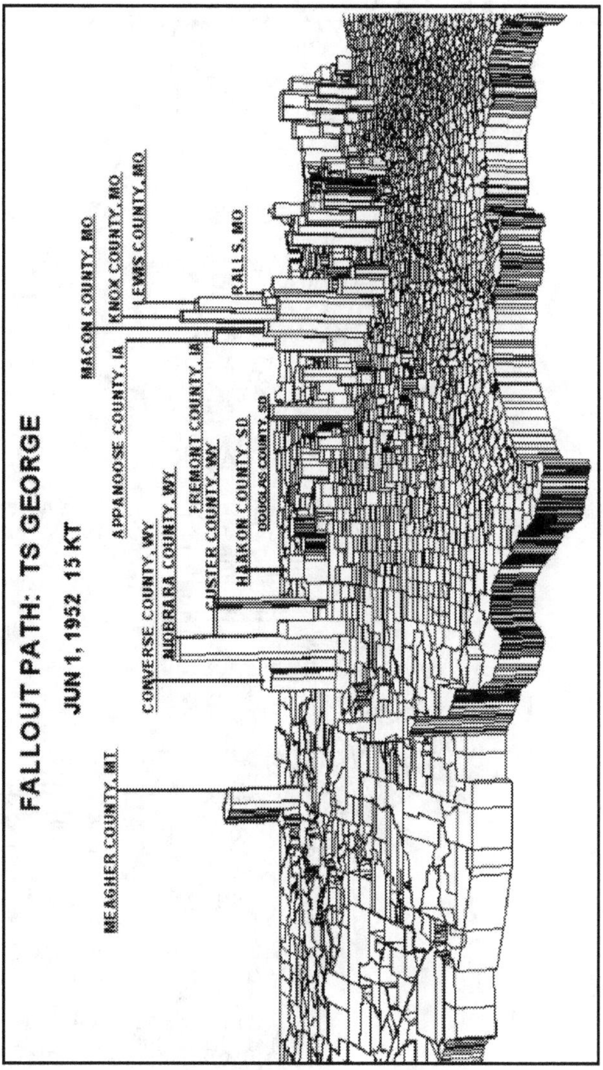
FALLOUT PATH: TS GEORGE
JUN 1, 1952 15 KT
MEAGHER COUNTY, MT
CONVERSE COUNTY, WY
NIOBRARA COUNTY, WY
CUSTER COUNTY, WY
APPANOOSE COUNTY, IA
FREMONT COUNTY, IA
HAAKON COUNTY, SD
DOUGLAS COUNTY, SD
MACON COUNTY, MO
KNOX COUNTY, MO
LEWIS COUNTY, MO
RALLS, MO

RELATIVE TOTAL FALLOUT: TS GEORGE

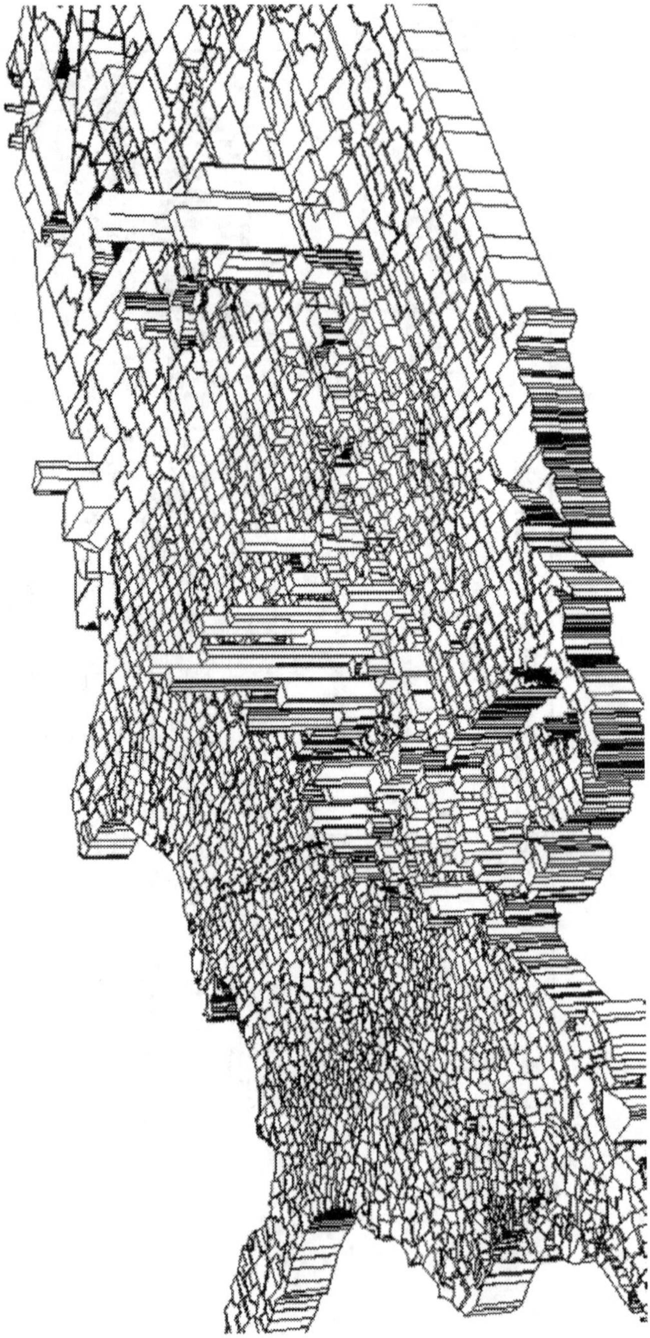

SHOT 8: TUMBLER-SNAPPER HOW (TS8)

TUMBLER-SNAPPER: HOW

Detonation Date: 5 Jun 1952 • *Detonation Time:* 3:55 A.M. • *Area:* 2 • *Sponsor:* Los Alamos • *Yield:* 14 kt • *Radiation Level at Ground Zero:* 1,500 R/hr • *Height of Burst:* 300 ft • *Cloud Top Height:* 41,800 ft

24,000-Foot Trajectory: NEVADA: Warm Springs, Round Mountain, Crescent Valley, Midas, Tuscarora • IDAHO: Riddle, Grasmere, Mayfield, Lowman • MONTANA: Darby, Bearmouth, Helmville, Greenfield, Power, Gildford, Kremlin • SASKATCHEWAN: Regina • NEW YORK: Youngstown, Model City, Niagara Falls, Tonawanda, Buffalo, Lancastor, Aurora, South Wales, Holland, Chaffee, Elton, North Cuba, Nile, Allentown • PENNSYLVANIA: Kinney, Gold, Carter Camp, Haneyville, Avis, Ranchtown, Cowan, New Berlin, Selinsgrove, Red Cross, Sharidan, Dayton, Jonestown, Bunker Hill, Avon, Iona, Clay, Ephrata, Groffdale, Gap, Atglen, Jennersville, Kelton, New London, Kemblesville • DELAWARE: Newark, Brookside, St. Georges, Taylors Bridge, Dupont Manor, Dover, Bowers, Slaughter Beach, Nassau, Cottonpatch Hill.

CHARACTERISTICS: TS HOW

This device was the first to use the beryllium neutron reflector surrounding the plutonium core. Tumbler-Snapper HOW is ranked 1st for percentage of both Am-241 and Sm-153 in nuclear fallout . In addition, perhaps due to rainouts soon after detonation, this shot resulted in some of the highest exposures of I-131 to the thyroid for Gem County, ID.

The deposition pattern for Tumbler-Snapper HOW is most similar to Plumbbob NEWTON (PB15) (R: 0.53; T: 34.85). Like Tumbler-Snapper GEORGE, the fallout was associated with rain. But unlike the earlier shot, the rainout occurred on the day of the shot, June 5, 1952. For that reason, the fallout level that day was among the highest calculated for any aboveground test. Counties with the highest total fallout deposition included Gem ID (estimated maximum. activity: 878.1 μCi/sq meter), Custer ID (estimated maximum activity 843.6 μCi/sq meter); Blaine ID (estimated maximum activity: 795.2 μCi/sq meter), Lemhi ID (estimated maximum activity: 578.2 μCi/sq meter) and Deer Lodge MT (estimated maximum activity: 517.4 μCi/sq meter).

This test deposited fallout over 3014 counties, or 97 percent of the U.S. The ratio of total fallout to I-131 for Tumbler-Snapper HOW is relatively low at 77.9 to 1. This shot is ranked 8th for deposition of Sr-90 across the United States and 9th for deposition of Cs-137.

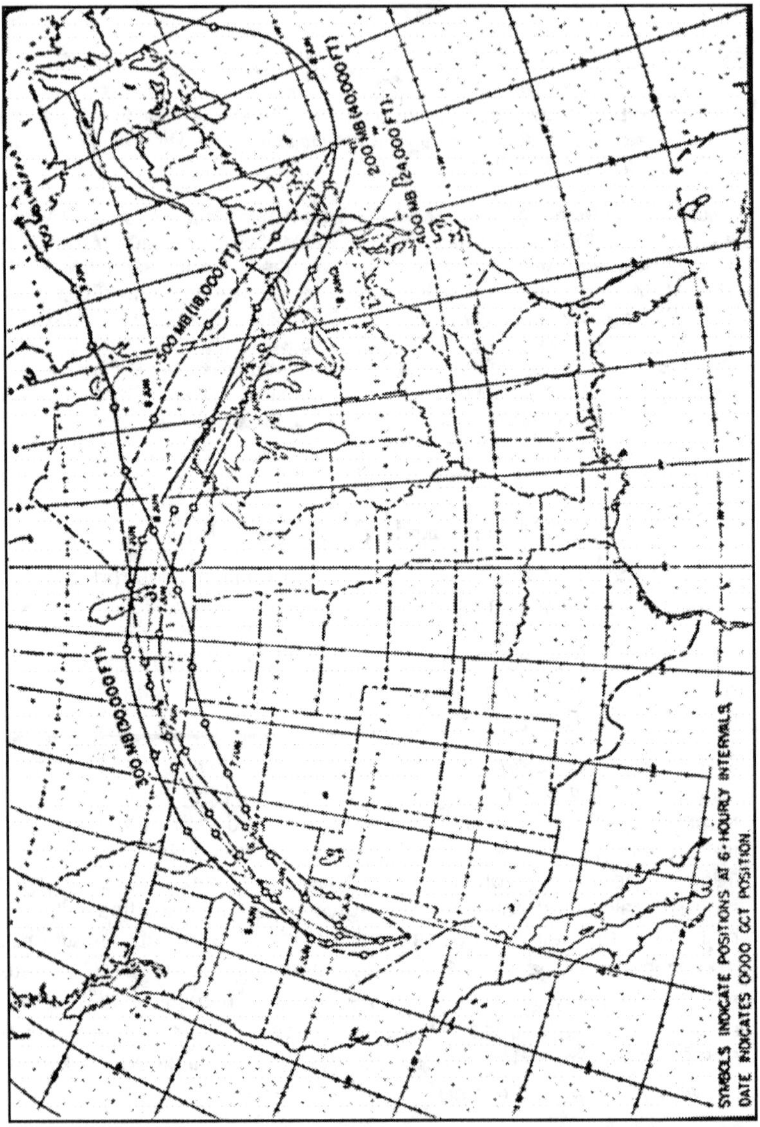

TRAJECTORY: TUMBLER-SNAPPER HOW (TS8)

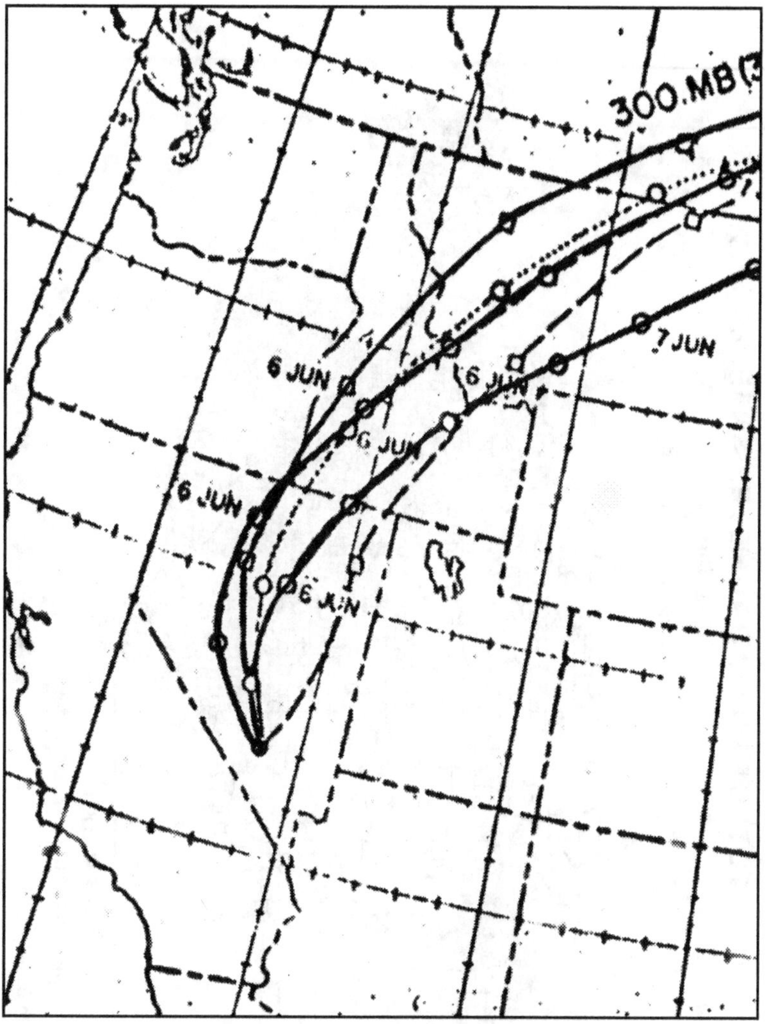

CLOSE-IN TRAJECTORY: TS HOW (TS8)

TOTAL FALLOUT: TUMBLER-SNAPPER HOW

Equal Intervals

nCi/square meter

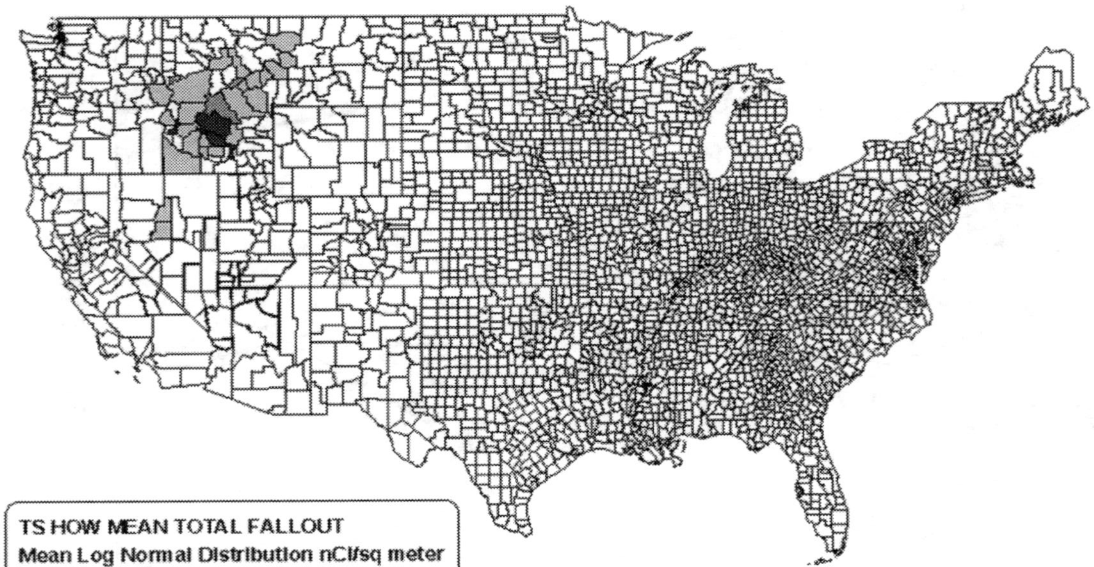

TS HOW MEAN TOTAL FALLOUT
Mean Log Normal Distribution nCi/sq meter

- 0.00 to 99841.95
- 99841.95 to 199683.90
- 199683.90 to 299525.85
- 299525.85 to 399367.80

6 Jun - 14 Jun, 1952

TOTAL FALLOUT: TUMBLER-SNAPPER HOW

Equal Number

nCi/square meter

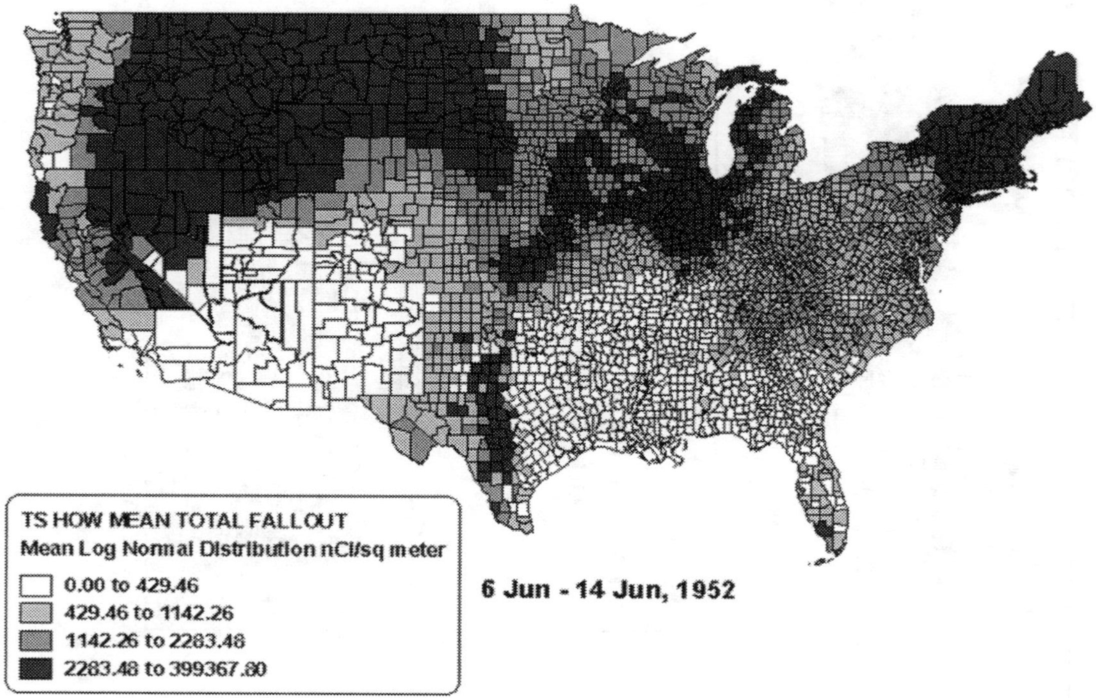

6 Jun - 14 Jun, 1952

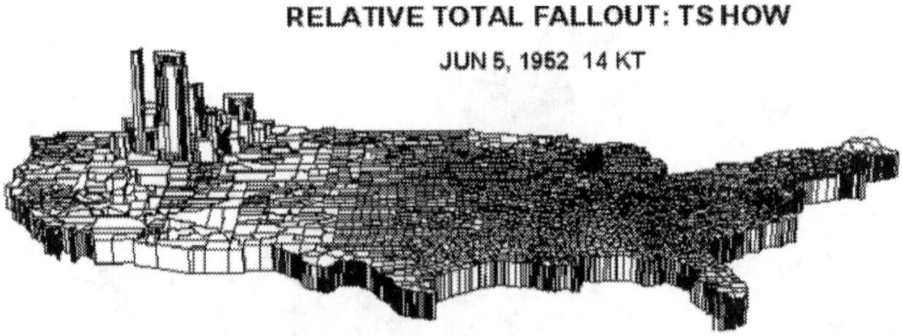

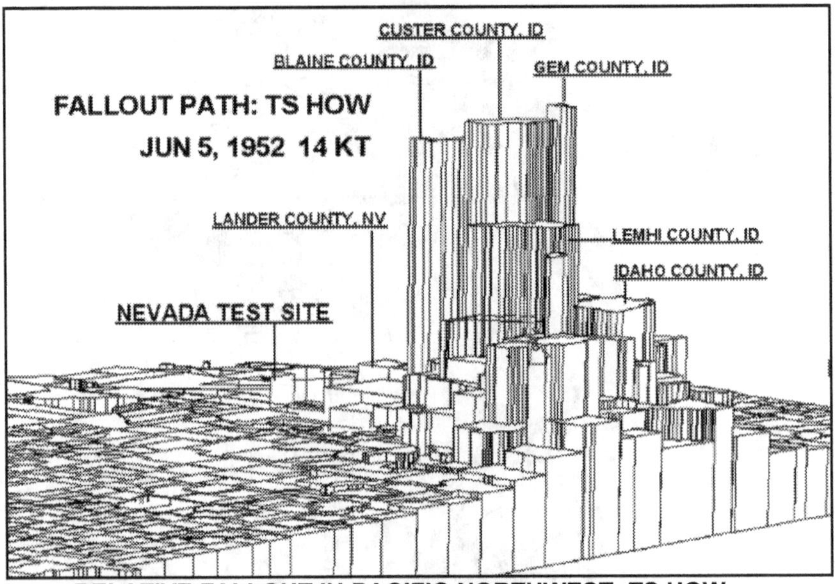

RELATIVE FALLOUT IN PACIFIC NORTHWEST: TS HOW

SECTION 4

UPSHOT-KNOTHOLE SERIES

Mar 17-Jun 4, 1953

SHOT 1: UPSHOT-KNOTHOLE ANNIE (UK1)

Upshot-Knothole: Annie

Detonation Date: 17 Mar 1953 • *Detonation Time:* 5:20 A.M. • *Area:* 3 • *Sponsor:* Los Alamos • *Yield:* 16 kt • *Radiation Level at Ground Zero:* 2,000 R/hr • *Height of Burst:* 300 ft • *Cloud Top Height:* 41,000 ft

10,000-Foot Trajectory: NEVADA: Alamo, Elgin, Carp • UTAH: Modena, Beryl, Paragonah, Spry, Panguitch • COLORADO: Grand Junction, De Begue, Parachute, Glenwood Springs, Boulder, Longmont, Greeley, New Raymer, Sedgwick, Ovid • NEBRASKA: Julesburg, Ogallala, Keystone, Stapleton, Gandy, Anselmo, Taylor, Neligh, Pierce, Carroll, Concord, Jackson • IOWA: Sioux City, Lawton, Kingsley, Diamond Center, Hanover, Nemaha, Rockwell City, Somers, Slifer, Stratford, Story City, Roland, Marshalltown, Chelsea, Haven, South Amana, Tiffin, Iowa City, Downey, Moscow, Pleasant Prairie, Bettendorf, Davenport • ILLINOIS: Rock Island, Moline, Buda, Tiskilwa, McNabb, Kangley, Streator, Dwight, Kankakee • INDIANA: Lake Village, Thayer, Wilders, Brems, Grovertown, Hamlet, Koontz Lake, Teegarden, Lakeville, Woodland, Jimtown, Elkhart, Bristol • MICHIGAN: Vistula, Sturgis, Findley, Coldwater, Litchfield, Pulaski, Vandercook, Grass Lake, Ann Arbor, Plymouth, Livonia, Oak Park, Highland Park, Detroit • NEW YORK: Westfield, Point Chautauqua, Jamestown • PENNSYLVANIA: Westline, Guffey, Mount Jewett, Howard, Castle Garden, Spring Garden, Unityville, Elk Grove, Lopez, Sugar Run, Camptown, Little Meadows. • NEW YORK: Binghamton, Oxford, Norwich, Shelburne, Utica, Grant, Saranac Lake, Redford, Dannemora, Jericho, Champlain • MAINE: Dickey, St. John, Fort Kent, Frenchville.

40,000-Foot Trajectory: NEVADA: Alamo, Ash Springs, Hiko • UTAH: Modena, Beryl, Paragonah, Spry, Panguitch, Mount Ellen • COLORADO: Gateway, Delta, Hotchkiss, Paonia, Crested Butte, Granite, Tarryall, Palmer Lake, Elbert, Hale • KANSAS: St. Francis • NEBRASKA: Danbury, Oxford, Mascot, Ragan, Hildneth, Holstein, Roseland, Hastings, Harvard, Lushton, McCool Junction, Beaver Crossing, Malcolm, Ashland, Smithfield, Bellevue, Omaha • IOWA: Council Bluffs, Hancock, Marne, Lorah, Monteith, Minburn, Alleman, Maxwell, Melbourne, Marshalltown, Dysart, Walker, Troy Mills, Farley, Centralia, Asbury, Dubuque • WISCONSIN: Kieler, Cuba City, Avon, Lamont, Argyle, Monticello, Evansville, Edgerton, Fort Atkinson, Hebron, Palmyra, North Prairie, Genesee, West Allis, Milwaukee • MICHIGAN: Holland, Grand Rapids, Cedar Springs, Trufant, Six Lakes, Wyman, Mount Pleasant, Leaton, Edenville, Estey, Rhodes, Bentley, Omer, Twining, Tawas City, Au Sable • NEW YORK: Waddington, Chase Mills, Norfolk, Dickenson Center, Duane Center, Hawkeye, Clintonville • VERMONT: Huntington Center, Waitsfield, Northfield, Corinth, Bradford • NEW HAMPSHIRE: Wentworth, Stinson Lake, Melvin Village, Wawbeck, Tuttonboro, Wolfeboro Center, Wakefield, Sanbornville • MAINE: Emery Mills, Springvale, Sanford, Kennebunkport.

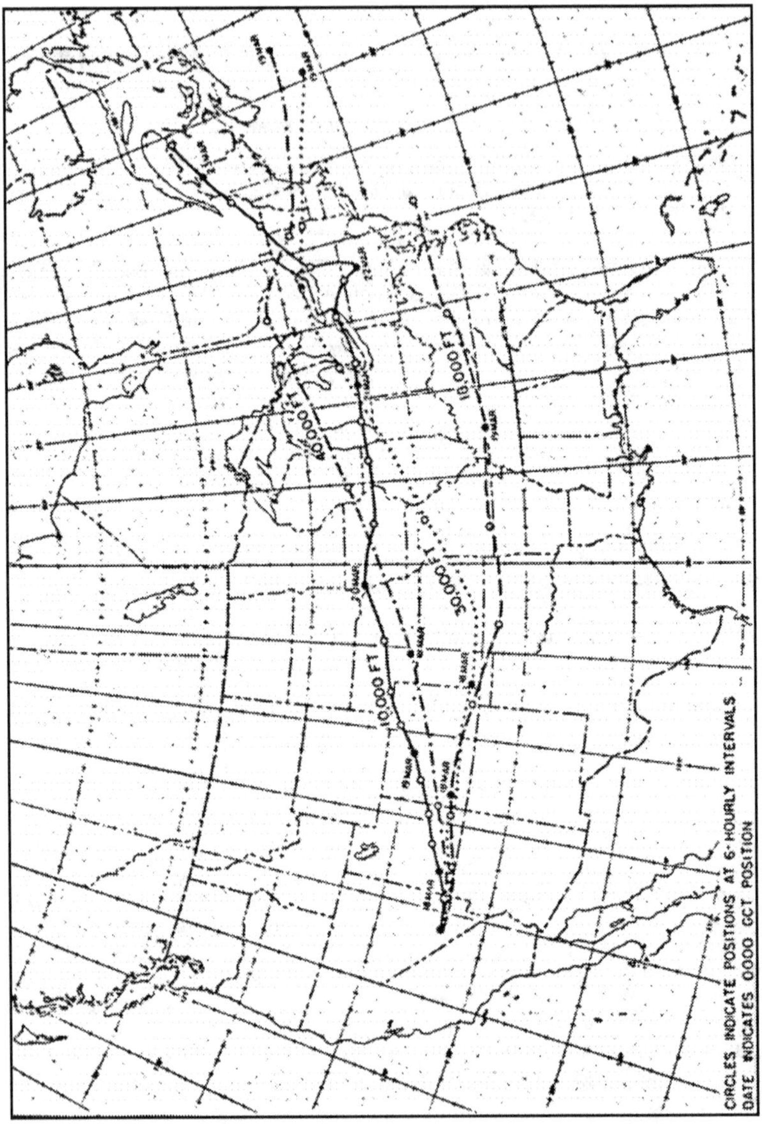

TRAJECTORY: UPSHOT-KNOTHOLE ANNIE (UK1)

CHARACTERISTICS: UK ANNIE

The device, designated as an "XR-3" design, weighed 2700 lbs and was detonated from atop a 300-foot tower. Upshot-Knothole ANNIE deposited fallout over 2937 counties, or approximately 95 percent of the total counties in the United States, thus ranking it 21st for widespread fallout. The deposition pattern for Upshot-Knothole ANNIE was most similar to Upshot-Knothole RUTH (Uk3) (R: 0.51; T: 32.89). The counties of highest fallout from this nuclear test included Washington, UT (maximum estimated activity: 2396 μCi/sq meter on Mar 17, 1953). Interestingly, on Mar 18, 1953 the fallout from this nuclear test also produced fallout hot spots on the East Coast, in the New Jersey counties of: Atlantic (estimated maximum activity: 306 μCi/sq meter); Monmouth (estimated maximum activity: 307 μCi/sq meter); Ocean (estimated maximum activity: 307 μCi/sq meter); Burlington (estimated maximum activity: *256 μCi/sq meter) and Cape May (estimated maximum activity: 230 μCi/sq meter).

TOTAL FALLOUT: UPSHOT-KNOTHOLE ANNIE
Equal Intervals
nCi/square meter

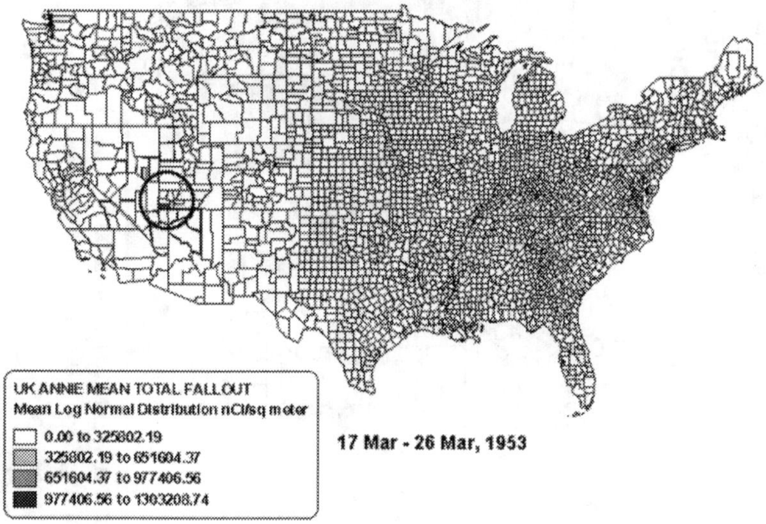

TOTAL FALLOUT: UPSHOT-KNOTHOLE ANNIE

Equal Number
nCi/square meter

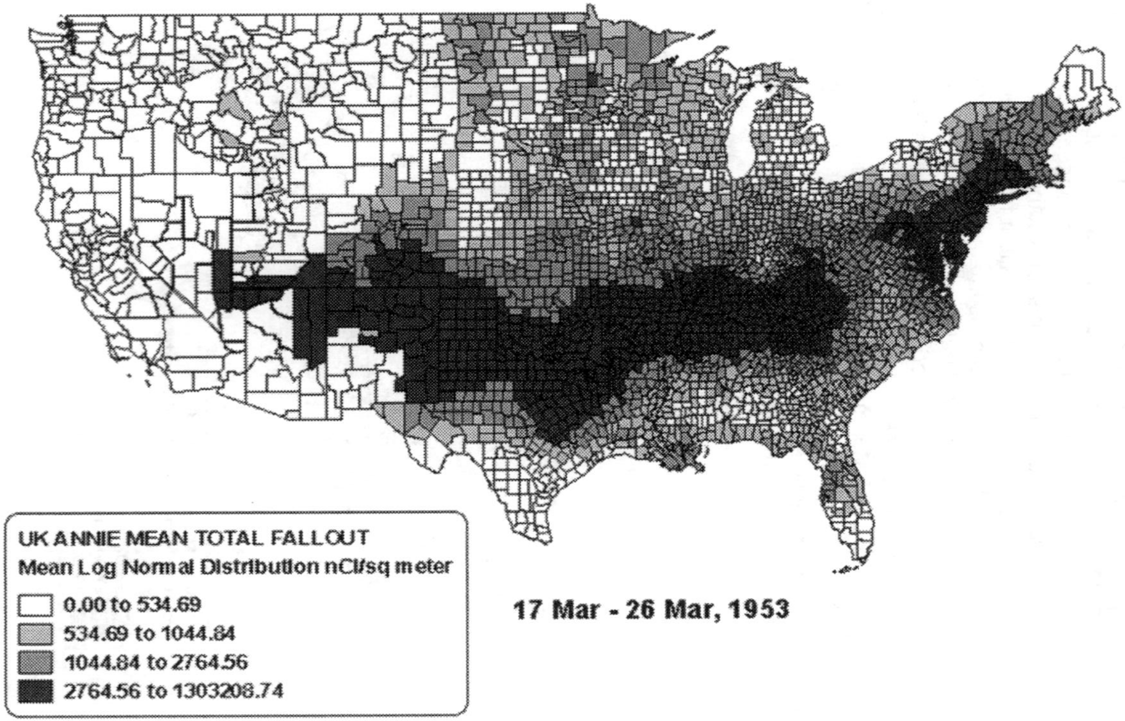

SHOT 2: UPSHOT-KNOTHOLE NANCY (UK2)

Upshot-Knothole: Nancy

Detonation Date: 24 Mar 1953 • *Detonation Time:* 5:10 A.M. • *Area:* 4 • *Sponsor:* Los Alamos • *Yield:* 24 kt • *Radiation Level at Ground Zero:* 2,000 R/hr • *Height of Burst:* 300 ft • *Cloud Top Height:* 41,500 ft

10,000-Foot Trajectory: NEVADA: (Summit Mountain), Dunphy, Tuscarora, Mountain City, Owyhee • IDAHO: Riddle, Grasmere, Mayfield, Lowman, Cobalt, Salmon • MONTANA: Butte, Boulder, Winston, Neihart, Windham, Benchland, Danvers, Hilger, Fergus, Roy, Mona, Andes • NORTH DAKOTA: Charleston, New Town, Plaza, Sawyer, Velva, Voltaire, Bergen, Balfour, Drake, Maddock, Oberon, Sheyenne, Warwick, Hamar, Tolna, Pekin, McVille, Kloten, Aneta, Northwood, Reynolds • MINNESOTA: Climax, Fertile, Rindal, Bejou, Naytahwaush, Akeley, Hackensack, Manhattan Beach, Fifty Lakes, Aitkin, Giese, Finlayson, Sandstone • WISCONSIN: Webster, Hertel, Barronett, Rice Lake, Canton, Cameron, Chetek, New Auburn, Cadott, Neillsville, Mather, Shennington, Cutler, New Lisbon, Lyndon Station, Baraboo, Prairie du Sac, Sauk City, Black Earth, Madison, Mount Horeb, Belleville, New Glarus, Monticello, Monroe, Juda • ILLINOIS: Rock Grove, Rock City, Ridott, Seward, Chana, Rockelle, Pawpaw, Earlville, Harding, Wedron, Marseilles, Seneca, Verona, Dwight, Reddick, Cabery, Stelle, Dunforth, Milford • INDIANA: Freeland Park, Pine Village, West Point, Romney, Kirkpatrick, Thorntown, Lebanon, Fayette, Royalton, Trader's Point, Indianapolis, Beech Grove, Acton, London, Fairland, Shelbyville, Waldron, St. Paul, Milford, Greensburg, Millhousen, Osgood, Farmers Retreat, East Enterprise, Quercus Grove, Florence • KENTUCKY: Warsaw, Glencoe, Poplar Grove, Mason, Corinth, Cynthiana, Millersburg, East Union, Sharpsburg, Owingsville, Sudith, Frenchburg, Mizel, Grassy Creek, Caney, Lakeville. Royalton, Blue River, Manton, Boldman, Pikeville, Chloe, Shelbiana • VIRGINIA: Prater, Mount Heron, Lynn Spring, Red Ash, Cedar Bluff, Asberrys, Wytheville, Austinville, Hillsville, Laurel Fork, Vesta, Spencer, Glenwood, Harmony, Mayo, Cluster Springs, Clarksville, Alberta, Dophpin, Stony Creek, Savedge, Williamsburg, Mobjack, Mathews, Diggs, Birdsnest, Nassawadox.

40,000-Foot Trajectory: NEVADA: Alamo, Ash Springs, Hiko • UTAH: Molena, Beryl, Lund, Beaver, Koosharem, Ferron • COLORADO: Palisade, Cedaredge, Rogers Mesa, Lazear, Crawford, Maher, LaGareta, Center, Mosca, Fort Garland • NEW MEXICO: (Laughlin Peak), Sofia, Clapham, Stead • TEXAS: Amarillo (near the Pantex Plant), Pullman, Palo Duro, South Brice, Parnell, Swearingen, Crowell, Thulia, Mabelle, Markley, Wizard Wells, Boonsville, Cottondale, Keeter, Springtown, Forth Worth, Hurst, Arlington, Grand Prairie, Dallas, Duncanville, Bristol, Palmer, Telico, Gun Barrel City, Caney City, Malakoff, Bradford, Nechas, Rusk, Wells, Lufkin, Jasper, Newton, Bon Wier • LOUISIANA: Merryville, Reeves, LeBlanc, Indian Village, Hathaway, Evangeline, Rayne, Crowley, Maurice, Youngsville, New Iberia, Jeanerette, Baldwin, Franklin, Morgan City, Houma, Chauvin, Golden Meadow, Leeville, Grand Isle • FLORIDA: Venice, South Venice, North Port, Port Charlotte, Lakeport, Port Mayaca, Jupiter.

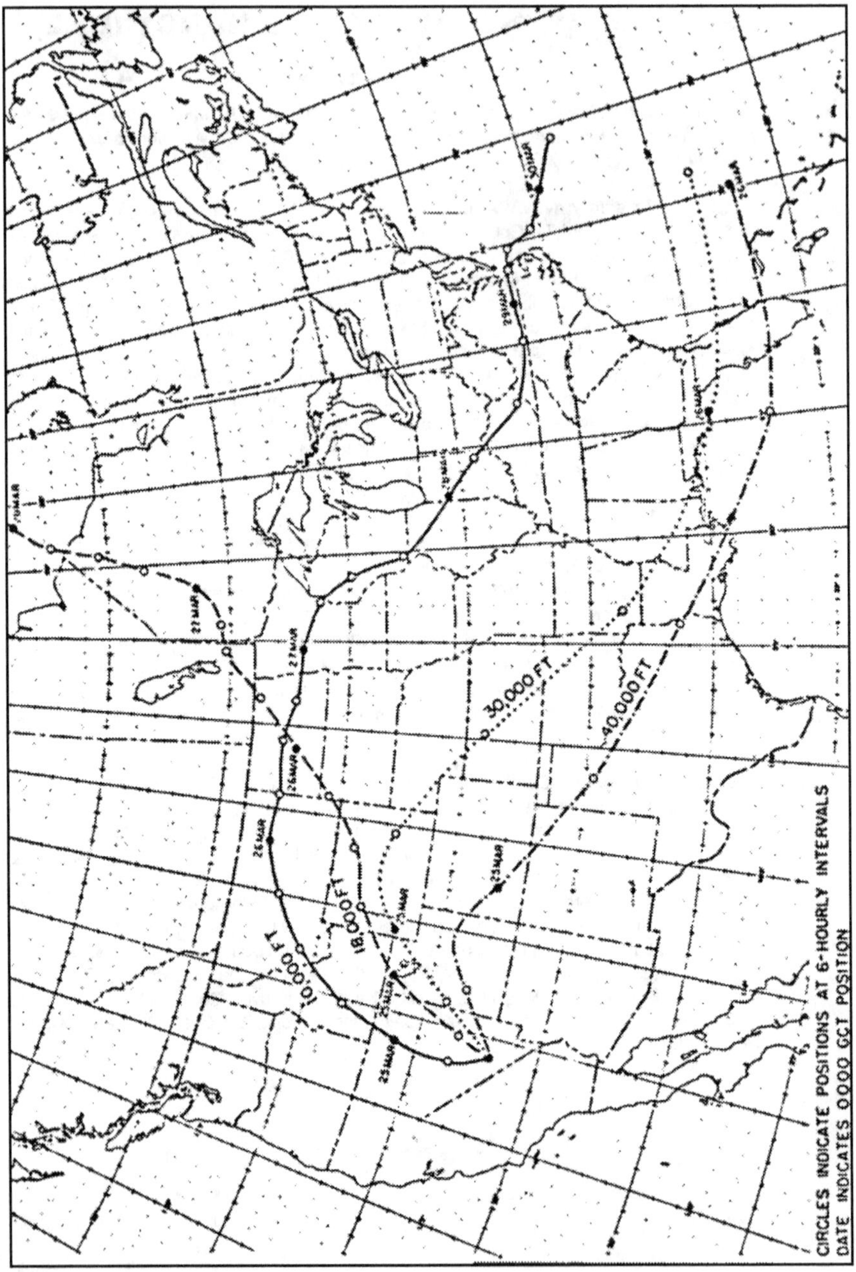

TRAJECTORY: UPSHOT-KNOTHOLE NANCY (UK2)

TOTAL FALLOUT: UPSHOT-KNOTHOLE NANCY

Equal Intervals

nCi/square meter

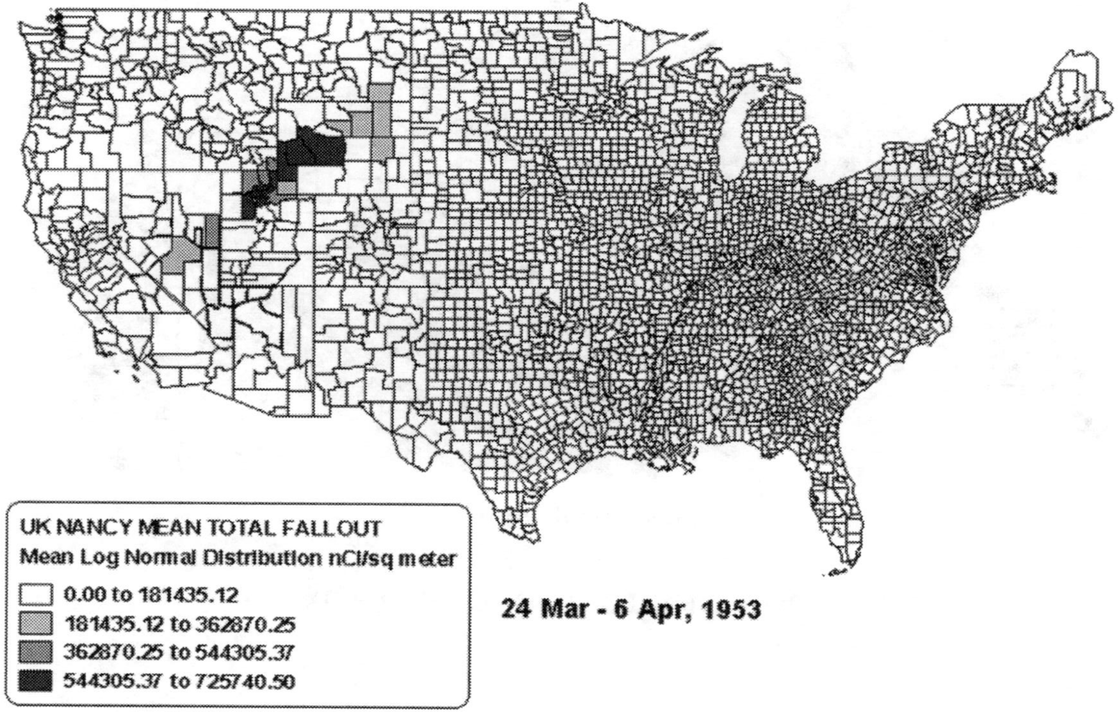

24 Mar - 6 Apr, 1953

TOTAL FALLOUT: UPSHOT-KNOTHOLE NANCY

Equal Number

nCi/square meter

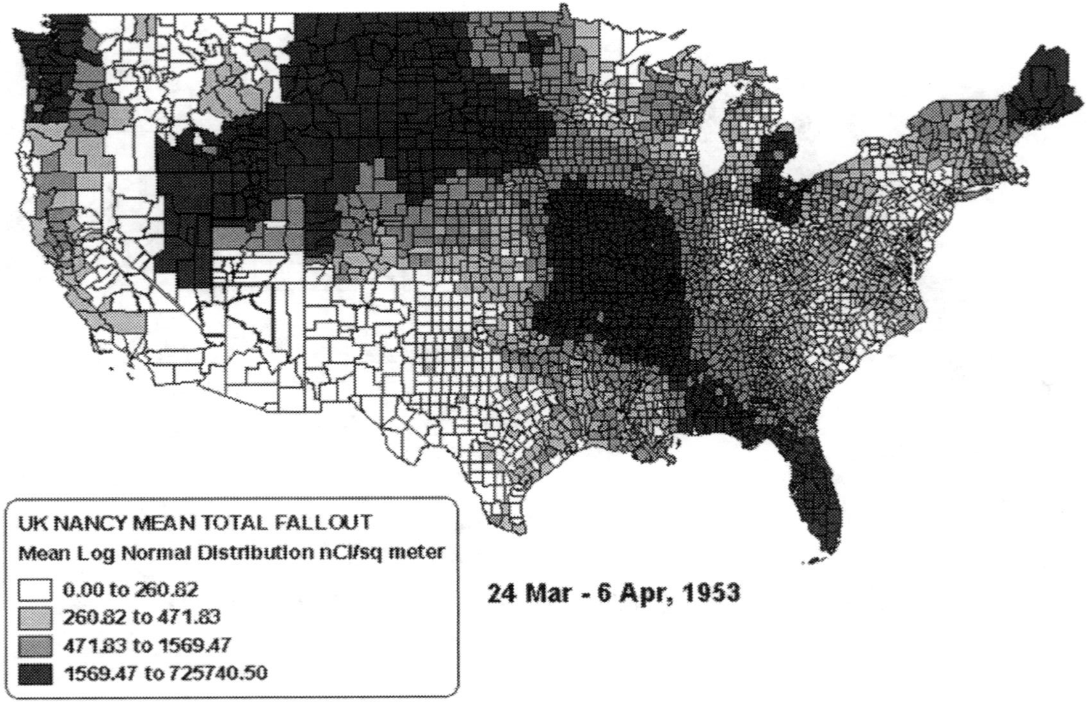

24 Mar - 6 Apr, 1953

RELATIVE TOTAL FALLOUT: UK NANCY

MAR 24, 1953 24 KT

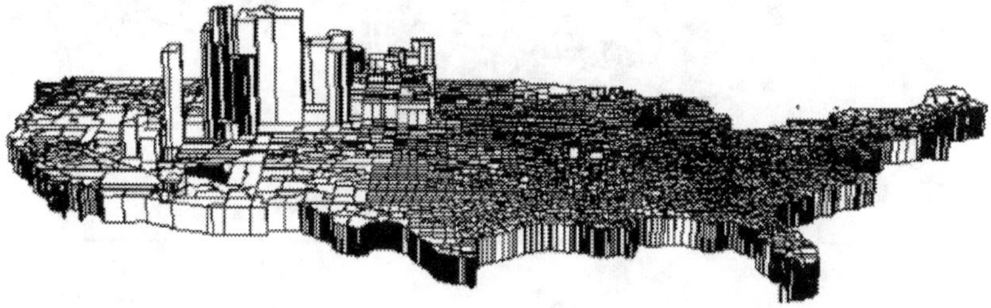

CHARACTERISTICS: UK NANCY

While the nuclear test was named "NANCY", the device that was the subject of the test was code-named "Nevada Zombie." A 3 ft diameter 10-foot long cylinder weighing 6.5-tons, Zombie was one of the biggest devices ever placed on top of a 300-foot tower. The shot was used to test a thermonuclear weapon design. Upshot-Knothole NANCY is ranked 3rd for percentage of Fe-59 in fallout, and 7th for percentages of Te-127, Te-127m and U-237. The deposition pattern for Upshot Knothole NANCY is most similar to Plumbbob SHASTA (PB8) (R: 0.49; T: 30.91). Fallout from this nuclear test was deposited on 97 percent of the U.S. counties (2997 counties) with an average per-county fallout deposition level of 14.6 μCi/sq meter.

Counties of highest fallout deposition include Morgan UT (estimated maximum activity: 1821 μCi/sq meter), Lincoln, WY (estimated maximum activity: 1670 μCi/Sq meter) and Sublette, WY (estimated maximum activity: 1658 μCi/sq meter).

The day one ratio of total fallout to I-131 is moderate at 91.6 to 1.

Upshot-Knothole NANCY also ranks 8th for deposition of the radionuclide Be-7, 9th for Co-60 and 10th among the nuclear tests for deposition of Cs-137

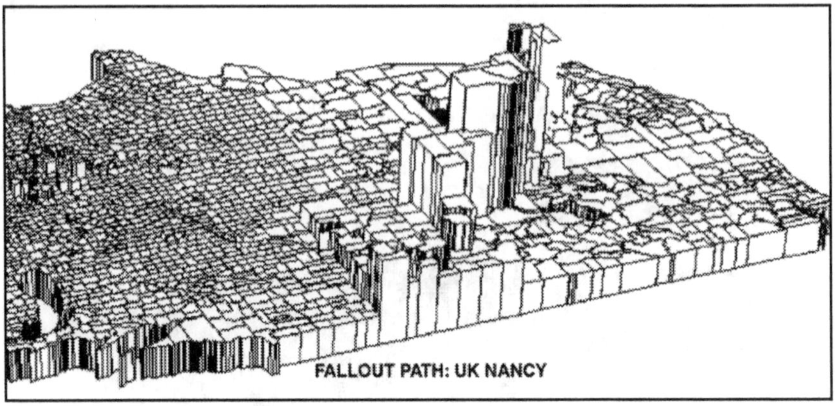

FALLOUT PATH: UK NANCY MAR 24, 1953

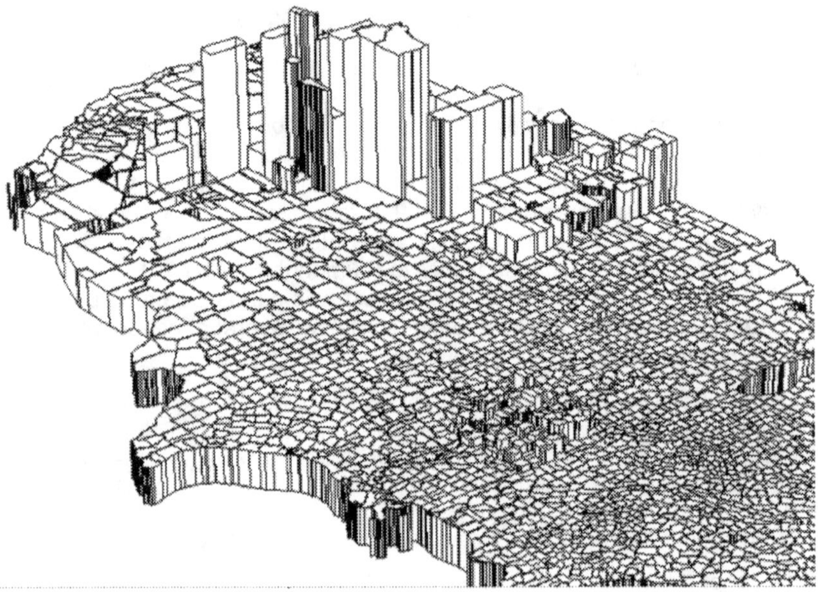

SHOT 3: UPSHOT-KNOTHOLE RUTH (UK3)

UPSHOT-KNOTHOLE: RUTH

Detonation Date: 31 Mar 1953 • Detonation Time: 5:00 A.M. • Area: 7-5a • Sponsor: Livermore Laboratory • Yield: 0.2 kt • Radiation Level at Ground Zero: 10 R/hr • Height of Burst: 304.69 ft • Cloud Top Height: 13,600 ft

10,000-Foot Trajectory: NEVADA: Indian Springs, Las Vegas, Winchester, Nelson, Cottonwood Cove, Riviera • CALIFORNIA: Needles, Vidal Junction, Rice, (Palen Mountains), (Chocolate Mountains), Glamis, (Sand hills) • ARIZONA: (Copper Mountains), (Sauceda Mountains), Casa Grande, Maricopa, Olberg, Palm Springs, Apache Junction, Tortilla Flat, Roosevelt, (Bear Mountain), Carrizo, Pinetop, Lakeside, (Greens Peak), Eagar, Alpine • NEW MEXICO: Luna, Reserve, (Tularosa Mountains), (Elk Mountain), Kingston, Las Cruces, La Mesa, Chamberino, La Union • TEXAS: El Paso, Horizon City, Clint, Andrews, Tarzan, Big Spring, Lomax, Otis Chalk, Silver, Robert Lee, Bronte, Maverick, Balinger, Voss, Mozelle, Trickham, Brookesmith, Indian Creek, Mullin, Golthwaite, Center City, Evant, Pearl, Arnett, Fort Gates, Flat, Moffatt, Pendelton, Moody, Cego, Mooresville, Perry, Otto, Ben Hur, Groesbeck, Personville, Point Enterprise, Fairfield, Red Lake, Montalba, Jacksonville, Reese, Black Jack, St. Clair City, New London, Kilgore, Easton, Darco, Longview, Karnack, Smithland, Gray • LOUISIANA: Trees, Vivian, Gilliam, Hosston, Mira, Bolinger, Ida • ARKANSAS: Springhill, Welcome, Taylor, Emerson, Plainfield, Atlanta, Wasson, Urbana, Gerdner, Strong, North Crossett, West Crossett, Crossett, Parkdale, Empire, Eudora • MISSISSIPPI: Chaltham, Hollandale, Darioue, Belzoni, Cruger, Acona, West, Beatty, French Camp, Weir, Ackerman, Sturgis, Bradley, Longview, Starkville, Columbus, Mayhew, Caledonia • ALABAMA: Fernbank, Kingville, Covin, Howard, Carbon Hill, Townby, Saragossa, Manchester, Macedonia, Cold Springs, Hanceville, Center Hill, Blountsville, Liberty, Brooksville, Nixons Chapel, Douglas, Horton, Boaz, Sardis City, Keener, Sand Rock, Leesburg, Cedar Bluff, Gaylesville • GEORGIA: Coosa, Garden Lakes, Rome, Halls, Cassville, Pine Log, Rydal, White, Waleska, Nelson, Ball Ground, New Holland, Lula, Raoul, Hollingsworth, Avalon, Martin, Lavonia • SOUTH CAROLINA: Anderson, Williamston, Cheddar, Pelzer, Fork Shoals, Owings, Enoree, Cross Keys, Buffalo, Monarch, Union, Chester, Eureka, Richburg, Fort Lawn, Grace, Lancaster, Pageland, Mount Crogham, Rugby, Chesterfield • NORTH CAROLINA: Gibson, Laurel Hill, Old Hundred, Laurinburg, Wakulla, Red Springs, Shannon, Rex, St. Pauls, Jerome, Parkerburg, Ingold, Magnolia, Rose Hill, Beulaville, Richlands, Petersburg, New Bern, Arapahoe, Merritt, Pamlico, Whortonville, Oracoke, Hatteras, Frisco, Buxton, Cape Hatteras.

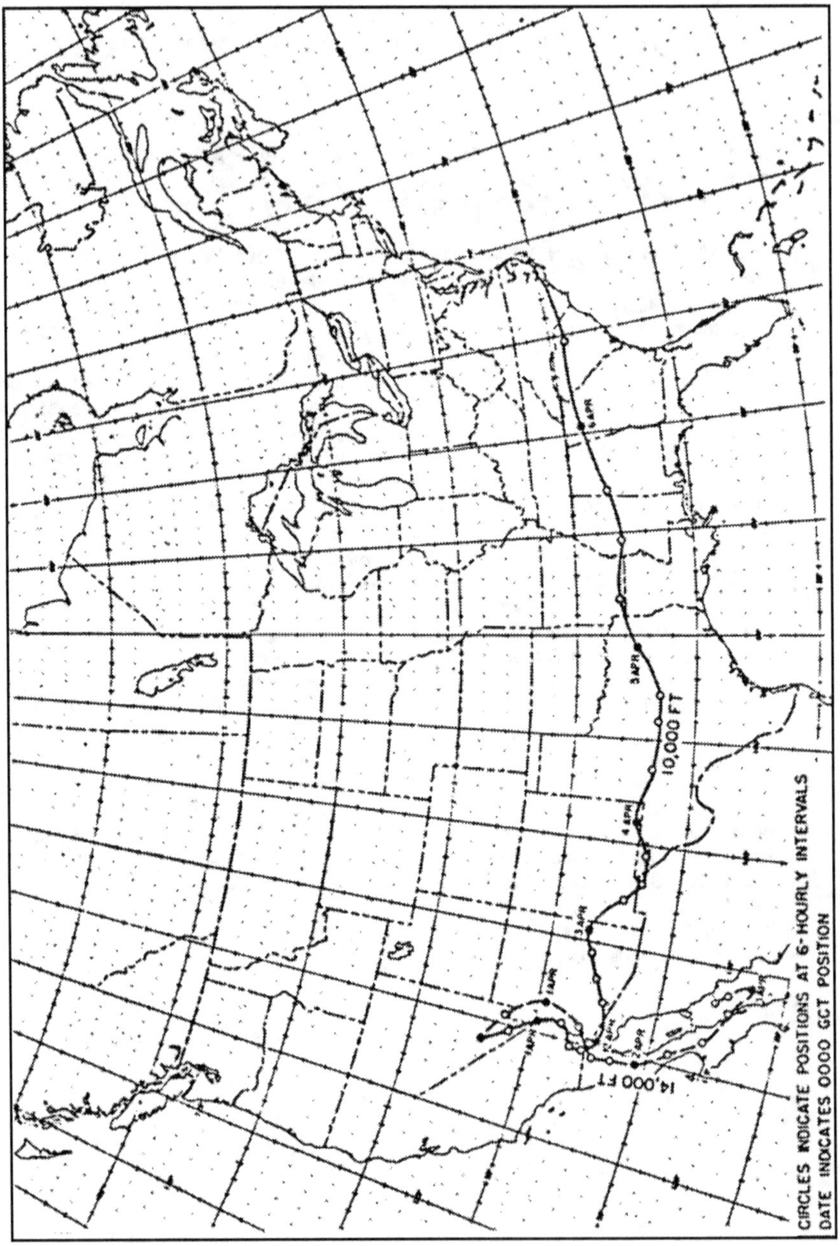

TRAJECTORY: UPSHOT-KNOTHOLE RUTH (UK3)

CHARACTERISTICS: UK RUTH

The device for the RUTH shot, code-named Hydride I, was the first University of California Radiation Lab bomb ever tested at the Nevada Test Site. Though huge (7400 lbs, 56 inches diameter, 66 inches long), the shot was a fizzle, vaporizing only the top 100 feet of the 300-foot shot tower. Among the nuclear tests, Upshot-Knothole RUTH is ranked 7th for Kr-85, Sr-90 and Y-90. The deposition pattern for Upshot-Knothole RUTH is most similar to Upshot-Knothole ANNIE (UK1) (R: 0.51; T: 32.89). Fallout from this nuclear test fell on 56 percent of the total U.S. counties (1733 counties out of 3098 designated by the NCI). Counties of highest deposition included Beckham OK, Greer OK, Harmon OK, Roger Mills OK, Armstrong TX, Carson TX Collingsworth TX, Donley TX and Parmer, TX. All received (estimated maximum activity) 83.58 μCi/sq meter on Apr 4, 1953.

Despite its low yield, low total fallout level, and the fact that it had been something of a fizzle, Upshot-Knothole RUTH has a relatively high day-one total fallout to I-131 ratio: 124 to 1.

TOTAL FALLOUT: UPSHOT-KNOTHOLE RUTH

Equal Intervals
nCi/square meter

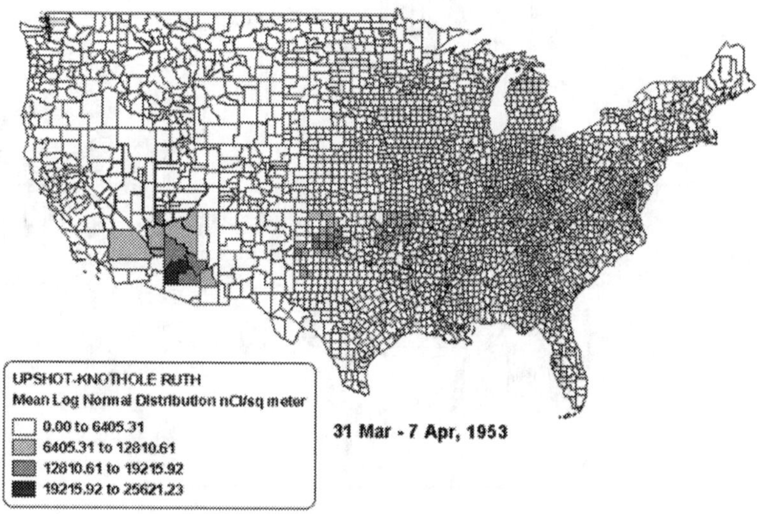

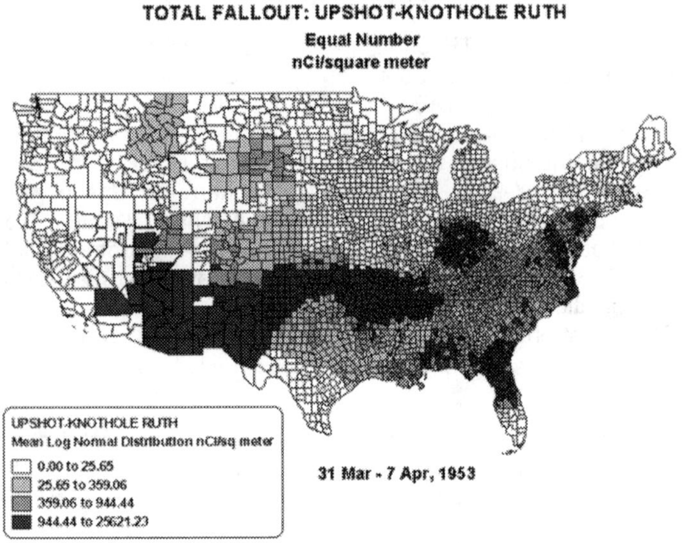
TOTAL FALLOUT: UPSHOT-KNOTHOLE RUTH
Equal Number
nCi/square meter
UPSHOT-KNOTHOLE RUTH
Mean Log Normal Distribution nCi/sq meter
0.00 to 25.65
25.65 to 359.06
359.06 to 944.44
944.44 to 25621.23
31 Mar - 7 Apr, 1953

FALLOUT PATH: UK RUTH

PATH OF FALLOUT: UK RUTH
MAR 31, 1953

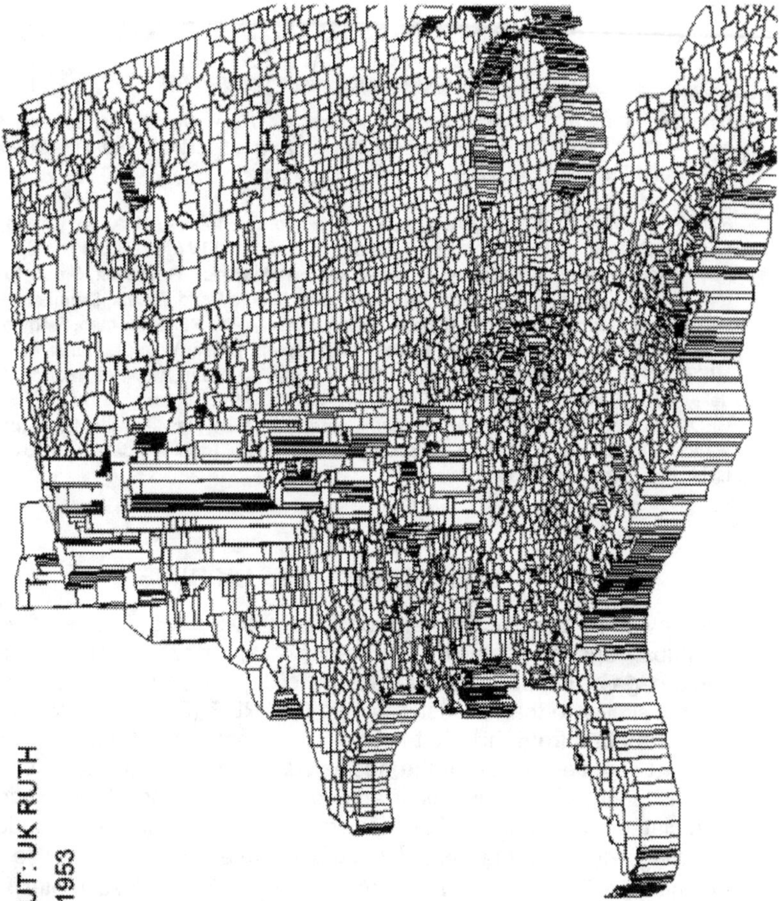

SHOT 4: UPSHOT-KNOTHOLE DIXIE (UK4)

UPSHOT-KNOTHOLE: DIXIE

Detonation Date: 6 Apr 1953 • *Detonation Time:* 7:30 A.M. • *Area:* 7–3 • *Sponsor:* Los Alamos • *Yield:* 11kt • *Radiation Level at Ground Zero:* 1.5 mR/hr • *Height of Burst:* 6,022 ft • *Cloud Top Height:* 45,000 ft

30,000-Foot Trajectory: NEVADA: Moapa, Logandale, Overton • ARIZONA: (Grand Canyon National Park), Flagstaff, Show Low, Nutrioso, Alpine • NEW MEXICO: Horse Springs, Negal, Hondo, Roswell, Caprock • TEXAS: Sundown, Lubbock, Roosevelt, Crosbytown, Croton, Truscott, Thalia, Electra, Wichita Falls, Terral, Illinois Bend, Dexter, Dennison, Ivanhoe, Paris, Reno, Detroit, Clarksville, Whaley, Texarkana • ARKANSAS: Homen, Patmos, Lamartine, Smackover, El Dorado, North Crossett, Crossett, Berlin, Wilmet, Indian, Readland • MISSISSIPPI: Glen Allan, Grace, Yazoo City, Vaughn, Carthage, McAffee, McDonald, Econdale • ALABAMA: Demopolis, Burnsville, Selma, Montgomery, Shorter, Tuskegee • GEORGIA: Fort Benning, Tazewell, Ideal, Montezuma, Elko, Hawkinsville, Gresston, Mount Vernon, Vidalia, Lyons, Santa Claus, Claxton, Pembroke, Bloomingdale • SOUTH CAROLINA: Port Wentworth, Forest Beach.

CHARACTERISTICS: UK DIXIE

The device for this shot (an airdrop) was a 10.5-ft x 5 ft cylinder weighing 3260 lbs. Upshot-Knothole DIXIE ranks 4th among the nuclear tests in the percentage of Co-58 in total fallout. The deposition pattern for UK DIXIE is most similar to that of Buster-Jangle CHARLIE (BJ2) (R: 0.25; T: 14.06). Nuclear debris from this shot fell on 50 percent of the total U.S. counties, or 1548 counties. Most of the nuclear fallout was deposited in the states of Massachusetts, Connecticut and Rhode Island and all deposition was associated with rain. Counties of highest deposition included the following: Plymouth Ma, (estimated maximum activity: 1260 μCi/sq meter), Washington RI (estimated maximum activity: 607 μCi/sq meter), Bristol MA (estimated maximum activity: 942 μCi/sq meter) and Middlesex CT (estimated maximum activity: 573 μCi/sq meter).

The ratio of total fallout to I-131for DIXIE is relatively high at 125 to 1.

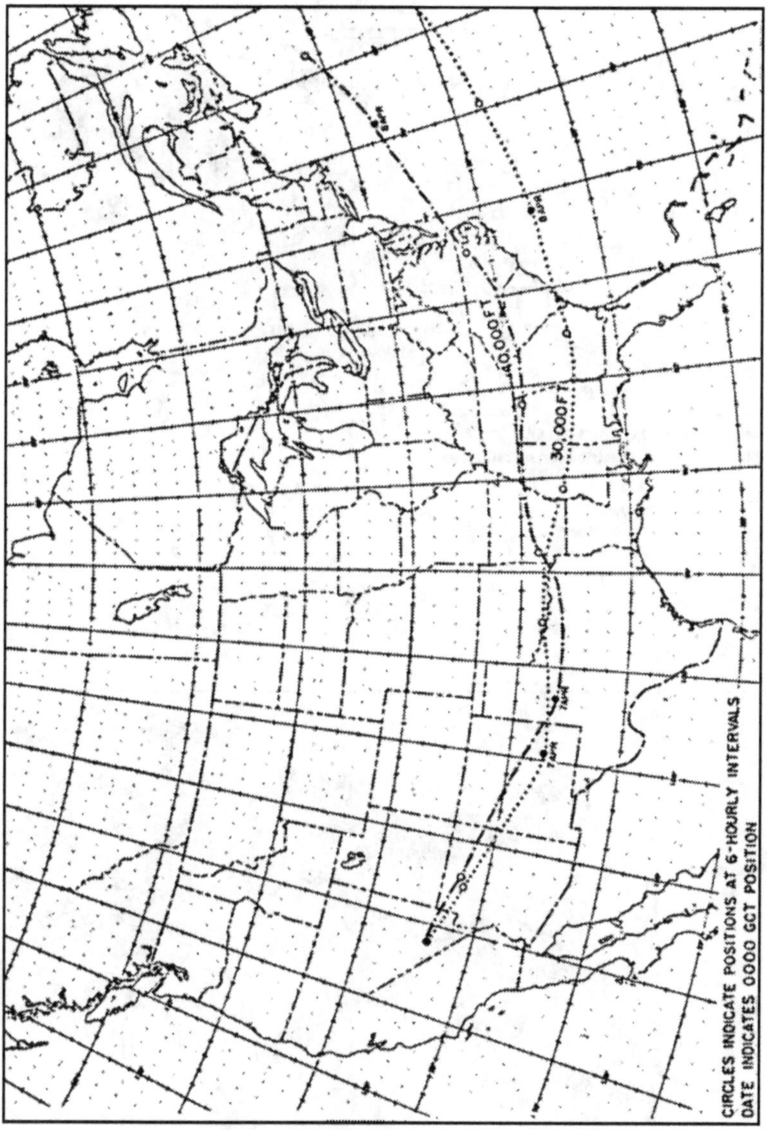

TRAJECTORY: UPSHOT-KNOTHOLE DIXIE (UK4)

TOTAL FALLOUT: UPSHOT-KNOTHOLE DIXIE

Equal Intervals
nCi/square meter

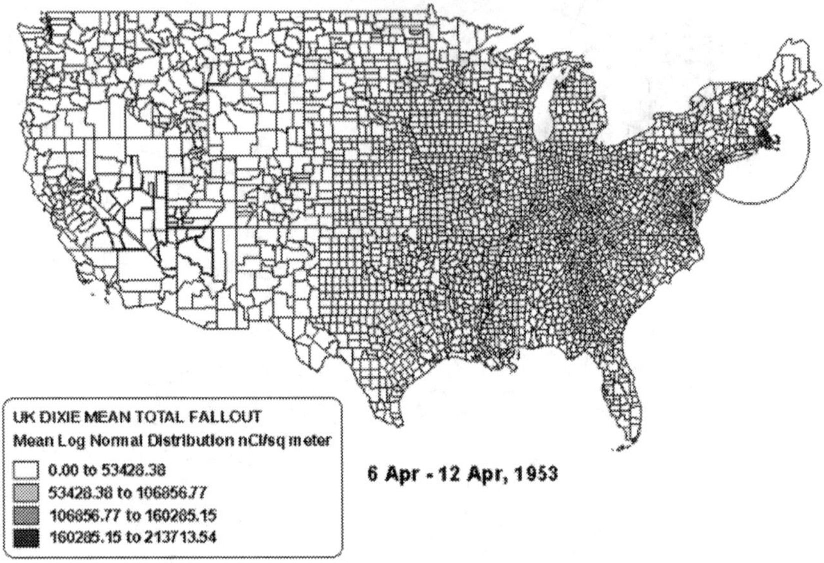

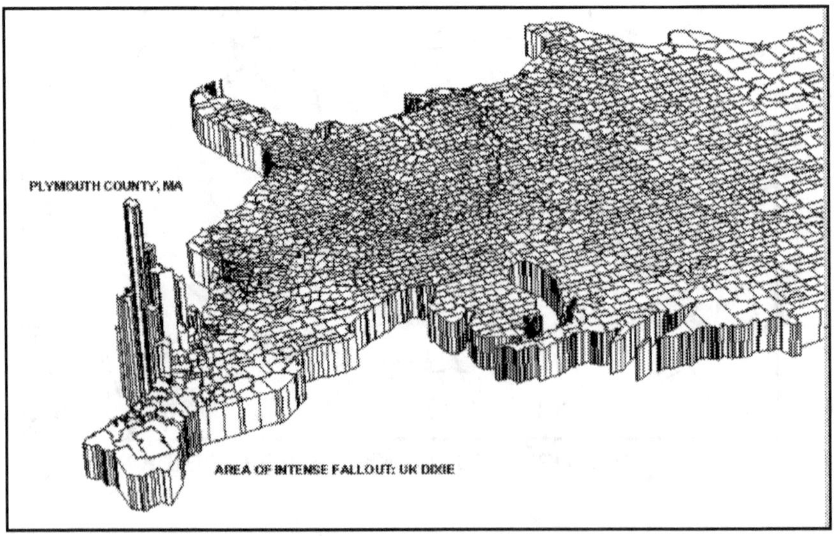

SHOT 5: UPSHOT-KNOTHOLE RAY (UK5)

UPSHOT-KNOTHOLE: RAY

Detonation Date: 11 Apr 1953 • *Detonation Time:* 4:45 A.M. • *Area:* 4a • *Sponsor:* Livermore Laboratory • *Yield:* 0.2 kt • *Radiation Level at Ground Zero:* 20 R/hr • *Height of Burst:* 100 ft • *Cloud Top Height:* 12,800 ft

10,000-Foot Trajectory: NEVADA: Lathrop Wells • CALIFORNIA: Death Valley Junction, (Greenwater Range), (Avawatz Mountains), (Bristol Mountains), Bagdad, Amboy, (Bristol Dry Lake), (Cadiz Dry Lake) (Palen Mountains), Blythe, East Blythé • ARIZONA: Ehrenburg, (Kofa Mountains), (Gila Bend Mountains), (Woolsey Peak), Maricopa Akchin Indian Reservation, Casa Grande, Stanfield, Eloy, Pichacho, Catalina, Mount Lemmon, (Lime Peak), Dragoon, (Chiricahua Peak) • NEW MEXICO: Rodeo, (Animas Mountains) • TEXAS: Socorro, Horizon City, (Hueco Mountains), (San Antonio Mountain) • NEW MEXICO: (Carlsbad Caverns National Park), (Sitting Bull Falls), Whites City, Loving, Carlsbad, Malaga, (Antelope Ridge), Bennett, Jal • TEXAS: Stanton, Greenwood, Sterling City, Water Valley, Carlsbad, Miles, Mereta, Millers View, Doole Fife, Placid, Hall, Algerita, San Saba, Lampasas, Watson, Prairie Dell, Bartlett, Granger, Thorndale, Navasota, Conroe, Moss Hill, Silsbee, Lumberton • LOUISIANA: Vinton, Edgerly, Sulphur, Westlake, Lake Charles, Iowa, Welsh, Jennings, Midland, Morse, Indian Bayou, Maurice, Youngsville, New Iberia, Jeanerette, Labadieville, Thibodaux, Raceland, Lockport, Larose, Lafite, Myrtle Grove, Magnolia, Port Sulphur, Empire, Buras, Triumph, Boothville, (North Pass) • FLORIDA: Marco, Goodland, Everglades City, Ochope, Monroe, Miccosukee Indian Reservation, Sweetwater, Miami, Miami Beach.

CHARACTERISTICS: UK RAY

Among the nuclear tests, Upshot-Knothole RAY is ranked 6th for percentages of Sr-89, Sr-90 and Y-90 in fallout. The deposition pattern for Upshot-Knothole RAY is most similar to that of Upshot-Knothole ANNIE (UK1) (R: 0.46; T: 28.46). Fallout from this shot was deposited on 2084 counties, or 67 percent of the total U.S. counties. The county of highest deposition was Yuma, AZ with an estimated maximum activity of 95.41 μCi/sq meter on Apr 11, 1953. The total fallout to I-131 ratio is relatively high at 121 to 1.

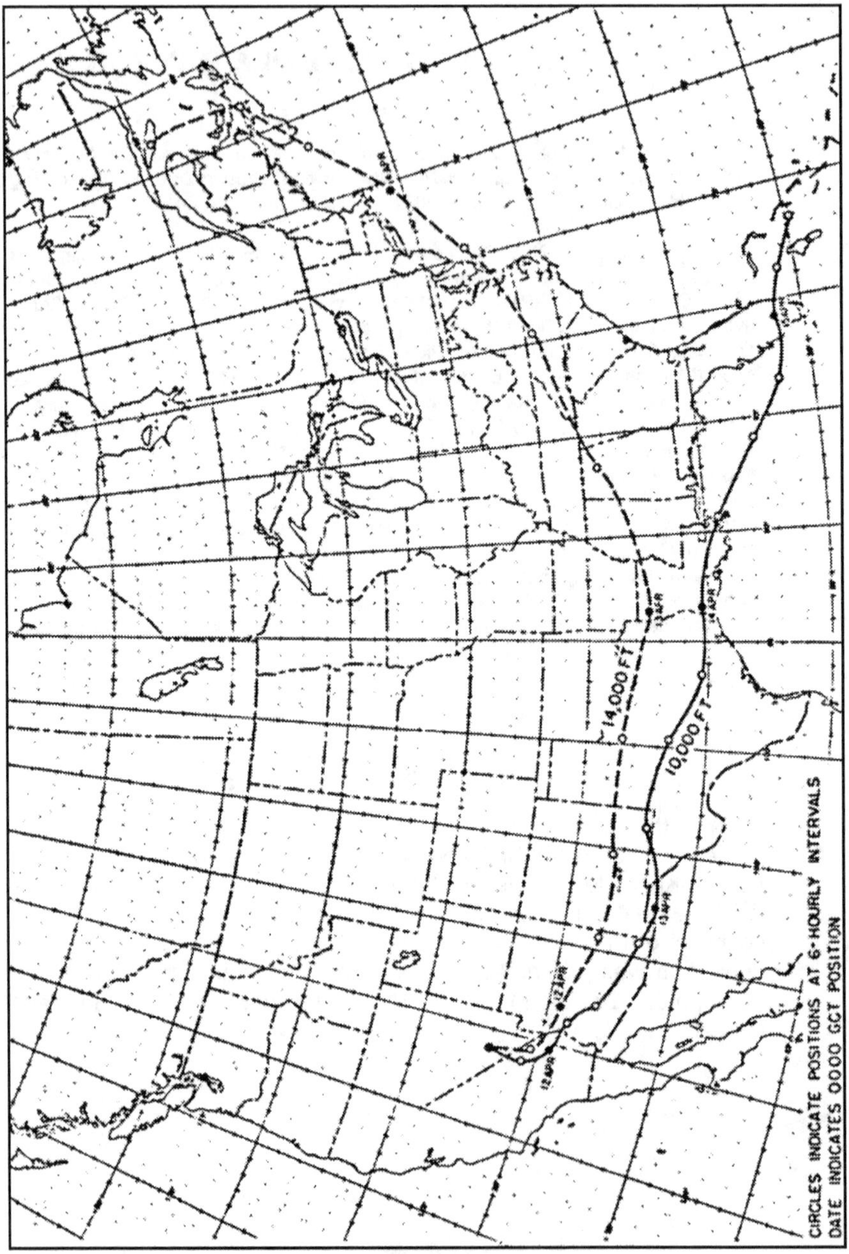

TRAJECTORY: UPSHOT-KNOTHOLE RAY (UK5)

TOTAL FALLOUT: UPSHOT-KNOTHOLE RAY

Equal Intervals

nCi/square meter

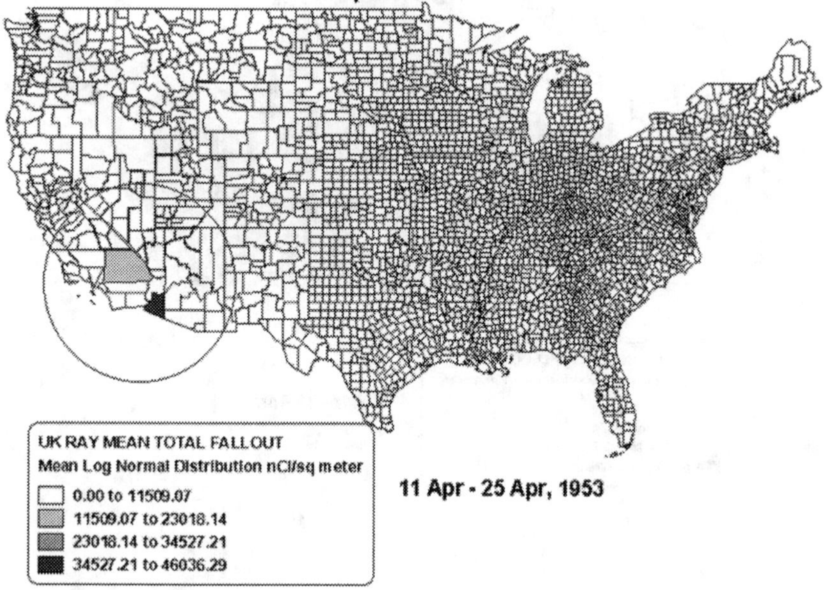

TOTAL FALLOUT: UPSHOT-KNOTHOLE RAY

Equal Number

nCi/square meter

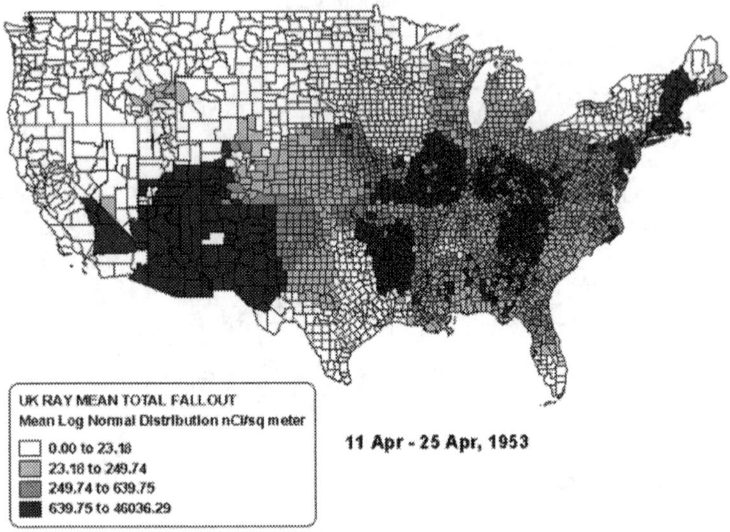

SHOT 6: UPSHOT-KNOTHOLE BADGER (UK6)

UPSHOT-KNOTHOLE: BADGER

Detonation Date: 18 Apr 1953 • *Detonation Time:* 4:35 A.M. • *Area:* 2 • *Sponsor:* Los Alamos • *Yield:* 23 kt • *Radiation Level at Ground Zero:* 1,200 R/hr • *Height of Burst:* 300 ft • *Cloud Top Height:* 36,000 ft

40,000-Foot Trajectory: NEVADA: Indian Springs, North Las Vegas, Henderson • ARIZONA: Hualpai Indian Reservation, Peach Springs, Nelson, Drake, Sedonia, Heber, Overgaard, Shumway, Vernon, Springerville • NEW MEXICO: Red Hill, Pie Town, Polvadera, Lemitar, Florida, Claunch, Ancho, Haystack Mountain, Elkins, Kenna, Milnesand, Pep, Lingo • TEXAS: Denver City, Higginbotham, Seminole, Three Leagues, Knott, Fairview, Big Spring, Otis Chalk, Edith, Tennyson, Orient, Rowena, Concho, Millersview, Lohn, Fife, Rochelle, Hall, San Saba, Lake Victor, Briggs, Youngsport, Little River, Heidenheimer, Temple, Westphalia, Durango, Barclay, Rosebud, Travis, Lott, Cedar Springs, Marlin, Oletha, Thronton, Donie, Buffalo, Oakwood, Elkhart, Palestine, Rusk, Gallatin, Ponta, New Salem, Laneville, Concord, Dotson, Gary, Carthage, Deadwood • LOUISIANA: Keatchie, Stonewall, Keithville, Gayles, McDude, Elm Grove, Janestown, Heflin, Arcadia, Bryceland, Grambling, Ruston, Vienna, Sibby, D'Arbonne, Rocky Branch, Monroe, Sterlington, Bastrop, Fairbanks, Collinston, Mer Rouge, Goodwill, Pioneer, Forest, Transylvania, Lake Providence • MISSISSIPPI: Cary, Fitler, Jonestown, Valley, Tinsley, Benton, Way, Vaughn, Camden, Ofahoma, Carthage, Edinburg, McAfee, Williamsville, Deemer, McDonald, Moscow, House, Daleville • ALABAMA: York, Livingston, Coatopa, Belmont, Demopolis, Prairieville, Suttle, Summerfield, Burnsville, Winslow, Independence, Booth, Prattville, Cobbs Ford, Ware, Milstead, Tuskegee, Society Hill, Uchee, Seale, Fort Mitchell • GEORGIA: Cusseta, Buena Vista, Murrays Crossroads, Oglethorpe, Montezuma, Unadilla, Hawkinsville, Hartford, Gresston, Plainfield, Mount Vernon, Alley, Higgston, Vidella, Lyons, Choopee, Collins, Bellville, Claxton, Pembroke, Blitchton, Eden • SOUTH CAROLINA: Pritchardville, Bluffton, Hilton Head Island, Folly Field Beach.

CHARACTERISTICS: UK BADGER

The device for the BADGER test was code-named "Buzzard." Like many other thermonuclear-styled designs, Buzzard was quite large, weighing 7400 lbs with a diameter of 56 inches. Among the nuclear tests, Upshot-Knothole BADGER is ranked 7th for percentages of Sn-123, Sb-125 and Te-127m in fallout. The deposition pattern for Upshot-Knothole BADGER is most similar to Plumbbob WHEELER-COULOMB B and LAPLACE (PB13) (R:0.38; T: 22.74). Fallout from this shot was deposited on 3000 counties, or 97 percent of the total U.S. counties. Counties of highest deposition included: Archuleta, CO (estimated maximum activity: 1136 μCi/sq meter) and Clark, NV (estimated maximum activity: 1103 μCi/sq meter), both on April 18, 1953.

The day one ratio of total fallout to I-131 is moderately high at 111 to 1.

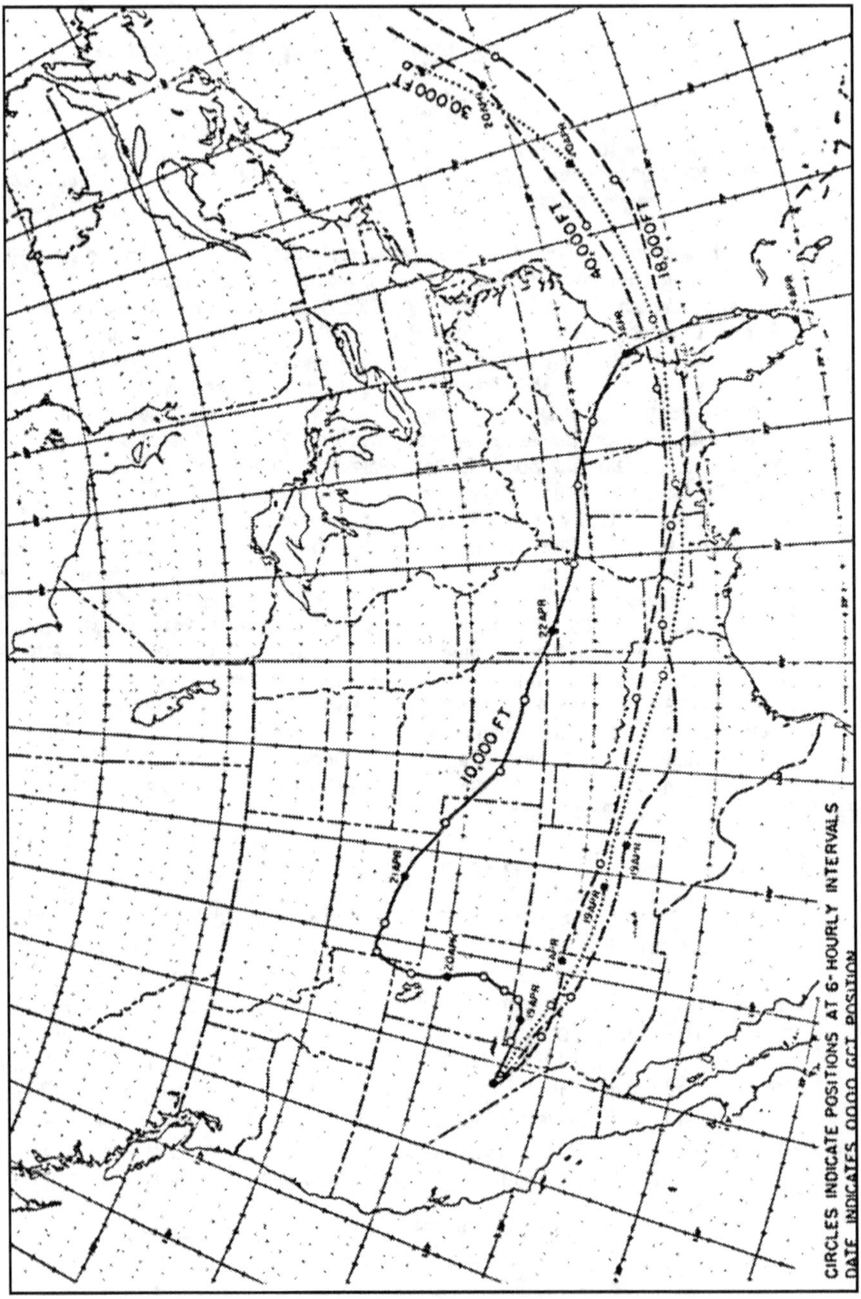

TRAJECTORY: UPSHOT-KNOTHOLE BADGER (UK6)

TOTAL FALLOUT: UPSHOT-KNOTHOLE BADGER

Equal Intervals

nCi/square meter

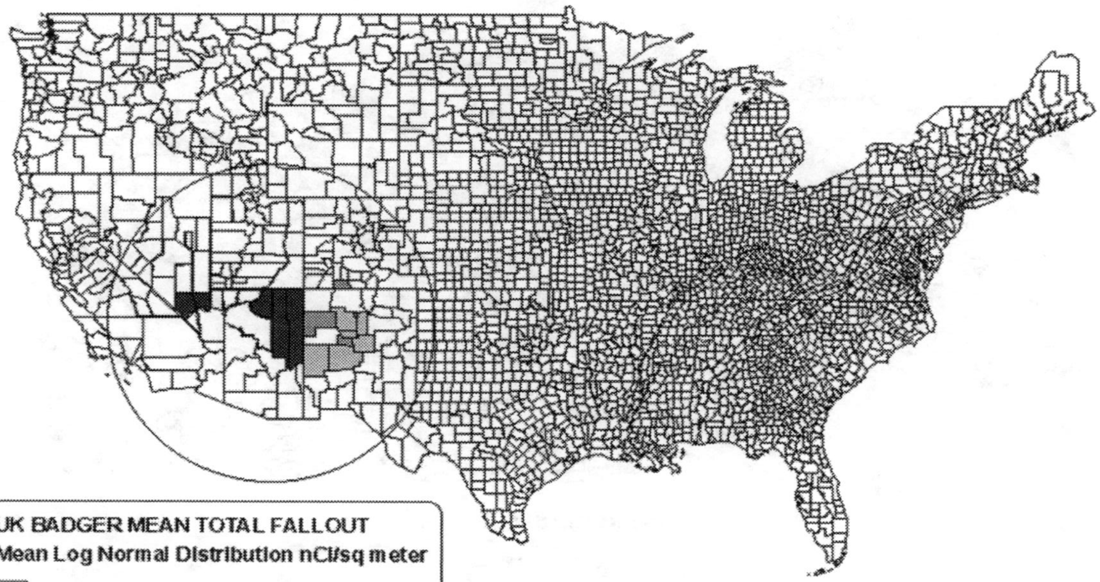

UK BADGER MEAN TOTAL FALLOUT
Mean Log Normal Distribution nCi/sq meter

- 0.00 to 82135.02
- 82135.02 to 164270.05
- 164270.05 to 246405.07
- 246405.07 to 328540.09

18 Apr - 25 Apr, 1953

TOTAL FALLOUT: UPSHOT-KNOTHOLE BADGER

Equal Number

nCi/square meter

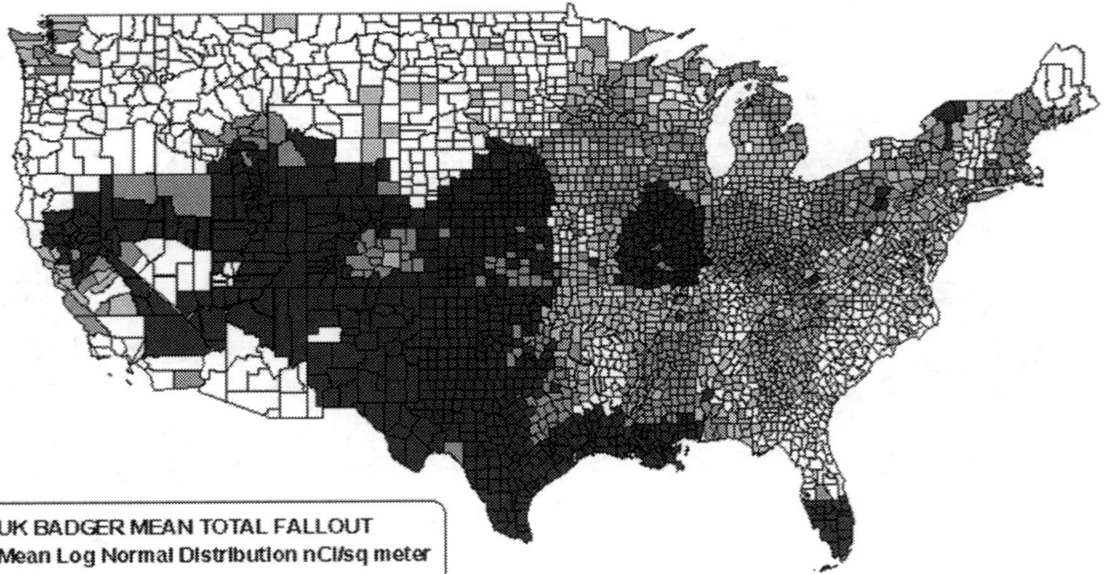

UK BADGER MEAN TOTAL FALLOUT
Mean Log Normal Distribution nCi/sq meter

- 0.00 to 237.39
- 237.39 to 407.30
- 407.30 to 1005.66
- 1005.66 to 328540.09

18 Apr - 25 Apr, 1953

FALLOUT PATH: UK BADGER

APR 18, 1953 23 KT

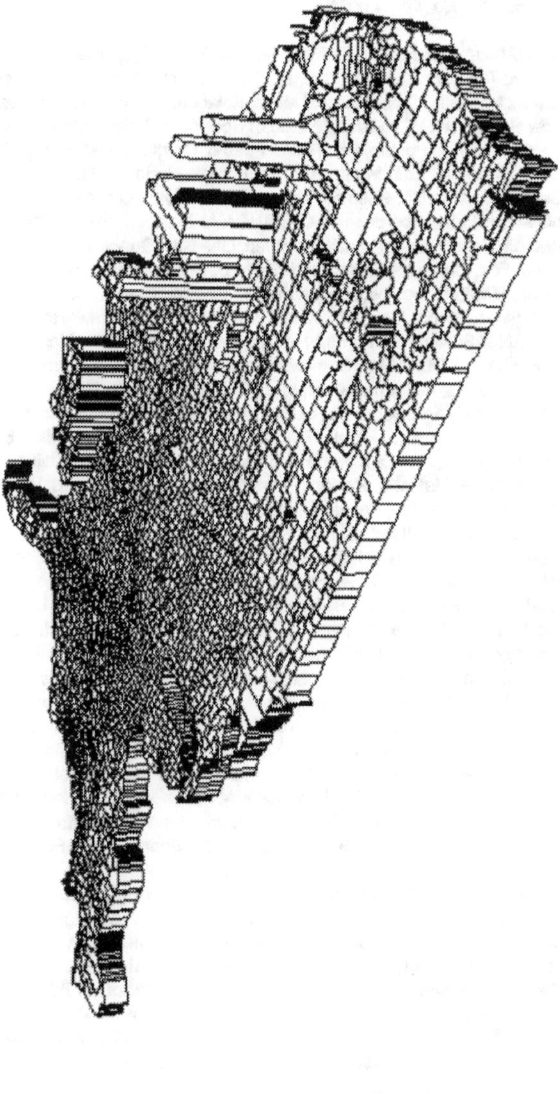

SHOT 7: UPSHOT-KNOTHOLE SIMON (UK7)

UPSHOT-KNOTHOLE: SIMON

Detonation Date: 25 Apr 1953 • *Detonation Time:* 4:30 A.M. • *Area:* 1 • *Sponsor:* Los Alamos • *Yield:* 43 kt • *Radiation Level at Ground Zero:* 300 R/hr • *Height of Burst:* 300 ft • *Cloud Top Height:* 44,000 ft

30,000-Foot Trajectory: NEVADA: Moapa Indian Reservation, Logandale, Overton • ARIZONA: Tusayan, Cameron, (Painted Desert), Navajo Indian Reservation, Hopi Indian Reservation, Hotevilla, Orabi, Second Mesa, Polacca, Keams Canyon, Nazlini, Fort Defiance, Window Rock • NEW MEXICO: Mexican Springs, Brimhall, Crown Point, Hospah, San Ysidro, Santo Domingo Pueblo, San Felipe Pueblo, Cerrillos, Madrid, Galisteo, Lamy, El Pueblo, Ribera, Serafina, Los Montoyas, Tucumcati, Porter • TEXAS: Adrian, Vega, Wildorado, Bushland, Amarillo, Pullman, Washburn, Conway, Claude, Goodnight, Howardwick, Samnorwood, Lutie, Aberdeen • OKLAHOMA: Madge, Vinson, Reed, Mangum, Hester, Blair, Warren, Snyder, Cache, Lawton, Hulen, Gas City, Duncan, Sunray, Velma, Countyline, Alma, Loco, Fox, Clemscot, Graham, Woodford, Newport, Springer, Gene Autry, Dickson, Baum, Ravia, Tishomingo, Nida, Armstrong, Caddo, Matoy, Cade, Bokchito, Boswell, Soper, Goodland, Hugo, Sawyer, Swink, Valient, Idabel, Haworth, Harris, Bokhoma, Tom • ARKANSAS: Arkinda, Fomby Homan, Stamps, Buckner, Magnolia, Calhoun, Macedonia, Emerson • LOUISIANA: Summerfield, Lillie, Spearsville, Farmerville, Linville, Spencer, Bastrop, Perryville, Fairbanks, Collinston, Oak Ridge, Epps, Monticello, Sondheimer • MISSISSIPPI: Blakely, Redwood, Bentonia, Ballard, Sharon, Offahoma, Carthage, Edinburg, Stallo, Plattsburg, McLeod, Prairie Point • ALABAMA: Pickensville, Beaver Town, Carrolton, Gordo, Echola, Port Birmingham, Bayview, Adamsville, Graysville, Fultondale, Gardendale, Mount Olive, Majestic, Pinson, Palmerdale, Remlap, Allgood, Whitney, Taits Gap, Gallant, Reece City, Gadsden, Keener, Leesburg, Sand Rock, Gaylesville, Ringgold • GEORGIA: Lylerly, Berryton, Holland, Crystal Springs, Summersville, Calhoun, Echota, Oakman, Whitestone, (Brasstown Bald, highest point in Georgia: 4,784 feet), Pillard • NORTH CAROLINA: Brasstown, Shooting Creek, Franklin, Iotla, Webster, Cullowhee, Sylva, Balsam, Hazelwood, Canton, Asheville, Leicester, Alexander, Jupiter, Barnardsville, Little Switzerland, Celo, Estatoe, Miraville, Spruce, Ingalls, Linville, Montezuma, Boone, Jefferson, Warrensville, Lansing, Grassy Creek • VIRGINIA: Independence, Fries, Sylvatus, Indian Valley, Alum Ridge, Copper Hill, Bent Mountain, Roanoke, Montvale, Big Island, Lowesville, Nellysford, White Hall, Free Union, Nortonsville, Ruckersville, Hood, Wolftown, Haywood, Boston, Viewtown, Amissville, Orlean, Old Tavern, Marshall, Rectorville, Halfway, Aldie, Oatlands, Hamilton, Waterford, Point of Rocks • MARYLAND: Tuscarora, Adamstown, Thurston, Buckeystown, Frederick, Mount Pleasant, Johnsville, Union Bridge, Ladiesburg, Uniontown, Middleburg, Tarrytown, Silver Run • PENNSYLVANIA: Hanover, Mount Royal, Newberrytown, Royalton, Hummelstown, Hershey, Shellsville, East Hanover, Bardnersville, Oak Grove, Ravine, Donaldson, Branch Dale, Buck Run, Gordon, Shenendoah, Fern Glen, Plymouth, Luzerne, Upper Exeter, Dalton, Lewisville, Royal, Winderdale • NEW YORK: Fish's Eddy, Shinhopple, Downsville, Lake Delaware, Bovina Center, Stamford, West Fulton, Schoharie, Schenectady, Rotterdam, Charlton, Factory Village, Saratoga Springs, Fort Howard, Hudson Falls, Kingsburg, Truthville, Poultney, Hampton • VERMONT: West Haven, Benson, Brandon, Forest Dale, Goshen, Alpine Village, Roxbury, Northfield, Berlin, Montpelier, Barre, North Montpelier, Plainfield, East Calais, Marshfield, Cabot, Walden, North Danville, Lyndon, Burke, East Haven, Lemington, Colebrook • NEW HAMPSHIRE: Stewartstown Hollow, Mount Pisgah, Magalloway Mountain • MAINE: Eustis, West Forks, Moosehead, Kokadjo, (Mount Katahdin, highest point in Maine), Patten, Crystal, Island Falls, Shin Pond, Knowles Corner, Bridgewater, Blaine, Mars Hill.

The 40,000-foot trajectory passed over the highest points in the states of Georgia and Maine.

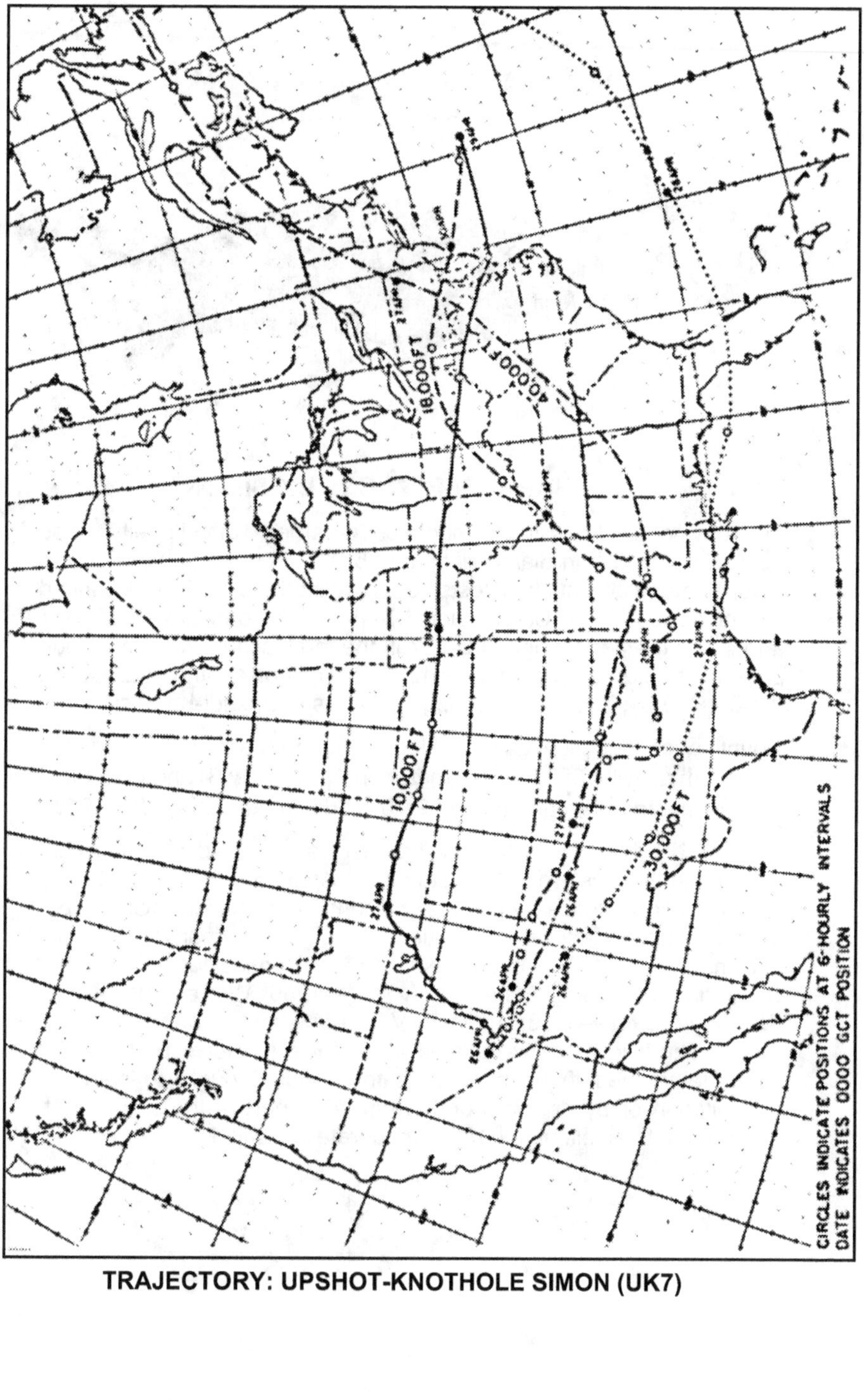

TRAJECTORY: UPSHOT-KNOTHOLE SIMON (UK7)

RELATIVE TOTAL FALLOUT: UK SIMON

APR 15, 1953 43 KT

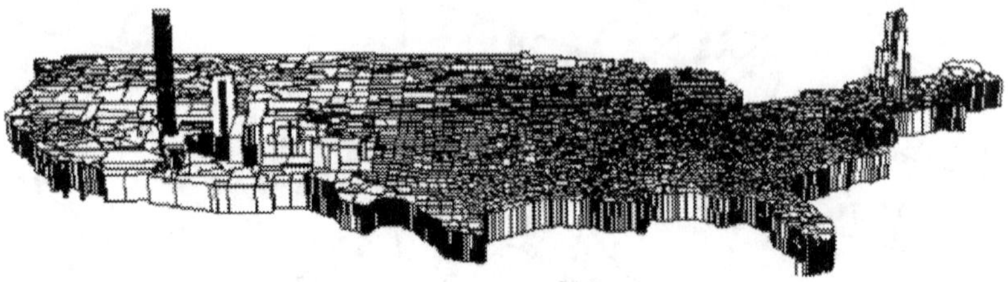

CHARACTERISTICS: UK SIMON

The device for the SIMON shot, code-named "Simultaneity", was approximately 3 ft in diameter and 18.5 ft in length and weighed 5.5 tons. Among the nuclear tests, Upshot-Knothole SIMON is ranked 2nd for deposition of Cs-137 and Nb-95, 3rd for percentage of Be-7 and 4th for the percentage of U-237 in the total fallout. The deposition pattern for this test is most similar to Upshot-Knothole ENCORE (UK8) (R: 0.51; T: 32.81). Among all the tests, Upshot-Knothole

Counties of highest close-in fallout include: Clark, NV (estimated maximum activity: 3882 μCi/sq meter on Apr 25, 1953) and Coconino, AZ (estimated maximum activity: 2197 μCi/sq meter on Apr 25, 1953).

Interestingly, fallout from this nuclear test was involved in a significant hot spot in the Eastern United States, on April 26, 1953. On that day the following counties received significant fallout (in microCuries per square meter, estimated maximum activity): Albany NY—894; Columbia NY—1058; Fulton NY—1769; Rensselaer, NY—2694; Saratoga, NY—1808; Schenectady, NY—1260; Warren, NY—3068; Washington, NY—3226; Addison, VT—1043; and Bennington, VT—1519. The fallout was associated with a rainout. However, it has not been determined if the nuclear material was brought to earth by precipitation or by dry downdrafts that were part of the storm front. The ratio of total fallout to I-131 is moderate at 98 to 1.

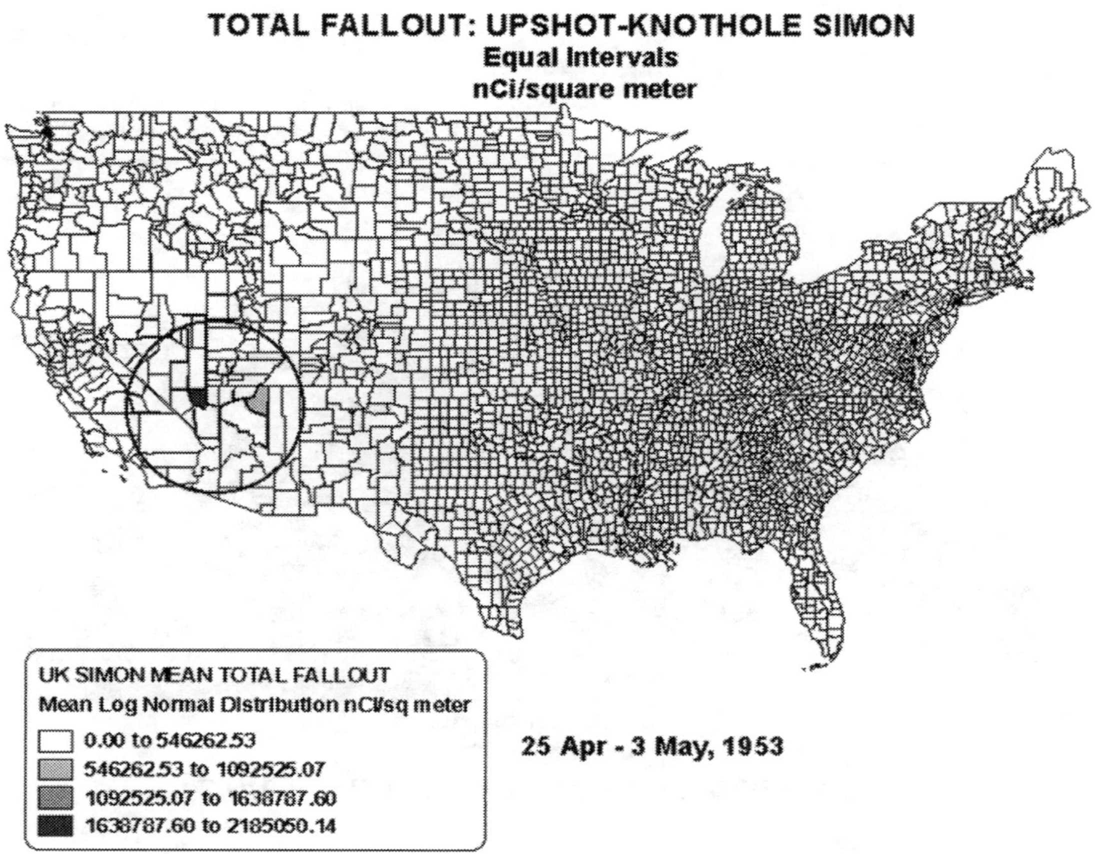
TOTAL FALLOUT: UPSHOT-KNOTHOLE SIMON
Equal Intervals
nCi/square meter
UK SIMON MEAN TOTAL FALLOUT
Mean Log Normal Distribution nCi/sq meter
0.00 to 546262.53
546262.53 to 1092525.07
1092525.07 to 1638787.60
1638787.60 to 2185050.14
25 Apr - 3 May, 1953

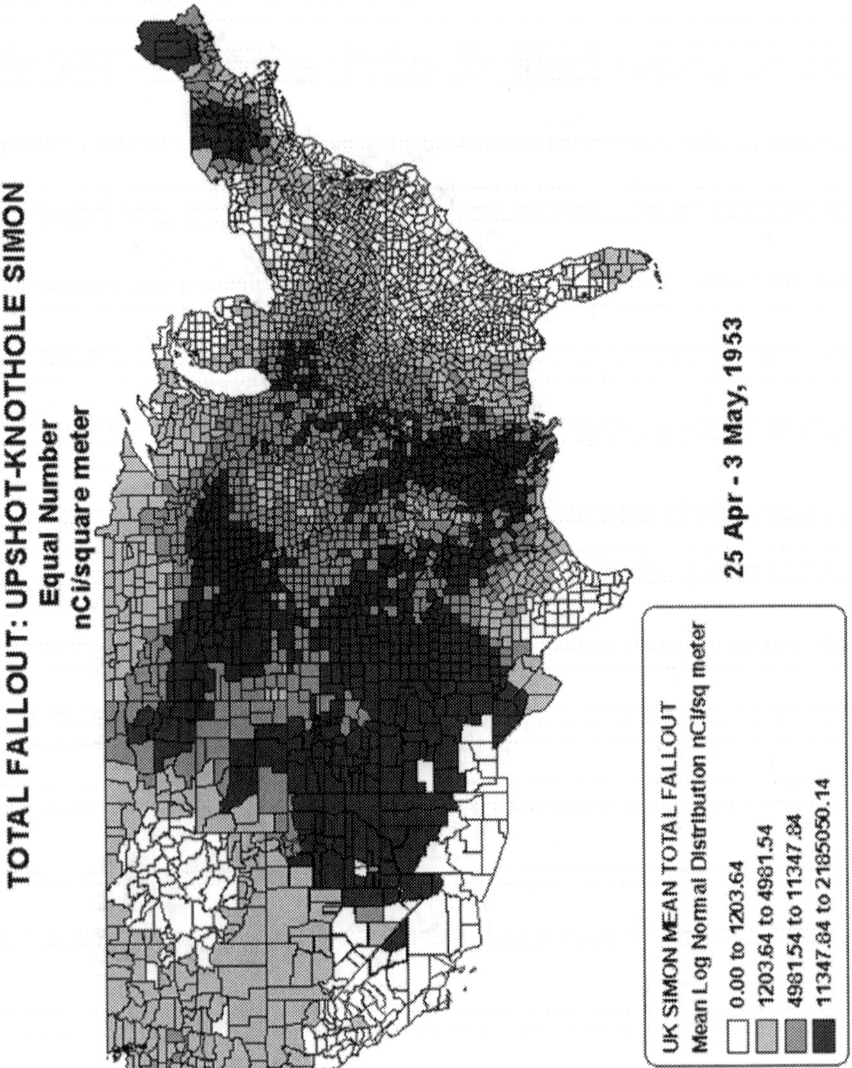
TOTAL FALLOUT: UPSHOT-KNOTHOLE SIMON
Equal Number
nCi/square meter
25 Apr - 3 May, 1953
UK SIMON MEAN TOTAL FALLOUT
Mean Log Normal Distribution nCi/sq meter
0.00 to 1203.64
1203.64 to 4981.54
4981.54 to 11347.84
11347.84 to 2185050.14

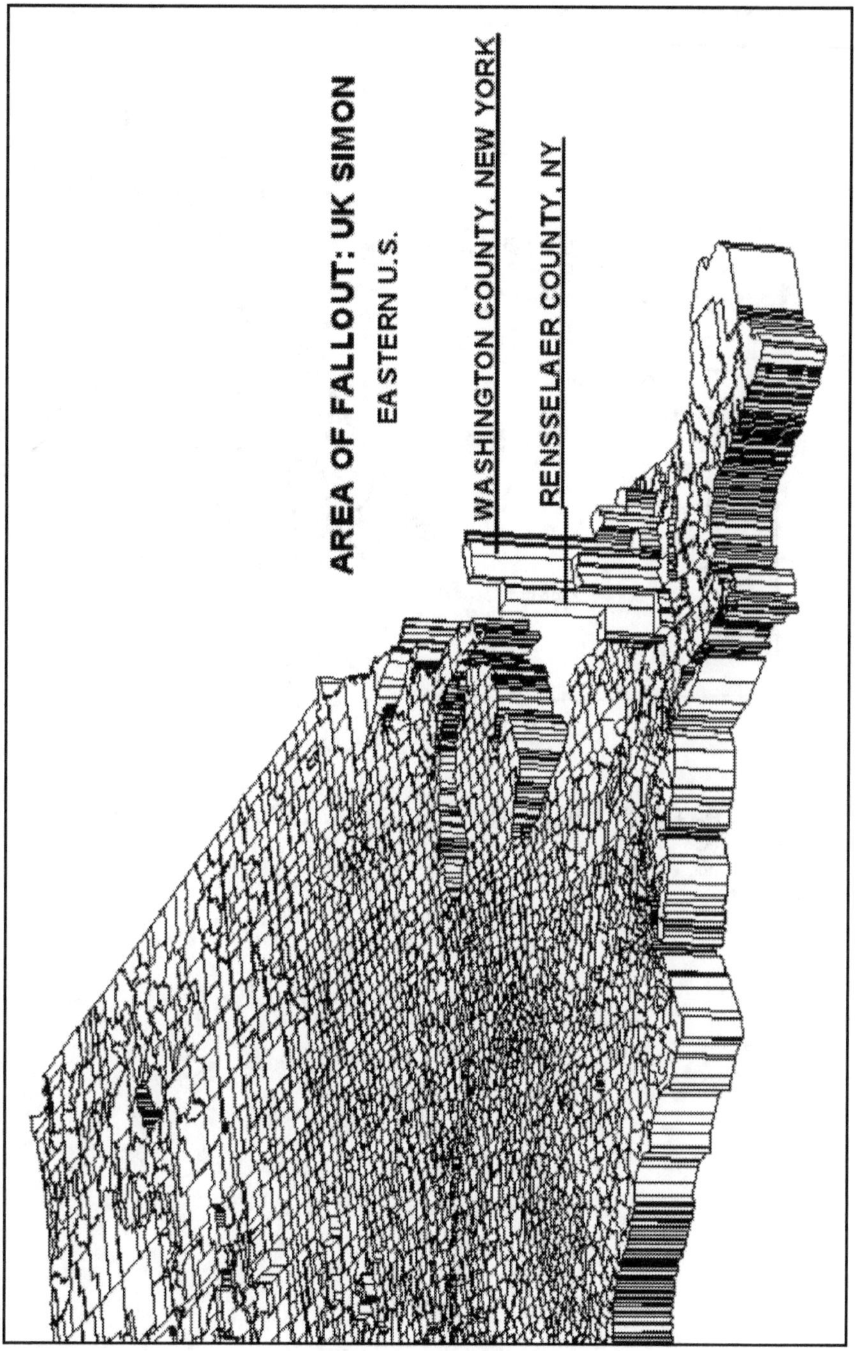
AREA OF FALLOUT: UK SIMON
EASTERN U.S.
WASHINGTON COUNTY, NEW YORK
RENSSELAER COUNTY, NY

FALLOUT IN EASTERN U.S. FROM UPSHOT-KNOTHOLE SIMON

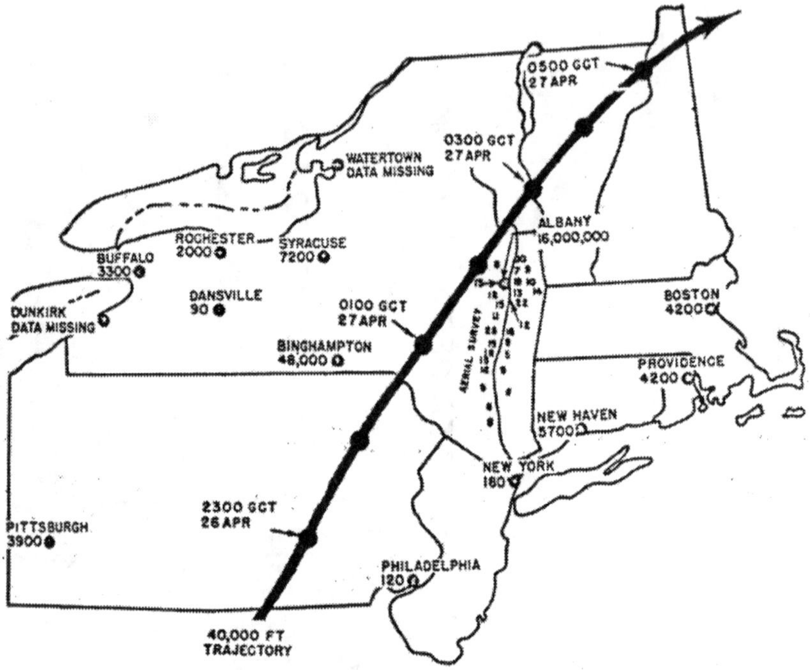

PATH OF THE SIMON CLOUD OVER NEW YORK STATE

APRIL 25, 1953

SHOT 8: UPSHOT-KNOTHOLE ENCORE (UK8)

UPSHOT-KNOTHOLE: ENCORE

Detonation Date: 8 May 1953 • *Detonation Time:* 7:30 A.M. • *Area:* Frenchman's Flat • *Sponsor:* Los Alamos, Department of Defense • *Yield:* 27 kt • *Radiation Level at Ground Zero:* 25 R/hr • *Height of Burst:* 2,423 ft • *Cloud Top Height:* 42,000 ft

10,000-Foot Trajectory: NEVADA: Moapa, Logandale, Overton • ARIZONA: (Mount Dellenbaugh), Havasupai Indian Reservation, North Rim, Cedar Ridge, Kaibito • UTAH: (Glen Canyon), (Cataract Canyon), (Canyonland National Park), (La Sal Mountains), Castleton, Mount Tomasaki • COLORADO: Gateway, Whitewater, Skyway, Grand Mesa, (Grand Mesa National Forest), Snowmass, Ruedi, Meredith, Norrie, (Tennessee Pass), Climax, Leadville, (Fremont Pass), Jefferson, Como, Deckers, Greenland, Palmer Lake, Peyton, Rush, Eads • KANSAS: Tennis, Wright, Windthorst, Brenham, Haviland, Wellsford, Coats, Sawyer, Isabel, Duquoin, Harper, Danville, Freeport, Argonia, Corbin, South Haven, Ashton, Arkansas City, Silverdale, Maple City, Cedar Vale, Wauneta, Elk City, Sycamore, Morehead, Galesburg, Erie, St. Paul, Walnut, Hepler, Farlington, Hiattville, Garland • MISSOURI: Deerfield, Richards, Horton, Prairie City, Appleton City, Ohio, Deepwater, Brownington, Tightwad, Coal, Leesville, Branden, Mora, Bahner, Otterville, Beaman, Pilot Grove, Bunceton, Bellair, Lone Elm, Lamine, Boonville, Wooldridge, Rocheport, New Franklin, Midway, Hinton, Harrisburg, Rucker, Sturgeon, Clark, Centralia, Paris, Stoutsville, Monroe City, Hassard, Hunnewell, Withers Mill, Palmyra, West Quincy • ILLINOIS: Quincy, Ursa, Mendon, Loraine, Bigneck, Stillwell, Bowen, Denver, Augusta, Plymouth, Colmar, Fondon, Macomb, Bardolph, New Philadelphia, Bushness, Marietta, Blyton, Ellisville, Fairview, London Mills, Rapatee, Farmington, Yates City, Elmwood, Oak Hill, Brimfield, Dunlap, Princeville, Edelstein, Speer, Camp Grove, Henry, Putnam, McNabb, Granville, Spring Valley, Oglesby, Cedar Point, Peru, LaSalle, Ottawa, Dayton, Wedrun, Norway, Serena, Sheridan, Newark, Sandwich, Plano, Yorkville, Bristol, Oswego, Montgomery, Autota, Batavia, Wheaton, Elmhurst, Bloomingdale, Roselle, Des Plaines, Highland Park, Glencoe, Winnettca, Wilmette • MICHIGAN: Graafschap, Beechwood, Holland, New Holland, Hudsonville, Jenison, Grandville, Wyoming, Walker, Grand Rapids, Cannonsburg, Bostwick Lake, Belding, Fenwick, Butternut, Vickeryville, Crystal, Sumner, Ithoca, Breckenridge, Wheeler, Merrill, Iva, Hemlock, Laporte, Freeland, Midland, Bay City, Essexville, Oakhurst, Bay Park, Sebewaing, Bay Port, Pigeon, Caseville, Pinnebog, Kinde, Port Austin, Grind Stone City • MAINE: Pittston Farm, Seboomook, North East Carry, (Caribou Lake), (Harrington Lake), (North Brother Mountain), (Mount Katahdin, highest point in Maine), Stacyville, Sherman Station, Sherman Mills, (Otter Lake Mountain), Orient.

CHARACTERISTICS: UK ENCORE

The Upshot-Knothole ENCORE device weighed 8330 lbs and was airdropped from a height of 19,000 feet. Among the nuclear tests, Upshot-Knothole ENCORE is ranked 7th for percentage of Co-57 in total fallout. The deposition pattern for Upshot-Knothole ENCORE is most similar to Upshot-Knothole SIMON (UK7) (R: 0.51; T: 32.81). Nuclear debris from this test was deposited on 2227 counties, or 72 percent of the total number of U.S. counties. County of highest deposition: Custer, MT (548 μCi/sq meter estimated maximum activity on May 8, 1953.) Other counties with high deposition amounts from this shot (all in μCi/sq meter, estimated maximum activity) included: Dawson MT—328; Fallon MT—328; McCone MT—328; Prairie MT—328; Richland MT—548; Roosevelt MT—328; Sheridan MT—328; Wibaux MT—328 and Billings ND—328

PRIMARY FALLOUT ZONE: UK ENCORE

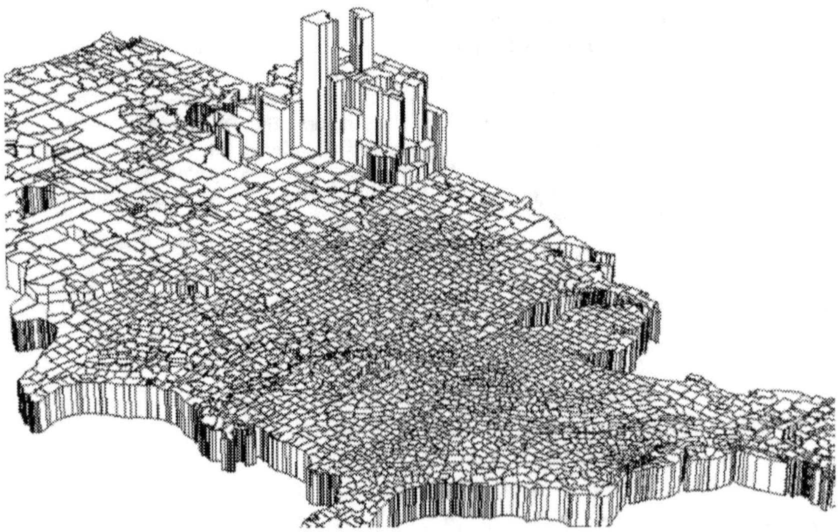

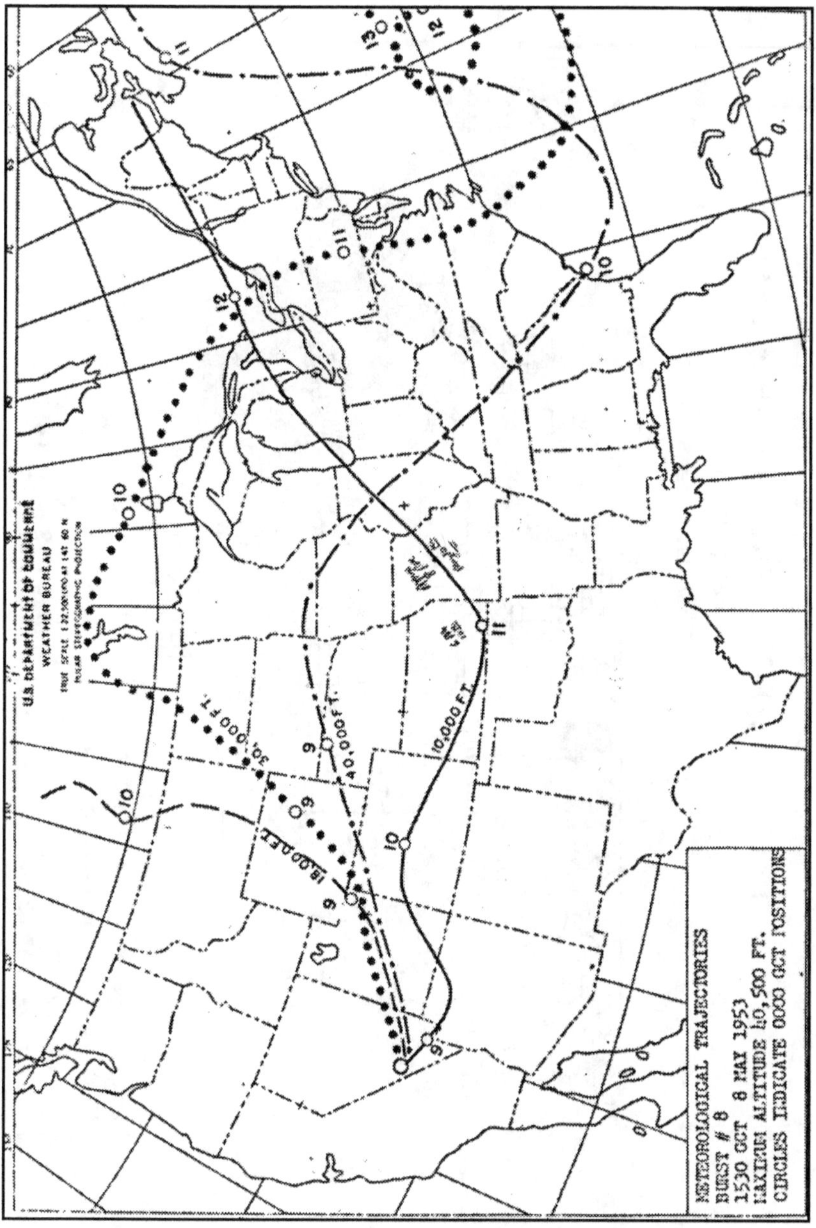

TRAJECTORY: UPSHOT-KNOTHOLE ENCORE (UK8)

TOTAL FALLOUT: UPSHOT-KNOTHOLE ENCORE

Equal Intervals

nCi/square meter

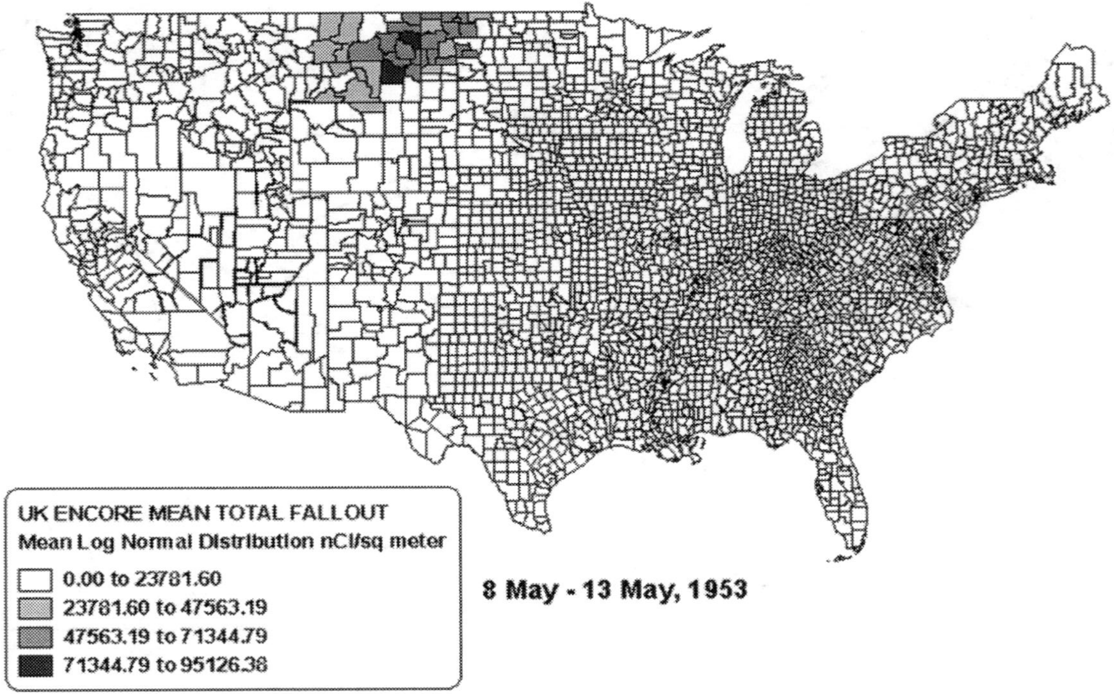

8 May - 13 May, 1953

SHOT 9: UPSHOT-KNOTHOLE HARRY (UK9)

UPSHOT-KNOTHOLE: HARRY

Detonation Date: 19 May 1953 • *Detonation Time:* 4:05 A.M. • *Area:* 3a • *Sponsor:* Los Alamos • *Yield:* 32 kt • *Radiation Level at Ground Zero:* 10 R/hr • *Height of Burst:* 300 ft • *Cloud Top Height:* 42,500 ft

10,000-Foot Trajectory: NEVADA: Preston, Ely, Tippett • UTAH: Callao, Grantsville, Salt Lake City, Layton, Ogden, Woodruff • WYOMING: Frontier, Diamondville, Kemmerer, Fontenelle, Farson, Yoder • NEBRASKA: Scottsbluff, Gering, McGrew, Bridgeport, Broadwater, Lisco, Oshkosh, Lewellen, Ogallala, Roscoe, Farnam, Kearney, Lowell, Prosser, Hansen, Trumbull, Harvard, Lushton, Cordova, Pleasantdale, Denton, Lincoln, Avoca, Union • IOWA: Shenandoah, Bingam, Yorktown, Clarinda, Sharpsburg, Gravity, Conway, Kellerton, Decatur City, Leon, High Point, Corydon, Promise City, Plano, Centerville, West Grove, Bloomfield, Hillsboro, Salem, Danville • ILLINOIS: Monmouth, Galesburg, Wataga, Victoria, Fayette, Toulon, Wyoming, Putnam, Oglesby, LaSalle, Ottowa, Wedron, Lisle, Chicago • MICHIGAN: Allendale, Wyoming, Grand Rapids, Alpine, Sparta, Sherman City, Farwell, St. Helen, Luzerne, Metz, Hawks, Rogers City.

18,000-Foot Trajectory: NEVADA: Elgin, Carp • UTAH: Santa Clara, St. George, Hurricane, Virgin, Rockville, Springdale, Mt. Carmel, Blanding • COLORADO: Pleasant View, Stoner, Spar City, Masonic Park, Center, Cedarwood, Rocky Ford, Las Animas, McClave, Big Bend, Wiley • KANSAS: Lydia, Rush Center, Timkin, Shaffer, Olmitz, Hoisington, Redwing, Odin, Lorraine, Falun, Mentor, Kipp, Holland, Pearl, Abiline, Junction City, St. Marys, Mayetta, Half Mound, Atchison • MISSOURI: St. Joseph, Rochester, Fairport, McFall, Blue Ridge, Goshen, Princeton, Mercer • IOWA: Seymour, Centerville, Ottumwa, Bladensburg, Abington, Brighton, Cranston, Muscatine, Fairport, Montpelier, Davenport, Bettendorf • ILLINOIS: Rock Island, Moline, Hillsdale, Lyndon, Rock Falls, Sterling, Dixon, Rochelle, Hillcrest, Clare, Genoa, Hampshire, Huntley, Algonquin, Libertyville, Waukeegan, North Chicago • MICHIGAN: Douglas, Saugatuck, Hamilton, Bentheim, Green Lake, Caldeonia, Clarksville, Lake Odessa, Portland, Fowler, St. Johns, Eureka, Bannister, Brant, St. Charles, Saginaw.

RELATIVE TOTAL FALLOUT UK HARRY

MAY 19, 1953 32 KT

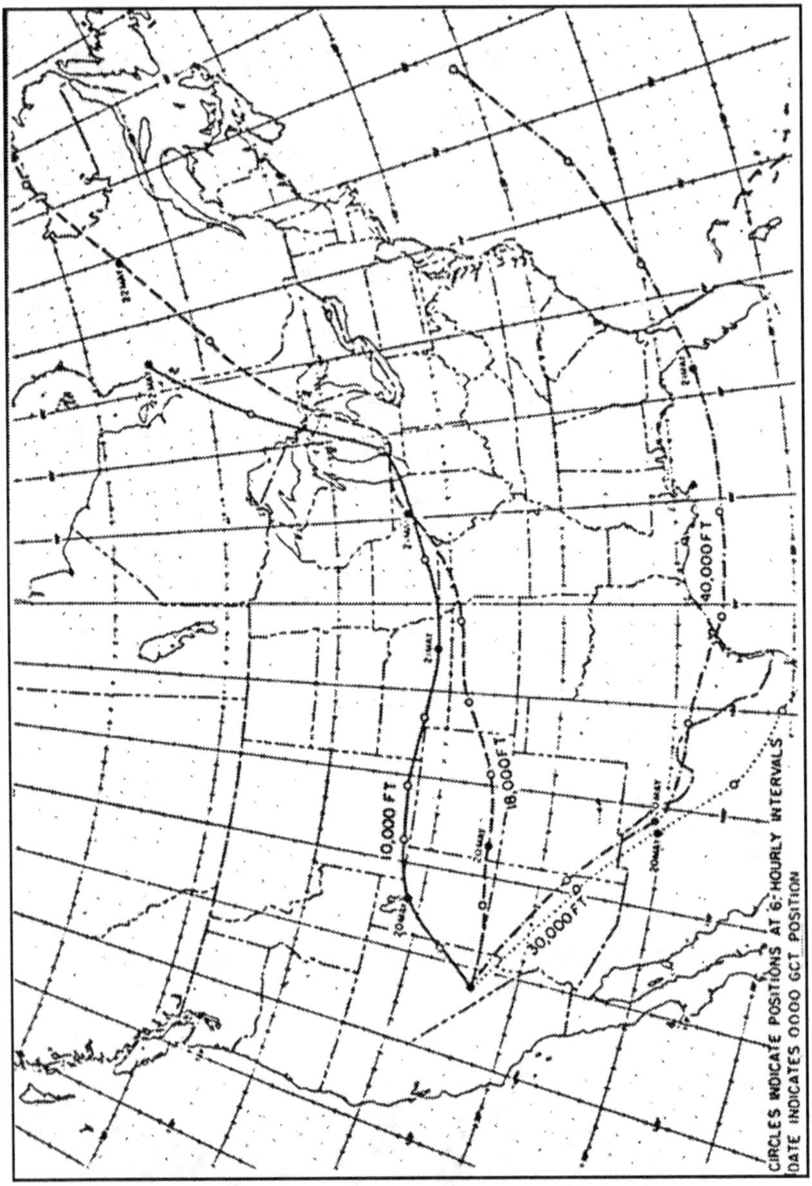

TRAJECTORY: UPSHOT-KNOTHOLE HARRY (UK9)

CHARACTERISTICS: UK HARRY

The device for shot HARRY was code-named "Hamlet." It was 56 inches in diameter, 66 inches in length, and weighed 4 tons. Among the nuclear tests, Upshot-Knothole HARRY is ranked 3rd for deposition of Cs-137; Nb-95; Sr-90; and Zr-95; 4th for deposition of Nb-95m and Pr-144; 5th for percentage of U-240 and Ru-106; 6th for percentage of I-130 and Te-127m, 8th for deposition of Co-60; 10th for deposition of Eu-155; 13th for percentage of Sr-89, Sr-90 and Y-90 and 16th for deposition of Be-7.

The deposition pattern for Upshot-Knothole HARRY is most similar to Upshot-Knothole CLIMAX (UK11) (R: 0.50; T: 32.26).

Fallout from this shot was deposited on 3046 counties, or 98 percent of the total counties in the U.S, ranking it the 6th for fallout dispersion.

The county with the highest deposition amount was also the county with one of the highest levels of fallout in the entire nuclear test series: Gunnison, CO. While this county received high amounts of fallout from a number of different nuclear tests, UK HARRY produced an exceptionally heavy fallout for this area: On May 19, 1953 Gunnison County, Colorado received an estimated maximum activity of 9609 μCi/square meter. Other counties in the southwest received slightly less (the following numbers in microCuries estimated maximum activity): Coconino, AZ—8037; Conejos CO—8036; Mineral CO—4224; Kane UT—6857; Washington(2) UT—7171.

RELATIVE TOTAL FALLOUT: UK HARRY

MAY 19, 1953 32 KT

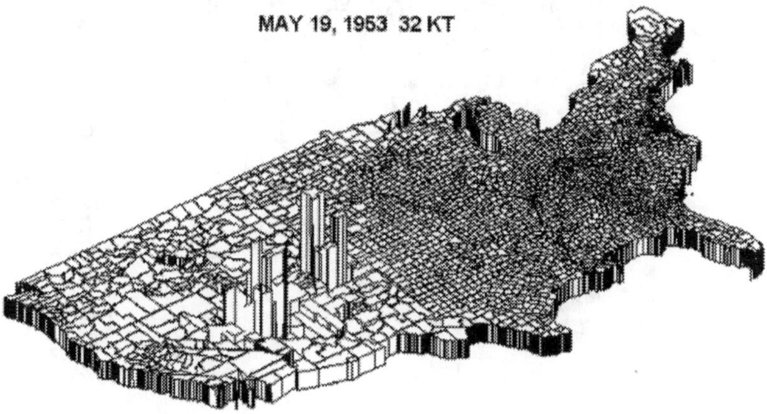

TOTAL FALLOUT: UPSHOT-KNOTHOLE HARRY

Equal Intervals

nCi/square meter

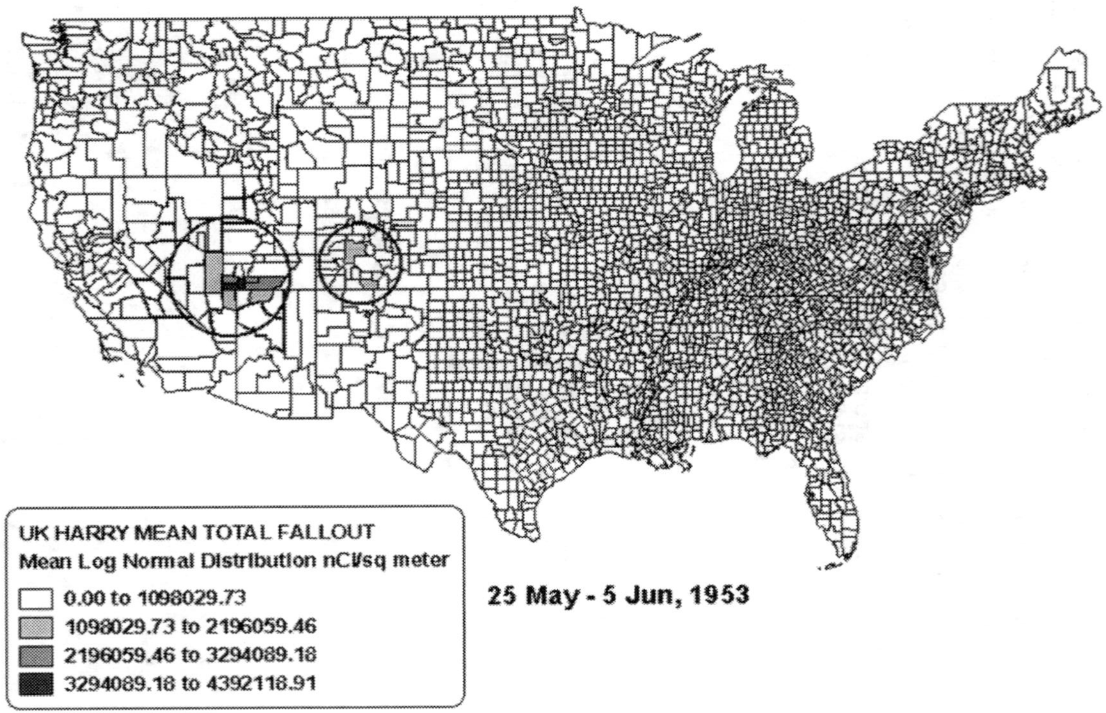

25 May - 5 Jun, 1953

TOTAL FALLOUT: UPSHOT-KNOTHOLE HARRY

Equal Number

nCi/square meter

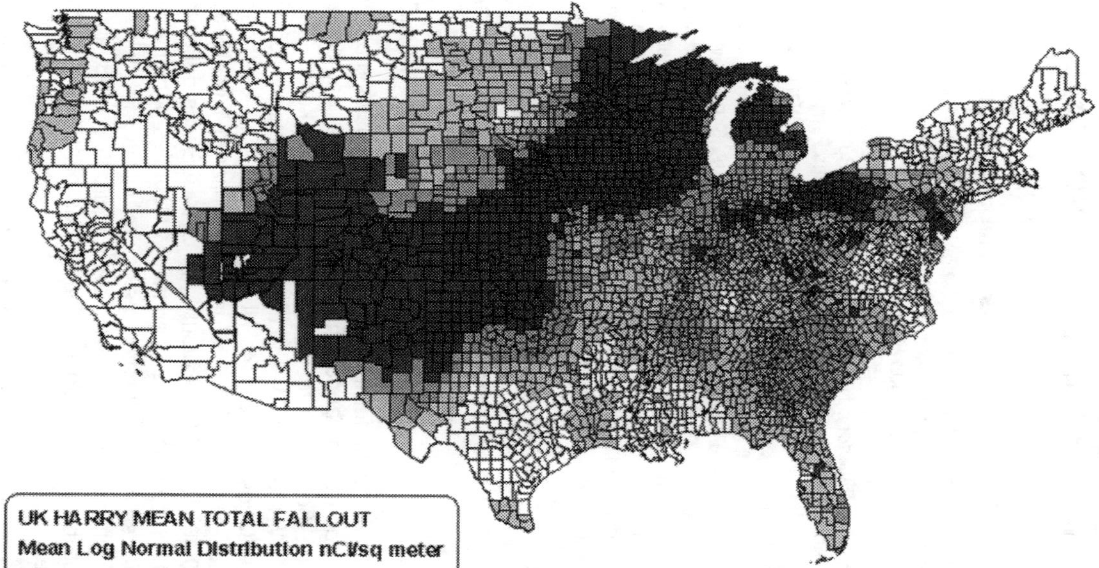

UK HARRY MEAN TOTAL FALLOUT
Mean Log Normal Distribution nCi/sq meter
0.00 to 2758.42
2758.42 to 5177.25
5177.25 to 11047.90
11047.90 to 4392118.91

25 May - 5 Jun, 1953

SHOT 10: UPSHOT-KNOTHOLE GRABLE (UK10)

UPSHOT-KNOTHOLE: GRABLE

Detonation Date: 25 May 1953 • *Detonation Time:* 7:30 A.M. • *Area:* Frenchman's Flat • *Sponsor:* Los Alamos • *Yield:* 15 kt • *Radiation Level at Ground Zero:* 10 R/hr • *Height of Burst:* 524 ft • *Cloud Top Height:* 35,000 ft

10,000-Foot Trajectory: NEVADA: (Grant Range), Ruth, Steptoe, Currie, (Snow Water Lake), Thousand Springs • IDAHO: Three Creek, Glenns Ferry, Mountain Home, Mayfield, Boise, Pearl, Montour, Payette, Weiser • OREGON: Pleasant Valley, Baker, Haines, North Powder, (Blue Mountains), Pilot Rock, Echo, Irrigon • WASHINGTON: Paterson, (Horse Heaven Hills), Yakima Indian Reservation, (Lincoln Plateau), (Mount Adams Wilderness Area) (Gifford Peak),(Gifford Pinchot National Forest), North Bonneville, Skamania • OREGON: Gresham, Boring, Eagle Creek, Estacada, (Goat Mountain), (Round Mountain), (Monument Peak), (Tombstone Pass), (Three Sisters Wilderness Area), Bend, (Bear Creek Butte), Post (Spanish Peak), Hamilton, (Meadow Creek Pass), (Blue Mountains), La Grande, Island City, Alicel, Imbler, Minam, Walowa • IDAHO: Waha, Lewiston, Lapwai, Spalding, Julietta, Southwick, Helmer, Boyill, Avery, Saltese, Mullan • MONTANA: Belknap, (Mount Headley), (Blue Mountain), Tray • IDAHO: Moyie Springs, Meadow Creek, (Moyie Falls), Nordman • WASHINGTON: (Molybdenite Mountain), Lost Creek, Arden, Addy, Cedonia, Hunters, Marlin, Moses Lake, Sunnyside, Grandview, Mabton, (Horse Heaven Hills) • OREGON: Arlington, Blalock, Rock Creek, Mikkalo, Kent, Shaniko, Antelope, Ashwood, (Grizzly Mountain).

CHARACTERISTICS: UK GRABLE

The device used for this shot was 54.4 inches long and weighed 803 lbs. It was fired from a cannon that weighed 85 tons and had a range of 20 miles. Among the nuclear tests, UK GRABLE is ranked 2nd for percentages of W-181, W-185 and W-188m 4th for percentages of Mn-54; and 6th for percentages of Ce-144, Nb-95 and Zr-95. The deposition pattern for Upshot-Knothole GRABLE is most similar to Tumbler-Snapper HOW (TS8) (R: 0.46; T: 28.61). Fallout from GRABLE was deposited on 98 percent of the U.S. counties (3025 counties). Counties of highest deposition included (in μCi/sq meter estimated maximum activity) Ferry WA—207; Bonner ID—109; Okanogan, WA—141; Pend Oreille WA—158; and Stevens WA—136. All of the aforementioned depositions of fallout occurred on a single day, June 5, 1953.

The ratio of total fallout to I-131 for this shot is relatively high at 124 to 1.

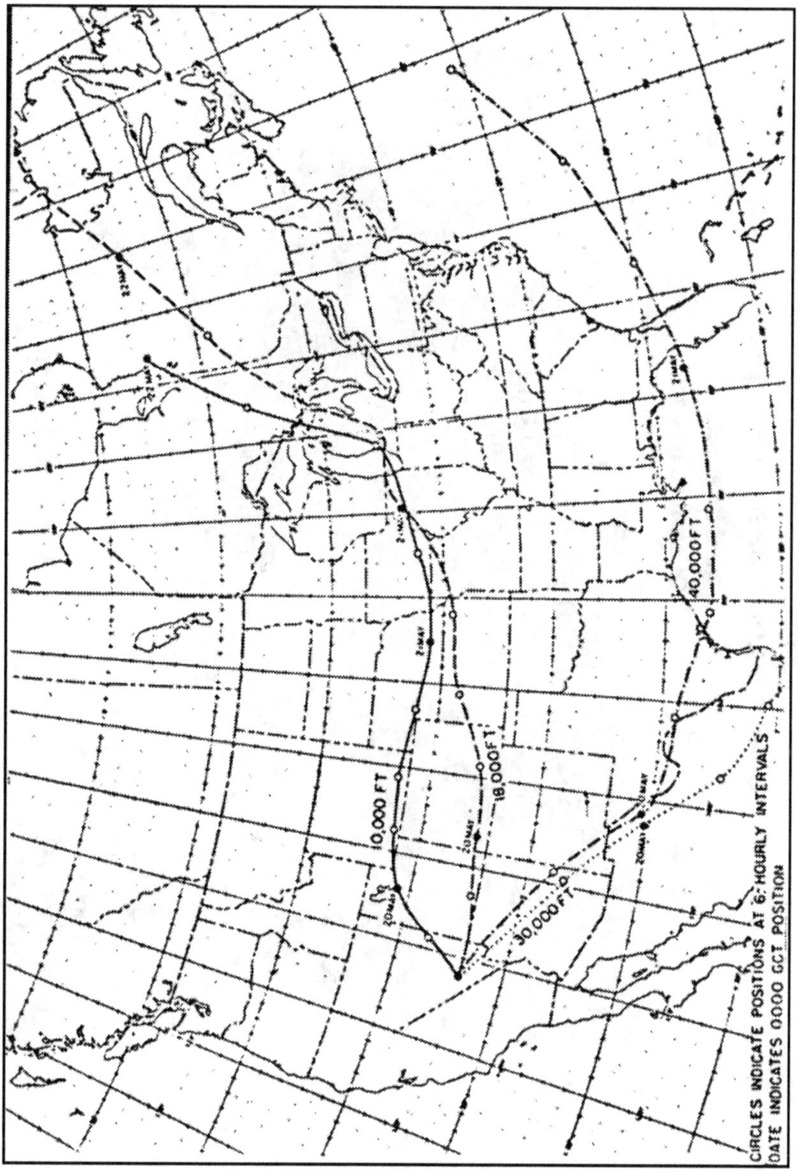

TRAJECTORY: UPSHOT-KNOTHOLE GRABLE (UK10)

TOTAL FALLOUT: UPSHOT-KNOTHOLE GRABLE

Equal Intervals

nCi/square meter

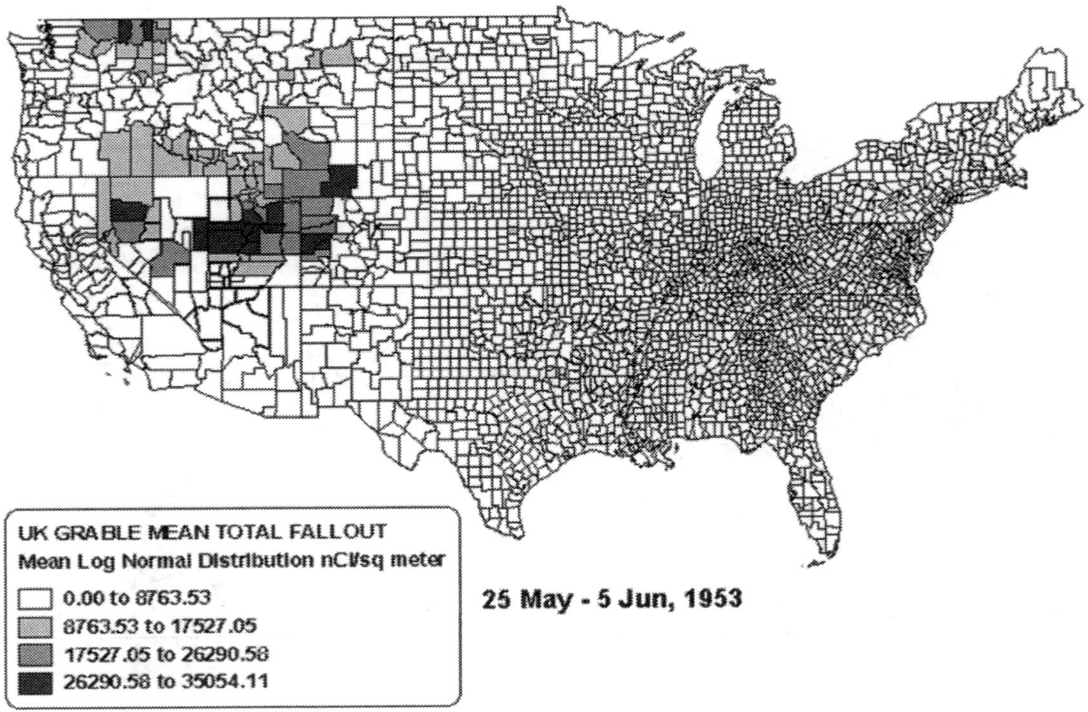

25 May - 5 Jun, 1953

SHOT 11: UPSHOT-KNOTHOLE CLIMAX (UK11)

UPSHOT-KNOTHOLE: CLIMAX

Detonation Date: 4 Jun 1953 • *Detonation Time:* 3:15A.M. • *Area:* 7-3 • *Sponsor:* Los Alamos • *Yield:* 61 kt • *Radiation Level at Ground Zero:* 10 R/hr • *Height of Burst:* 1,334 ft • *Cloud Top Height:* 42,700 ft

40,000-Foot Trajectory: NEVADA: (Pintwater Range), Carp • UTAH: (Snow Canyon), St. George, Washington, Hurricane, Rockville, Mount Carmel Junction, (Cottonwood Canyon), (White Canyon), (Cataract Canyon), (Angel Arch), La Sal Junction, Caselton • COLORADO: Gateway, Glade Park, Grand Junction, Orchard Mesa, Fruitvale, Clifton, De Beque, Parachute, Rulison, Rio Blanco, Buford, (Williams Fork Mountains), Haydon, Clark, Hahns Peak • WYOMING: (Blackhall Mountain), Riverside, Ryan Park, (Medicine Bow Peak), McFaddin, Rock River, Esterbrook, Douglas, Orin, Newcastle, Four Corners • SOUTH DAKOTA: Cheyenne Crossing, Savoy, Central City, Whitewood, Spearfish, St. Onge, Fruitdale, Nisland, Newell, Zeona, Bison, Meadow, Morristown, Watanga • NORTH DAKOTA: Shields, Breien, Fort Rice, Huff, Moffitt, Sterling, Steele, Dawson, Crystal Springs, Predina, Goldwin, Buchanan, Pingree, Spiritwood Lake, Courtenay, Wimbledon, Hannaford, Dazey, Sibley, Luverne, Blabon, Hope, Clifford, Galesburg, Blanchard, Hillsboro, Kelso • MINNESOTA: Climax, Shelly, Halstad, Lackhart, Gary, Rindal, Bejou, Ebro, Bagley, Shelvin, Solway, Bemidji, Turtle River, Tenstrike, Hines, Blackduck, Bergville, Effie, Craigville, Bois Fort, Nelt Lake, Ash Lake, Cusson, Crane Lake.

CHARACTERISTICS: UK CLIMAX

The device for the CLIMAX test was 30.6 inches wide, 15 ft long, and weighed 1840 lbs. Among the nuclear tests, Upshot-Knothole CLIMAX is ranked 1st in percentages of Ce-141 and Ce-144 in fallout and 2nd for percentages of Pr-144 in fallout. The deposition pattern for UK CLIMAX is most similar to Teapot TESLA (TP3) (R: 0.53; T: 35.09). Fallout from Upshot-Knothole CLIMAX was deposited on 2726 counties, or 88 percent of the total number of counties in the United States.

Counties of highest deposition included Coconino, AZ at 289 μCi/sq meter (estimated maximum activity) and Clark NV at 143 μCi/sq meter (estimated maximum activity)—both on May 25, 1953. The day one total fallout to I-131 ratio for CLIMAX is in the upper moderate range at 121 to 1.

TRAJECTORY: UPSHOT-KNOTHOLE CLIMAX (UK11)

TOTAL FALLOUT: UPSHOT-KNOTHOLE CLIMAX

Equal Intervals

nCi/square meter

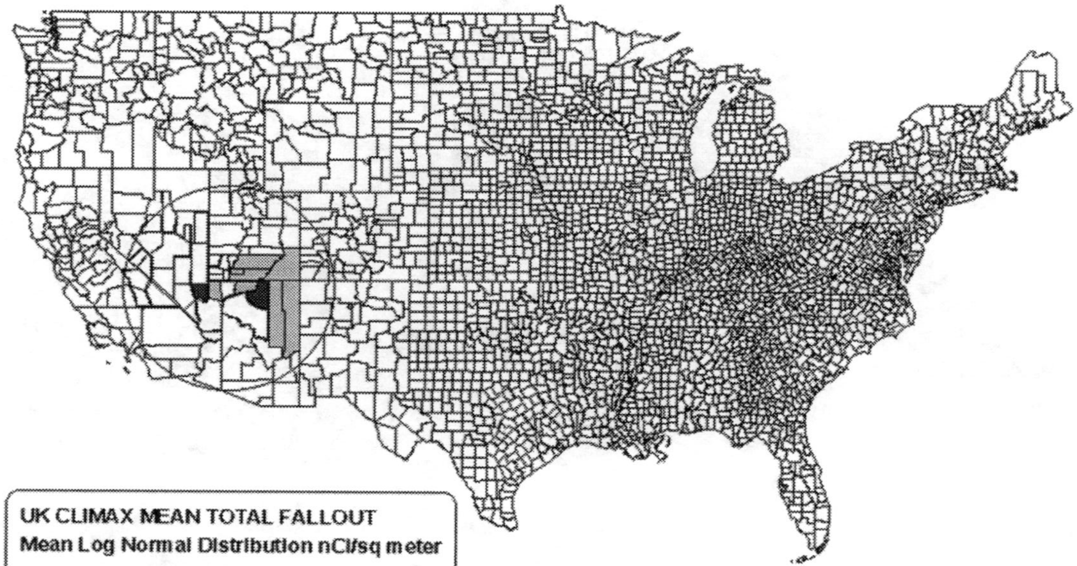

UK CLIMAX MEAN TOTAL FALLOUT
Mean Log Normal Distribution nCi/sq meter

- 0.00 to 28809.87
- 28809.87 to 57619.73
- 57619.73 to 86429.60
- 86429.60 to 115239.46

4 Jun - 11 Jun, 1953

TOTAL FALLOUT: UPSHOT-KNOTHOLE CLIMAX

Equal Number
nCi/square meter

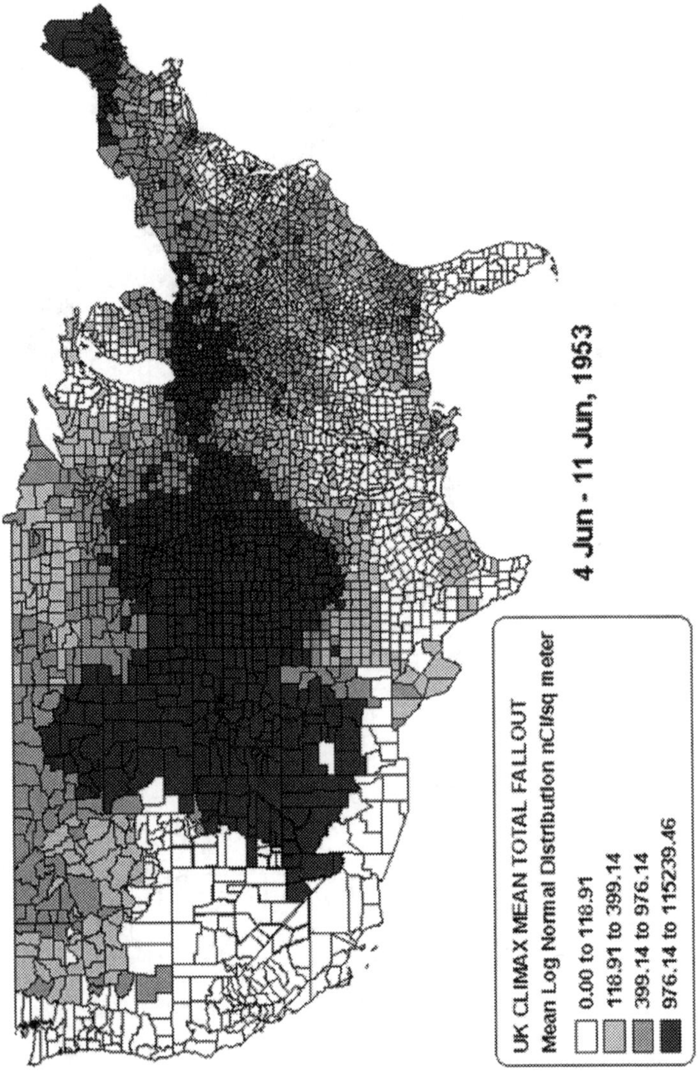

SECTION 5

TEAPOT SERIES

Feb 18-May 15, 1955

SHOT 1: TEAPOT WASP (TP1)

TEAPOT: WASP

Detonation Date: 18 Feb 1955 • *Detonation Time:* 12:00 noon • *Area:* T-7-4 • *Sponsor:* Los Alamos • *Yield:* 1 kt • *Radiation Level at Ground Zero:* 10 R/hr • *Height of Burst:* 762 ft • *Cloud Top Height:* 21,500 ft

18,000-Foot Trajectory: NEVADA: (Sheep Range), Moapa Indian Reservation, (Valley of Fire State Park), Overton • ARIZONA: (Grand Wash Cliffs), (Poverty Mountain), Havasupai Indian Reservation, (Grant Canyon), Tusayan, Cameron, Gray Mountain, Steamboat Canyon, Sunrise Springs, Cornfield, Ganado, Cross Canyon, Fort Defiance, Sawmill • NEW MEXICO: Mexican Springs, Tohatchi, Brimhall, Standing Rock, Pueblo Pintado, La Jara, Regina, Gallina, Coyote, Vallecitos, Las Tables, Tres Piedras, Cerro, Amalia • COLORADO: La Valley, Stonewall, Gulpare, Aguilar, Timpas, La Junta, Swink, North La Junta, Cheraw, Eads, Cheyenne Wells, Arapahoe • KANSAS: (Mount Sunflower, highest point in Kansas: 4,039 feet), Edson, Brewster, Achilles, Traer, Cedar Bluffs • NEBRASKA: Danbury, Cambridge, Holbrook, Elwood, Lexington, Berwyn, Almeria, Rose, Bassett, Burton.

At this point, the 16,000-foot trajectory split into two sections; one traveled east; the other west.

East Trajectory: SOUTH DAKOTA: Burke, Gregory, Oacoma, Reliance, Lower Brule, Harrold, Gettysburg, Lowry, Java, Artas • NORTH DAKOTA: Hague, Napoleon, Dawson, Lake Williams, Sykeston, Cathay, Sheyenne, Devils Lake, Webster, Hampden, Loma, Langden (into Manitoba).

West Trajectory: SOUTH DAKOTA: Wewela, Winner, Witten, Draper, Hajes, Cherry Creek, Howes, Marcus, Redowl, Stoneville, Fairpoint, (Mud Buttes) • WYOMING: Beulah, Aladdin, Sundance, Moorcroft, (Devils Tower), Rozet, Sussex, Powder River, Natrona, (Rattlesnake Hills), (Ferris Mountains), Rawlins, (Medicine Bow National Forest) • COLORADO: Columbine, Clark, Milner, Toponas, McCoy, Bond, Aron, Golman, Redcliff, Leadville, Texas Creek, Fairview, San Isabel, Green Towers, Walsonburg, Pryor, Aguilar, Ludlow, El Moro, Trinidad, (Raton Pass) • NEW MEXICO: Folsom, Des Moines, Capulin, Sofia, Mount Dora, Clapham, Stoad, Sedan • TEXAS: Perico, Dalhart, Cactus, Gruver, Spearman, Waka, Farnsworth, Perryton, Twitchell, Booker, Darrouzett • OKLAHOMA: Catesby, May, Camp Houston • KANSAS: Hardtner, Sharon, Attica, Harper, Duquoin, Milton Violla, Clonmel, Clearwater, Shulte, Wichita, Andover, Towanda, El Dorado, Thrall, Madison, Hartford, Strawn, Sharpe, Halls Summit, Waverly, Agricola, Williamsburg, Homewood, Ottawa, LeLoup, Edgerton, Wellsville, Gardner, Lenexa, Overland Park, Shawnee, Edwardsville, Kansas City • MISSOURI: Kansas City, Parkville, Gladstone, Kearney, Holt, Elmira, Mirabile, Kingston, Hamilton, Breckenridge, Lock Springs, Jamesport, Trenton, Dunlap, Galt, Osgood, Harris, Newton, Lucerne, Powersville • IOWA: Numa, Centerville, Brazil, Mystic, Rathbun, Moravia, Avery, Eddyville, Cedar, Wright, Keomah, Rose Hill, What Cheer, Thornburg, Millersburg, Williamsburg, West Ammana, Watkins, Norway, Atkins, Palo, Center Point, Waller, Quasqueton, Winthrop, Manchester, Lamart, Dundee, Strawberry Point, Osborle, Elkader, St. Olaf, Farmersburg, Freelich, Monona, Watson, Vallen, Waterville, Lansing. • WISCONSIN: De Soto, Red Mound, Victory, Genoa, Chaseburg, Coon Valley, La Crosse, Middle Ridge, Bangor, Rockland, West Salem, Cataract, Melrose, Shamrock, Black River Falls, Alma Center, Merrillan, Humbird, Willard, Thorp, Lublin, Kennan, Catawba, (Butternut Lake), Mallan, High Bridge, Marengo, Bad River Indian Reservation, Ashland, Washburn, La Pointe, Bayfield, Red Cliff, (York Island), (Bear Island) • MINNESOTA: Little Marais, (Nine Mile Lake), (Snowbank Lake), Orleans, Noyes, Humbolt, Northcote, Bowesmont • NORTH DAKOTA: Joliette, Bowesmont, Auburn, Oakwood, Minto, Warsaw, Ardoch, Johnstown, Manvel, Makinock, Holmes, Reynolds, Buxton, Mayville, Cummings, Kelso, Grandin, Gardiner • MINNESOTA: Georgetown, Moorhead, Dilworth, Lyndon, Rollag, Cormorant, Vergas, Perham, Sebeka, Nimrod, Poplar, Pequot Lakes, Breezy Point, Hassman, McGregor, Tamerack, Wright, Cromwell, Cloquet, Twig, Hermantown, Arnold, Lax Lake, Finland, Little Marais, Taconite Harbor, Schroeder, Tofte, Lutsen, Grand Marais, Grand Portage Indian Reservation.

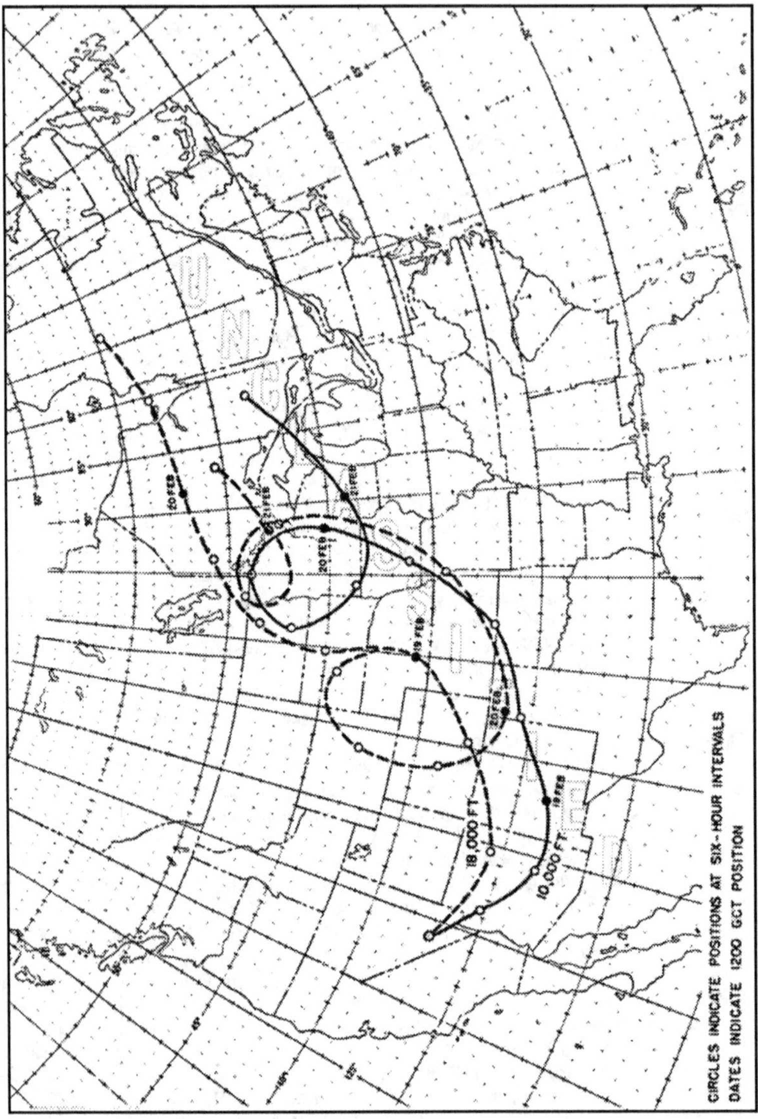

TRAJECTORY: TEAPOT WASP (TP1)

CHARACTERISTICS: TEAPOT WASP

The WASP device was relatively small, consisting of a 120 lb, 22-inch diameter sphere packed inside a bomb case. The total weight of the unit was approximately 1500 lbs. Among the nuclear tests, Teapot WASP is ranked 2nd for percentages of Pm-147 and 3rd for Tc-99m and Eu-155. The deposition pattern for Teapot WASP is most similar to Tumbler-Snapper BAKER (TS2) (R: 0.10; T: 5.86). County of highest deposition: San Bernadino, CA.

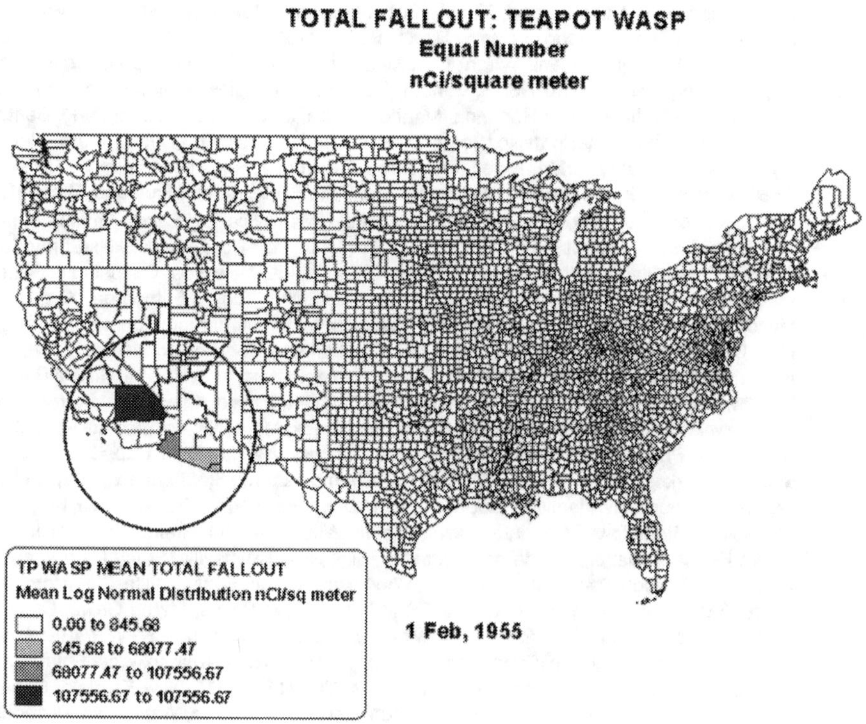

SHOT 2: TEAPOT MOTH (TP2)

TEAPOT: MOTH

Detonation Date: 22 Feb 1955 • *Detonation Time:* 5:45 A.M. • *Area:* 3 • *Sponsor:* Los Alamos • *Yield:* 2 kt • *Radiation Level at Ground Zero:* 500 R/hr • *Height of Burst:* 300 ft • *Cloud Top Height:* 24,200 ft

24,000-Foot Trajectory: NEVADA:(Spotted Range), Indian Springs, Las Vegas, Winchester, N. Las Vegas, Henderson, Boulder City • ARIZONA: Willow Beach, Dolan Springs, Chloride, Prescott, Mayer, (Mazatlan Mountains), Tonto Basin, (Rockinstraw Mountain), San Carlos Indian Reservation, Morenci, Clifton, (Macerick Hill) • NEW MEXICO: Mule Creek, Buckhorn, Kingston, Hillsboro, Caballo, (Caballo Mountains), Alamogordo, Sunspot, Sacramento, Weed, Dunken, Hope, Atoka, Loco Hills, Humble City, Hobbs • TEXAS: Seminole, Lamesa, Arvana, Mesquite (in Borden County), Gail, Fluvanna, Dermott, Rotan, Hamlin, Tuxedo, Stamford, Avoca, Woodson, South Bend, Eliasville, Graham, Bryson, Jacksboro, Wizard Wells, Chico, Alvord, Greenwood, Slidell, Pilot Point, Tioga, Dorchester, Sherman, Ambrose, Ivanhoe, Telephone, Elwood, Riverby, Direct, Belk, Chicota, Arthur City, Albion • OKLAHOMA: Ord, Frogville, Valiant, Wright City, Millerton, Garvin, Golden, Glover, Broken Bow, Eagleton, DeQueen • ARKANSAS: Gilham, New Hope, Langley, Salem, Glenwood, Bonnerdale, Hempwallace, Lake Hamilton, Magnet Cove, (Hot Springs National Park), Benton, Little Rock, Shannon, Wrightville, Scott, Lonoke, Carlisle, Des Arc, Cotton Plant, Hunter, Wynne, Colt, Gieseck, Parkin, Turrell, Frenchmans Bayou, Joiner, Pecan Point • TENNESSEE: Millington, Kerrville, Rosemark, Brownsville, Belle Eagle, Alamo, Fruitland, Gibson, Milan, Cades, Atwood, Trezevant, McKenzie, Mansfield, Manlyville, Big Sandy, Faxon, Tennessee Ridge, Erin, Shiloh, Marian, Cunningham, Fredonia, Port Royal, Cedar Hill, Springfield, Orlinda, Cross Plains, Mitchellville • KENTUCKY: Rapids, New Roe, Chapel Hill, Scottsville, Maynard, Cedar Springs, Austin, Roseville, Eighty Eight, Edmonton, Gradyville, Milltown, Casy Creek, Creston, Pricetown, Yosemite, Middleburg, King Mountain, Gum Sulphur, Conway, Morrill, Kerby Knob, Drip Rock, Old Landing, Zoe, Vortex, Malaga, Elsie, Leatha, Wheelersburg, Volga, Sitka, Nippa, Lowmansville, Clifford • WEST VIRGINIA: Radnor, Dunlow, Sias, Sumerco, Ruth, Rand, Charleston, Bomant, Maysel, Clay, Tesla, Sutton, Newville, Cleveland, Czar, Cassity, Beverly, Elkins, Red Creek, Maysville, Points, Paw Paw, Great Cacapon • PENNSYLVANIA: Lashley, Big Cove Tannery, Webster Mills, Cito, Cove Gap, LeMasters, Markes, Fort Loudon, St. Thomas, Chambersburg, Stoufferstown, Fayetteville, Wenksville, Laurel, Bendersville, Goodyear, Latimore, Clear Spring, Franklintown, Maytown, Lewisburg, Middletown, Deodate, Lawn, Mount Wilson, Mount Gretna, Iona, Richland, Mount Pleasant, Dauberville, Windsor Castle, Klinesville, Krumsville, Stony Run, Saegersville, Laurys Station, Pennsville, Bearsville, Wind Gap, Roseto, Bossandville, Slaterford, Delaware, East Stroudsburg • NEW JERSEY: Millbrook, Five Points, Myrtle Grove, Crandon Lakes, Augusta, Branchville, Pettettown, McAfee, Hamburg, Vernon • NEW YORK: Amity, Warwick, Edenville, Bellvale, Sugar Loaf, Central Valley, Arden, Harriman, West Point, Garrison Lake, Carmel, Lake Carmel, Putnam Lake • CONNECTICUT: New Fairfield, Candlewood Shores, Brookfield, Roxbury Falls, Hotchkissville, Minortown, Watertown, Plymouth, Pequabuck, Whigville, Unionville, Bloomfield, Windsor Locks, Kings Corner, Enfield, Hazardville • MASSACHUSETTS: Hampden, Monson, Brimfield, Spencer, Leicester, Worcester, Morningdale, South Berlin, Berlin, Marlborough, Hudson, Maynard, West Concord, Concord, Beford, Pinehurst, North Reading, Middleton, Topsfield, Ipswich, Little Neck, (Plum Island State Park).

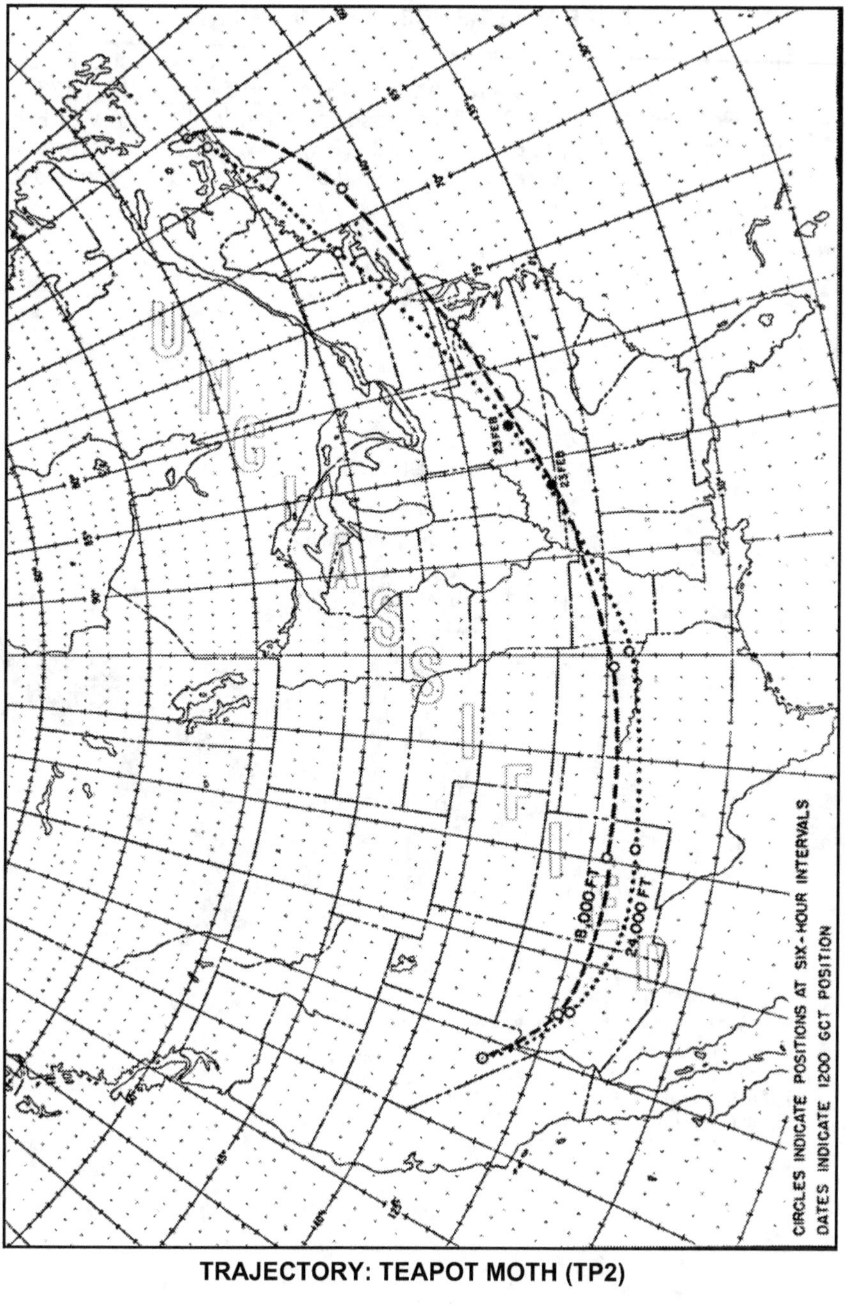

TRAJECTORY: TEAPOT MOTH (TP2)

CHARACTERISTICS: TEAPOT MOTH

The MOTH device weighed 445 lbs. Among the nuclear tests, Teapot MOTH is ranked 2nd for Cs-137 and Ba-137m and 5th for I-132, I-133, I-135, Rh-103m and Ru-103. The deposition pattern for Teapot MOTH is slightly similar to Teapot BEE and ESS (TP6), however, the Spearman R is only 0.05 with a T score of 2.53. County of highest (and only) recorded deposition: Clark, NV.

TOTAL FALLOUT: TEAPOT MOTH

Equal Number

nCi/square meter

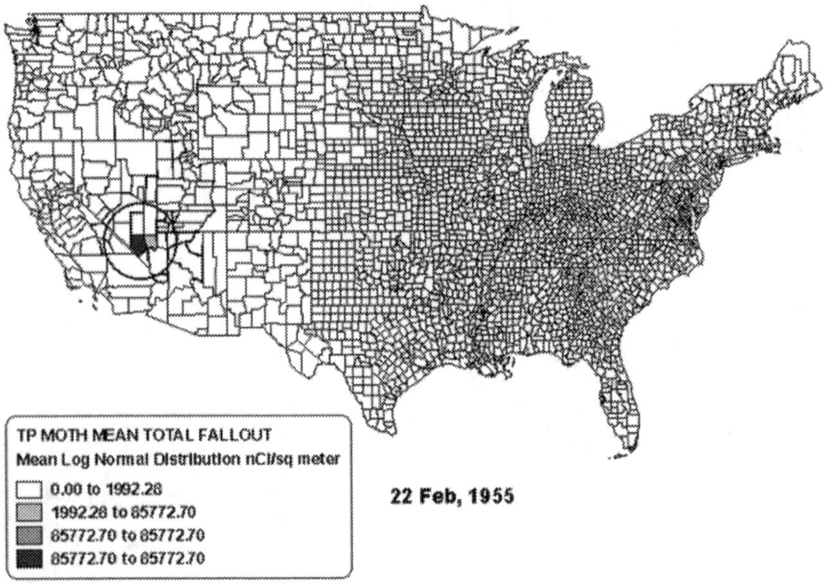

SHOT 3: TEAPOT TESLA (TP3)

Teapot: Tesla

Detonation Date: 1 Mar 1955 • *Detonation Time:* 5:30 A.M. • *Area:* 9b • *Sponsor:* Livermore Laboratory • *Yield:* 7 kt • *Radiation Level at Ground Zero:* 750 R/hr • *Height of Burst:* 300 ft • *Cloud Top Height:* 30,000 ft

24,000-Foot Trajectory: NEVADA: (Spotted Range), (Sheep Range), (Morman Peak), Mesquite • ARIZONA: Littlefield, Colorado City, Cane Beds, Moccason, Kaibab Indian Reservation, Kaibab, Fredonia, Page, (Tower Butte), (Monument Valley), Mexican Water • NEW MEXICO: Shiprock, (Shiprock Peak: 7,178 feet), Waterflow, Fruitland, Flora Vista, Aztec, Manero, Rutherton, Brazos, Eusenada, Tierra, Amarilla, (San Juan Mountains), Tres Piedras, Cerro, Questa, Red River, (Wheeler Peak, highest point in New Mexico: 13,161 feet), Maxwell, Grenville, Seneca • OKLAHOMA: Felt, Griggs, Guymon, Hardesty, Rosston, Laverne, Buffalo, Selman, Camp Houston, Capron, Burlington, Driftwood, Amorita, Byron, Wakita, Renfrow • KANSAS: Caldwell, Hunnewell, Ashton, Dexter, Oak Valley, Elk City, Sycamore, Morehead, Parsons, Strauss, McCune, Cherokee • MISSOURI: Mindenmines, Nashville, Kenoma, Golden City, Lockwood, Greenfield, Everton, Pennsboro, Ash Grove, Springfield, Willard, Walnut Grove, Glidewell, Fordland, Seymour, Mansfield, Norwood, Mountain Grove, Pine Crest, Mountain View, Birch Tree, Bartlett, Winna, Fremont, Van Buren, Chicopee, Williamsville, Hendrickson, Asherville, Wappapello, Puxico, Acornridge, Aguilla, Salcedo, Sikeston, Miner, Bertrand, Charleston, Anniston, Wyatt • KENTUCKY: Bardwell, Arlington, Fancy Farm, Mayfield, Brewers, Dexter, Fairdealing, Golden Pond, Cadiz, Caledonia, Hopkinsville, Mewstead, Pembroke, Elkton, Trenton, Daysville, Olmstead, Middleton, Franklin, Mount Aerial, Chapel Hill, Petroleum, Holland, Scottsville, Flippin, Mud Lick, Grandview, Thompkinsville, Judio, Littrell, Kettle, Mount Pisgah, Co-Operative, Pine Knot, Whitley City, Hollyhill, Mountain Ash, Gatliff, Middlesboro, Harrogate • TENNESSEE: Kyle's Ford, Okolona, Mount Carmel, Church Hill, Kingsport, Blountville, Avoca, Bluff City, Sadie, Buladeen, Shouns, Mountain City • NORTH CAROLINA: Creston, Clifton, Warrensville, Jefferson, Index, Laurel Springs, McGrady, Traphill, Austin, Elkin, Jonesville, Boonville, Smithtown, Flint Hill, Pfafftown, Enon, Lewsville, Winston-Salem, Union Cross, Jamestown, High Point, Archdale, Climax, Pleasant Gardens, Julian, Liberty, Pittsboro, Bynum, New Hill, Holly Springs, Fuquay-Varina, Willow Springs, Smithfield, Four Oaks, Goldsboro, Seymour Johnson Air Force Base, Seven Springs, Rivermont, Wise Forks, Trenton, Rhems, Pollocksville, Riverdale, Croatan, Hewelock, Newport, Morehead City, Otway, Beaufort, Parkers Island, (Shackleford Banks), (Cape Lookout).

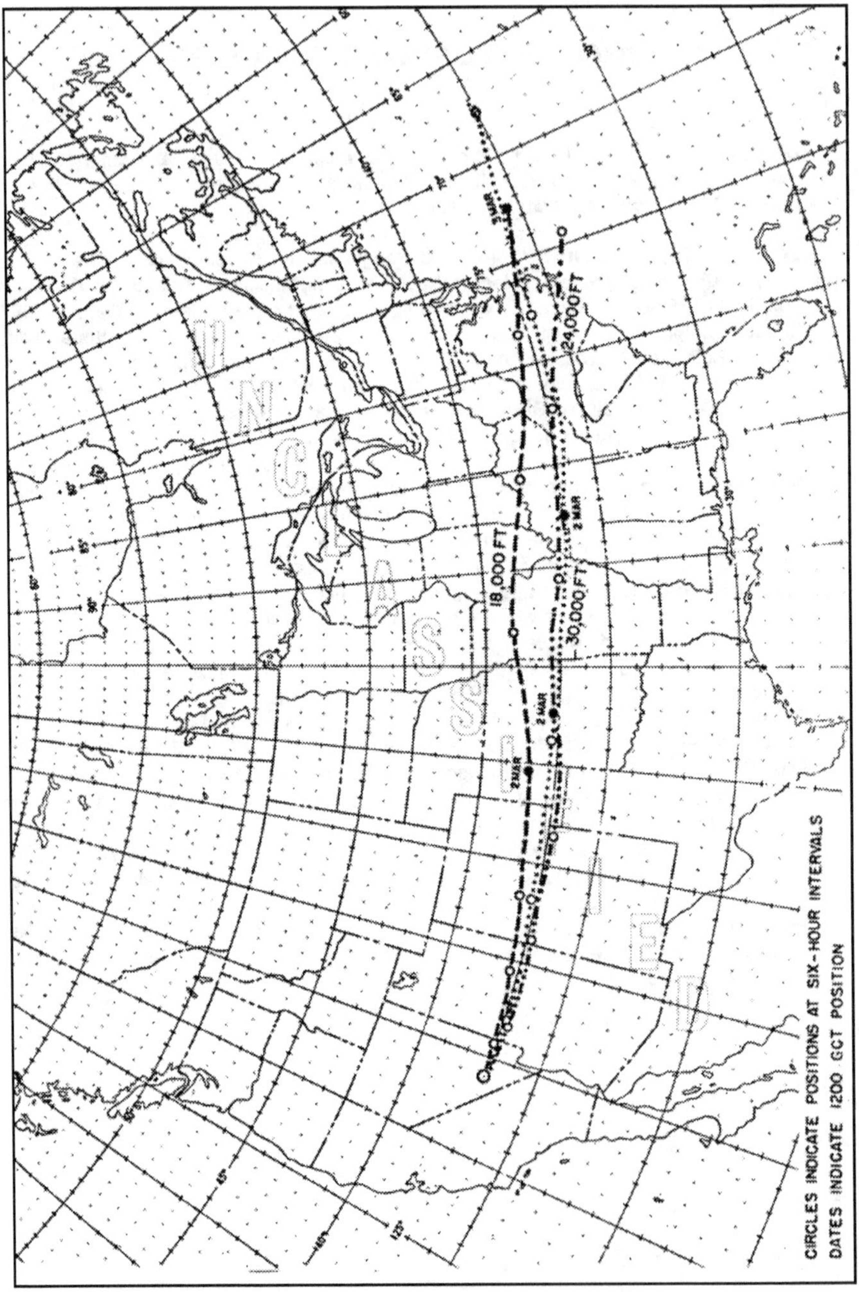

TRAJECTORY: TEAPOT TESLA (TP3)

CHARACTERISTICS: TEAPOT TESLA

The device for the TESLA shot was code-named "Cleo." Among the nuclear tests, Teapot TESLA is ranked 1st for percentages of Ba-137m, Cs-137, Rh-103m and Ru-103; 2nd for percentage of I-132 and 6th for percentage of Ag-109m. The deposition pattern for Teapot TESLA is most similar to that of Teapot TURK (TP4) (R:0.66; T: 48.46). The Spearman correlation between these two deposition patterns is among the highest in the entire series of aboveground tests.

Teapot TESLA deposited fallout on 1196 counties, equal to 39 percent of the total number of U.S. counties. Counties of highest estimated maximum activity included: Washington UT (415 μCi/sq meter estimated maximum activity) followed by the Colorado counties of Adams, Arapahoe, Boulder, Clear Creek, Douglas, Eagle, Gilpin, Grand, Jefferson and La Plata—all received a 246 μCi/sq meter fallout (estimated maximum activity) on March 2, 1955. The ratio of total fallout to I-131 for this shot is relatively low at 79 to 1.

RELATIVE TOTAL FALLOUT TP TESLA

MAR 1, 1955 7 KT

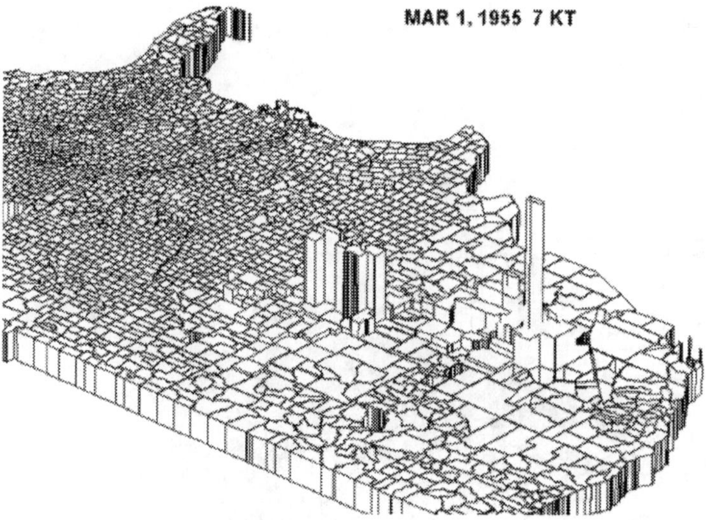

TOTAL FALLOUT: TEAPOT TESLA
Equal Intervals
nCi/square meter

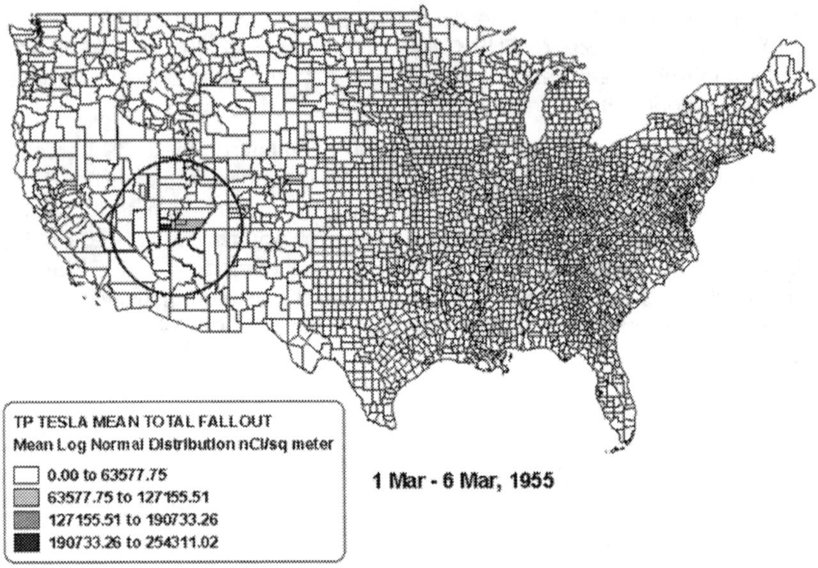

TOTAL FALLOUT: TEAPOT TESLA
Equal Number
nCi/square meter

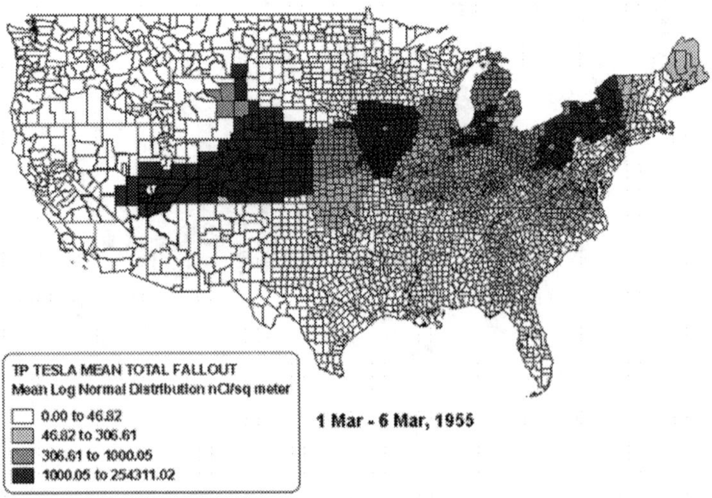

SHOT 4: TEAPOT TURK (TP4)

Teapot: Turk

Detonation Date: 7 Mar 1955 • *Detonation Time:* 5:20 A.M. • *Area:* 2 • *Sponsor:* Livermore Laboratory • *Yield:* 43 kt • *Radiation Level at Ground Zero:* 1,000 R/hr • *Height of Burst:* 500 ft • *Cloud Top Height:* 44,700 ft

40,000-Foot Trajectory: NEVADA: (Mormon Peak) • UTAH: (Snow Canyon), St. George, Washington, Rockville, Kanab, (Vermillion Cliffs), Glen Canyon, Navajo Indian Reservation, (Monument Valley), Mexican Hat • COLORADO: Ute Mountain Indian Reservation, Red Mesa, Southern Ute Indian Reservation, Tiffany, Allison, Arboles, Juanita, Chrome, Ortiz • NEW MEXICO: Chama, San Miguel, Las Pinos, Amalia, Costilla, (Ortiz Peak), Capulin, Grenville, (Mount Dora), Clayton • TEXAS: Texline, Stratford, McKibben, Notla, Glazier, Canadian • OKLAHOMA: Durham, Crawford, Roll, Thomas, Geary, Calumet, El Reno, Yukon, Bethany, Oklahoma City, Midwest City, Woods, Dale, Shawnee, Little, Butner, Wetumka, Vernon, Indianola, Russellville, Quinton, Lewisville, Kinta, Wells, Wister, Poteau, Howe, Monroe • ARKANSAS: Midland, Hartford, Waldron, Harvey, Gravelly, Story, Buckville, Jessieville, Owensville, (Hot Springs National Park), Benton, Redfield, Wright, Sherrill, Altheimer, Lodge Corner, DeWitt, DeLuce, Crumrod, Snow Lake • MISSISSIPPI: Perthshire, Deeson, Shelby, Drew, Minter City, Philipp, La Flora, Duck Hill, Lodi, Tomnolen, Reform, Sturgis, Brooksville, Macon, Prairie Point • ALABAMA: Dancy, Clinton, Stewart, Akron, Havana, Heiberger, Sprott, Plantersville, Winslow, Autaugaville, Hanter, Montgomery, Pile Road, Mathews, Fitzpatrick, Enon, Three Notch, Springhill • GEORGIA: Troutman, Westran, Bronwood, Oakfield, Ashburn, Sycamore, Mystic, Lax, Upton, Douglas, Bickley, Dixie Union, Blackshear, Nahonta, Atkinson, Brunswick, (St. Simons Island).

CHARACTERISTICS: TEAPOT TURK

The device for shot TURK, code-named "Linda," weighed 2325 lbs and was 30.5 inches by 61.3 inches. Among the nuclear tests, Teapot TURK is ranked 6th for percentage of Br-82 in fallout and 8th for percentage of Be-7 in fallout. The deposition pattern for Teapot TURK is most similar to Teapot TESLA (R: 0.66; T: 48.46). Fallout from TURK was deposited over 3008 counties. The county of highest deposition: Inyo, CA with an estimated maximum activity of 1043 μCi/sq meter total fallout. Other areas receiving high fallout were the Colorado counties of Adams, Arapahoe, Boulder, Clear Creek, Douglas, Eagle, Gilpin, Grand, Jefferson, and LaPlata. All received 922 μCi/sq meter (estimated maximum activity) on March 8, 1955.

The total fallout to I-131 ratio for Teapot TURK was relatively low at 79 to 1.

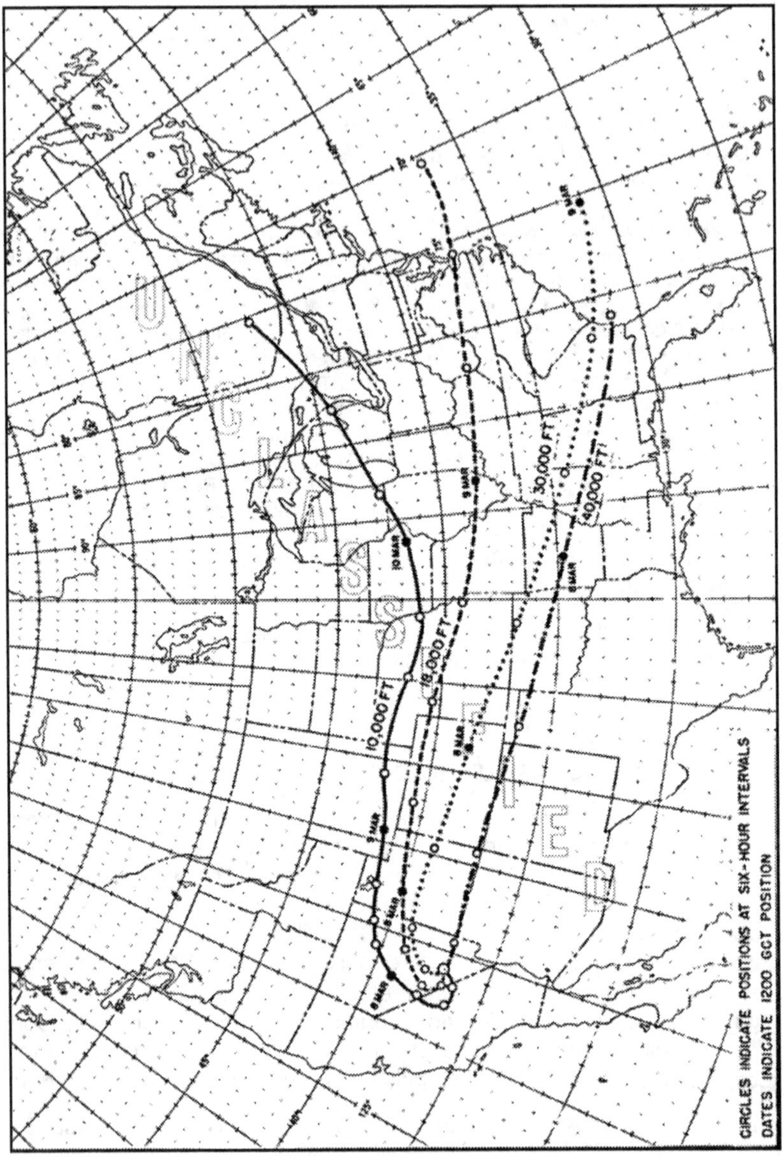

TRAJECTORY: TEAPOT TURK (TP4)

TOTAL FALLOUT: TEAPOT TURK
Equal Intervals

nCi/square meter

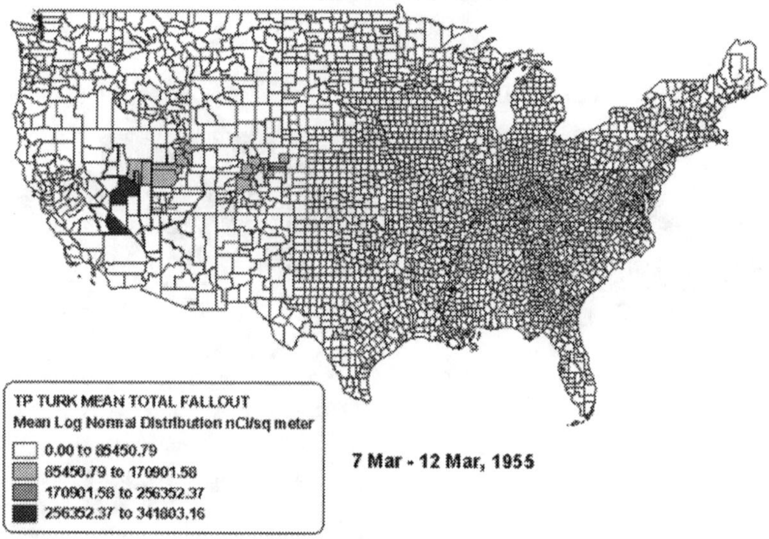

TOTAL FALLOUT: TEAPOT TURK
Equal Number

nCi/square meter

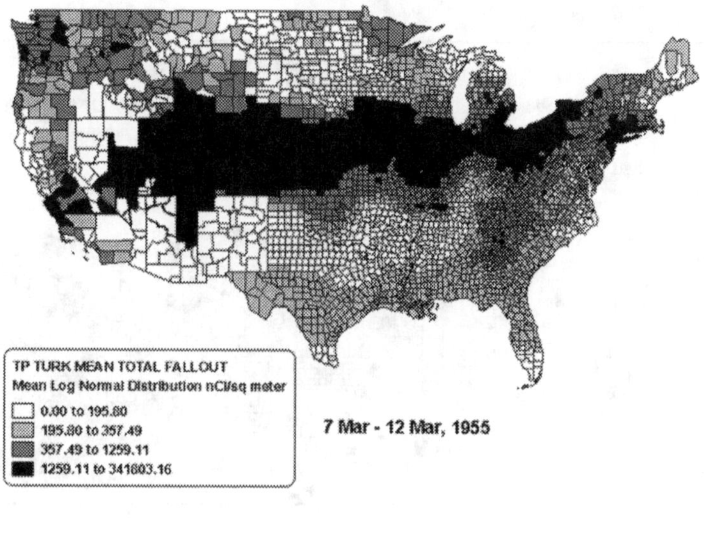

RELATIVE TOTAL FALLOUT TP TURK

MAR 7, 1955 43 KT

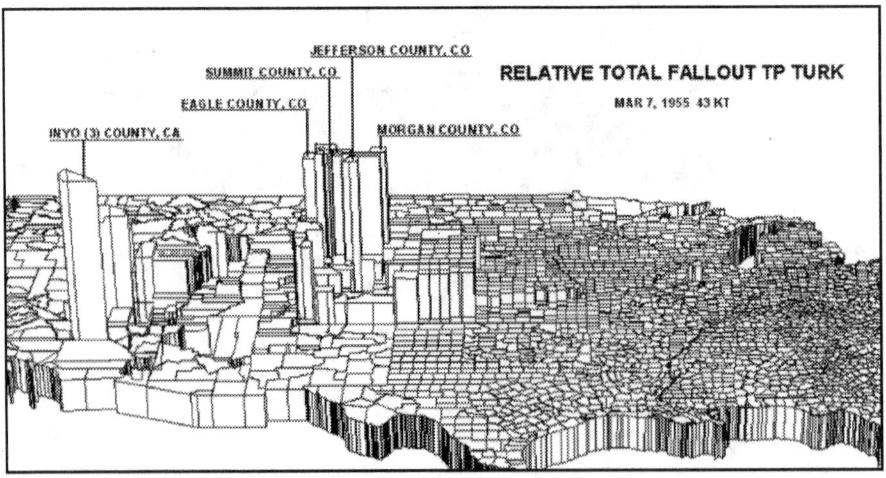

SHOT 5: TEAPOT HORNET (TP5)

Teapot: Hornet

Detonation Date: 12 Mar 1955 • *Detonation Time:* 5:20 A.M. • *Area:* 3a • *Sponsor:* Los Alamos • *Yield:* 4 kt • *Radiation Level at Ground Zero:* 2,500 R/hr • *Height of Burst:* 300 ft • *Cloud Top Height:* 37,000 ft

40,000-Foot Trajectory: NEVADA: (Papoose Range), Carp • UTAH: (Snow Canyon), Gunlock, Shivwits, Hurricane, La Verkin, Pintura, (Zion National Park), Glendale, Orderville, Mount Carmel Junction, (Valley of the Gods) • COLORADO: Squaw Point, Pleasant View, Silverton, Olney Springs, Numa, Ordway, Sugar City, Chivington, Sheridan Lake, Towner • KANSAS: Horace, Tribune, Selkirk, Lepti, Marienthal, Modor, Scott City, Manning, Grigston, Dighton, Shields, Utica, Ransom, Brownell, McCracken, Liebenthal, Loretta, Galatia, Susank, Beaver, Dubuque, Ellsworth, Carneiro, Brookville, Salina, Kipp, Abilene, Enterprise, Pear, Skiddy, Alta Vista, Hessdale, Keene, Topeka, Pauline, Big Spring, Williamstown, McLouth, Fairmount, Kansas City • MISSOURI: Kansas City, Raytown, Independence, Blue Springs, Wellington, Mayview, Higginsville, Alma, Mount Leonard, Marshall, Malta Bend, Slater, Forest Green, Roanoke, Mount Airy, Moberly, Middle Grove, Paris, Perry, Center, New London, Saverton • ILLINOIS: New Canton, New Salem, Valley City, Chapin, Literberry, Prentice, Pleasant Plains, Salisbury, Cantrall, Williamsville, Mount Pulaski, Chestnut, Kenny, Lane, Clinton, Weldon, De Laud, Mansfield, Fisher, Dewey, Rantoul, Ludlow, Clarence, East Lynn, Hoopeston • INDIANA: Ambra, Talbot, Fowler, Reynolds, Norway, Sitka, Royal Center, Lucerne, Metea, Perrysburg, Twelve Mile, Denver, Chili, Roann, Urbana, Andrews, Huntington, Zanesville, Ossian, Poe, Hoagland, Williams, Dixon, Monmouth • OHIO: Dixon, Cavett, Scott, Kalida, Pandora, Benton Ridge, Vanlue, Carey, Tymochtee, Belle Vernon, Sycamore, Benton, Chatfield, Tiro, Ganges, Epworth, Paradise Hill, Ashland, England Station, Rowsburg, New Pittsburg, Wooster, Orrville, Massillon, Perry Heights, Canton, Fairhope, Paris, Bayard, Hanoverton, West Point, Frederickstown • PENNSYLVANIA: Blackhawk, Fallston, East Rochester, Fernway, Wexford, Dorseyville, Oakmart, Plum, Sardis, Export, Delmont, Crabtree, New Derry, Derry, Wilpen, Tire Hill, Paint, Scalp Level, Ogletown, Weyant, Imber, New Enterprise, Riddlesburg, Kearney, Enid, Gracey, Clear Ridge, Fort Littleton, Fennettsburg, Metal, Creenvillage, Arendtsville, Table Rock, New Chester, New Oxford, Abbotstown, Spring Grove, Stoverstown, Rye, Felton, Pleasant Grove, Kyleville, Little Britain, Oxford, Strickersville • DELAWARE: Newark, Christiana, Bear, Red Lion, Delaware City • NEW JERSEY: New Castle, Harrisonville, Pointers, Salem, Renton, Alloway, Aldine, Centerton, Norma, Vineland, East Vineland, Milmay, Dorothy, Bargaintown, West Atlantic City, Atlantic City.

CHARACTERISTICS: TEAPOT HORNET

The HORNET device weighed 500 lbs and was detonated atop a 300-foot tower at the Nevada Test Site Area 2. Among the nuclear tests, Teapot HORNET is ranked 6th in percentages of Cd-115, In-115m and In-117 in total fallout. The deposition pattern for Teapot HORNET is most similar to that of Teapot APPLE-2 (TP10) (R: 0.43; T: 26.26). Fallout from Teapot HORNET was deposited on 2926 U.S. counties. County of highest deposition: Clark, NV.

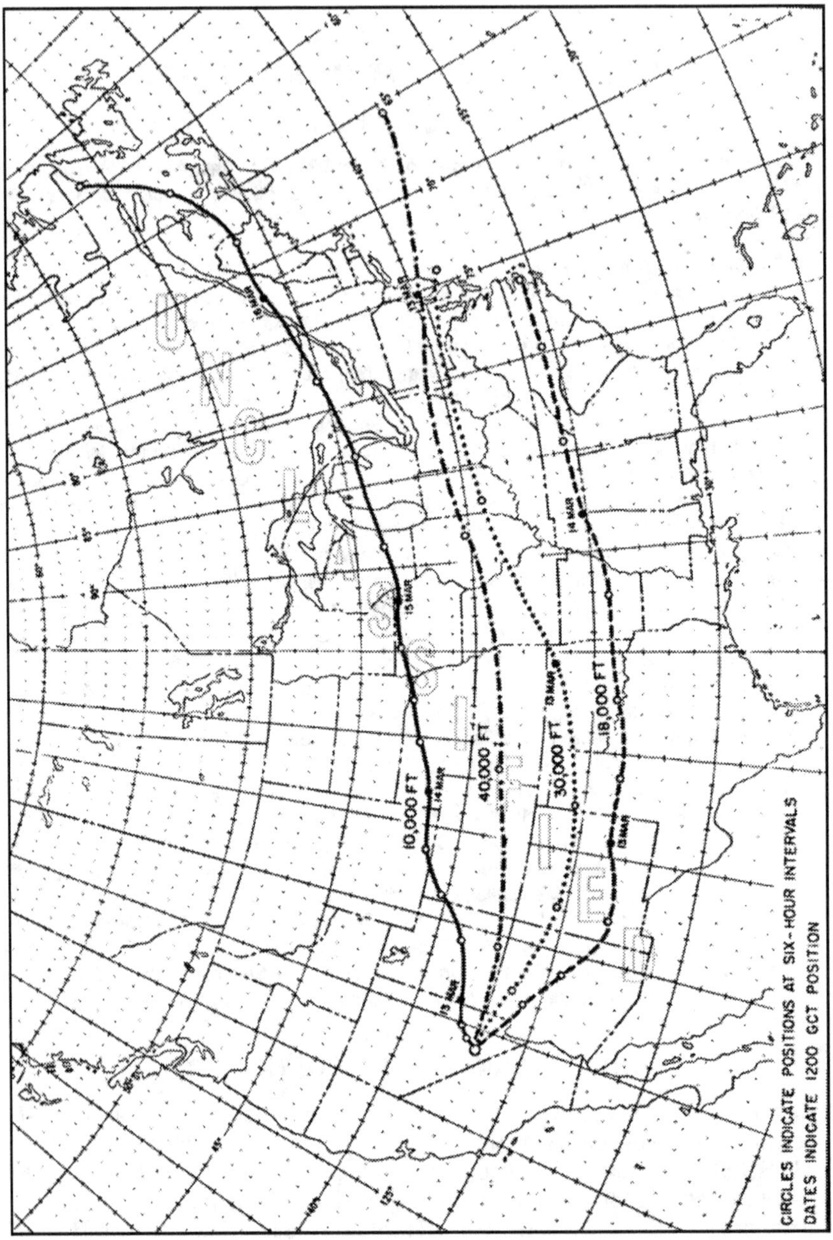

TRAJECTORY: TEAPOT HORNET (TP5)

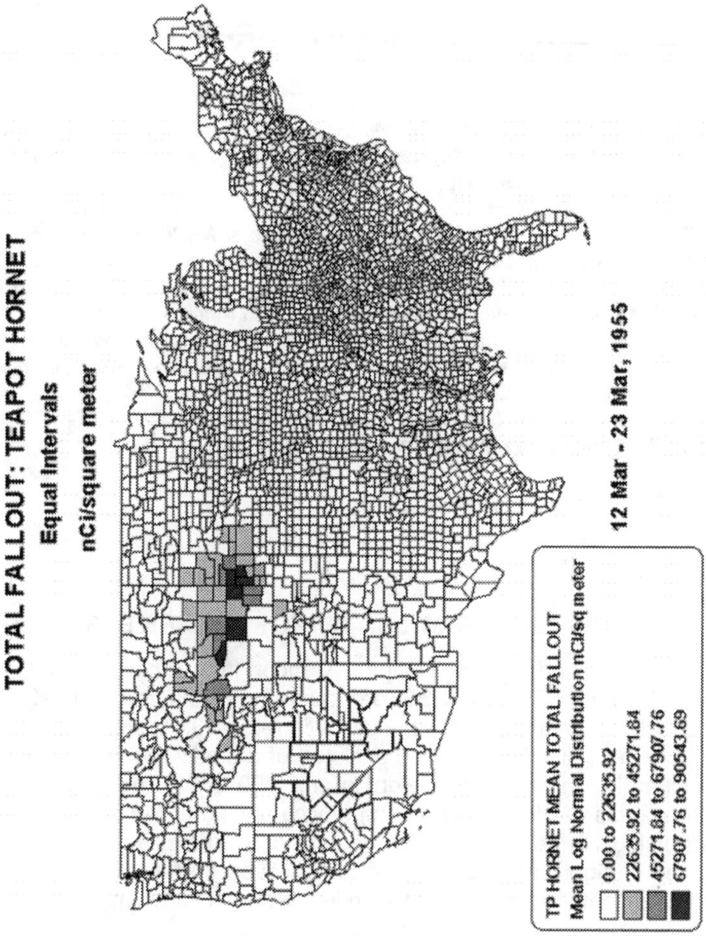

RELATIVE TOTAL FALLOUT: TP HORNET

MAR 7, 1955 43 KT

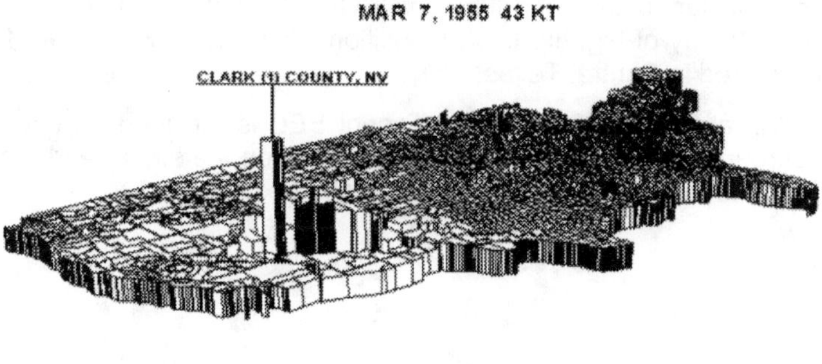

SHOT 6a: TEAPOT BEE (TP6A)

Teapot: Bee

Detonation Date: 22 Mar 1955 • *Detonation Time:* 5:05 A.M. • *Area:* 7–1a • *Sponsor:* Los Alamos • *Yield:* 8 kt • *Radiation Level at Ground Zero:* 2,000 R/hr • *Height of Burst:* 500 ft • *Cloud Top Height:* 39,700 ft

18,000-Foot Trajectory: NEVADA: Cottonwood Cove • ARIZONA: Bullhead City, Yucca, Congress, Wickenburg, Sun City, Phoenix, Scottsdale, Mesa, Mammoth, San Manuel, Cochise, Portal • NEW MEXICO: Rodeo, (Alamo Hueco Mountains) • TEXAS: Alpine, Sanderson, Dryden, Loma Alta, Campwood, Hondo, San Antonio, Union Valley, Shiner, Rock Island, Eagle Lake, Houston, Crosby, Shiloh, Devers, Beaumont, Orange • LOUISIANA: Vinton, Sulphur, Lake Charles, Church Point, Grand Coteau, Baton Rouge, Livingston, Hammond, Covington • MISSISSIPPI: Picayune, Caesar, Necaise, Wade, Nurley • ALABAMA: Mobile, Malbais, Robertsdale, Lillian • FLORIDA: Pensacola, Panama City, Highland Park, Chipola Park, Tallahassee, Waukeena, Lamont, Taylor, Jacksonville.

CHARACTERISTICS: TEAPOT BEE AND ESS

Though combined by the National Cancer Institute in the 1997 I-131 study, Teapot BEE and ESS are remarkably different shots, both in radionuclide content and in total fallout-to-I-131 ratios. The nuclear component of the BEE device consisted of a small sphere approximately 17 inches in diameter and weighing only 130 lbs. The ESS device, by contrast, was a 4-ton, six foot-long device placed at the bottom of a steel shaft drilled 70 feet beneath the desert floor. The BEE test undoubtedly resulted in significant dispersal of radioactive iron fume, while the fallout from ESS contained materials found in the desert alluvium.

Among the nuclear tests, Teapot BEE ranks 2nd in percentages of Pb-203 and 6th in I-133 and I-135. Teapot ESS is ranked first in Np-239, 3rd in U-237 and 7th in Na-24. The deposition pattern for the two combined shots is most similar to that of Plumbbob BOLTZMANN (PB1) ((R: 0.41; T: 25.16). County of highest total deposition: Clark, NV. Significant fallout also occurred in Central Texas.

The total fallout to I-131 ratio for Teapot BEE is the third lowest in the aboveground test series at 73 to 1. Shot ESS, on the other hand, has the third ***highest*** total fallout to I-131 ratio at 199 to 1.

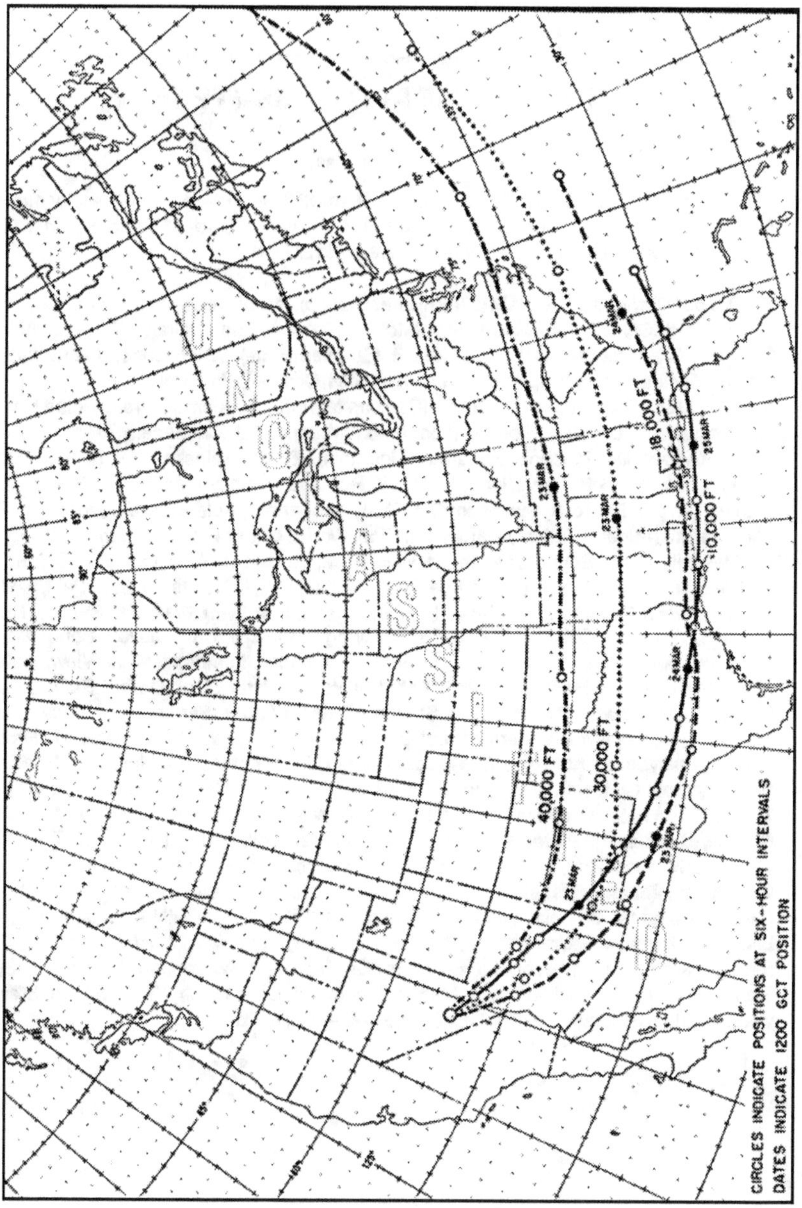

TRAJECTORY: TEAPOT BEE (TP6A)

SHOT 6B: TEAPOT ESS (TP6B)

TEAPOT: ESS

Detonation Date: 23 Mar 1955 • *Detonation Time:* 12:30 P.M. • *Area:* T-10a • *Sponsor:* Los Alamos, Department of Defense • *Yield:* 1 kt • *Radiation Level at Ground Zero:* 6,000 R/hr • *Height of Burst:* – 67 ft • *Cloud Top Height:* 12,000 ft

10,000-Foot Trajectory: NEVADA: (Pintwater Range), Mopa, Logandale, Overton, (Virgin Peak) • ARIZONA: (Virgin Mountains), (Mount Trumbull), Supai, (Grand Canyon), Tusayan, Cameron, Gray Mountain, (Painted Desert), (Stephen Butte), White Cone, Steamboat Canyon, Cornfields, Sunrise Springs, Ganada, (Cross Canyon), Fort Defiance, St. Michaels • NEW MEXICO: Fence Lake, Trechado, (Gallinas Mountains), Alamo Navajo Indian Reservation, Polvadera, Lemitar, Florida, (Los Pinos Mountains), (Chupadera Mesa), Claunch, Ancho, (Arroyo Del Macho), (Haystack Mountain), Elkins, Milnesand • TEXAS: Griffith, Lehman, Morton, Pettit, Levelland, Lubbock, Roosevelt, Posey, Cap Rock, Canyon Valley, Kalgary, Jayton, Peacock, Swenson, Old Glory, Aspermont, Rule, Haskell, Woodson, Eliasville, Graford, Mineral Wells, Weatherford, Fort Worth, Grand Prairie, Arlington, Irving, Dallas, Mesquite, Balch Springs, Lawrence, Wills Point, Edgewood, Alba, Quitman, Golden, Forest Hill, Hainesville, Rhonesboro, Grice, Thomas, LaFayett, Lone Star, Daingerfield, Red Hill, Atlanta, Queen City, Cass • ARKANSAS: Brightstar, Fort Lynn, Fouke, Lewisville, Falcon, Rosston, Waterloo, Childester, Amy, Holly Springs, Ramsey, Ivan, Staves, Randall, Ladd, Moscow, Bayou, Lodge Corner, DeWitt, Ethel, St. Charles, Lambrook, Lake View, Wabash, West Helena, Friars Point • MISSISSIPPI: Hill House, Rena Lara, Sherard Baugh, Farrell, Rudyard, Jonestown, Darling, Curtis Station, Pleasant Grove, Sardis, Harmontown, Looxahoma, Tyro, Thyatyra, Waterford, Lows Hill, Potts Camp, Winborn, Ashland, Peoples, Falkner, Chalybeate, Kossuth, Wenasoga, Corinth, Farmington, Kendrick • TENNESSEE: Michie, Acton, Childers Hill, Nixon, Burnt Church, Houston, Center, Crossroads, Ethridge, Webber City, Waco, Mooresville, Lewisburg, Shelbyville, Manchester, Ragsdale, Viola, Irving College, (Cumberland Plateau), Hitchcox, Mount Crest, Cold Spring, (Walden Ridge), Grandview, Roddy, Rockwood, Kingston, Solway, Ball Camp, Knoxville, Mascot, Blaine, Jefferson City, Shiloh, Talbott, Russellville, Whitesburg, St. Clair, Ballytown, McCloud, Fall Branch, Sullivan Gardens, Colonial Heights, Blountsville, Bristol • VIRGINIA: Bristol, Oscoala, Lodi, Adwolf, Attoway, Crockett, Wytheville, Draper, Graysontown, Christianburg, Alleghany Spring, Bent Mountain, Roanoke, Stewartsville, Bedford, New London, Lynchburg, Spout Spring, Appomattox, Guinea Mills, Macon, Richmond, Old Church, King William, Stevensville, Morattico, Burgess, Fair Port, Lillan, Reedville, (Smith Point).

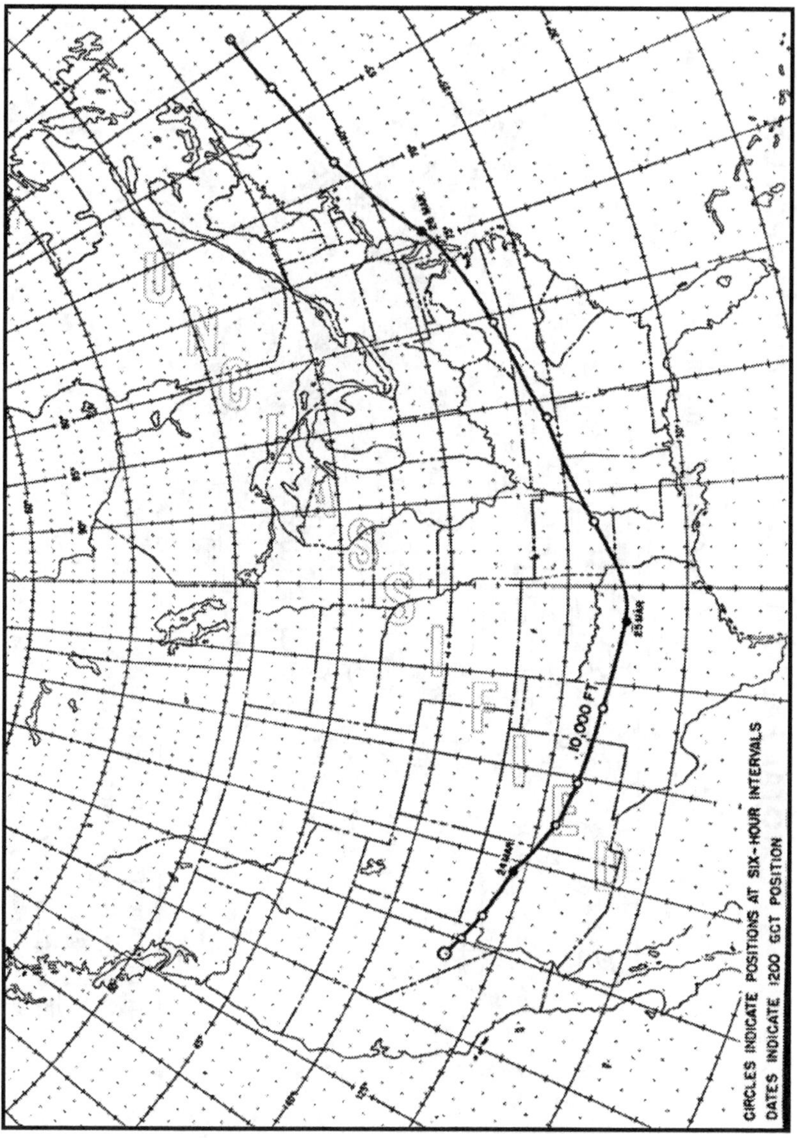

TRAJECTORY OF TEAPOT ESS (TP6B)

TOTAL FALLOUT: TEAPOT BEE AND TEAPOT ESS

Equal Intervals

nCi/square meter

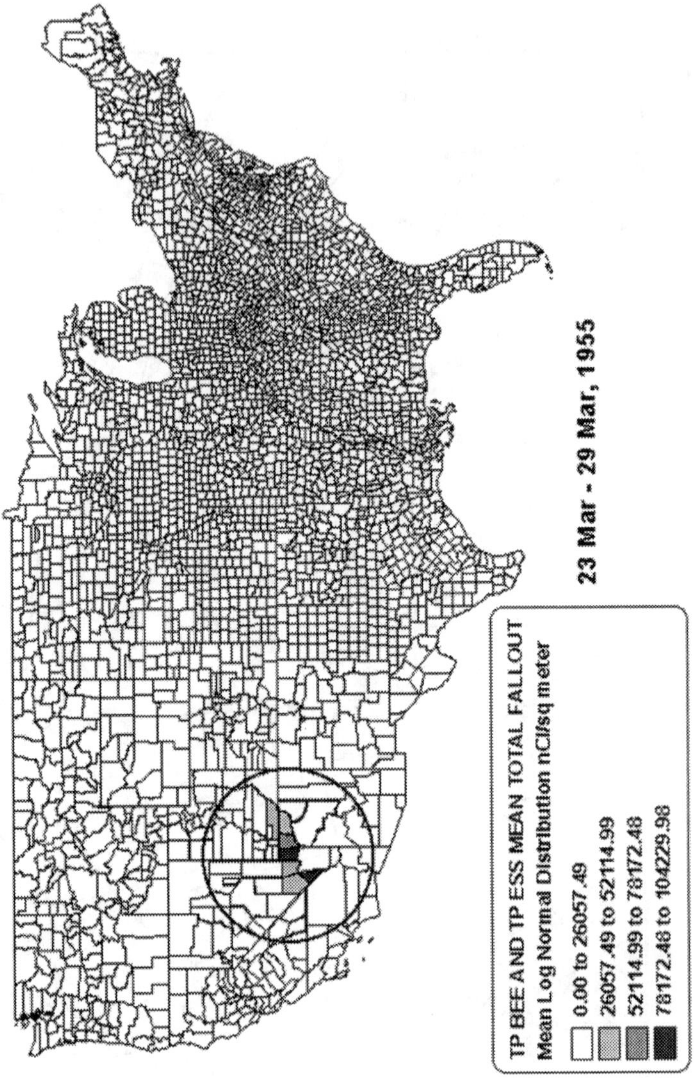

TOTAL FALLOUT: TEAPOT BEE AND TEAPOT ESS

Equal Number

nCi/square meter

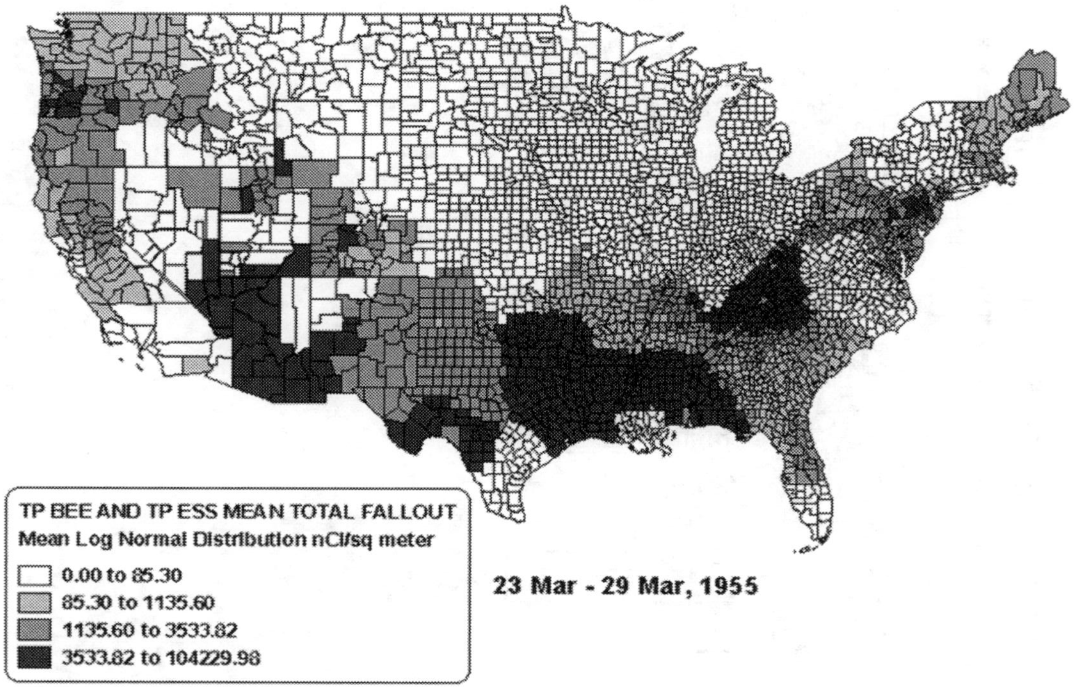

23 Mar - 29 Mar, 1955

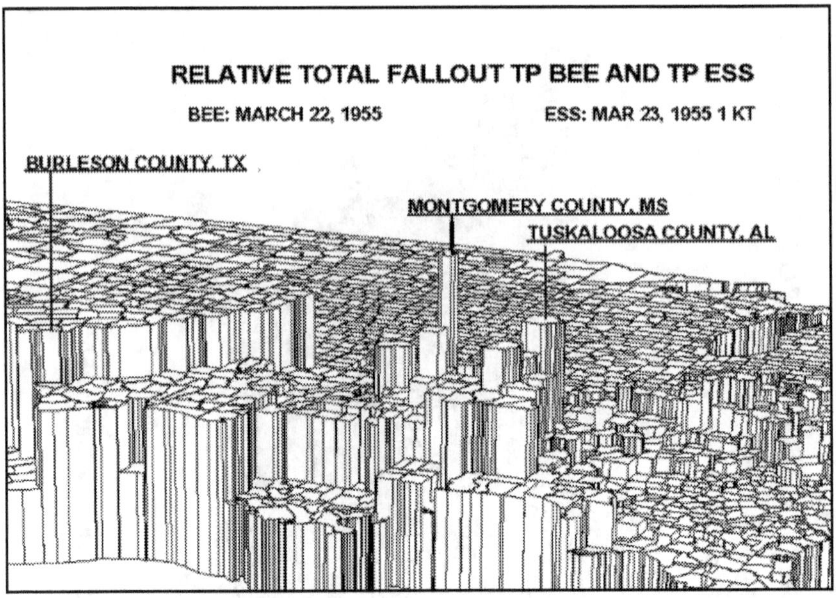
RELATIVE TOTAL FALLOUT TP BEE AND TP ESS
BEE: MARCH 22, 1955
ESS: MAR 23, 1955 1 KT
BURLESON COUNTY, TX
MONTGOMERY COUNTY, MS
TUSKALOOSA COUNTY, AL

SHOT 7: TEAPOT APPLE-1 (TP7)

TEAPOT: APPLE-1

Detonation Date: 29 Mar 1955 • *Detonation Time:* 4:55 A.M. • *Area:* 4 • *Sponsor:* Los Alamos • *Yield:* 14 kt • *Radiation Level at Ground Zero:* 500 R/hr • *Height of Burst:* 500 ft • *Cloud Top Height:* 32,000 ft

10,000-Foot Trajectory: NEVADA: (Papoose Range), (Groom Lake), (Timpahute Range), Lake Valley, Baker • UTAH: Garrison, Grantsville, Salt Lake City, (Antelope Island), (Castle Rock), Tremonton, Newton, Plymouth, Cornish • IDAHO: Snowville, Woodruff, Samaria, Malad City, Virginia, McCammon, Swan Valley, Driggs, Felt, Tetonia • WYOMING: (Mount Moran), (Mount Gancock), (Yellowstone Park: Heart Lake, Yellowstone Lake), Pahaska, (Windy Mountain), (Bear Tooth Pass) • MONTANA: Red Lodge, Washoe, Bearcreek, Bridger, Pryor, Crow Indian Reservation, Hardin, Hathaway, Miles City, Locate, Ismay, Ollie • NORTH DAKOTA: Galva, (Tracy Mountain), Southeart, Dickinson, Marshall, Dodge, Halliday, Twin Buttes, White Shield, Roseglen, Ryder, Douglas, Sawyer, Logan, Surrey, Minot, Norwich, Granville, Towner, Berwick, Barton, Mylo, Agate, Porth, Rocklake, Calvin, Hannah.

CHARACTERISTICS: TEAPOT APPLE-1

The nuclear component of the APPLE-1 device was 29.5 inches wide, 74.6 inches long and weighed 2300 lbs. Among the nuclear tests, Teapot APPLE-1 is ranked 6th in percentage of U-240 in fallout. The deposition pattern for Teapot APPLE-1 is most similar to that of Teapot ZUCCHINI (TP11) (R: 0.43, T: 26.37). Fallout from Teapot APPLE 1 was deposited on 3062 counties.

Counties of highest deposition (all figures in μCi/sq meter, estimated maximum activity): Texas, OK—940.7; Stevens, KS—940.7; Oldam, TX, Potter, TX—940.7; Lincoln, NV—388.9; all on Mar 29, 1955.

With regard to radionuclide content, Teapot APPLE-1 appears to represent a "standard" atomic test. In deposition of radionuclides, Teapot APPLE 1's rank among all aboveground tests is as follows:

Co-60—26; Nb-95m—30; Nb-95—28; Cs-137—28; Be-7—25; Mn-54—28; Pr-144—29; Ru-106—30; Te-127m—29; Sr-90—24.

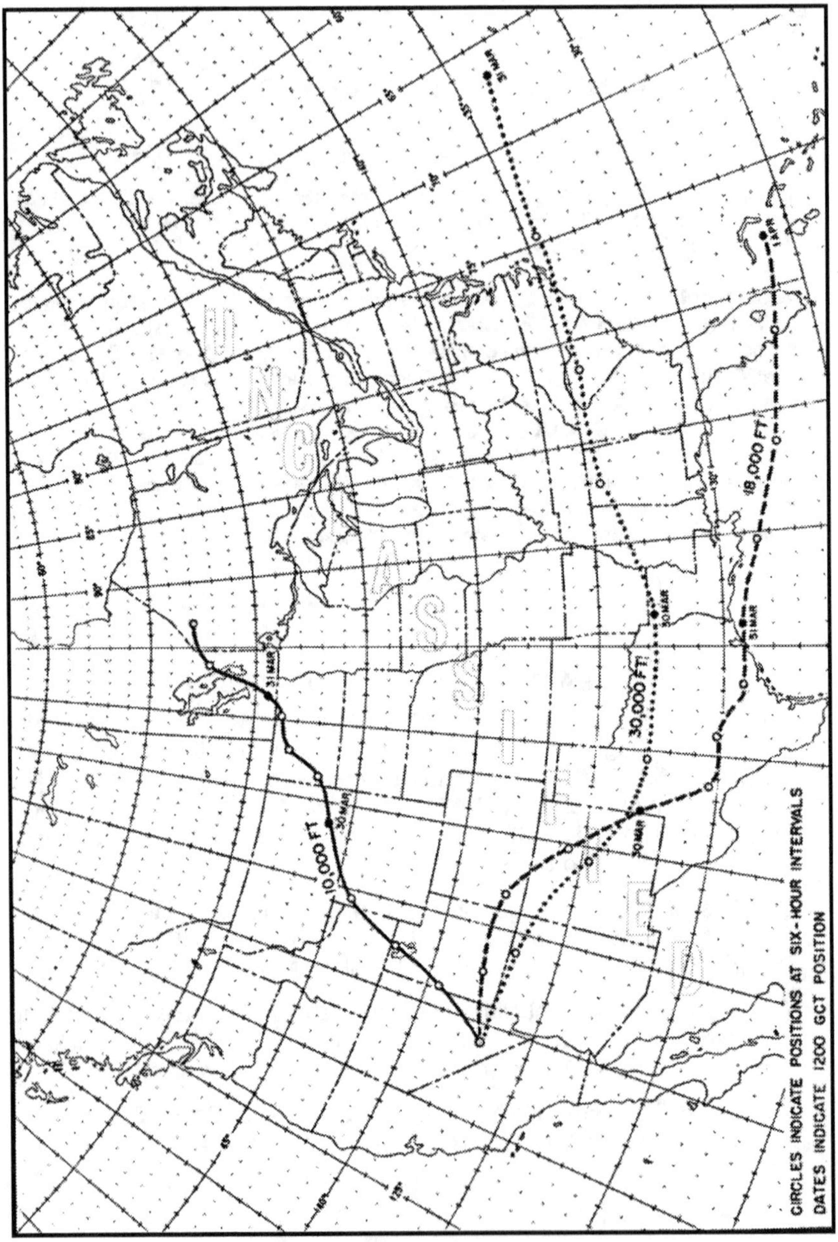

TRAJECTORY: TEAPOT APPLE-1 (TP7)

TOTAL FALLOUT: TEAPOT APPLE-1

Equal Intervals

nCi/square meter

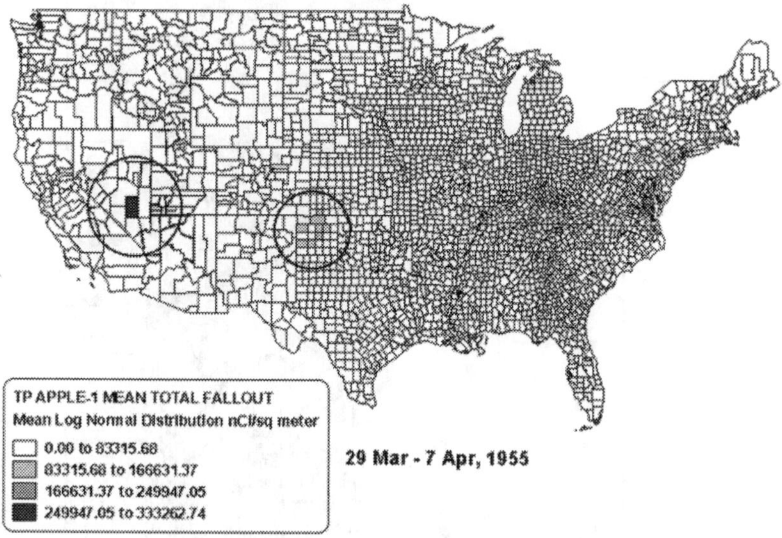

TOTAL FALLOUT: TEAPOT APPLE-1

Equal Number

nCi/square meter

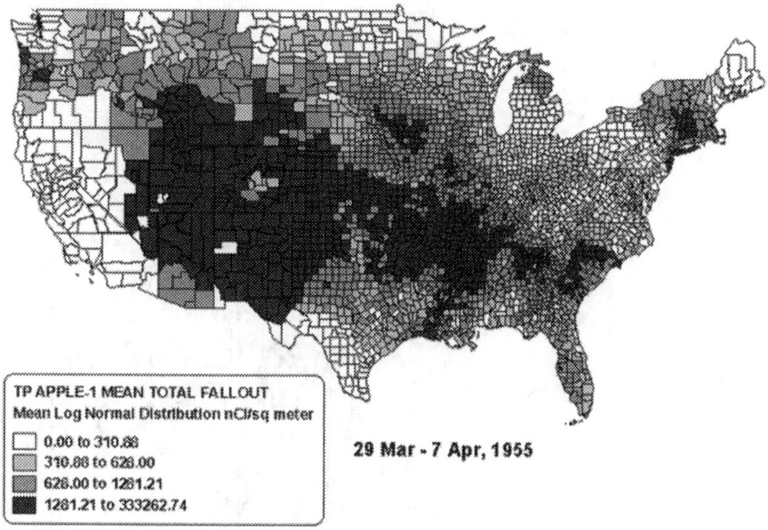

RELATIVE TOTAL FALLOUT: TP APPLE-1
MAR 29, 1955 14 KT
LINCOLN COUNTY, NV
POTTER COUNTY, TX
OLDHAM COUNTY, TX
TEXAS COUNTY, OK

SHOT TEAPOT WASP PRIME

(Not Included in NCI Study)

TEAPOT: WASP PRIME

Detonation Date: 29 Mar 1955 • *Detonation Time:* 10:00 A.M. • *Area:* T-7-4 • *Sponsor:* Los Alamos • *Yield:* 3 kt • *Radiation Level at Ground Zero:* 10 R/hr • *Height of Burst:* 739 ft • *Cloud Top Height:* 32,000 ft

18,000-Foot Trajectory: NEVADA: (Spotted Range), (Pintwater Range), Elgin • UTAH: Enterprise, Newcastle, Cedar City, Enoch, Summit, Parowan, Paragonah, Circleville, Junction, Greenwich, Fremont, (Thousand Lake Mountain), (San Rafael Knob), Willowblind, Uintah and Quray Indian Reservation, Bonanza • COLORADO: Dinosaur, Blue Mountain • WYOMING: Baggs, (Medicine Bow National Forest), Saratoga, (Medicine Bow Peak), Bosler, Iron Mountain, Meriden, Albin • NEBRASKA: Potter, Sidney, Colton, Sunol, Lodgepole, Chappell, Big Springs, Brule, Ogallala, Keystone, Dunning, Almeria, Taylor, Elyria, Ord, North Loup, Scotia, Cotesfield, Elba, Cushing, Palmer, Archer, Central City, Hordville, Marquette, Polk, Arborville, Benedict, Bradshaw, York, Waco, Utica, Beaver Crossing, Cordova, Crete, Wilber, Kramer, Clatonia, Pickrell, Crab Orchard, Tecumseh, Elk Creek, Howe, Stella, Shubert, Barada • MISSOURI: Corning, Craig, Maitland, Graham, Barnard, Guilford, Skidmore, Arkoe, Conception Junction, Ravenwood, Parnell, Grant City, Sheridan • IOWA: Redding, Delphos, Benton, Mount Ayr, Tingley, Arispe, Talmadge, Lorimore, Barney, East Peru, Hanley, Winterset, Patterson, Des Moines, Bonderant, Clyde, Rhodes, Melbourne, Van Cleve, Haverhill, Marshalltown, Le Grand, Clutier, Elberon, Van Horne, Newhall, Atkins, Covington, Cedar Rapids, Mount Vernon, Tipton, Bennett, New Liberty, Sunbury, Maysville, Walcott, Davenport, Bettendorf • ILLINOIS: Moline, Oak Grove, Arion, Andove, Nekoma, Bishop Hill, Galva, LaFayette, Speer, Edelstein, Chillicothe, Rome, Metamora, Eureka, Hudson, Normal, Holder, Bellflower, Blue Ridge, Urbana, Champaign, Philo, Sidney, Allerton, Sidell, Ridge Farm, Scotland • INDIANA: Cayuga, Lodi, Newport, Annapolis, Marshall, Milligan, Morton, Clinton Falls, Brick Chapel, Fillmore, Stilesville, Hazelwood, Mooresville, Smith Valley, Stones Crossing, Greenwood, Whiteland, Rocklane, Boggstown, Needham, Shelbyville, Waldron, St. Paul, Sandusky, Oldenburg, St. Peters, South Gate, St. Leon, New Trenton, West Harrison, Harrison • OHIO: Harrison, New Baltimore, Miamitown, Groesbeck, Monfort Heights, Brentwood, Deer Park, Montgomery, Madeira, Cincinnati, Norwood, Milford, Owensville, Crosstown, Greenbush, Bulford, Gath, Mowrystown, May Hill, Locust Grove, Mount Joy, Sedan, Fire Brick, Eifort, Blackfork, Gallia, Rodney, Gallipolis • WEST VIRGINIA: Point Pleasant, Mount Alto, Cottageville, Liverpool, Reedy, Mount Zion, Sand Ridge, Copen, Cedarville, Napier, Arlington, Rock Cave, Alton, Cassity, Valley Bend, (Spruce Knob, highest point in West Virginia), Judy Gap, Fort Seybert • VIRGINIA: Broadway, Tenth Legion, Alma, Stanley, Brightwood, Leon, Burr Hill, Todds Tavern, Massaponax, Guinea, Woodford, Sparta, Beulaville, Mangohick, Aglett, Walkerton, King William, Port Richmond, West Point, Barhamsville, Williamsburg, Fort Eustis, Lawson, Rescue, Newport News, Norfolk, Chesapeake, Virginia Beach.

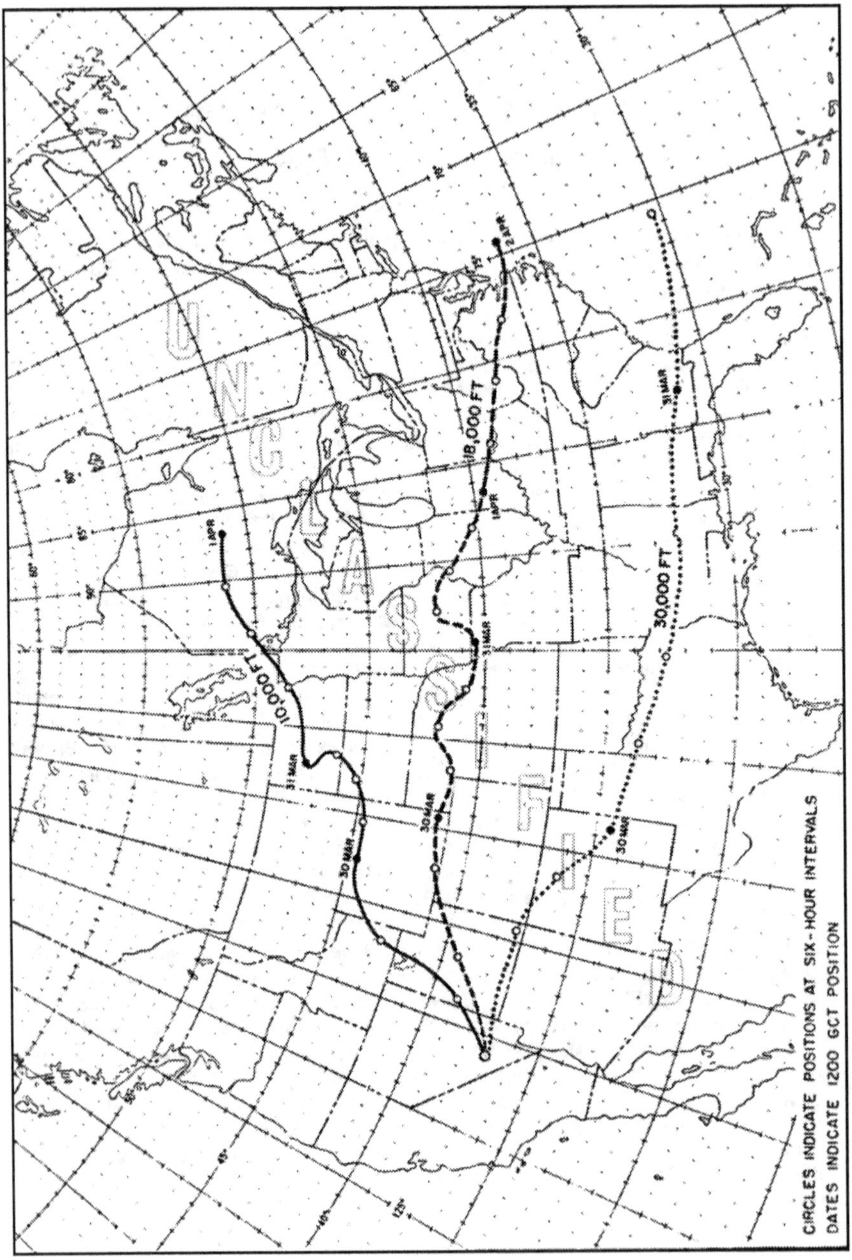

TRAJECTORY: TEAPOT WASP PRIME

SHOT TEAPOT HA (HIGH ALTITUDE)

(Not Included in the NCI Study)

Teapot: HA

Detonation Date: 6 Apr 1955 • *Detonation Time:* 10:00 A.M. • *Area:* T-5 • *Sponsor:* Department of Defense • *Yield:* 3 kt • *Radiation Level at Ground Zero:* None detected • *Height of Burst:* 32,582 + / – 100 ft • *Cloud Top Height:* 55,000 ft

45,000-Foot Trajectory: NEVADA: (Sheep Range), Mesquite, Bunkerville • ARIZONA: Littlefield, (Mount Bangs), (Kaibab Plateau), (Marble Canyon), Cedar Ridge, (Wildcat Peak), Tonalea, Cow Springs, Chilchinbito, Rough Rock, Many Farms, Tsaile • NEW MEXICO: Sanostee, Toadlena, Newcomb, Burnham, Blanco, Nageezi, Ojito, Gavilan, Llaves, El Rito, Vallecitos, Petaca, La Madera, Taos, Pueblo, Talpa, Arroyo Seca, Angel Fire, (Aqua Fria), (Christo Range), (Philmont Boy Scout Ranch), Cimarron, Miami, Springer, Maxwell, (Laughlin Peak), Grenville, Sierra Grande, Moses • OKLAHOMA: Wheeless, Boise City, Keyes, Tyrone, Lookout, Manchester • KANSAS: Elkard, Liberal, Hardtner, Kiowa, Caldwell, Arkansas City, Silverdale, Maple City, Hewins, Elgin, Chautauqua, Caney, Tyro, Coffeyville, Edna, Chetopa, Melrose, Baxter Springs, Treece, Lowell • MISSOURI: Joplin, Fidelity, Sarcoxie, Wentworth, Pierce City, Freistatt, Verona, Marionville, Aurora, Crane, Hurley, Jamesville, Union City, Highlandville, Chadwick, Garrison, Ava, Squires, Gentryville, Drury, Vanzant, (Cedar Knobs), Pomona, Olden, White Church, Peace Valley, Thomasville, Wilderness, Handy, Eastwood, Grandlin, Poplar Bluff, Ash Hill, Fisk, Dudley, Dexter, Essex, Matthews, Whiting, E. Prairie, Wolf Island • KENTUCKY: Hickman, Moscow, Oakton, Cayce, Crutchfield, Water Valley, Cuba, Tri City, Lynn Grove, Murray, Hamlin, La Fayette, Oak Grove, Henleytown, Tiny Town, Keysburg, Adairville, Prices Mill, Providence, Rapids, New Roe, Adolphus, Akersville, Gamaliel, Hestand, Vernan, Littrell, Albany, Static, Sunnybrook, Mount Pisgah, Co-Operative, Pine Knot, Strunk, Saxton, Pearl, Pruden, Fonde, Edgewood, Middleboro, (Cumberland Gap), Weber City, Morrison City • TENNESSEE: Kingsport, Lynn Garden, Bloomingdale, Orebank, Bristol, Sutherland, Laurel Bloomery • NORTH CAROLINA: Lansing, Grassy Creek, Piney Creek, Twin Oaks, Sparta, Lowgap, Bottom, Mount Airy, Bannertown, Francisco, Lawsonville, Sandy Ridge, Stoneville, Eden, Ruffin, Pelham, Yancyville, Concord, Leasburg, Gentry's Store, Allensville, Stovall, Williamsboro, Middleburg, Manson, Norlina, Macon, Warrenton, Vaughan, Littleton, Sunny Side, Roanoke Rapids, Weldon, Jackson, Conway, Petecasi, Murfreesboro, Winston, Union, Tunis, Cofield, Gatesville, Sandy Cross, Trotville, Hobbsville, Morgan's Corner, Shawboro, Gregory, Belcross, Camden, Currituck, Maple, Barco, Waterlily, Aydlett, Carolia.

CHARACTERISTICS: TEAPOT HA

The HA device was airdropped by a B36 from an altitude of over 32,000 ft and allowed to descend via parachute to the detonation height of 739 ft. The device, including bomb casing, weighted 1085 lbs.

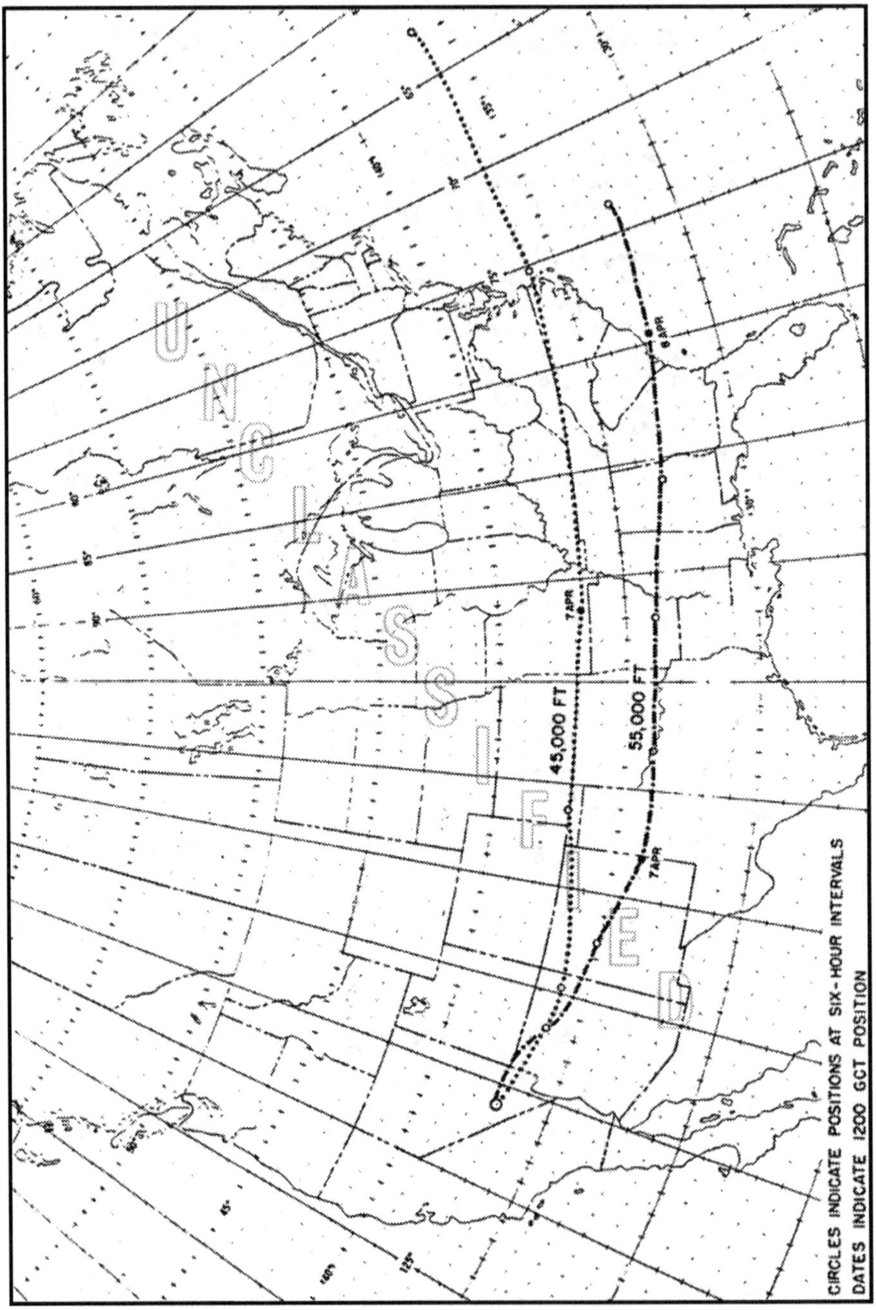

TRAJECTORY: TEAPOT HA (HIGH ALTITUDE)

SHOT 8: TEAPOT POST (TP8)

Teapot: Post

Detonation Date: 9 Apr 1955 • *Detonation Time:* 4:30 A.M. • *Area:* 9c • *Sponsor:* Livermore Laboratory • *Yield:* 2 kt • *Radiation Level at Ground Zero:* 1,000 R/hr • *Height of Burst:* 300 ft • *Cloud Top Height:* 15,500 ft

14,000-Foot Trajectory: NEVADA: Sandy, Goodsprings, Jean, Searchlight, Laughlin • ARIZONA: Kingman, Drake, Flagstaff, Winona, White Cone, Steamboat Canyon, Nazlini • NEW MEXICO: Crystal, Naschitti, Brazos, Ensenada, Tierra Amarilla, Cerro, Questa, Red River, Capulin, Moses • OKLAHOMA: Boise City, Keyes, Eva, Optima, Adams, Beaver, Laverne, Freedom, Cherokee, Vining, Medford • KANSAS: Arkansas City, Cambridge, Grenola, Moline, Fredonia, Thayer, St. Paul, Girard, Frontenac • MISSOURI: Lamar, Dadeville, Eudora, Elkland, Grovespring, Success, Oscar, Centerville, Lesterville, Cobalt City, Patton, Sedgewickville, Daisy, New Wells, Pocahontas • ILLINOIS: Alto Pass, Carbondale, Carterville, Energy, Johnston City, Akin, Diamond City, Springerton, Burnt Prairie, Bellmont, St. Francisville • INDIANA: Vincennes, Bicknell, Westphalia, Bloomfield, Bloomington, Unionville, Helmsburg, Spearsville, Marietta, Shelbyville, Rushville, Harrisburg, Richmond • OHIO: Gettysburg, Verona, Nashville, Tipp City, North Hampton, Springfield, Catawba, Tradersville, Columbus, Whitehall, Jacksontown, Gratiot, Hopewell, Zanesville, Opperman, Mount Ephraim, Calais, Ozark, Clarington • WEST VIRGINIA: Wileyville, Fairmont, Newburg, Arthur, Rig, Lost City • VIRGINIA: Edinburg, Mount Jackson, Sperryville, Culpepper, Fredericksburg, Woodford, Bowling Green, Millers Tavern, Dutton.

CHARACTERISTICS: TEAPOT POST

The nuclear device used for shot POST was 34.2 inches long and weighed 322 lbs. Among the nuclear tests, Teapot POST is ranked 1st in percentage of Rh-106, Ru-106, Te-127 and Te-127m and ranked 2nd for percentage of Ag-109m, Pd-111m and Sb-126 in fallout. The deposition pattern for Teapot POST is most similar to Plumbbob GALILEO (PB12) (R: 0.39; T: 23.35). Fallout from this shot was deposited over 2914 counties (94 percent of the total counties). Counties of highest deposition: Tooele, UT; Wasatch, UT; Morgan, UT; White Pine, NV and Salt Lake, UT. All received less than 100 µCi/sq meter of total fallout (estimated maximum activity).

The total fallout to I-131 ratio for this shot was one of the lowest of the aboveground test program at 72 to 1. Only Plumbbob KEPLER had a lower total fallout-to –I-131 ratio at 71 to 1.

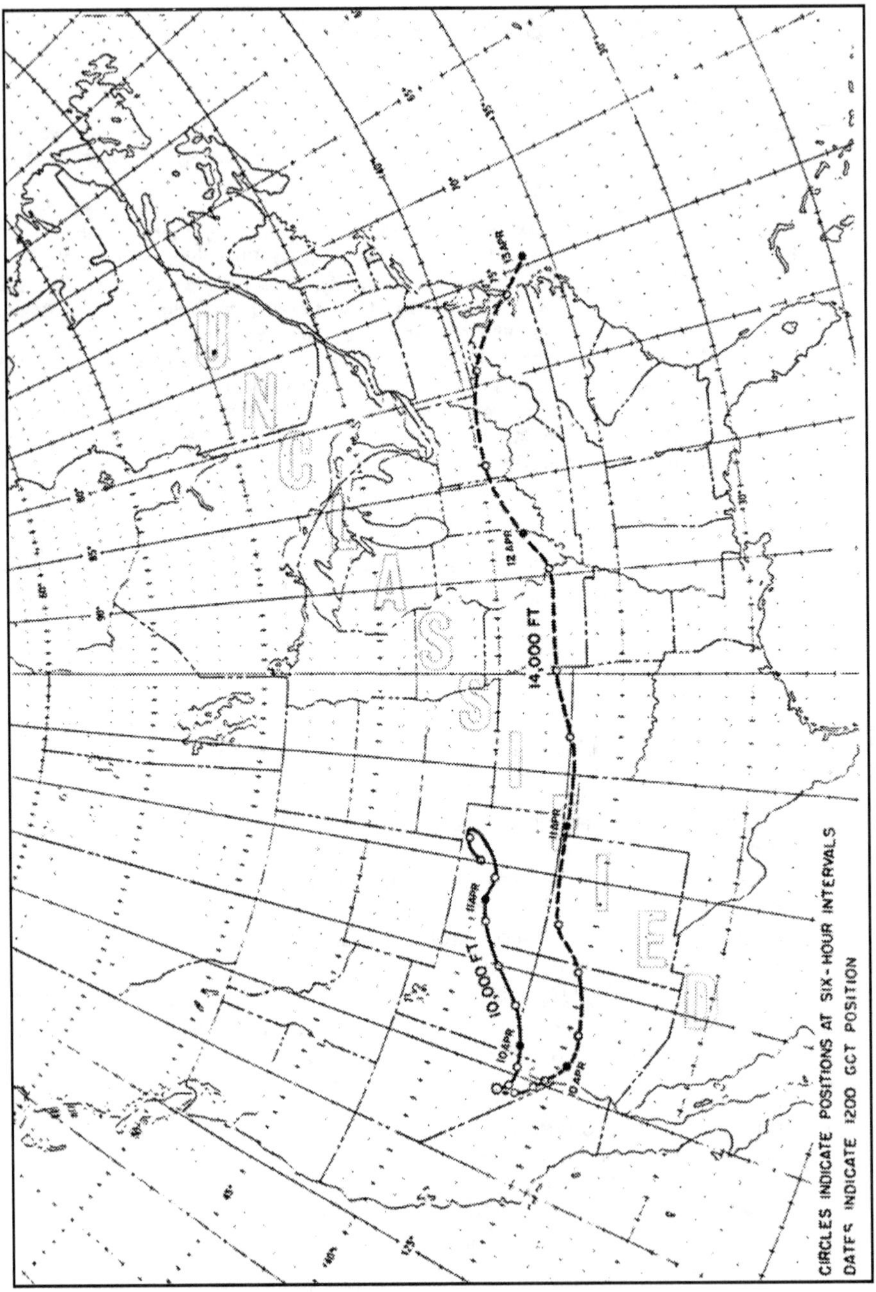

TRAJECTORY: TEAPOT POST (TP8)

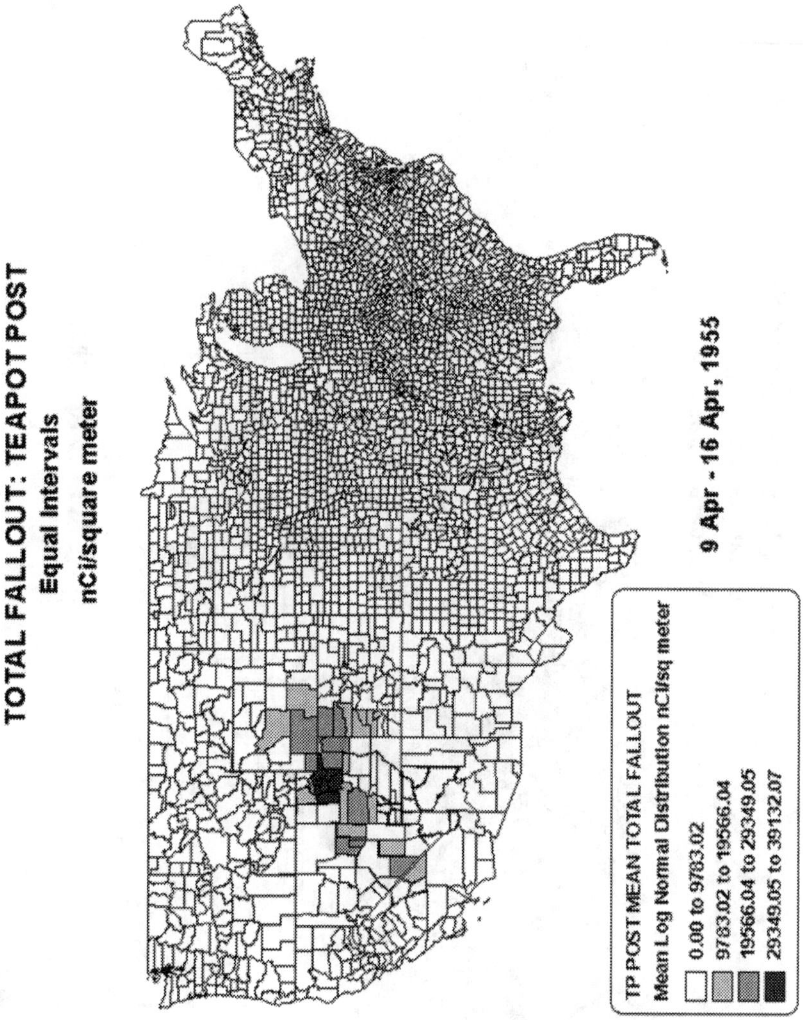
TOTAL FALLOUT: TEAPOT POST
Equal Intervals
nCi/square meter
9 Apr - 16 Apr, 1955
TP POST MEAN TOTAL FALLOUT
Mean Log Normal Distribution nCi/sq meter
0.00 to 9783.02
9783.02 to 19566.04
19566.04 to 29349.05
29349.05 to 39132.07

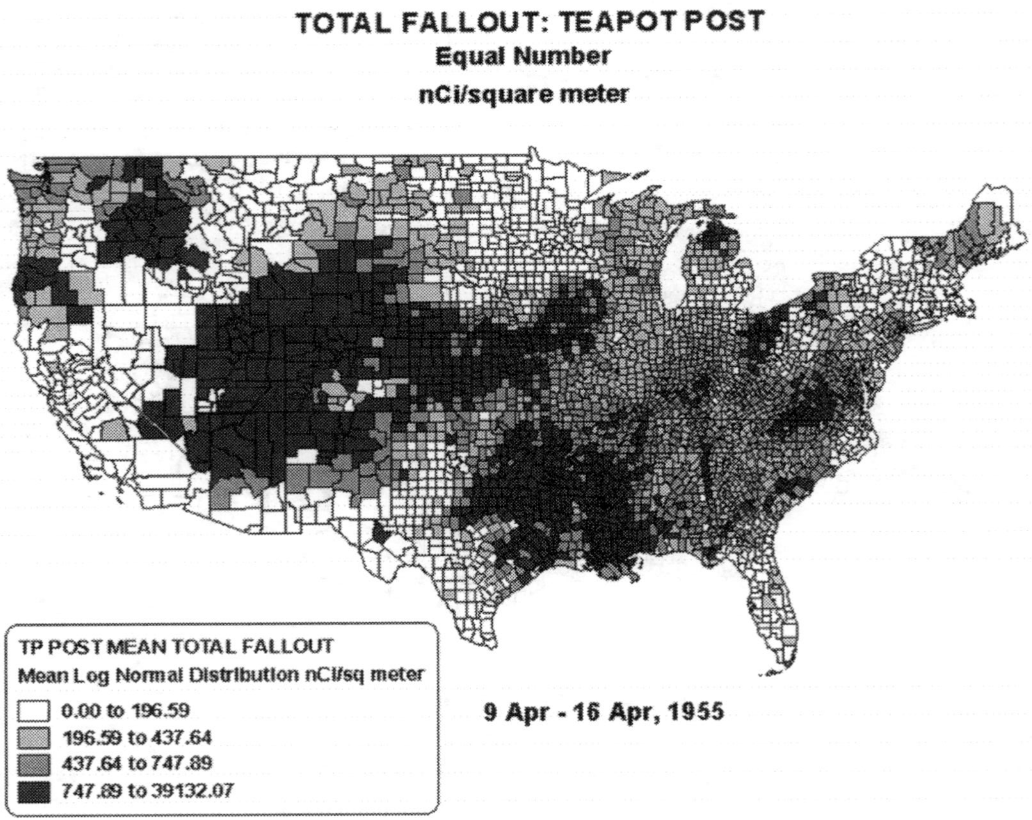
TOTAL FALLOUT: TEAPOT POST
Equal Number
nCi/square meter
TP POST MEAN TOTAL FALLOUT
Mean Log Normal Distribution nCi/sq meter
0.00 to 196.59
196.59 to 437.64
437.64 to 747.89
747.89 to 39132.07
9 Apr - 16 Apr, 1955

SHOT 9: TEAPOT MET (TP9)

Teapot: Met

Detonation Date: 15 Apr 1955 • *Detonation Time:* 11:15 A.M. • *Area:* Frenchman's Flat • *Sponsor:* Los Alamos, Department of Defense • *Yield:* 22 kt • *Radiation Level at Ground Zero:* 10 R/hr • *Height of Burst:* 400 ft • *Cloud Top Height:* 40,300 ft

40,000-Foot Trajectory: NEVADA: (Pintwater Range), (Pahranagat Range), (Delmar Mountains), Elgin • UTAH: Uvada, (Escalante Desert), Circleville, Junction, Greenwich, Fremont, (San Rafael Knob), (Windowblind Peak), (East Tavaputs Plateau), (Sweetwater Canyon) • COLORADO: (Douglas Pass), Buford, Phippsburg, Oak Creek, (Muddy Pass), Rand, Gould, (Cameron Pass), Poudre Park, Livermore, Wellington, Hereford • NEBRASKA: (Johnson Township, highest point in Nebraska), Kimball, Dix, Potter, Gurley, Oshkosh, Arthur, (Three Mile Lake), Halsey, Purdum, Emmet, O'Neill, Page, Winnetoon, Bazile Mills, Bloomfield, Bow Valley, Hartingen, Obert, Maskell • SOUTH DAKOTA: Vermillion, Burbank, Hub City, Junction City, Big Springs • IOWA: Elk Point, LeMars, Remsen, Peterson, Cornell, Gillett Grove, Ayrshire, Rodman, Whittemore, Hobarton, Algona, Sexton, Wesley, Hutchins, Britt, Duncan, Miller, Ventura, Clear Lake, Emery, Central Heights, Mason City, Nora Springs, Floyd, Colwell, Alata Vista, Jerico, Little Turkey, Fort Atkinson, Ossian, Frankville, Rossville, Volney, Waukon • WISCONSIN: Prairie du Chien, Wauzeka, Woodman, Ridgeway, Dodgeville, Daleyville, Paoli, Belleville, Edgerton, Indianford, Milton, Lima Center, Burlington, Union Grove, Racine, Kenosha • MICHIGAN: Hagar Shores, Riverside, Milburg, Watervliet, Keeler, Dowagiac, Volina, Wakelee, Vandalia, Jones, Constantine, Mottville • INDIANA: Bristol, Middlebury, Shipshewana, Lagrange, Plato, Stroh, Woodruff, Hudson, Ashley, Hamilton • OHIO: Edgerton, Farmer, Brunersburg, Defiance, Ayersville, New Bavaria, Holgate, Prentiss, Belmore, Townwood, McComb, Findlay, Upper Sandusky, Wyandott, Caledonia, Denmark, Mount Gilead, Pulaskiville, Waterford, Fredericktown, Howard, Millwood, Newcastle, New Guilford, West Bedford, Cooperdale, New Moscow, Adams Mills, Conesville, Bloomfield, Cambridge, Greenwood, Whigville, Calais, Mittonsburg, Woodsfield, Laings, Round Botton, Hannibal • WEST VIRGINIA: New Martinsville, Porters Falls, Reader, Folsom, Sedalia, Clarksburg, Rangoon, Belington, Elkins, Red Creek, Harmon, Onego, Milam • VIRGINIA: (Shenandoah Mountains), Bergton, Fulks Run, Timberville, Broadway, New Market, Tenth Legion, Alma, Haywood, Brightwood, Madison, Aroda, Rapidan, Paytes, Post Oak, Snell, Thornburg, Milford, Owenton, Stephens Church, Bruington, Truhart, Saluda, Glenns, Hartfield, Hudgins, Diggs, (Chesapeake Bay), Eastville, Cheriton, (Wreck Island), (Cobb Island).

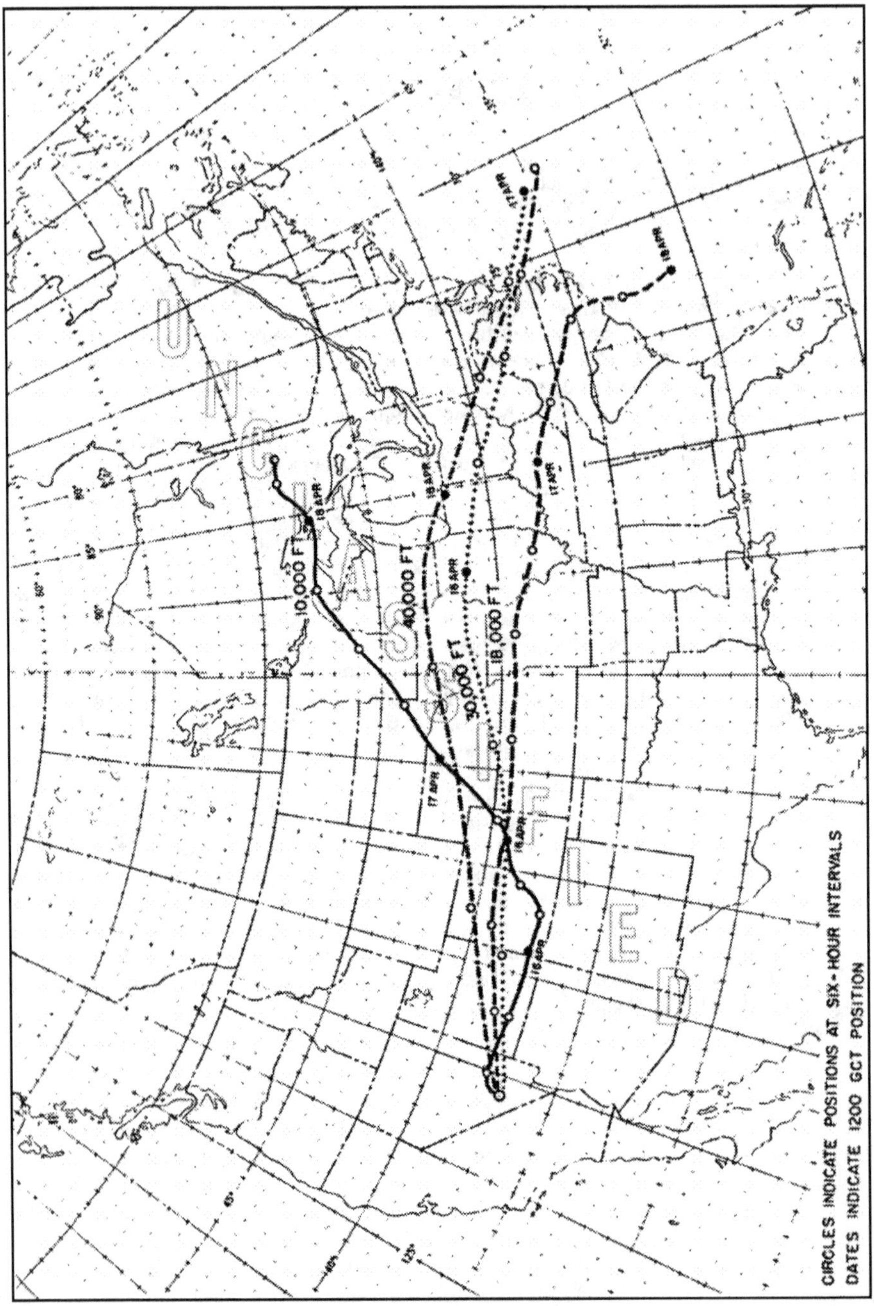

TRAJECTORY: TEAPOT MET (TP9)

CHARACTERISTICS: TEAPOT MET

The nuclear component for the MET ("Military Effects Test") shot was an 800 lb, 30-inch diameter sphere. The design was based on the earlier Buster Jangle EASY design. Teapot MET was a somewhat unique shot, ranking ***first*** among the nuclear tests for percentages of Br-82, Kr-85, Sr-89, Sr-90, Y-90 and Y-91 in fallout. The deposition pattern for Teapot MET is most similar to Plumbbob NEWTON (PB15) (R: 0.38; T: 22.92).

Fallout from this nuclear test was deposited over 3018 counties, 98 percent of the total counties in the U.S. The counties of highest estimated maximum activity included:

- Beaver UT (1057 μci/sq meter);
- Bowman ND (733 μCi/sq meter);
- Mesa CO (820 μCi/sq meter);
- Emery UT (846 μCi/sq meter); and
- Grand UT (846 μCi/sq meter), all on Apr 15, 1955.

The total fallout to I-131 ratio for shot Met is moderate at 88 to 1.

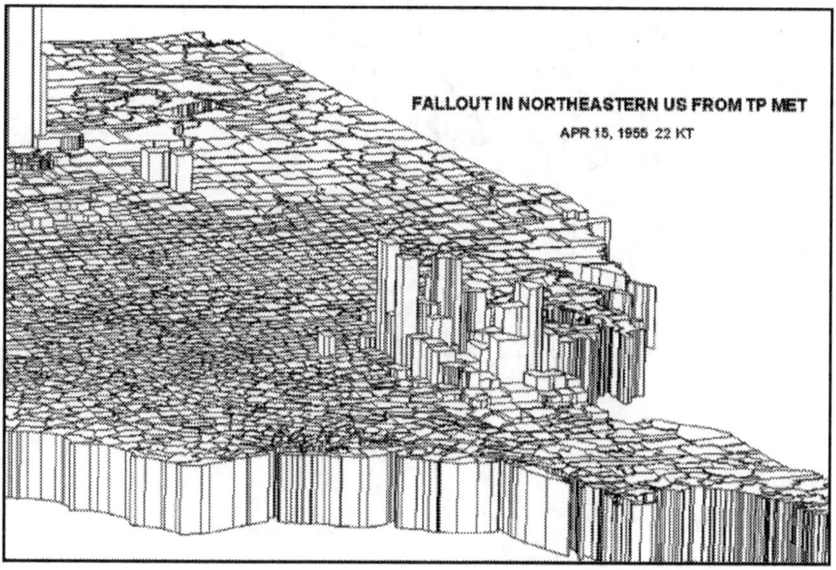

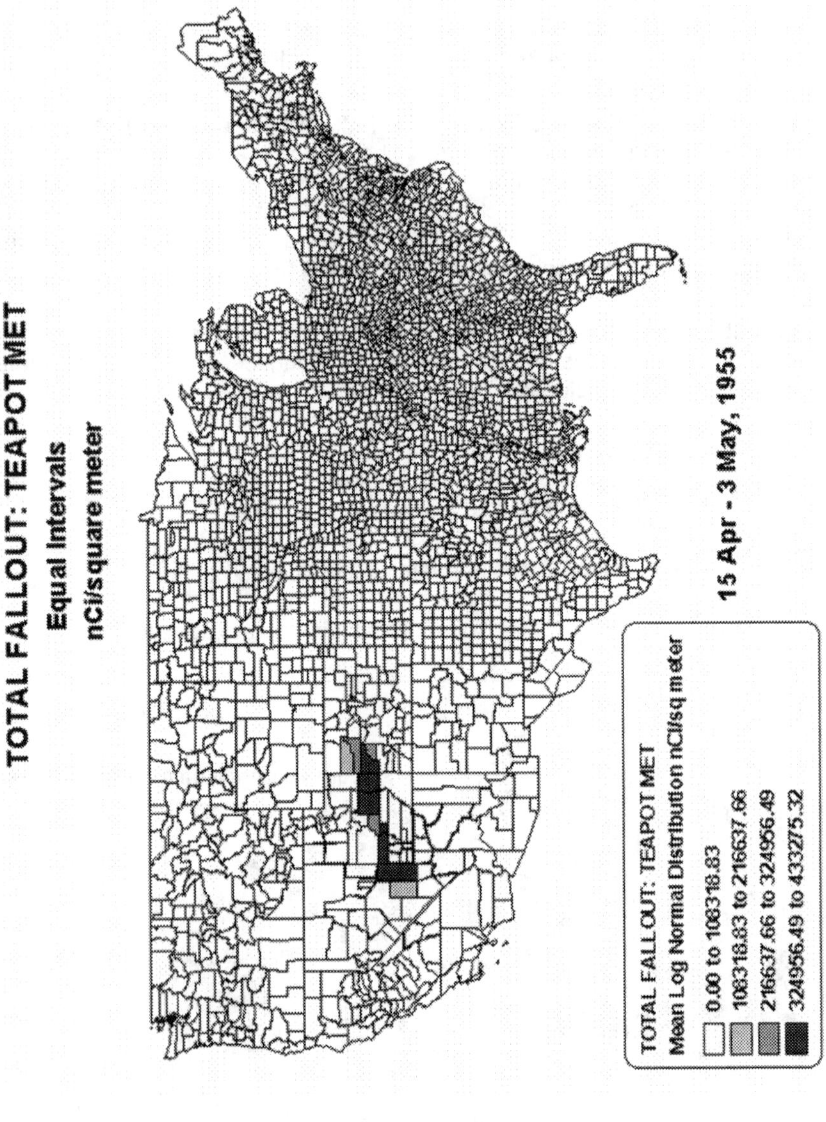
TOTAL FALLOUT: TEAPOT MET
Equal Intervals
nCi/square meter
15 Apr - 3 May, 1955
TOTAL FALLOUT: TEAPOT MET
Mean Log Normal Distribution nCi/sq meter
0.00 to 108318.83
108318.83 to 216637.66
216637.66 to 324956.49
324956.49 to 433275.32

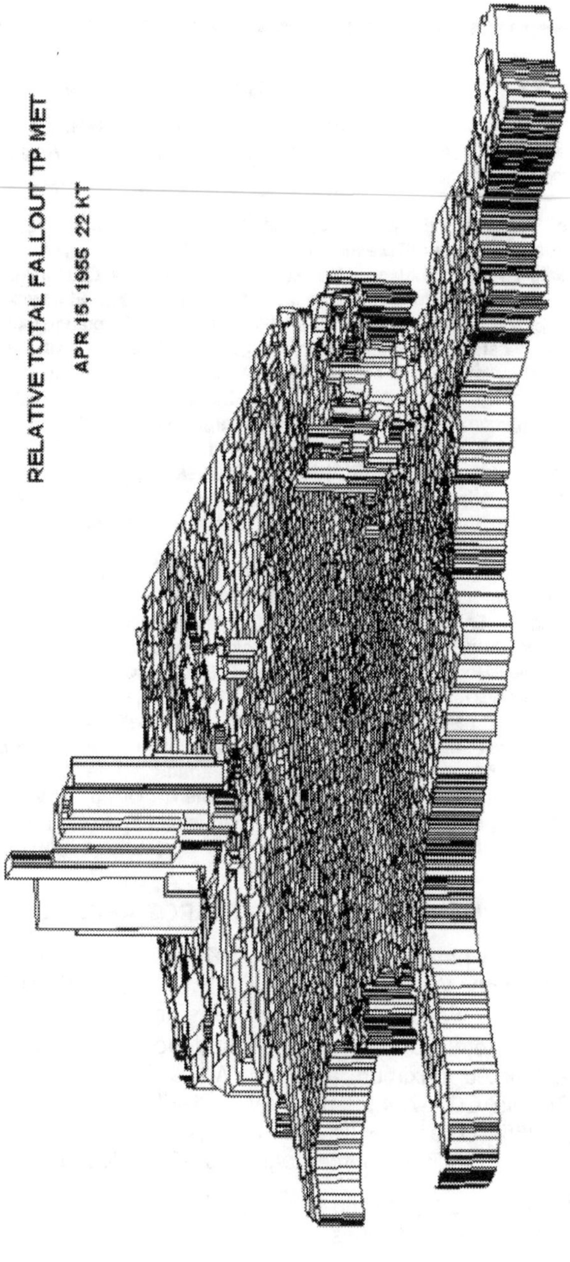
RELATIVE TOTAL FALLOUT TP MET
APR 15, 1955 22 KT

SHOT 10: TEAPOT APPLE-2 (TP10)

TEAPOT: APPLE-2

Detonation Date: 5 May 1955 • *Detonation Time:* 5:10 A.M. • *Area:* 1 • *Sponsor:* Los Alamos • *Yield:* 29 kt • *Radiation Level at Ground Zero:* 500 R/hr • *Height of Burst:* 500 ft • *Cloud Top Height:* 51,000 ft

30,000-Foot Trajectory: NEVADA: (Shoshone Peak), (Belted Range), (Egan Range), (Snake Range), (Humbolt National Forest) • UTAH: Trout Creek, Faust, Vernon, Fairfield, Oram, Provo, (Daniels Pass), Fruitland, Upalco, Myton, Randlett • COLORADO: Rangely, Rio Blanco, New Castle, Glenwood Springs, Cardiff, Cattle Creek, El Jezebel, Basalt, Snowmass, Woody Creek, Aspen, (Mount Elbert, highest point in Colorado), Twin Lakes, (Weston Pass), Fairplay, Estabrook, Pine, Foxton, South Platte, Leuviers, Littleton, Bennett, Strasburg, Woodrow, Akron, Platner • NEBRASKA: Champion, Wellfleet, Lexington, Overton, Elm Creek, Kearney, Minden, Norman, Bladen, Superior, Hardy • KANSAS: Republic, Rydal, Wayne, Agenda, Clyde, Vining, Clifton, Morganville, Fort Riley, Manhattan, Ogden, Miller, Osage City, Melvern, Agricola, Waverly, Harris, Garnett, Bush City, Blue Mound, Mantey, Harding, Fulton • MISSOURI: Richard, Deerfield, Nevada, Milo, Montevallo, Jerico Springs, Arcola, Dadeville, Walnut Grove, Ash Grove, Williard, Springfield, Nichols, Battlefield, Nixa, Linden, Ozark, Sparta, Chadwick, Bradleyville, Brownbranch, Longrun, Theodosia, Isabella, Howards Ridge • ARKANSAS: Gemalel, Viola, Mitchell, Wiseman, Sidney, Mount Pleasant, Cave City, Saffell, Dowdy, Egypt, Cash, Otwell, Bay, Caraway, Trumann, Keiser, Marie, Driver, Osceola • TENNESSEE: Garland, Covington, Stanton, Hillville, Uptonville, Henderson, Enville, Saltillo, Clifton, Center, Webber City, Campbellsville, Wales, Frankewing, McBurg, Fayetteville, Skinem, Coldwater, Bellview, Lincoln, Elora, Anderson, Richard City, New Hope • ALABAMA: Bryant • GEORGIA: New England, (Lookout Mountain), Chickamauga, Tunnel Hill, Elton, Crandell, Cherry Log, Robertstown, Helen, Batesville, Turnerville, Tallulah Falls • SOUTH CAROLINA: Long Creek, Westminster, Seneca, Clemson, Utica, Central, Pelzer, Woodville, Owings, Gray Court, Cross Anchor, Sedalia, Delta, Whitmire, Tuckertown, Shelton, Salemn Crossroads, White Oak, Winnsboro Mills, Langtown, DeKalb, Lucknow, Una, Lydia, Oats, Florence, Quinby, Peedee, Blue Brick, Marion, Zion, Nichols, Mullins • NORTH CAROLINA: Tabor City, Mollie, Nakina, Cruise Island, Winnabow, Bolivia, Boiling Spring Lakes, Seabreeze, Carolina Beach, Kure Beach.

CHARACTERISTICS: TEAPOT APPLE-2

The deposition pattern for Teapot APPLE 2 is most similar to that of Plumbbob shots WHEELER-COULOMB B and LAPLACE (PB13) (R: 0.58; T: 39.53). Fallout from this test was deposited over 3045 counties (98.4% of the total). Counties of highest deposition included (all in μCi/sq meter, estimated maximum activity:): White Pine NV— 533.6 on May 5, 1955; Costilla CO—734.29 on May 6, 1955 and Shelby MO—372 on May 7, 1955. Teapot APPLE-2 ranks 4th in the percentage of Sr-90 produced and 11th in the percentage of Be-7 produced.

The total fallout to I-131 ratio for the APPLE-2 shot is relatively low at 84 to 1. That is only slightly less than that for the APPLE-1 shot at 85 to 1.

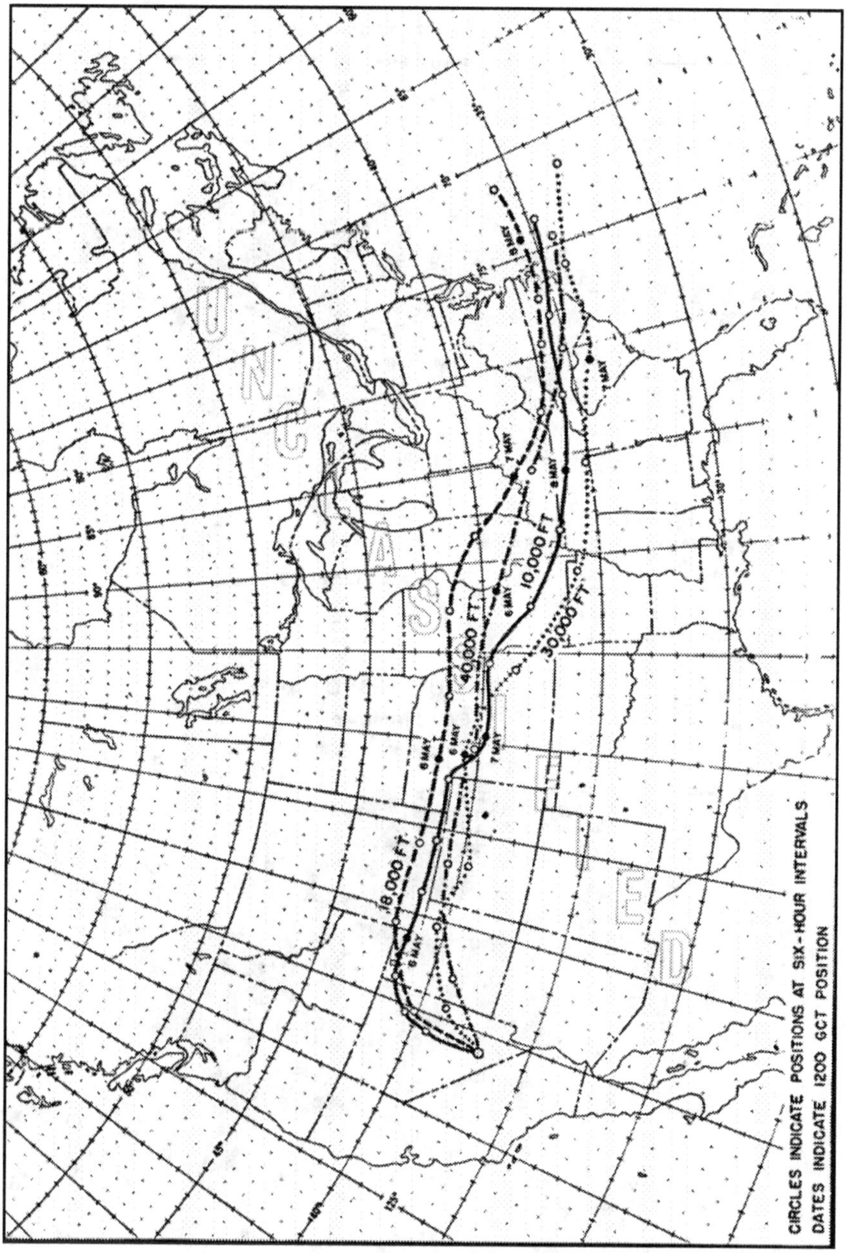

TRAJECTORY: TEAPOT APPLE-2 (TP10)

TOTAL FALLOUT: TEAPOT APPLE-2

Equal Intervals

nCi/square meter

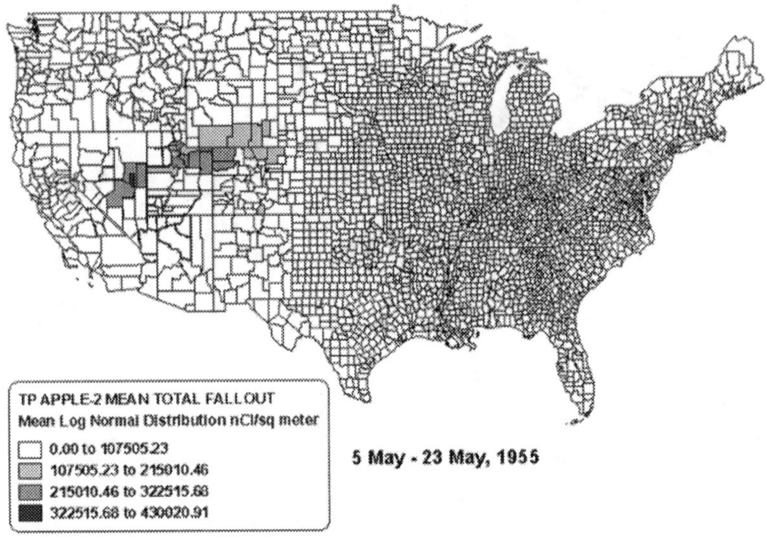

TOTAL FALLOUT: TEAPOT APPLE-2

Equal Number

nCi/square meter

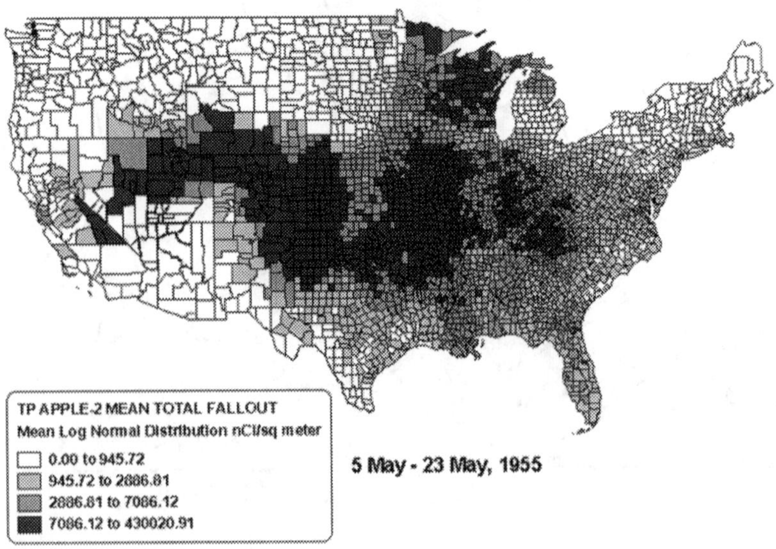

SHOT 11: TEAPOT ZUCCHINI (TP11)

Teapot: Zucchini

Detonation Date: 15 May 1955 • *Detonation Time:* 5:00 A.M. • *Area:* 7–1a • *Sponsor:* Los Alamos • *Yield:* 28 kt • *Radiation Level at Ground Zero:* 500 R/hr • *Height of Burst:* 500 ft • *Cloud Top Height:* 40,000 ft

40,000-Foot Trajectory: NEVADA: (Delmar Mountains) • UTAH: (Morman Range), Central, New Harmony, Cedar City, Summit, Parowan, Panguitch, Teasdale, Grover, Torrey, Bicknell, (Temple Mountain), Green River, (East Tavaputs Plateau), Bonanza • COLORADO: Dinosaur, (Dinosaur National Monument), Greystone, (Vermillion Bluffs) • WYOMING: Table Rock, Wamsutter, Red Desert, (Great Divide Basin), (Green Mountains), Jeffrey City, Gas Hills, Moneta, Lyside, Lost Cabin, Hyattsville, Greybull, Shell, (Bald Mountain), Kane, Lovell, (Horseshoe Bend) • MONTANA: (Bighorn Lake), Crow Indian Reservation, Billings, Lockwood, Huntley, Shepherd, Warden, Klein, Roundup, Forestgrove, Hilger, Christina, Suffolk, Iliad, Eagleton, Rocky Boy • MINNESOTA: Greenbush, Strathcona, (Mud Lake), Goodridge, High Landing, Trail, Gully, Lengby, Ebro, Roy Lake, Ponsford, Snellman, Osage, Wolf Lake, Midway, Evergreen, Hillview, Bluffton, Wrightstown, Eagle Bend, Clotho, Little Sauk, Ward Springs, Melrose, Greenwald, St. Martin, Roscoe, Paynesville, Crow River, Grove City, Litchfield, Greenleaf, Corvuso, Cedar Mills, Hutchinson, Stewart, Brownton, Ferando, New Auburn, Winthrop, Gaylord, Norseland, St. Peter, Kasota, Mankato, Eagle Lake, Smiths Mill, St. Clair, Alma City, Waldorf, Matawan, Minnesota Lake, Freeborn, Manchester, Albert Lea, Twin Lakes, Emmons, Glenville • IOWA: Silver Lake, Kensett, Mankey, Rock Falls, Mason City, Nora Springs, Rockford, Marble Rock, Greene, Packard, Clarksville, Shell Rock, Waverly, Finchford, Cedar Falls, Waterloo, Garrison, Van Horne, Newhall, Norway, Walford, East Amana, Amana, Tiffin, Iowa City, Coralville, Hills, Lone Tree, Conesville, Fredonia, Wapello, Toolesboro, Oakville, Northfield Kingston • ILLINOIS: Oquawka, Gulf Port, Media, Bushnell, New Philadelphia, Ipava, Summum, Bath, Chandlerville, Curran, Chatlam, Glenarm, Pawnee, Morrisonville, Wenonah, Nokomis, Ramsey, St. Elmo, Altamont, Edgewood, La Clede, Iola, Louisville, Clay City, Wynoose, Samsville, Bone Gap, Gards Point, Mount Carmel • INDIANA: East Mount Carmel, Johnson, Owensville, Haubstadt, Warrenton, Elberfeld, Daylight, Chandler, Boonville, Pelzer, Yankeetown, Hatfield, Eureka, Richland, Patronville • KENTUCKY: Owensboro, Philpot, Whitesville, Fordsville, Short Creek, Leitchfield, Grayson Springs, Millerstown, Jonesville, Pikes View, Campbellsville, Hatcher, Yuma, Casy Creek, Pellyton, Dunnville, Mintonville, (Daniel Boone National Forest), Lidu, Bush, Bluehole, Ogle, Brightshade, Roark, Chappell, Gilley, Benham • VIRGINIA: Keokee, Big Stone Gap, East Stone Gap, Fort Blackmore, Nickelsville, Collinwood, Benhams • NORTH CAROLINA: Baldwin, Fleetwood, Moravian Falls, Love Valley, Union Grove, Turnersburg, Harmony, Cooleemee, Churchland, Cid, Denton, Farmer, Utah, Seagrove, Highfalls, Parkwood, Spout Springs, Spring Lake, Fayettevile, Vanda, Stedman, Autreyville, Parkersburg, Ingold, Harrells, Penderlea, Watha, Holly Ridge.

CHARACTERISTICS: TEAPOT ZUCCHINI:

The device used for ZUCCHINI was 40 inches in diameter, 80.5 inches long, weighed 2935 lbs and was detonated atop a 500-ft steel tower. Teapot ZUCCHINI is ranked 2nd for percentages of Br-82, Kr-85 and Sr-89 in the total fallout. The deposition pattern for Teapot ZUCCHINI is most similar to Plumbbob BOLTZMANN (PB1) (R: 0.58; T: 39.44). Fallout from this shot was deposited on 3018 counties, or 98 percent of the total U.S. Counties of highest deposition included Iron UT (estimated maximum activity: 857 μCi/sq meter on May 15 1955) and Clark NV (estimated maximum activity: 406.4 μCi/sq meter on May 15, 1955).

There was some speculation that a rainout associated with debris cloud from ZUCCHINI occurred in Harney County, Oregon shortly after May 15, 1955. However, according to the NCI data, the fallout level at Harney County reached only 30 μCi/sq meter, estimated maximum activity.

The total fallout to I-131 ratio for Teapot ZUCCHINI is relatively low at 85 to 1.

TOTAL FALLOUT: TEAPOT ZUCCHINI

Equal Intervals
nCi/square meter

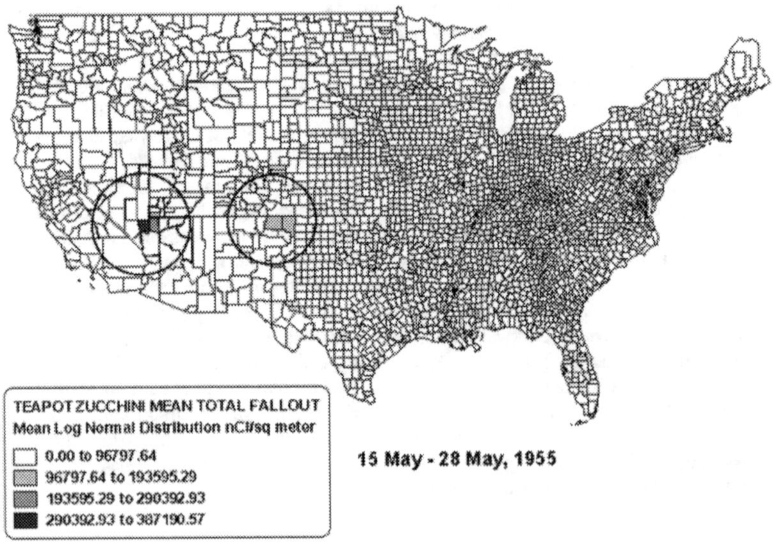

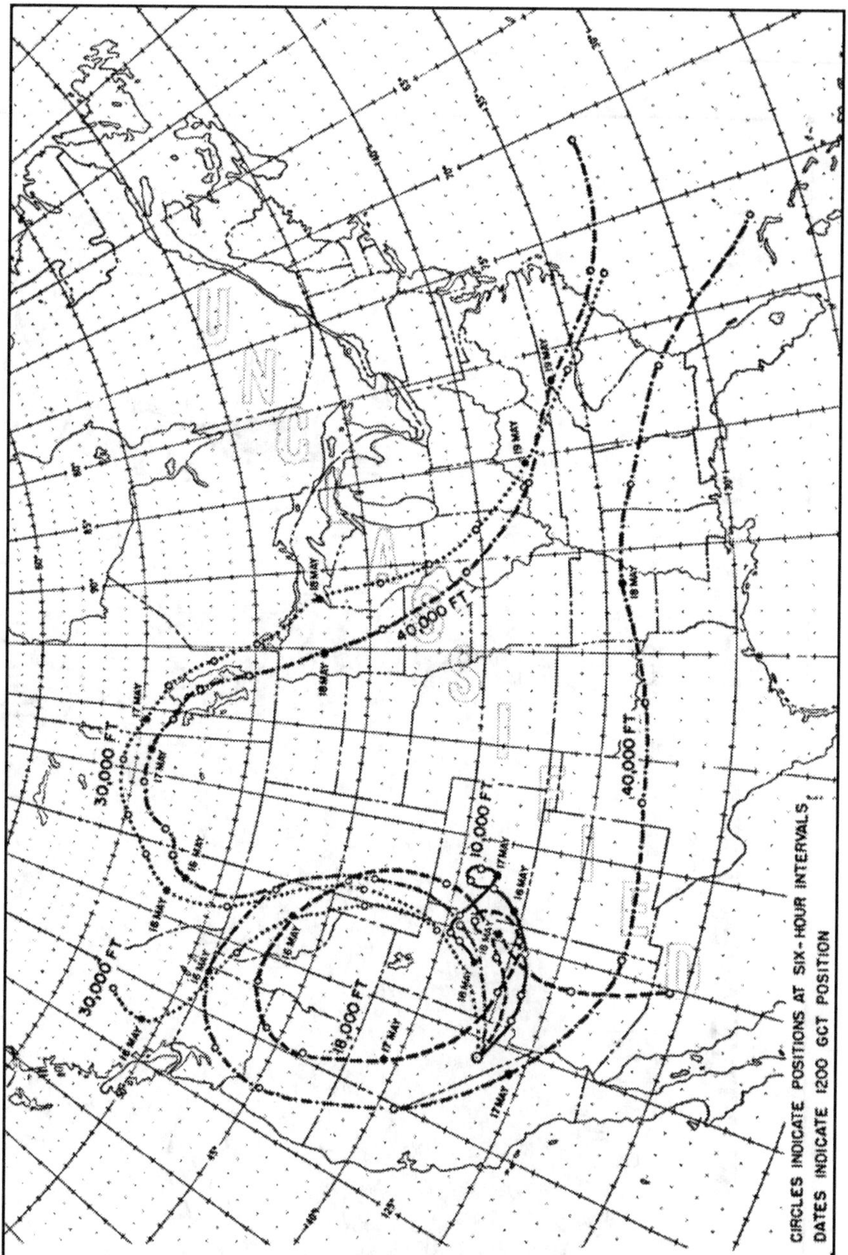

TRAJECTORY: TEAPOT ZUCCHINI (TP11)

RELATIVE AVERAGE TOTAL FALLOUT TP ZUCCHINI
MAY 15, 1955 28 KT
UNION COUNTY, NM
COLFAX COUNTY, NM
CLARK (1) COUNTY, NV

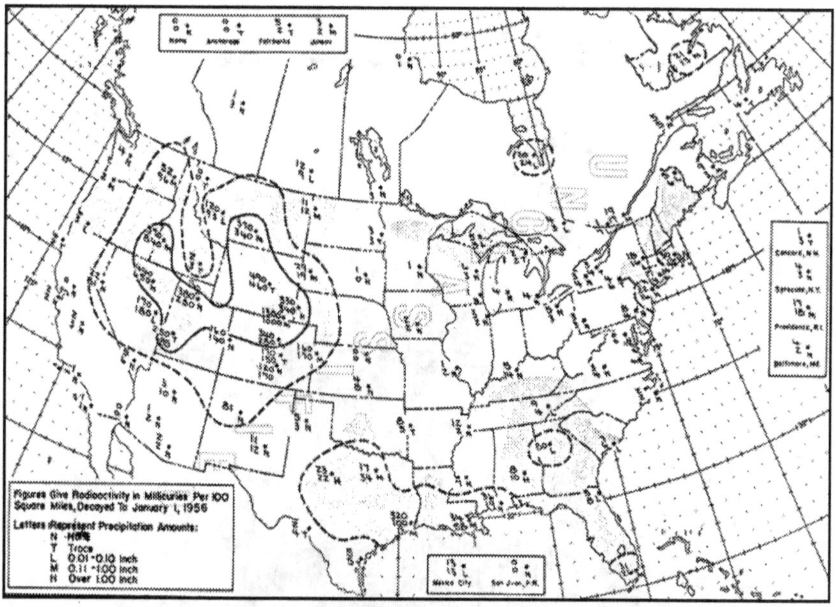

PRECIPITATION AMOUNTS DURING TEAPOT SERIES

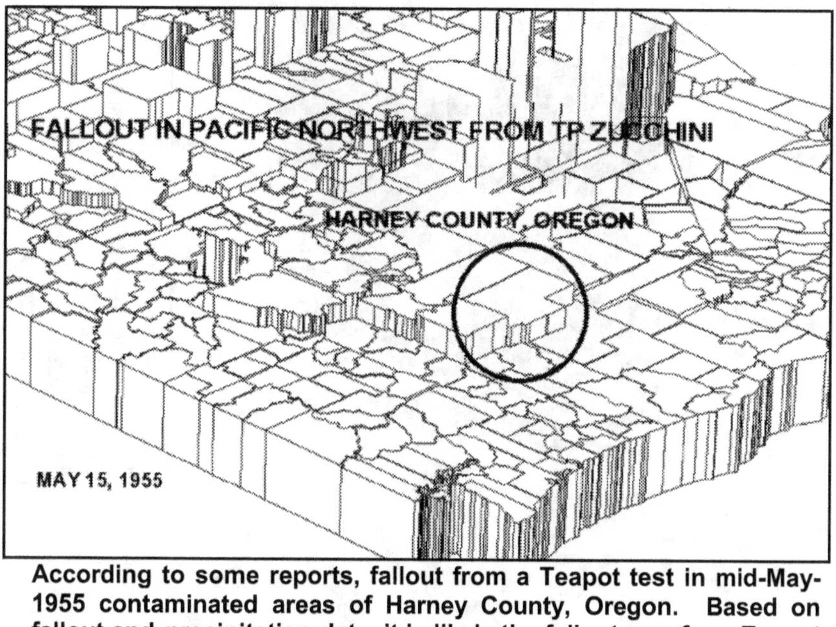

According to some reports, fallout from a Teapot test in mid-May-1955 contaminated areas of Harney County, Oregon. Based on fallout and precipitation data, it is likely the fallout was from Teapot ZUCCHINI.

TOTAL FALLOUT: TEAPOT ZUCCHINI

Equal Number

nCi/square meter

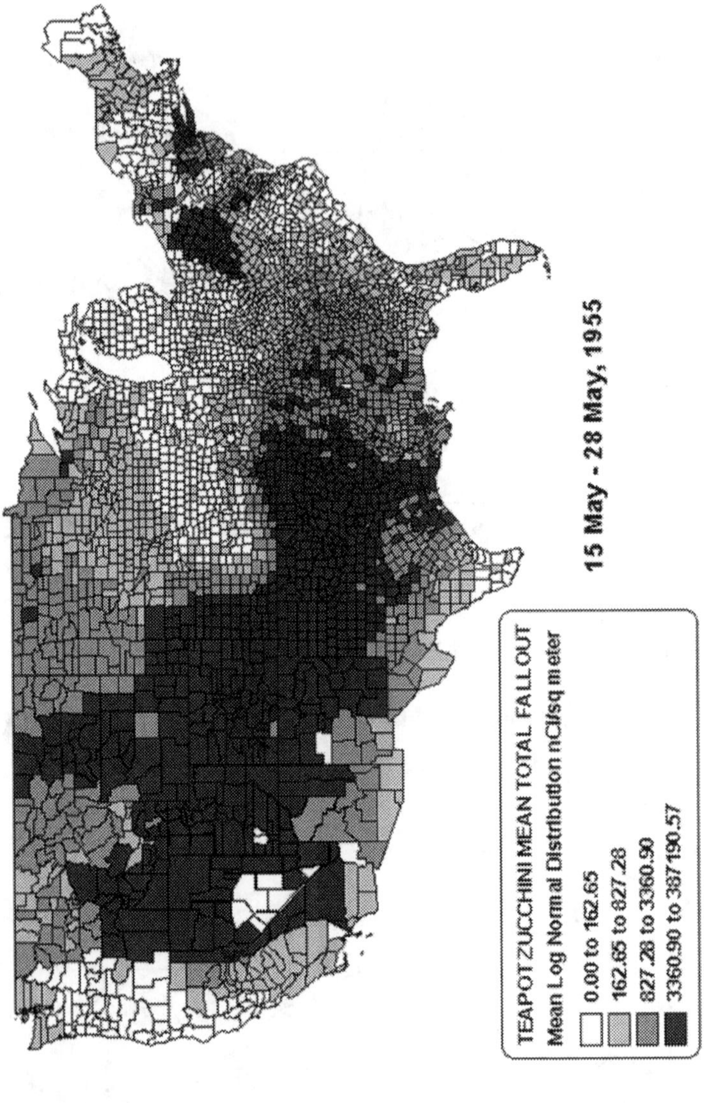

15 May - 28 May, 1955

SECTION 6

PLUMBBOB SERIES

May 28-Oct 7, 1957

SHOT 1: PLUMBBOB BOLTZMANN (PB1)

Plumbbob: Boltzmann

Detonation Date: 28 May 1957 • *Detonation Time:* 5:55 A.M. • *Area:* 7c • *Sponsor:* Los Alamos • *Yield:* 12 kt • *Radiation Level at Ground Zero:* 20 R/hr • *Height of Burst:* 500 ft • *Cloud Top Height:* 33,000 ft

10,000-Foot Trajectory: NEVADA: Warm Springs, Manhattan, Luning, Hawthorne, Wellington, Lake Tahoe • CALIFORNIA: Tahoe Pines, (Squaw Valley Ski Area), Placerville, Loomis, Lincoln, Trobridge, Marysville, Yuba City, Hamilton City, Richfield, Red Bluff, Cottonwood, Redding, Weed, Edgewood, Gazelle, Yreka, Hornbrook, Hilt • OREGON: Ashland, Medford, Shady Grove, Drew, Cottage Grove, Springfield, Albany, Salem, Brooks, Newburg, Beaverton, Hillsboro, Burlington, Columbia City, Rainier • WASHINGTON: Longview, Lexington, Ryderwood, Curtis, Chehalis, Fords Prairie, Littlerock, Shelton, Port Angeles.

RELATIVE TOTAL FALLOUT: PB BOLTZMANN

MAY 28, 1957 10 KT

CHARACTERISTICS: PLUMBBOB BOLTZMANN

The device for the BOLTZMANN shot was 31.6 inches long, 18 inches wide and weighed 295 lbs. The nuclear component itself weighed only 144.6 lbs. Reportedly, the shot cab atop the 500-ft tower was weighted with 9 tons of sand and 15 tons of paraffin as a means to shield instruments. Among the nuclear tests, Plumbbob BOLTZMANN ranks 5th in percentage of Cs137 in fallout The deposition pattern for Plumbbob BOLTZMANN is most similar to Teapot ZUCCHINI (TP11) (R: 0.58; T: 39.44). Fallout from this nuclear test was deposited over 2781 U.S. counties. The counties of highest deposition included Nye NV (estimated maximum activity: 1419 μCi/sq meter); Lander NV (estimated maximum activity: 598 μci/sq meter) and Mineral NV (estimated maximum activity: 585 μCi/sq meter) all on May 28, 1957. The total fallout to I-131 ratio for this shot was the 4th lowest of all above-ground nuclear tests at 74:1.

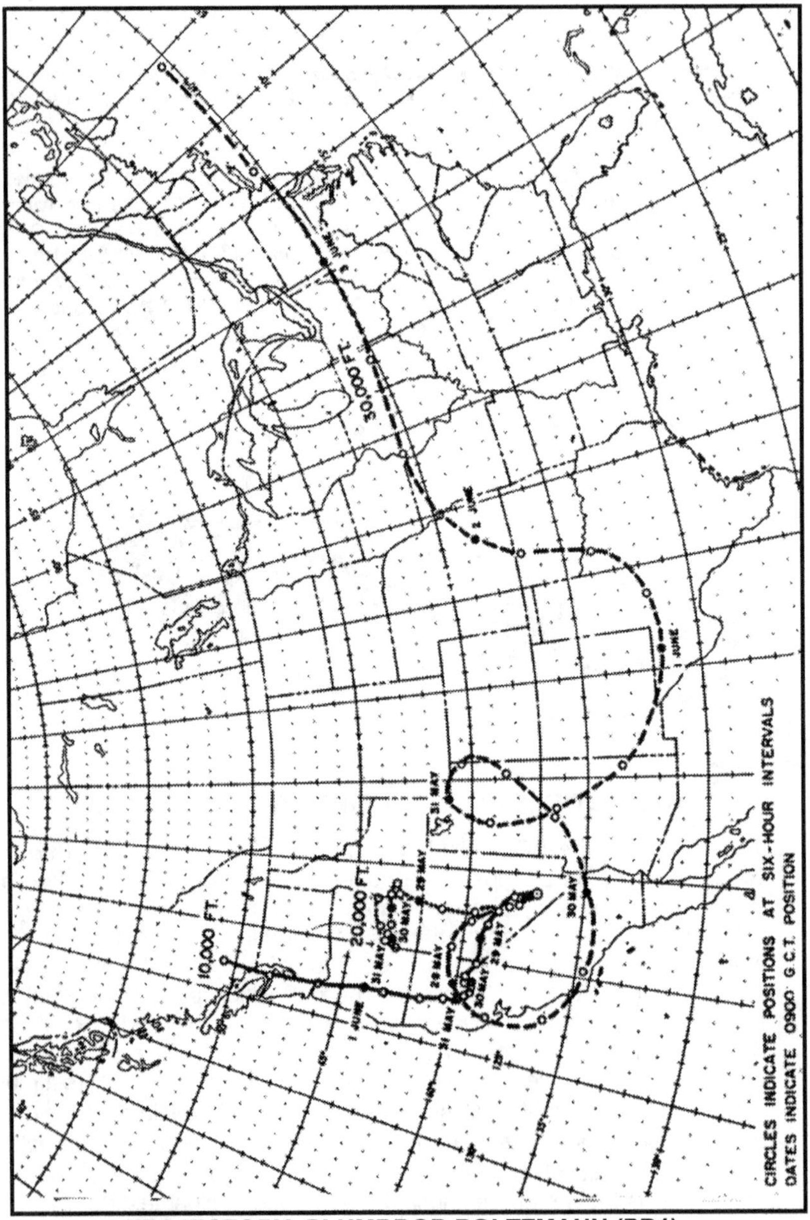

TRAJECTORY: PLUMBBOB BOLTZMANN (PB1)

TOTAL FALLOUT: PLUMBBOB BOLTZMANN

Equal Intervals

nCi/square meter

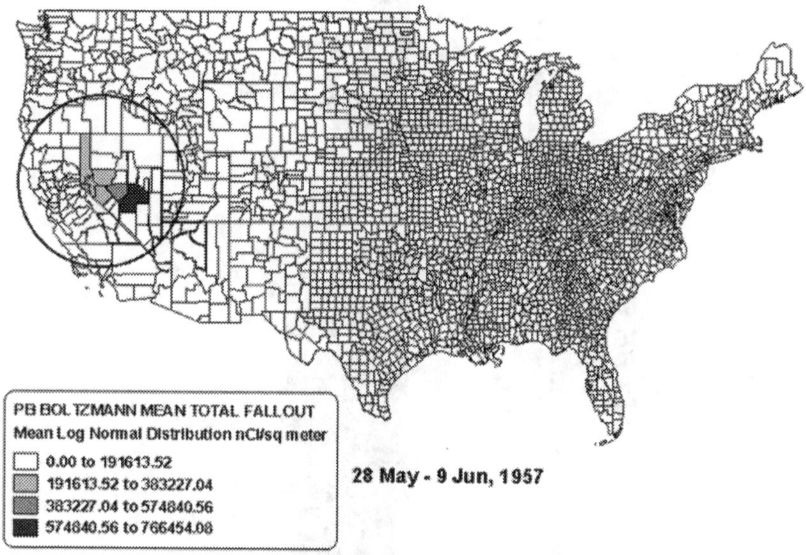

TOTAL FALLOUT: PLUMBBOB BOLTZMANN

Equal Number

nCi/square meter

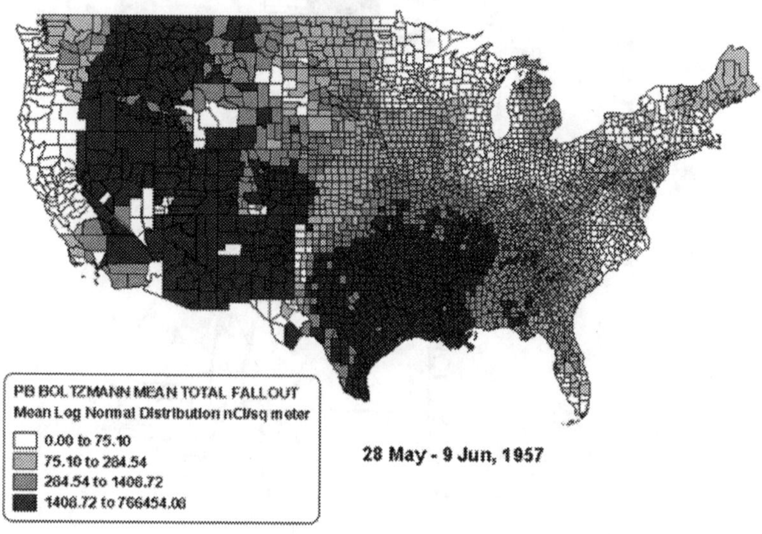

RELATIVE TOTAL FALLOUT: PB BOLTZMANN

MAY 28, 1957 10 KT

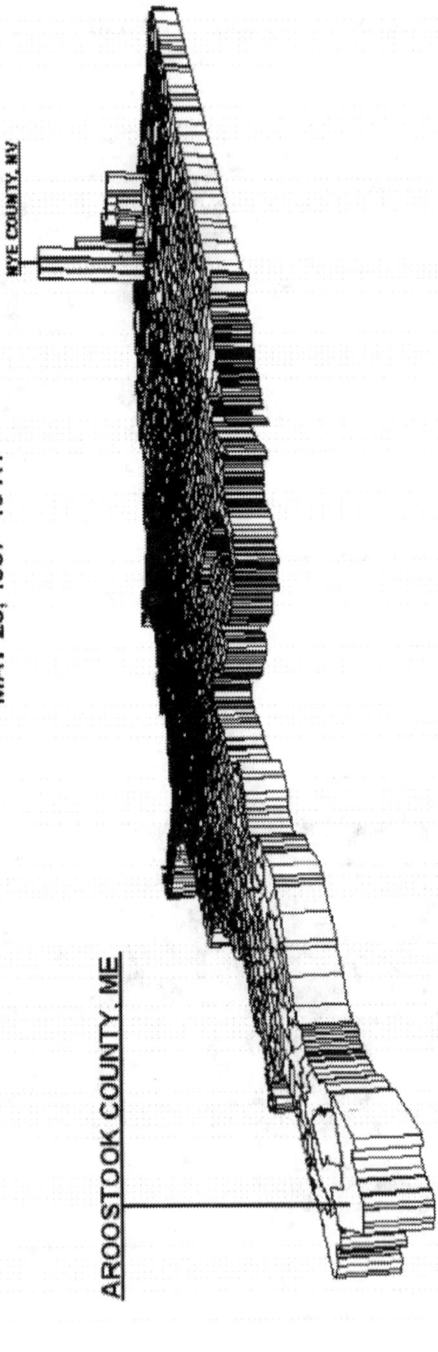

SHOT PLUMBBOB FRANKLIN

(Not included in NCI Study)

PLUMBBOB: FRANKLIN

Detonation Date: 2 Jun 1957 • *Detonation Time:* 4:55 A.M. • *Area:* 3 • *Sponsor:* Los Alamos • *Yield:*140 tons • *Radiation Level at Ground Zero:* 35 R/hr • *Height of Burst:* 300 ft • *Cloud Top Height:* 16,700 ft

16,000-Foot Trajectory: NEVADA: (Belted Range), (Egan Range) • UTAH: (Frisco Peak), Minersville, Enoch, Cedar City, (Zion National Park), Rockville, Hilldale • ARIZONA: Colorado City, (Mount Trumbull), Hualapai Indian Reservation, Nelson, Yucca • CALIFORNIA: Needles, (Sada Mountains), (Tiefort Mountains), Argos, Borosolvay, Dunmovin, Olancha, Cartago, (Mount Whitney), (Mount Woodworth), (Florence Lake), Mono, Hot Springs, (Ritter Range), (Eagle Peak), (Sweetwater Mountains) • NEVADA: Wellington, Smith, Wabuska, Silver Springs, Fernley, Nixon, (Seven Troughs Range), (Jackson Mountains), (Disaster Peak) • OREGON: (Trout Creek Mountains), Rome, Arock, Danner • IDAHO: Melba, Kuna, Bowmont, Mora, Boise, Idaho City, Stanley, Sunbeam, Bonanza, Challis, May, Leadore, (Continental Divide) • MONTANA: Bell, Lima, (Beaverhead National Forest), Raynolds Pass, West Yellowstone • WYOMING: (Mount Holmes), Canyon, Pelican Cone, (Sleeping Giant Mountain), (Buffalo Bill Dam), Cody, (Elk Butte), Manderson, Ten Sleep, Sussex, (Thunder Basin National Grassland) • NEBRASKA: Chadron, Hay Springs, (Big Hill), Broken Bow, Litchfield, Poole, Wood River, Diniphan, Hansen, Trumbull, Harvard, Saronville, Sutton, Geneva, Milligan, Tobias, Swanton, Plymouth, Beatrice, Filley, Elk Creek, Howe, Stella, Shubert, Barada • MISSOURI: Craig, Corning, Maryville, Skidmore, Arkoe, Conception Junction, Gentry, Bethany, Modena, Spickard, Osgood, Milan, Green Castle, Novenger, Pure Air, Kirskville, Hardland, Edina, Knox City, La Belle, Monticello, Canton • ILLINOIS: Meyer, Mendon, Camp Point, Clayton, Mount Sterling, Ripley, Bluff Spring, Philadelphia, Ashland, Pleasant Plains, Springfield, Riverton, Mechanicsburg, Decatur, Elwin, Mt. Zion, Lake City, Lovington, Arthur, Arcola, Bushton, Kansas, Paris, Grandview, Elbridge, Vermillion, Dennison • INDIANA: St. Mary-of-the-Woods, North Terre Haute, Brazil, Harmony, Reelsville, Manhattan, Belle Union, Crown Center, Hall, Bethany, Waverly, Stones Crossing, New Whiteland, Whiteland, Needam, Shelbyville, Rays Crossing, Blue Ridge, New Salem, Milroy, Richland, Andersonville, Laurel, Blooming Grove, Bath, Mixerville • OHIO: Morning Sun, West Elkton, Greenbush, Franklin, Springboro, Lytle, Ridgeville, Corwin, Kingman, Reeseville, Sabina, Staunton, Good Hope, Austin, Greenland, Andersonvile, Kinnickinnick, South Bloomingville, Hue, Mount Pleasant, Orland, New Plymouth, The Plains, Athens, Canaanville, Stewart, Little Hocking • WEST VIRGINIA: Parkersburg, Cedar Grove, Belmont, Wick Mountain, Alma, Lima, Alvy, Smithfield, Mannington, Arnettsville, Westover, Morgantown, Hazelton • PENNSYLVANIA: Markleysburg, Addison, Springs, West Salisbury, Boynton, Salisbury, Pocahontas, Pleasant Union, Wellersburg, Purcell, Inglesmith, Buck Valley, Plum Rush, Sylven, Claylick, Welsh Run, Shady Grove, Waynesboro, Fountain Dale, Green Mounds, Barlow • MARYLAND: Silver Run, Union Mills, Hampstead, Whitehouse, Butler, Sparks, Sunnybrook, Sweet Air, Perry Hall, White Marsh, Chase, Newtown, Melitota, Chestertown, Barclay, Roberts, Mount Zion, Henderson • DELAWARE: Petersburg, Sandtown, Felton, Masten Corner, Chestnut Knoll, Milford, Lincoln, Milton, Nassau, Harbeson, Angola, Seabreeze, Cottonpatch Hill.

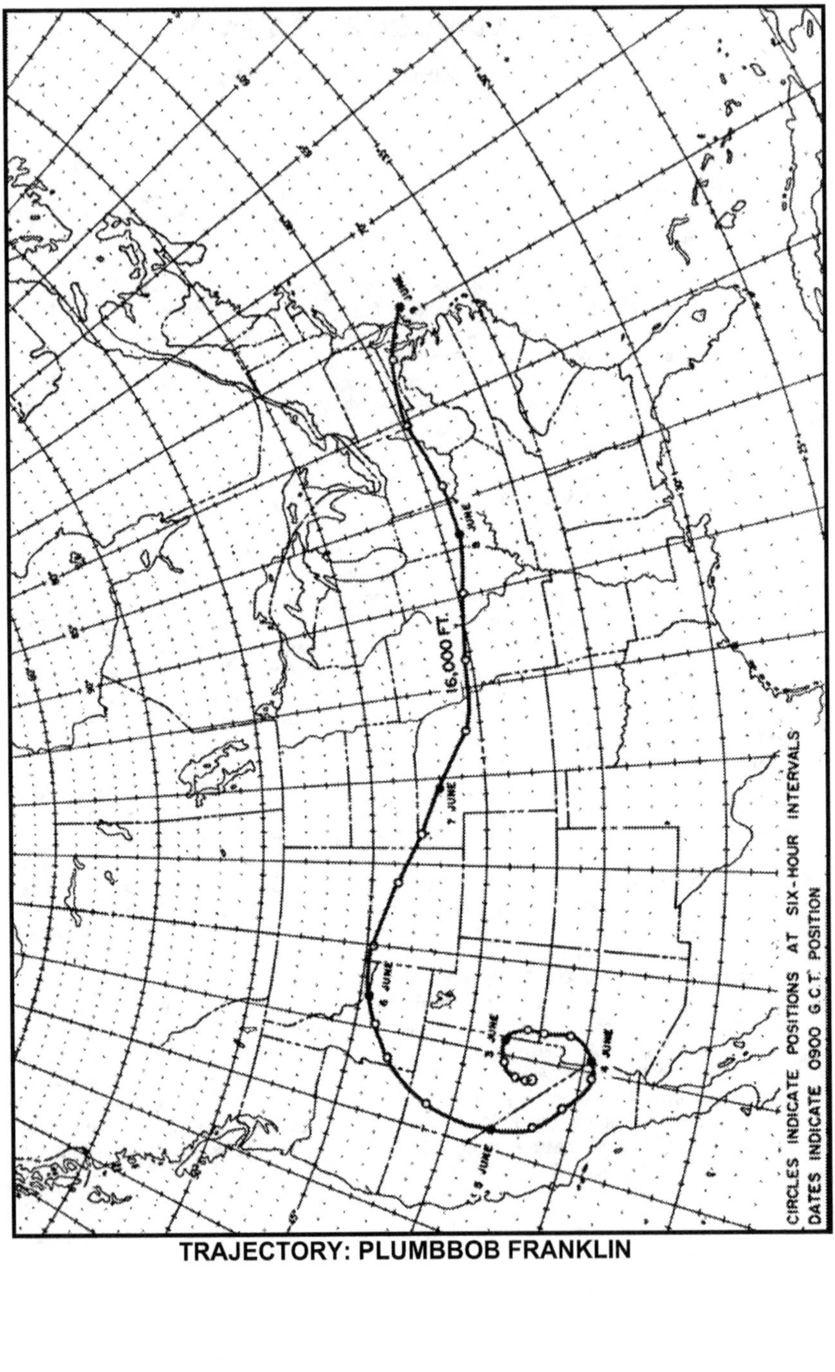

TRAJECTORY: PLUMBBOB FRANKLIN

SHOT 2: PLUMBBOB WILSON (PB2)

Plumbbob: Wilson

Detonation Date: 18 Jun 1957 • *Detonation Time:* 4:55 A.M. • *Area:* 9a • *Sponsor:* Livermore Laboratory • *Yield:* 10 kt • *Radiation Level at Ground Zero:* 20 R/hr • *Height of Burst:* 500 ft • *Cloud Top Height:* 35,000 ft

10,000-Foot Trajectory: NEVADA: Lathrop Wells, (Funeral Mountains) • CALIFORNIA: (Death Valley), (Last Chance Range) • NEVADA: (Magruder Mountain), Lida, (Montezuma Peak), Tonopah, (Pinion Peak), (Pancake Range), (Mount Hamilton), Steptoe, Tippett • UTAH: Ibapah, (Great Salt Lake Desert), Grantsville, Salt Lake City, Layton, Syracuse, Clearfield, Roy, Ogden, Laketown • WYOMING: Cokeville, Calpert, La Barge, (Continental Peak), (Great Divide Basin), Bairvil, Muddy Gap, (Emigrant Gap), Casper, Clareton, Newcastle, Osage, Four Corners • SOUTH DAKOTA: Cheyenne Crossing, Deadwood, Lead, Sturgis, Stoneville, Faith, Red Elm, Glencross, Trail City, Mobridge, Glenham, Eureka • NORTH DAKOTA: Venturia, Ashley, Fredonia, Jud, Nortonville, Millarton, Montpelier, Ypsilanti, Eckelson, Rogers, Dazey, Sibley, Cooperstown, Sharon, Aneta, Northwood, Kempton, Karimore, Arvilla, Honeyford, Gilby, McKinock, Manvel, Ardoch • MINNESOTA: Oslo, Aluorado, Argyle, Strandquist, Strathcona, Wannaska, Pencer, Roosevelt, Wheelers Point.

CHARACTERISTICS: PLUMBBOB WILSON

Among the nuclear tests, Plumbbob WILSON is ranked 2nd for percentage of Cm242 in fallout. The deposition pattern for this shot is most similar to that for Upshot-Knothole CLIMAX (UK11) (R: 0.26, T: 15.21). Fallout from this test was deposited over 2689 counties (87% of the total). This shot ranked 8th in amount of Mn54 deposition across the U.S. The total fallout to I-131 ratio was moderately high at 101:1.

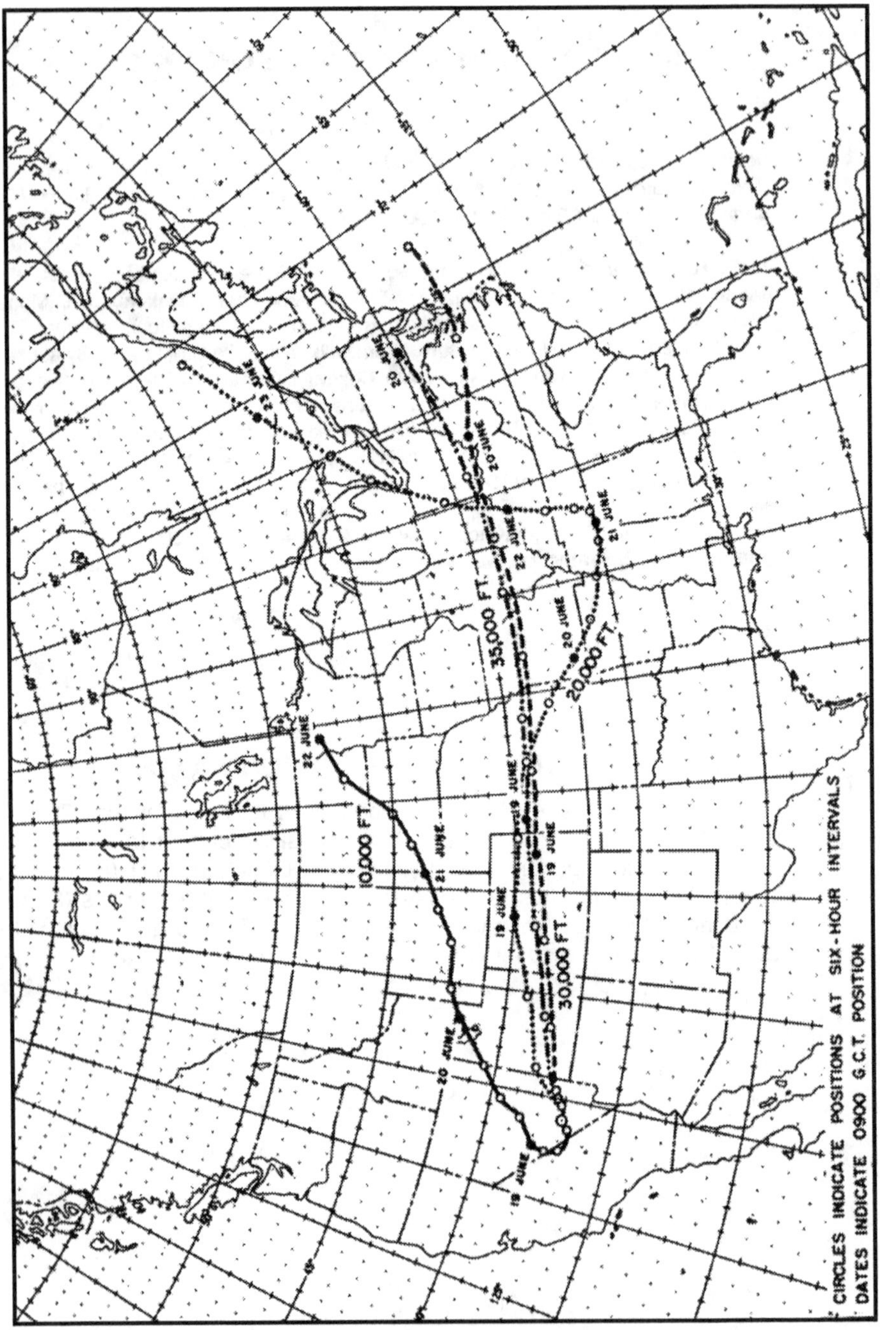

TRAJECTORY: PLUMBBOB WILSON (PB2)

RELATIVE TOTAL FALLOUT: PB WILSON

JUN 18, 1957 10 KT

FALLOUT FROM PB WILSON JUN 18, 1957

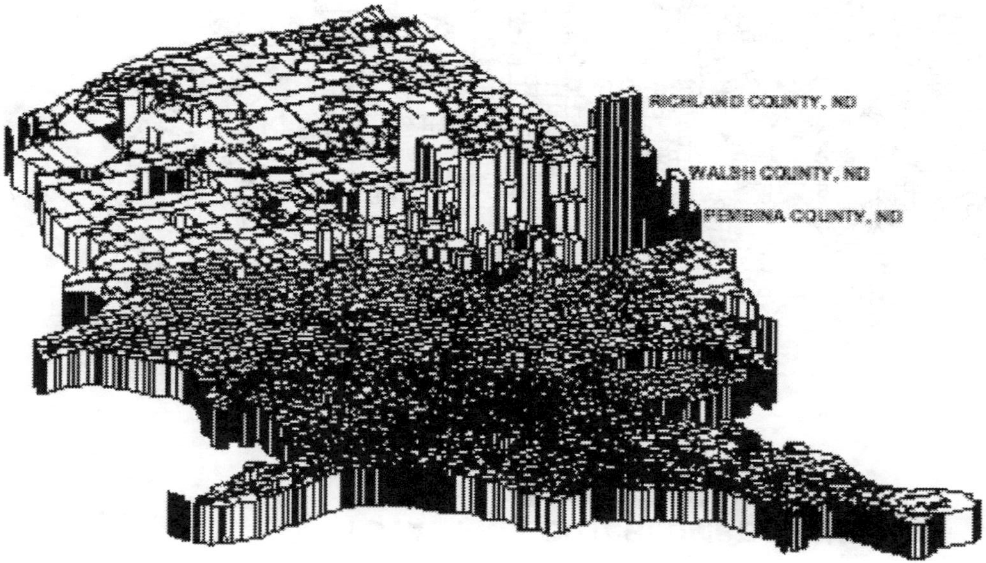

TOTAL FALLOUT: PLUMBBOB WILSON

Equal Intervals

nCi/square meter

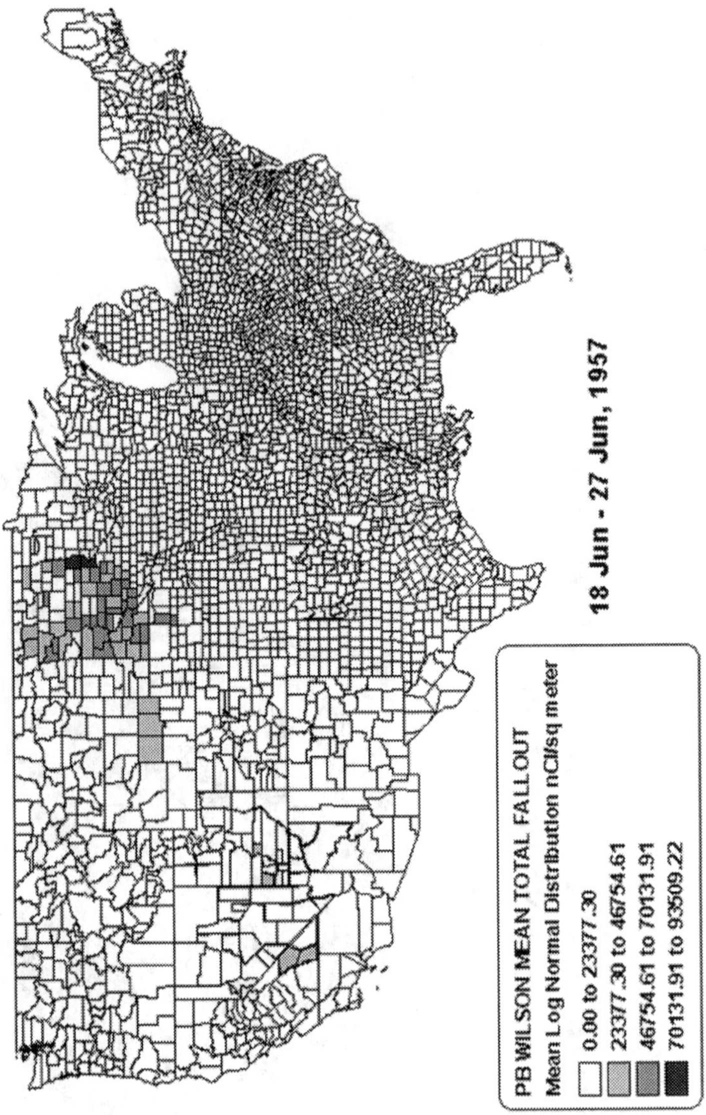

18 Jun - 27 Jun, 1957

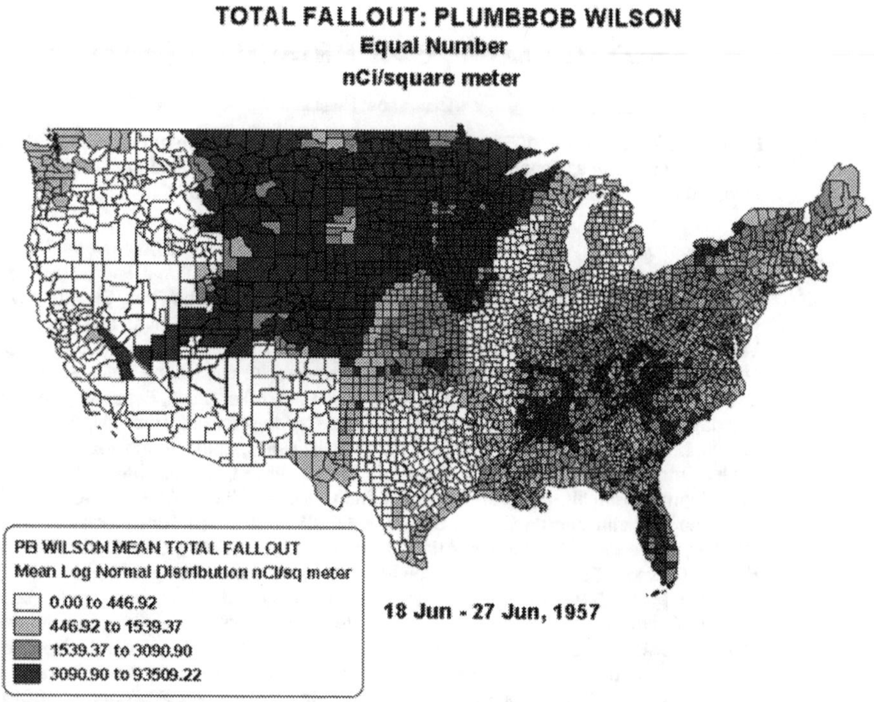
TOTAL FALLOUT: PLUMBBOB WILSON
Equal Number
nCi/square meter
PB WILSON MEAN TOTAL FALLOUT
Mean Log Normal Distribution nCi/sq meter
0.00 to 446.92
446.92 to 1539.37
1539.37 to 3090.90
3090.90 to 93509.22
18 Jun - 27 Jun, 1957

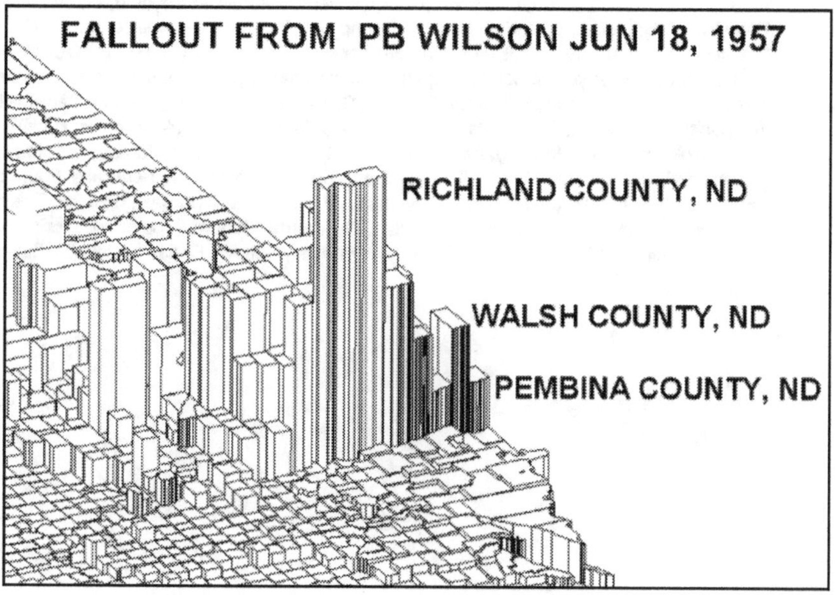
FALLOUT FROM PB WILSON JUN 18, 1957
RICHLAND COUNTY, ND
WALSH COUNTY, ND
PEMBINA COUNTY, ND

SHOT 3: PLUMBBOB PRISCILLA (PB3)

PLUMBBOB: PRISCILLA

Detonation Date: 24 Jun 1957 • *Detonation Time:* 6:30 A.M. • *Area:* Frenchman's Flat • *Sponsor:* Los Alamos, Department of Defense • *Yield:* 37 kt • *Radiation Level at Ground Zero:* 500 R/hr • *Height of Burst:* 700 ft • *Cloud Top Height:* 43,000 ft

40,000-Foot Trajectory: NEVADA: Alamo, Ash Springs, Caliente • UTAH: Uvada, Moderna, Beryl, Circleville, Junction, Loa, Lyman, Bicknell, (Canyons National Park), Dead Horse Point, La Sal • COLORADO: Bedrock, Mount Wilson, Lizard Head Pass, San Juan Mountains, Conejos Peak, (Rio Grande National Forest), Ortiz • NEW MEXICO: Los Pinos, Cerro, Valdez, Eagle Nest, Arroyo Ceso, Angel Fire, Black Lake, Ocate, Wagon Mound, Roy, Solano, Mosquero, Galegos • TEXAS: Glenrio, Hereford, Arney, Happy, Tulia, Rock Creek, Silverton, Gasoline, Flomot, White Flat, Paducah, Sneedville, Chalk, Benjamin, Rhineland, Munday, Throckmorton, Woodson, Crystal Falls, Caddo, Frankell, Huckaby, Stephenville, Selden, Duffau, Iredell, Meridian, Cranfill's Gap, Valley Mills, China Spring, Waco, Bellmead, Riesel, Otto, Ben Hur, Groesbeck, Thornton, Oletha, Jewett, Donie, Buffalo, Crockett, Latexo, Kennard, Ratcliff, Clawson, Central, Lufkin, Redland, Etoile, Chireno, Norwood, Penning, San Augustine, Fords, Corner, Sexton • LOUISIANA: Zwole, Natchitoches, Robaline, Natchez, Hazlewood, St. Maurice, Atlanta, Packton, Urania, Kelley, Olla, Clarks, Rosefield, Holum, Grayson, Fort Necessity, Extension, Jigger, Chase, Newlight, Somerset, Afton, Mound • MISSISSIPPI: Edwards, Oakley, Bolton, Raymound Forest, Jackson, Pelhatchie, Morton, Kalem, Forest, Conehatla, Lake, Decatur, Duffee, Collinsville, Susqualena, Martin, Obadiah, Daleville, Lauderdale • ALABAMA: Sumpterville, Epes, Forkland, Greensboro, Heiberger, Stanton, Maplesville, Fairview, Coopers, Verbena, Mountain Creek, Equality, Nixburg, Cottage Grove, Jacksons Gap, Dadeville, Lafayette, Buffalo, Fredonia • GEORGIA: La Grange, Stovall, Greenville, Gay, Woodbury, Molena, Meansville, Zebulon, Aldora, Milner, Barnesville, Forsyth, Juliette, East Juliette, Round Oak, Wayside, Devereaux, Sparta, Calverton, Jewell, Dearing, Boneville, Harlem, Grovetown, Martinez • SOUTH CAROLINA: Augusta, Graniteville, Aiken, Vaucluse, Wagener, New Holland, Thor, Pelion, Gaston, Gadsden, Eastover, Statesburg, Horatio, Dalzell, Oswego, St. Charles, Lynchburg, Atkins, Cartersville, Timmonsville, Florence, Quinby, Peedee, Blue Brick, Sellers, Floyd Dale, Kemper • NORTH CAROLINA: Orrum, Bladenburg, Abbottsburg, Butters, Elizabethtown, Tomahawk, Harrells, Delway, Wallace, Teachey, Rose Hill, Greenovers, Chinquapin, Lyman, Beulaville, Fountain, Catherine Lake, Richlands, Petersburg, Belgrade, Pollocksville, Riverdale, Croatan, Rhems, New Bern, Minnesott Beach, Arapahoe, Olympia, Merritt, Stonewall, Bayboro, Mesic, Florence, Hoboken, Pamlico Point, Bluff Point, New Holland, Gulrock, Rodantha, Waves, Salvo.

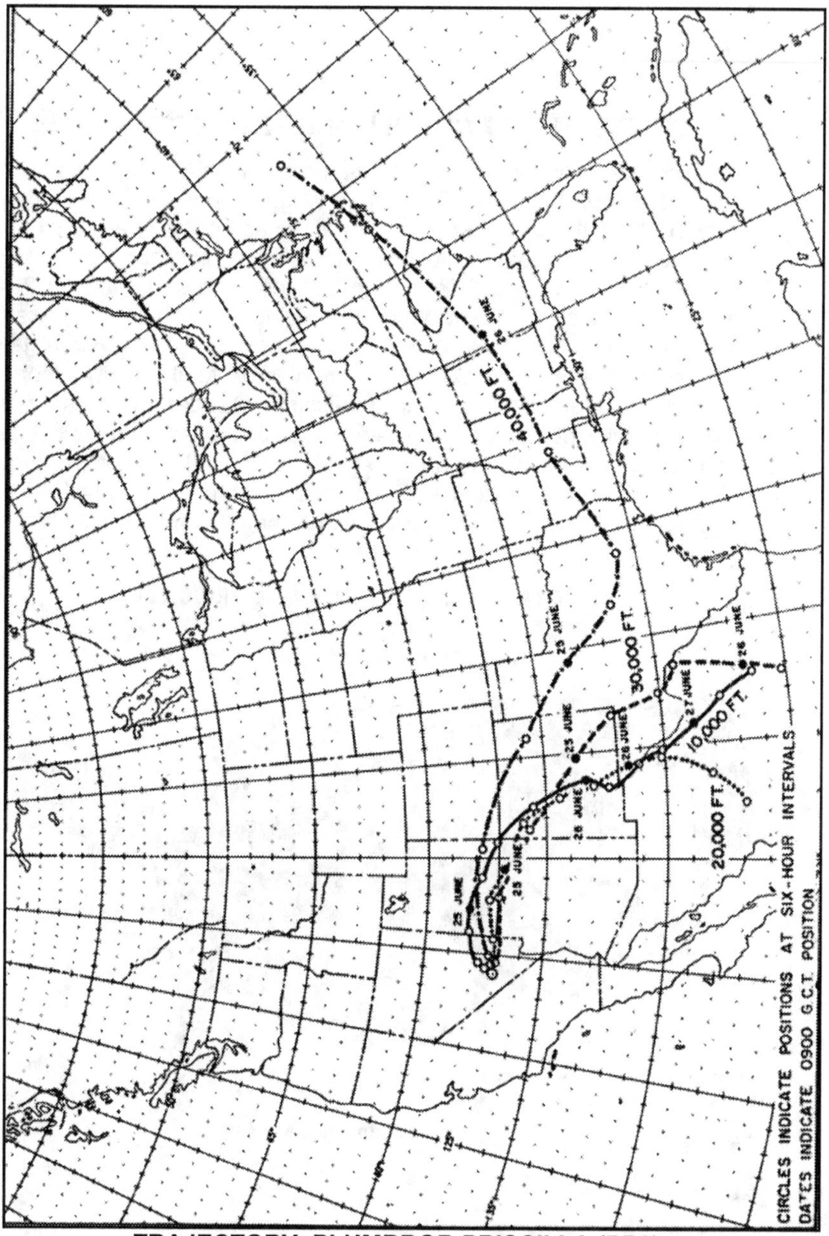

TRAJECTORY: PLUMBBOB PRISCILLA (PB3)

CHARACTERISTICS: PLUMBBOB PRISCILLA

The device used in the PRISCILLA test weighed 581.4 lbs, was 27.1 inches in diameter, and was 27.1 inches long. Among the nuclear tests, Plumbbob PRISCILLA is ranked 2nd for percentage of Na24 and Ce144 in fallout. The deposition pattern for this shot is most similar to Teapot MET (TP9) (R: 0.09, T: 5.05). Fallout from this test was deposited on 2941 counties. Counties of highest deposition included Avoyelles LA, Acadia LA, Wilkinson MS, West Feliciana LA, Assumption, LA and Beauregard, LA .(all approximately 350 μCi/sq meter *average* total fallout). Hottest shot days included the following Louisiana counties with estimated *maximum* activity of 702 μCi/sq meter: Acadia, Assumption, Avoyelles, Beuregard, Calcasieu, Cameron, East Baton Rouge, Evangeline, Grant and Iberia. Fallout from this shot covered 2941 counties (94% of the total). The total fallout to I-131 ratio for this shot was moderately high at 115:1.

TOTAL FALLOUT: PLUMBBOB PRISCILLA
Equal Number
nCi/square meter

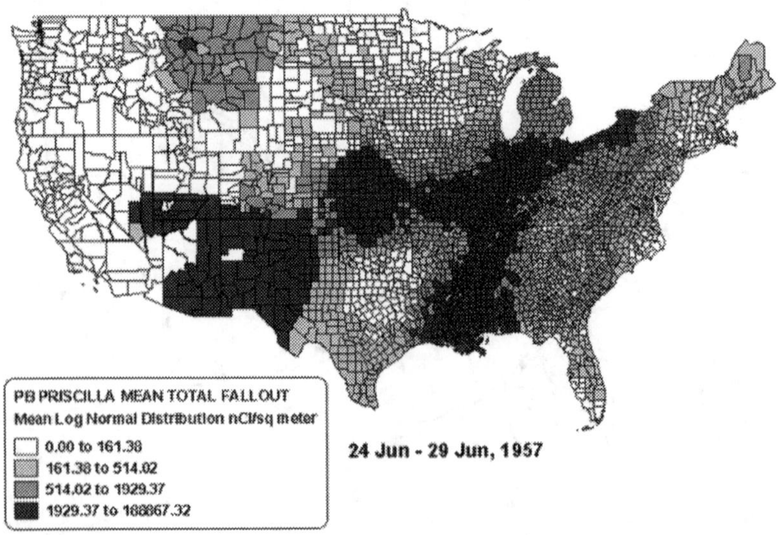

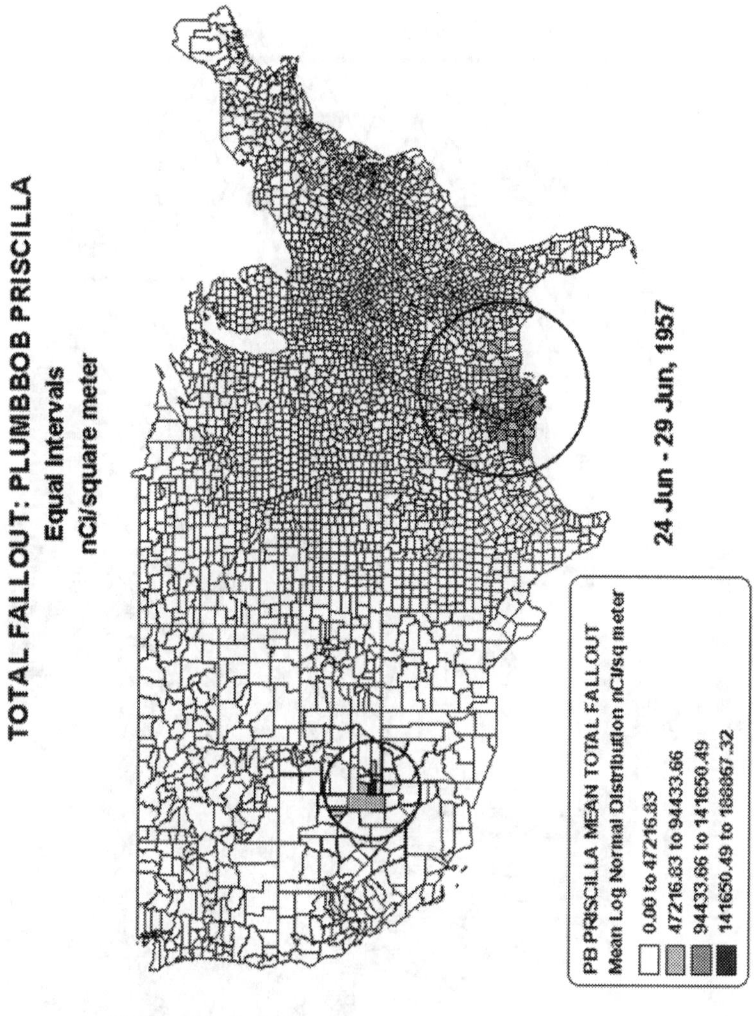

RELATIVE TOTAL FALLOUT: PB PRISCILLA

JUN 24, 1957 37 KT

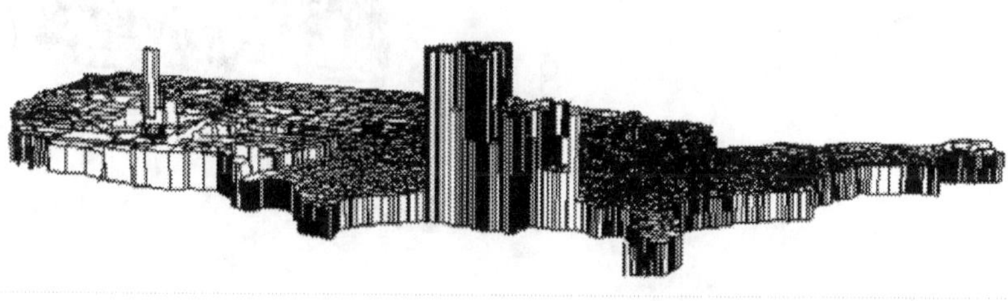

HIGH FALLOUT DEPOSITION FROM PB PRISCILLA

JUNE 24, 1957

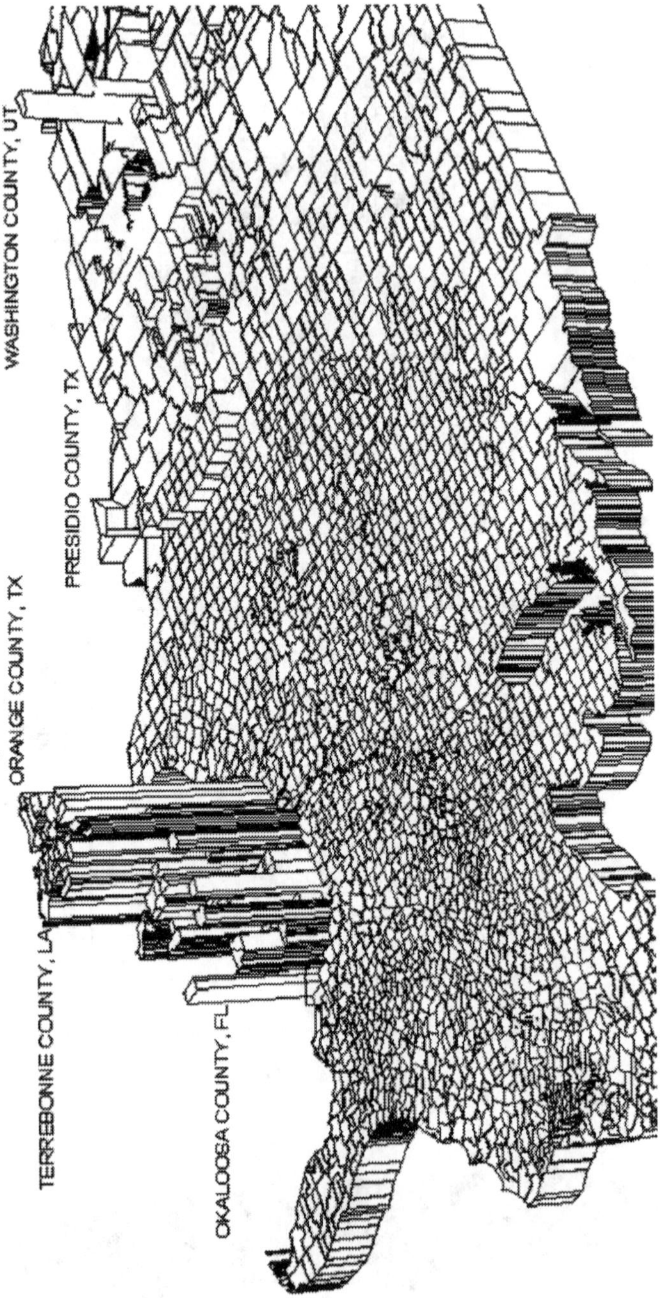

SHOT 4: PLUMBBOB HOOD (PB4)

Plumbob: Hood

Detonation Date: 5 Jul 1957 • *Detonation Time:* 4:40 A.M. • *Area:* 9a • *Sponsor:* Livermore Laboratory • *Yield:* 74 kt • *Radiation Level at Ground Zero:* 100 R/hr • *Height of Burst:* 1,500 ft • *Cloud Top Height:* 48,000 ft

10,000-Foot Trajectory: NEVADA: (Railroad Valley), Duckwater, Duckwater Indian Reservation, (Butte Mountains), Currie, Wendover • UTAH: (Great Salt Lake Desert), (Dolphin Island), Snowville • IDAHO: Woodruff, Samaria, Dayton, Clifton, Banida, Mink Creek, Bern, Bennington • WYOMING: Smoot, (Salt River Range), Halfway, Merna, (Gannett Peak, highest point in Wyoming: 13,806 feet), (Wind River Range), Burris, Hamilton Dome, Winchester, Worland, (Powder River Pass), Upton, Four Corners • SOUTH DAKOTA: Nemo, Piedmont, Blackhawk, Hereford, Milesville, Hayes, Pierre, Lower Brule Indian Reservation, Crow Creek Indian Reservation, Wessington Springs, Lane, Artesian, Carthage, Roswell, Ramona, Madison, Nunda, Rutland, Ward • MINNESOTA: Verdi, Ruthton, Garvin, Godahl, La Salle, Garden City, Good Thunder, Mapleton, Waldorf, New Richland, Blooming Prairie, Oslo, Hayfield, High Forest, Stewartville, Chatfield, Peterson, Whalan, Houston, Bratsberg, Brownsville • WISCONSIN: Stoddard, Genoa, Chaseburg, Newry, Westby, Bloomgingdale, Rockton, Valley, Hillsboro, Union Center, Wonewoc, Lake Delton, Wisconsin, Dells, Briggsville, Friesland, Cambria, Fox Lake, Waupun, Brownsville, Knowles, Le Roy, Le Mira, Kewaskum, Random Lake, Belgium, Cedar Grove • MICHIGAN: Roosevelt Park, Norton Shores, Cloverville, Conklin, Alpine, Grand Rapids, Saranac, Portland, Eagle, Grand Lodge, Lansing, Okemos, Haslett, Williamston, Bell Oak, Oak Grove, Fowlerville, Howell, Hartland, Highland, Milford, Walled Lake, Commerce, Farmington Hills, Southfield, Ferndale, Oak Park, Highland Park, Detroit, Grosse Point Woods, (Lake St. Clair) • ONTARIO: Windsor, Tecumseh, Belle River, S. Woodslee, Comber, Wheatley • OHIO: (Hubbard Homestead), Ashtabula, North Kingsville, Bashnell, Kellogsville, Monroe Center • PENNSYLVANIA: Pennside, Springboro, Hickernell, Mosiertown, Saegertown, New Richmond, Blooming Valley, Townville, Troy Center, Gresham, Titusville, E. Titusville, Pleasantville, West Hickory, Starr, Guitonville, Marienville, Roses, Halfton, Portland Mills, Brandy Camp, Toby, Byrnedale, Force, Weedville, Caledonia, Karthaus, Pine Glen, Clarence, Howard, Jacksonville, Nittany, Lamar, Clintondale, Tylerville, Livonia, Hartleton, Millmont, White Springs, Dice, New Berlin, Kratzerville, Shamokin Dam, Sunbury, Selinsgrove, Edgewood, Shamokin, Gowen City, Ranshaw, Kulpmont, Locust Gap, Helfenstein, Gordon, Primrose, Minersvile, Jonestown, Marlin, Pottsville, Mount Carbon, Schuylkill Haven, Adamsdale, Orwigsburg, Deer Lake, Eckville, Port Clinton, Albany, Lenhartsville, Klinesville, Grimville, Maxatawney, Trexlertown, Breinigsville, Alburtis, Mertztown, Macungie, Shimerville, Old Zionsville, Zionsville, Spinnerstown, Milford Square, Rich Hill, Hargersville, Blooming Glen, Dublin, Plumsteadville, Gardenville, Danbord, Fountainville, Mechanicsville, Lahaska, Buckingham, Doylestown, Furlong, Pineville, Penns Park, Washington Crossing, Yardley, Marrisville • NEW JERSEY: Ewing, Trenton, White Horse, Groveville, Creamridge, Hornerstown, Lakehurst, Ridgeway, Pine Lake Park, Toms River, Gilford Park, Pine Beach, Bay Shore, Island Heights, (Pelican Island), Seaside Heights, Seaside Park, South Seaside Park.

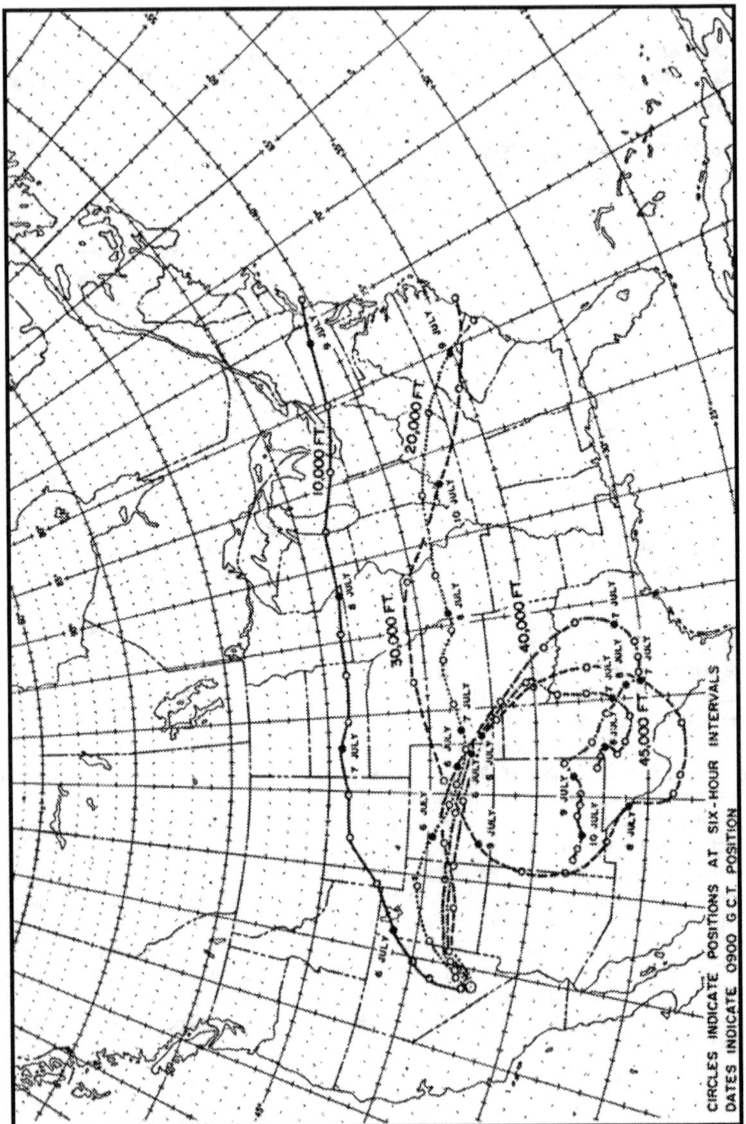

TRAJECTORY: PLUMBBOB HOOD (PB4)

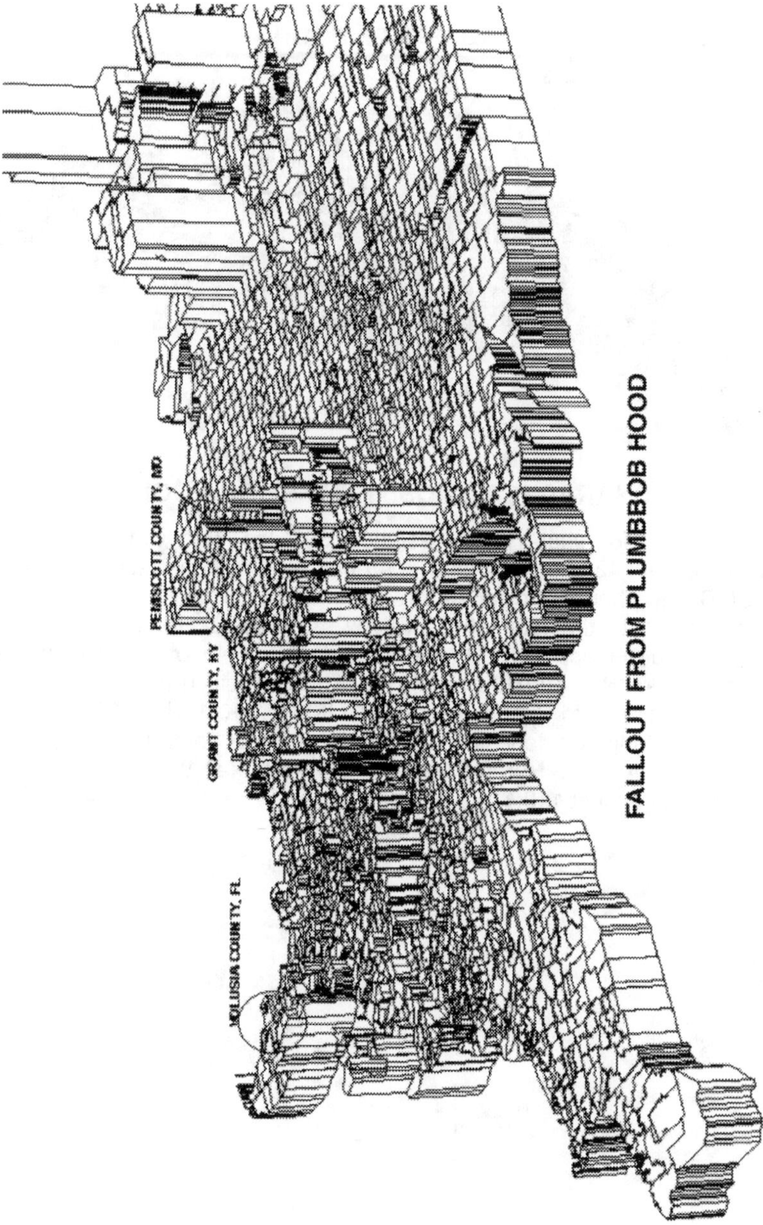

FALLOUT FROM PLUMBBOB HOOD

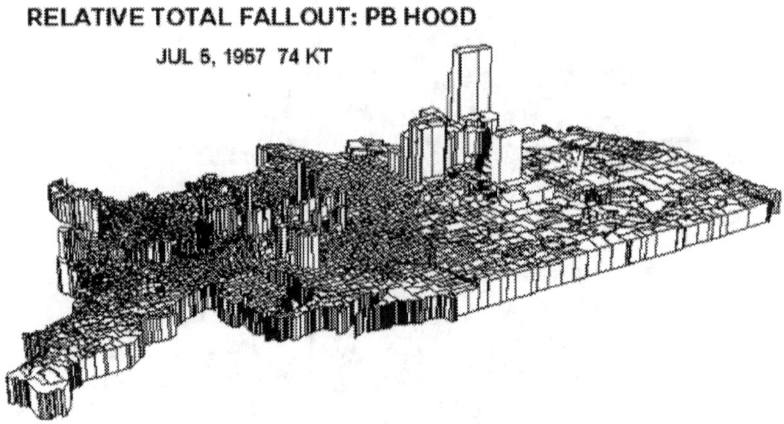

CHARACTERISTICS: PLUMBBOB HOOD

The device for shot HOOD weighed 395 lbs, and was 12.1 inches diameter by 42.2 inches long. Among the nuclear tests, Plumbbob HOOD, a thermonuclear shot, was very similar in radionuclide composition to the later shot SEDAN. It was 2nd for percentages of Au198, Au199 and Be7 (1st being SEDAN). No other shot, in fact, produced Au198 or Au199 in fallout. The deposition pattern for HOOD was most similar to Upshot-Knothole CLIMAX (UK11) (R: 0.58, T: 39.44). Counties of highest deposition included Pima AZ, Santa Cruz AZ, Osage MO, Colfax NM and Harding NM. HOOD's total fallout to I-131 ratio was moderately high at 110:1.

TOTAL FALLOUT: PLUMBBOB HOOD

Equal Intervals

nCi/square meter

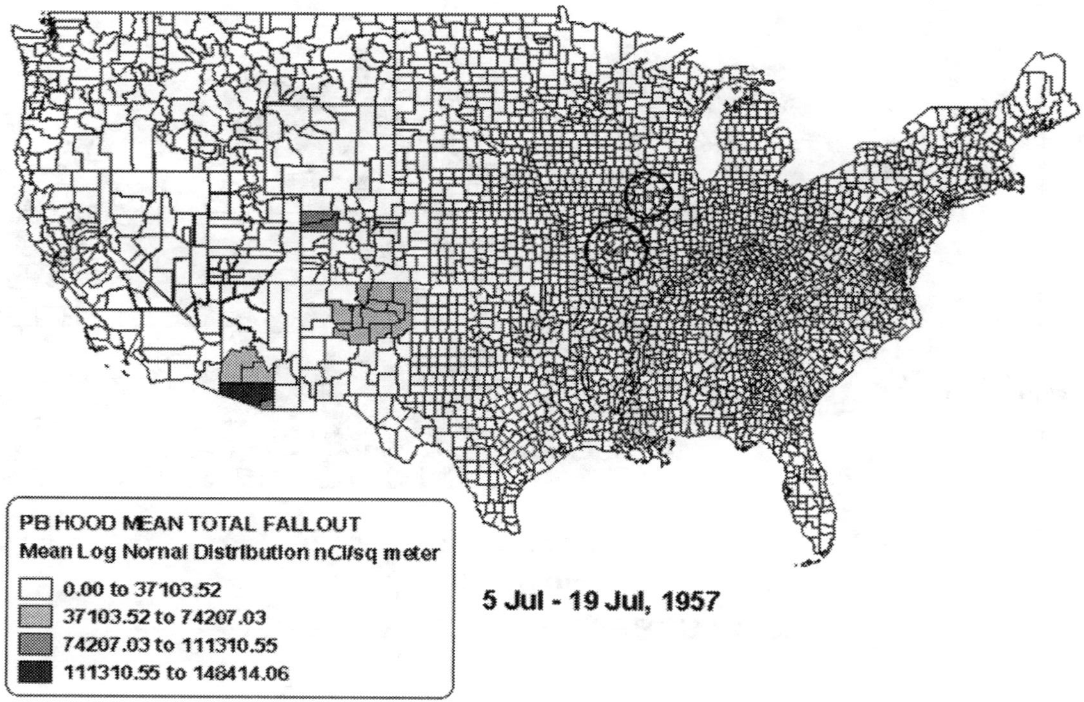

TOTAL FALLOUT: PLUMBBOB HOOD

Equal Number

nCi/square meter

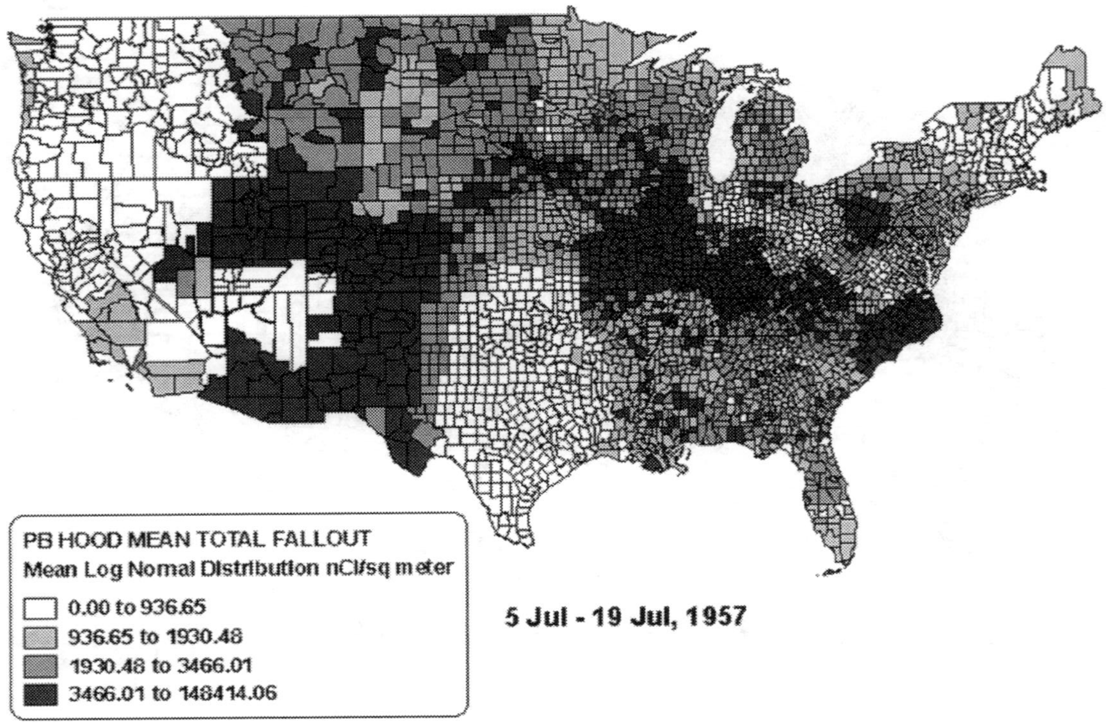

5 Jul - 19 Jul, 1957

FALLOUT FROM PB HOOD JULY 5 1957
PEMISCOTT COUNTY, MO
GREEN COUNTY, WI
GRANT COUNTY, KY
VOLUSIA COUNTY, FL

SHOT 5: PLUMBBOB DIABLO (PB5)

Plumbbob: Diablo

Detonation Date: 15 Jul 1957 • *Detonation Time:* 4:30 A.M. • *Area:* 2b • *Sponsor:* Livermore Laboratory • *Yield:* 17 kt • *Radiation Level at Ground Zero:* 200 R/hr • *Height of Burst:* 500 ft • *Cloud Top Height:* 32,000 ft

30,000-Foot Trajectory: NEVADA: Indian Springs, Las Vegas, Overton, Logandale, Mesquite • UTAH: Santa Clara, Pine Valley, New Harmony, Kanarraville, Cedar City, Enoch, Summit, Parowan, Paragonah, Circleville, Junction, Marysvale, Monroe, Annabella, Glenwood, Sigurd, Salina, Price, Helper, Kenilworth, Bridgeland, Roosevelt, Cedarview, Whiterocks, Tridell, Dry Fork, (Mount Lena) • WYOMING: (Richard's Mountain), Red Desert, Creston, Shirley Basin, (Laramie Mountains), Douglas, Orpha, Lance Creek • SOUTH DAKOTA: Pringle, Fairburn, Hermosa, Owanka, Wall, Quinn, Ottumwa, Hayes, Fort Pierre, Pierre, Blunt, Harrold, Holabird, Highmore, Ree Heights, Miller, St. Lawrence, Veyland, Wessington, Bonilla, Hitchcock, Broadland, Carpenter, Willow Lake, Bryant, Vienna, Lake Norden, Hayti, Dempster, Estelline, Toronto, Astoria • MINNESOTA: Hendricks, Wilno, Ivanhoe, Marshall, Lynd, Revere, Lamberton, Sanborn, Comfrey, Sveadahl, Darfur, St. James, Grogan, Lewisvile, Winnebago, Huntley, Delavan, Frost, Bricelyn, Kiester • IOWA: Scarville, Lake Mills, Leland, Ventura, Clear Lake, Thornton, Latimer, Iowa Falls, New Providence, St. Anthony, Clemons, Melbourne, Newton, Rasnor, Otley, Flagler, Pershing, Attica, Maysville, Weller, Melrose, Iconium, Mystic, Brazil, Centerville, Numa, Cincinnati • MISSOURI: Hartford, Pure Air, Melrose, Ethel, New Cambria, Kaseyville, Prairie Hill, Salisbury, Forest Green, Glasgow, Lisbon, Boonesboro, Boonville, Pilot Grove, Bellair, Syracuse, Stover, Climax Springs, Urbana, Louisburg, Halfway, Brighton, Pleasant Hope, Glidewell, Springfield, Clever, Aurora, McDowell, Ridgley, Exeter • ARKANSAS: Pea Ridge, Bentonville, Rogers, Highfill, Robinson, Siloam Springs • OKLAHOMA: West Siloam Springs, Kansas, Daks, Leach, Peggs, Yonkers, Wagover, Red Bird, Coweta, Leonard, Mounds, Slick, Beggs, Nuyaka, Henryetta, Pharoah, Vernon, Hanna, Indianola, Krebs, Alderson, Dow, Haileyville, Hartshorne, Daisy, Jumbo, Miller, Farris, Boehler, Sunkist, Cade, Bokchito, Utika, Achille, Hendrix • TEXAS: Denison, Bells, Ida, Tom Bean, Sherman, Gunter, Tioga, Pilot Point, Sanger, Boliver, Krum, Ponder, Argyle, Justin, Roanoke, Bartonville, Grand Prairie, Arlington, Irving, Midlothian, Waxahachie, Maypearl, Italy, Frost, Milford, Martens, Emmett, Irene, Malone, Mount Calm, Axtell, Mart, Riesel, Ben Hur, Otto, Perry, Marlin, Cedar Springs, Rosebud, Baileyville, Burlington, Ben Arnold, Cameron, Minerva, Milano, Rockdale, Tanglewood, Old Dime Box, Lincoln, Dime Box, Hills, Giddings, Winchester, West Point, Plum, Muldoon, High Hill, Engle, Moravia, Breslau, Shiner, Sweet Home, Halletsville, Speaks, Cordele, Louise, Danevang, Bay City.

CHARACTERISTICS: PLUMBBOB DIABLO

The device for the DIABLO shot was 16.2 inches in diameter, 68.4 inches long and weighed 1352 lb. Among the nuclear tests, Plumbbob DIABLO was ranked 6th for percentage of Pb203 in fallout. The deposition pattern for this shot is most similar to Upshot-Knothole GRABLE (UK10) (R: 0.42; T: 25.85). Plumbbob DIABLO was ranked 3rd in the amount of Te127m deposited nationwide and 6th in the amount of Ru106. Counties of highest deposition included Nye NV, White Pine NV, Lincoln NV, Stark ND, Teton ID, Utah UT, Dunn ND. The ratio of total fallout to I-131 was moderately low at 91:1.

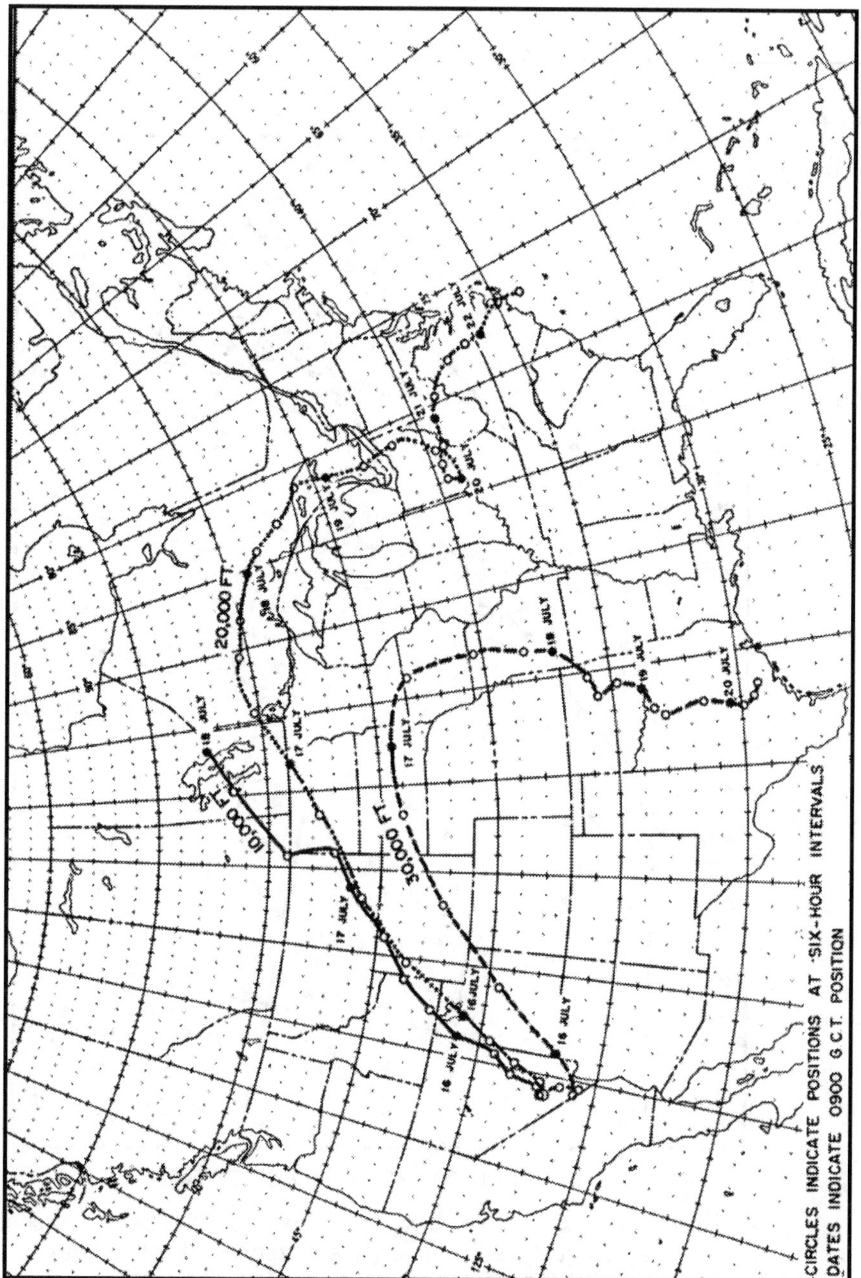

TRAJECTORY: PLUMBBOB DIABLO (PB5)

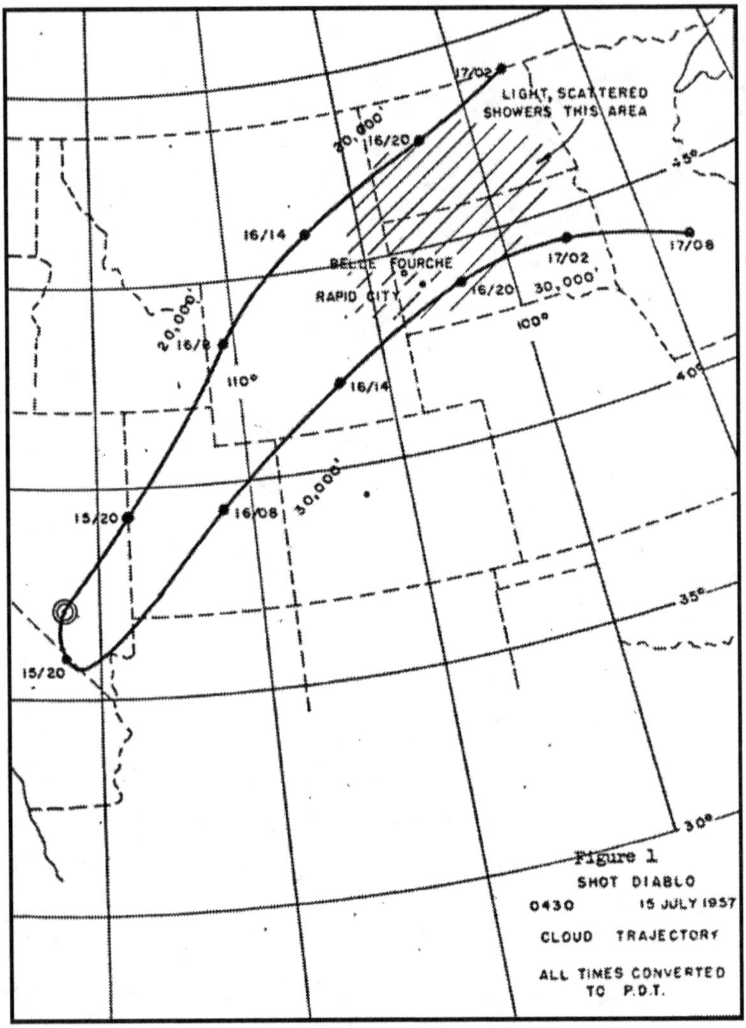

THE BELLE FOURCHE INVESTIGATION

Relatively high levels of radioactivity were reported in the Belle Fourche, SD area in late August, 1957. Investigators from the Atomic Energy Commission suggested that the increased levels of radioactivity may have come from either shot DIABLO or shot Stokes. This map was taken from the report of the investigation.

TOTAL FALLOUT: PLUMBBOB DIABLO

Equal Intervals

nCi/square meter

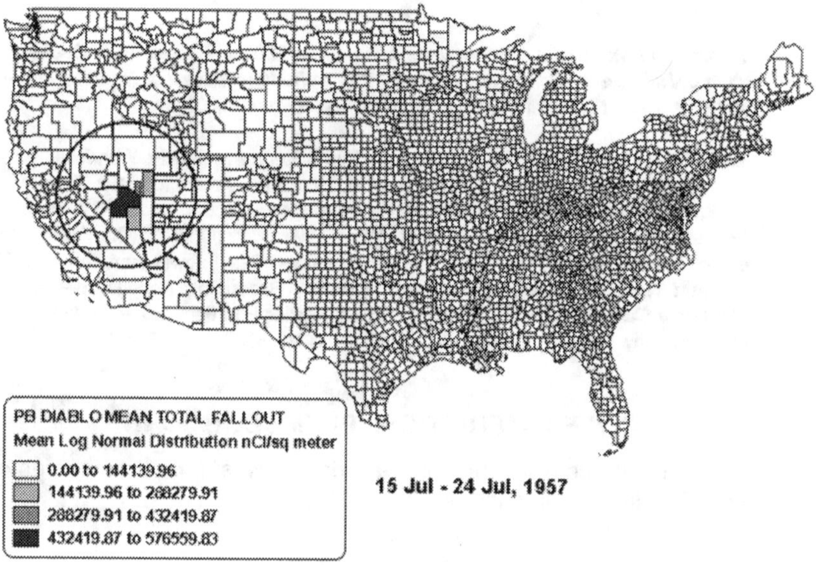

TOTAL FALLOUT: PLUMBBOB DIABLO

Equal Number

nCi/square meter

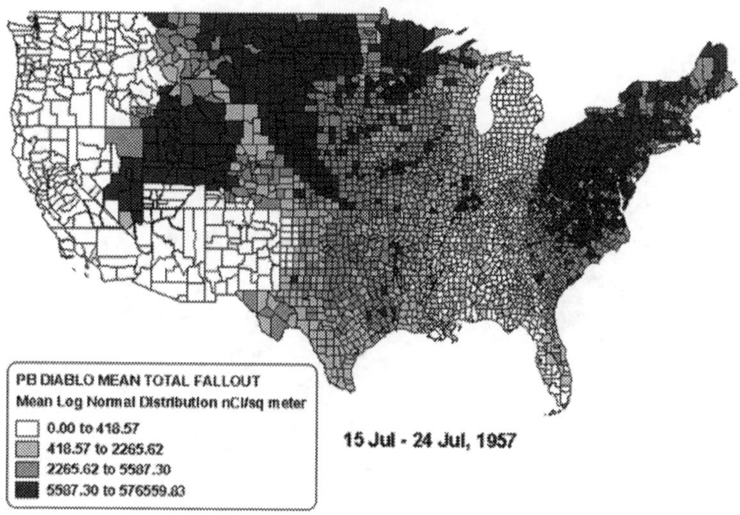

PLUMBBOB JOHN

(Not Included in the NCI Study)

PLUMBBOB: JOHN

Detonation Date: 19 Jul 1957 • *Detonation Time:* 7:00 A.M. • *Area:* 10 • *Sponsor:* Department of Defense • *Yield:* Approximately 2 kt • *Radiation Level at Ground Zero:* None detected • *Height of Burst:* 20,000 ft • *Cloud Top Height:* 44,000 ft

40,000-Foot Trajectory: NEVADA: Currant, Ruth, Steptoe • UTAH: Wendover, Park Valley, Clear Creek • IDAHO: Naf, Bridge, Sublett, Rockland, American Falls, Pocatello, Chubbock, Blackfoot, Basalt, Shelly, Idaho Falls, Ammon, Lincoln, Ucon, Ririe, Heise, Archer, Rexburg, Teton, Drummond, Marysville, Warm River •. WYOMING: (Old Faithful in Yellowstone National Park), (Craig Pass), Canyon, (Dunraven Pass), Tower Junction • MONTANA: (Gallatin National Forest), Limestone, Greycliff, Reedpoint, Rapelje, Belmont, Lavina, Flatwillow, Cat Creek, Mosby, (Fort Peck Lake), Frazier, Fort Peck Indian Reservation, Plentywood, Westby.

CHARACTERISTICS: PLUMBBOB JOHN

The plutonium core for this air-to-air device was 17 inches in diameter, was 17.25 inches long and weighed 221 lbs.

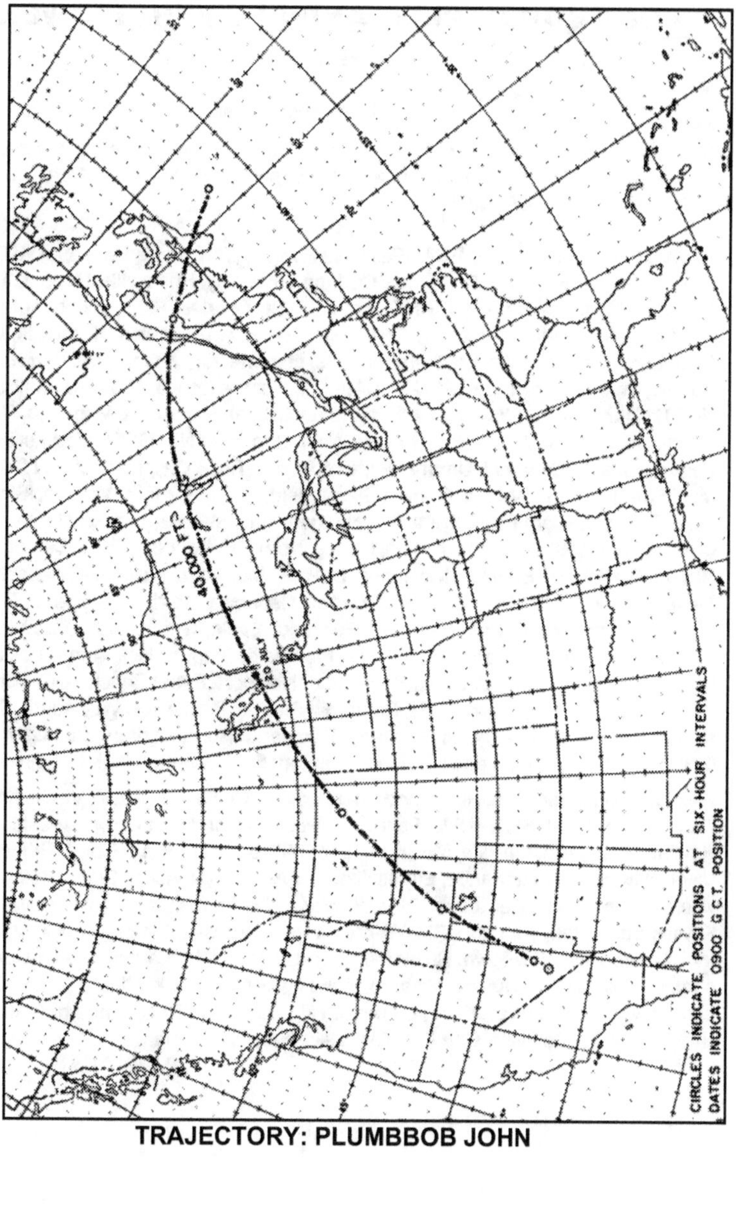

TRAJECTORY: PLUMBBOB JOHN

SHOTS 6A AND 6B

PLUMBBOB KEPLER AND PLUMBBOB OWENS (PB6)

PLUMBBOB: KEPLER

Detonation Date: 24 Jul 1957 • *Detonation Time:* 4:50 A.M. • *Area:* 4 • *Sponsor:* Los Alamos • *Yield:* 10 kt • *Radiation Level at Ground Zero:* 100 R/hr • *Height of Burst:* 500 ft • *Cloud Top Height:* 28,000 ft

20,000-Foot Trajectory: NEVADA: (Papoose Range), (Timpahute Range), • UTAH: Garrison, (Cedar Mountains), (Great Salt Lake), Corinne, River City, Honeyville, Garland, Fielding, Newton, Clarkston, Lewiston • IDAHO: Franklin, St. Charles, Paris, Bloomington, Montpelier • WYOMING: Daniel, Pinedale, Crowheart, Wind River Indian Reservation, (Owl Creek Mountains), Thermopolis, East Thermopolis, Lucerne, Kirby, Ten Sheep, (Powder River Pass), Buffalo, Weston, Colony • SOUTH DAKOTA: Hoover, Zeona, Meadow, Bullhead, Walker • NORTH DAKOTA: Fort Yates, Strasburg, Wishek, Lehr, Merricourt, Monango, Crete, Gwinner, Milnor, De Lamere, Wyndmere, Barney, Wahpeton • MINNESOTA: Breckenridge, Everdell, Fergus Falls, Underwood, Urbank, Perkers Prairie, Rose City, Flensburg, Little Falls, Buckman, Ramey, Morrill, Milaca, Grandy, Stanchfield, Stark, Harris • WISCONSIN: Atlas, Trade Lake, Frederic, Clam Falls, Barranett, Haugen, Brill, Mikana, Ladysmith, Jump River, Little Chicago, Wausau, Schofield, Ringle, Hatley, Norrie, Wittenberg, Pella, Caroline, Embarrass, Navarino, Nichols, Briarton, Seymour, DePere, Morrison, Wayside, Reedsville, Whitelaw, Manitowoc • MICHIGAN: Whitehall, Dalton, Casnovia, Harrisburg, Sparta, Belmont, Rockford, Cannonsburg, Lowell, Saramac, Portland, Eagle, Grand Lodge, Lansing, Millett, Dimondale, Golt, Mison, Dansville, Millville, Plainfield, Gregory, Hell, Ann Arbor, Dixboro, Ypsilanti, Carleton, Rockwood, South Rockwood, Newport, Estral Beach • OHIO: Lorain, Sheffield Lake, North Ridgeville, Elyria, Columbia Station, Brunswick, Weymouth, Sharon Center, Norton, Barberton, Liberty, Greensburg, Marshand, North Canton, East Canton, Mapleton, Oneida, Pattersonville, Wattsville, North Canton, East Canton, Mapleton, Oneida, Pattersonville, Wattsville, Bergholz, Richmond, Steubenville, Colliers • WEST VIRGINIA: Follansbee, Wellsburg • PENNSYLVANIA: Penowa, Avella, West Middletown, Washington, Gabby Heights, Vestaburg, LaBelle, Millsboro, East Millsboro, Isabella, Filbert, Hibbs, New Salem, Uledi, Newcomer, Fairchance, Elliotsville • MARYLAND: Asher Glade, Selbysport, Friendville, Accident, Bitlinger, Luke, Bloomington, Westernport, McCoole • WEST VIRGINIA: Keyser, Burlington, Junction, Rio, Wardensville • VIRGINIA: Mount Olive, Maurertown, Detrick, Bentonville, Viewtown, Rixeyville, Brandy Station, Richardsville, Flat Run, Fredericksburg, Four Mile Fork, Guinea, Villboro, Bowling Green, Spark, Millers Tavern, Center Cross, Laneview, Church View, Urbanna, Locust Hill, Hartfield, Dixie, Amburg, Deltaville, Gwynn, Hudgins, Diggs, Eastville, Cheriton, Bay View, (Ship Shoal Island).

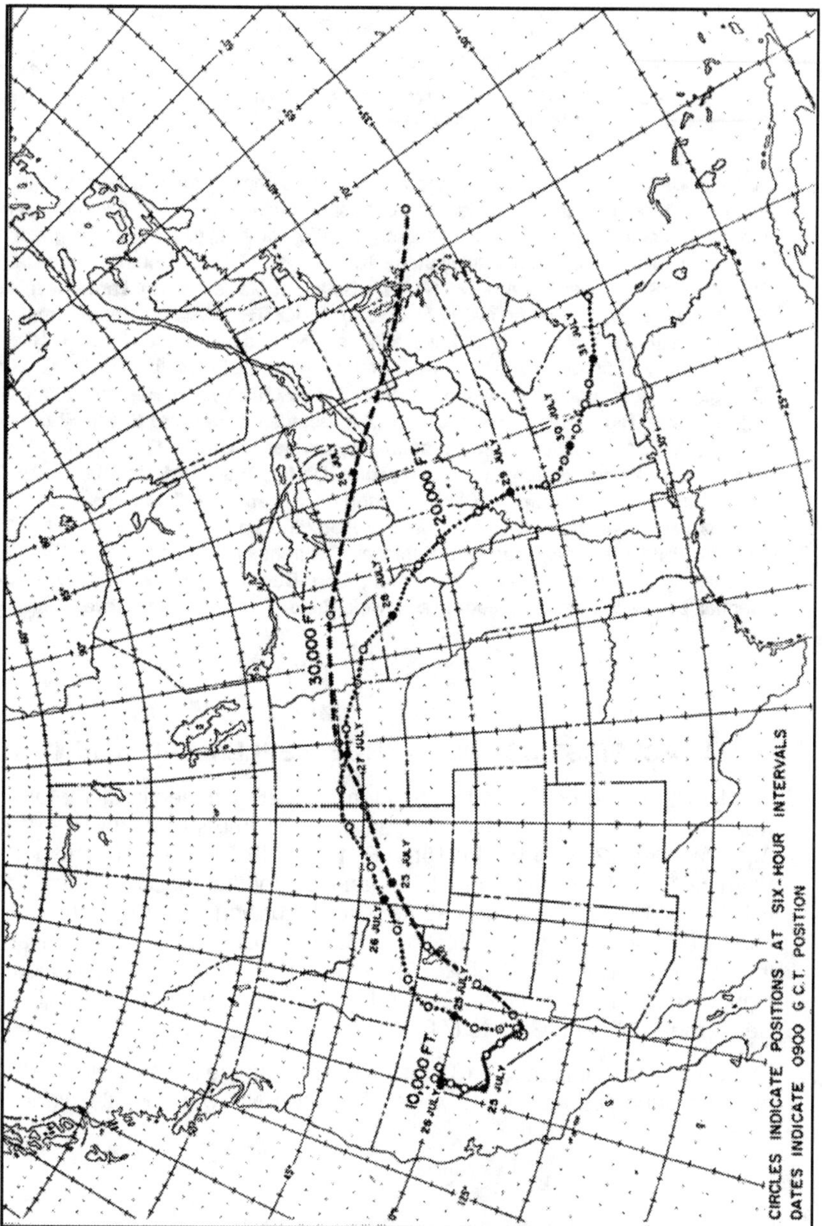

TRAJECTORY: PLUMBBOB KEPLER (PB6A)

Plumbbob: Owens

Detonation Date: 25 Jul 1957 • *Detonation Time:* 6:30 A.M. • *Area:* 9b • *Sponsor:* Livermore Laboratory • *Yield:* 9.7 kt • *Radiation Level at Ground Zero:* 1,000 R/hr • *Height of Burst:* 500 ft • *Cloud Top Height:* 35,000 ft

10,000-Foot Trajectory: NEVADA: Currant, (Pancake Summit), Ruby Valley, Lamoille, (Secret Pass), (Humboldt Range), Wells, Metropolis, Thousand Springs, Contact, Jackpot • IDAHO: Rogerson, Kimberly, Eden, Dietrich, Picabo, Gannett, (Borah Peak, highest point in Idaho), Small • WYOMING: (Yellowstone National Park), West Thumb, Pahaska, (Windy Mountain), (Bear Tooth Pass) • MONTANA: Belfry, St. Xavier, Busby, Lame Deer, Ashland, San Labre Mission, (Custer National Forest) • SOUTH DAKOTA: Camp Crook, Zeona, Maurine, Isabel, Firested, Timber Lake, Wakpale, Mobridge, Herreid, Artas • NORTH DAKOTA: Zeeland, Wishek, Nortonville, Jud, Grand Rapids, La Moure, Verona, Stirum, Gwinner, Cayuga, Sisseton Indian Reservation • SOUTH DAKOTA: Veiben, Hammer, Claire City, Sisseton, Peever, Big Stone City • MINNESOTA: Beardsley, (Hartford Beach State Park), Ortonville, Odessa, Louisburg, Milan, Watson, Montevideo, Wegdahl, Granite Falls, North Redwood, Evan, Cobden, Sleepy Eye, Leavenworth, Darfur, Butterfield, Mountain Lake, Bergen, Alpha, Jackson, Petersburg • IOWA: Huntington, Superior, Terril, Dickens, Greenville, Rossio, Peterson, Cherokee, Quimby, Washta, Pierson, Climbing Hill, Sloan • NEBRASKA: Walthill, Pender, Wisner, Clarkson, Schuyler, Columbus, Bellwood, Rising City, Surprise, Cresham, Waco, Lushton, Grafton, Sutton, Ong, Davenport, Ruskin, Hardy • KANSAS: Webber, Lovewell, Formoso, Montrose, Jewell, Randall, Hunter, Ash, Grove, Lucas, (Wilson Lake).

CHARACTERISTICS: PLUMBBOB KEPLER AND OWENS

The KEPLER device was 44 inches long, ws 28.5 inches in diameter and weighed 1517 lbs. Among the nuclear tests, Plumbbob KEPLER is 1st for Ag109m, Ag112, Cd115, In115m, Pd111m, Pd112, Sb126 and I-130. Plumbbob OWENS is 1st for Tb161 and 2nd for Ag112, Cd115, In115m, In117, Pd112, Sm153 and Tc99m. The deposition pattern for these two shots (as combined by the National Cancer Institute) is most similar to that of Plumbbob DOPPLER (PB9) (R: 0.39: T: 23.55.) Counties of highest deposition include: Esmeralda NV, Mineral NV, Churchill NV, Beaverhead MT and Granite MT.

The ratio of total fallout to I-131 for KEPLER was the lowest among all the above-ground tests at 81:1, while shot OWENS was slightly higher at 86:1.

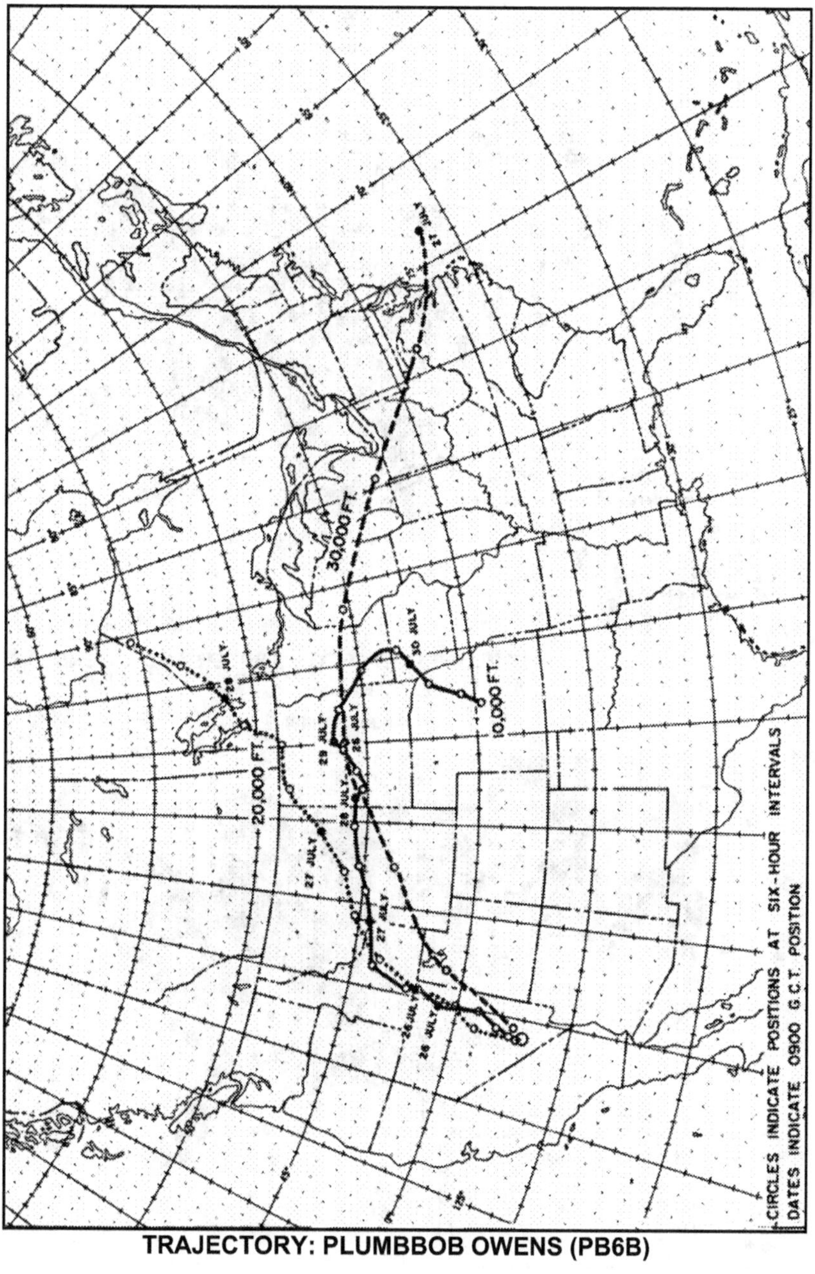

TRAJECTORY: PLUMBBOB OWENS (PB6B)

TOTAL FALLOUT PLUMBBOB SHOTS KEPLER AND OWENS

Equal Intervals

nCi/square meter

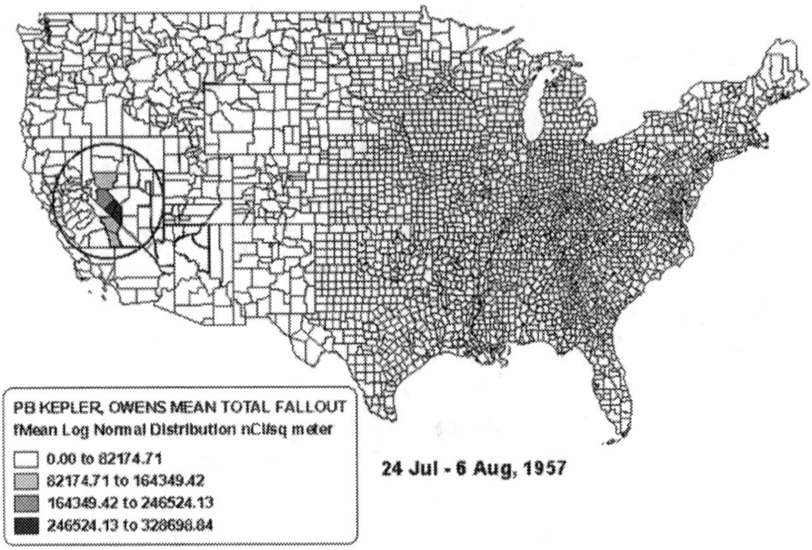

TOTAL FALLOUT PLUMBBOB SHOTS KEPLER AND OWENS

Equal Number

nCi/square meter

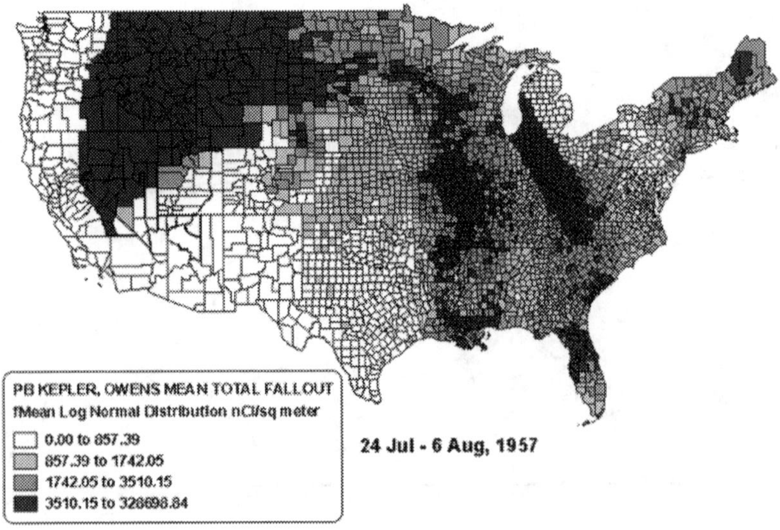

SHOT 7: PLUMBBOB STOKES (PB7)

PLUMBBOB: STOKES

Detonation Date: 7 Aug 1957 • *Detonation Time:* 5:25 A.M. • *Area:* 7b • *Sponsor:* Los Alamos • *Yield:* 19 kt • *Radiation Level at Ground Zero:* 50 R/hr • *Height of Burst:* 1,500 ft • *Cloud Top Height:* 37,000 ft

30,000-Foot Trajectory: NEVADA: Ely, Tippett • UTAH: Ibapah, Brigham City, Logan, Pickleville • WYOMING: Geneva, Afton, Bondurant, Dubois, Meeteetse, Lovell, Kane • MONTANA: (Bighorn Canyon), Saint Xavier, (Custer Battlefield), Castle Rock, Hathaway, Fort Keough, Miles City, Wibaux • NORTH DAKOTA: Grassy Butte, Parshall, Makoti, Logan, Granville, Denbigh, Towner, Rugby, Wolford, Bisbee, Egeland, Munich, Langdon, Leroy, Pembina • CANADA: Montreal • NEW YORK: Moopers Forks, West Chazy, Beekmantown, Plattsburgh, Clintonville, Westport, Mineville, Crown Point, Ticonderoga, Putnam Station, Clemons, Whitehall, South Hartford, Cossayuna, Salem, Greenwich, Johnsonville, West Steventown, New Concord, Hillsdale, West Copake, Hope Farm, Wingdale, Holmes, Brewster, Pleasantville, White Plains, New Rochelle, New York City.

CHARACTERISTICS: PLUMBBOB STOKES

The device for the STOKES test was 44 inches long, had a diameter of 28.5 inches, and weighed 1517 lbs. The Stokes shot ranked 2nd for (theoretical) percentage of Zr95 in fallout. HOWever, the NCI Plumbbob STOKES data suggests this test deposited no fallout. But an investigation by researchers for the Atomic Energy Commission strongly suggested that either the STOKES ore the earlier DIABLO shot deposited significant fallout near the town of Belle Fourche, South Dakota.

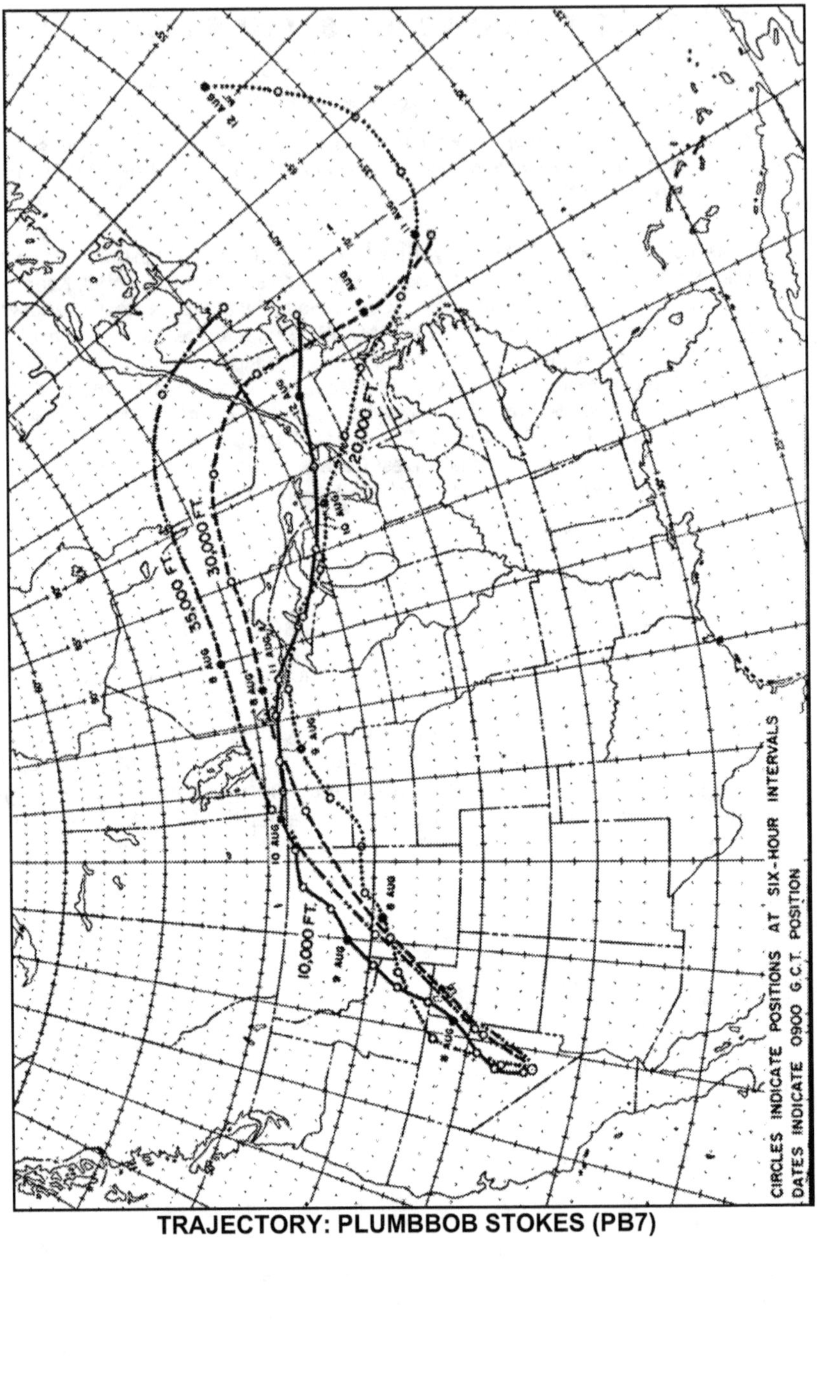

TRAJECTORY: PLUMBBOB STOKES (PB7)

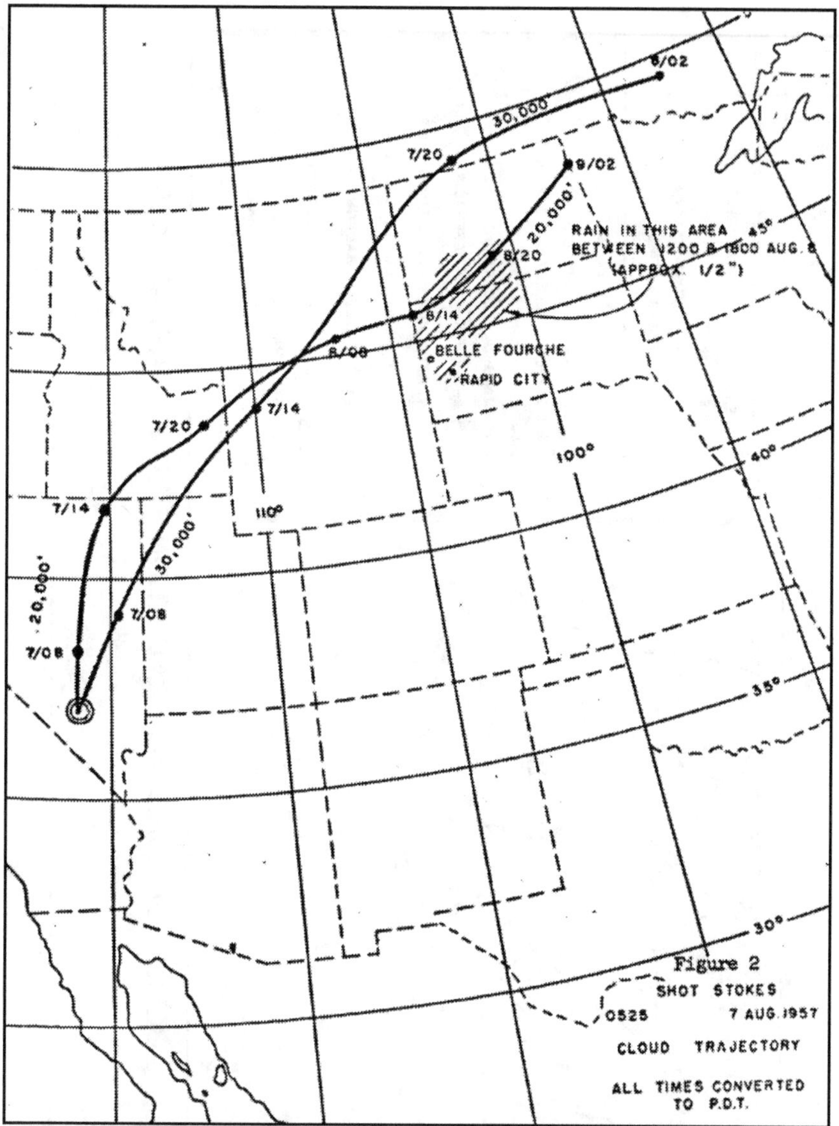

TRAJECTORIES POSSIBLY ASSOCIATED WITH THE BELLE FOURCHE RADIOACTIVITY

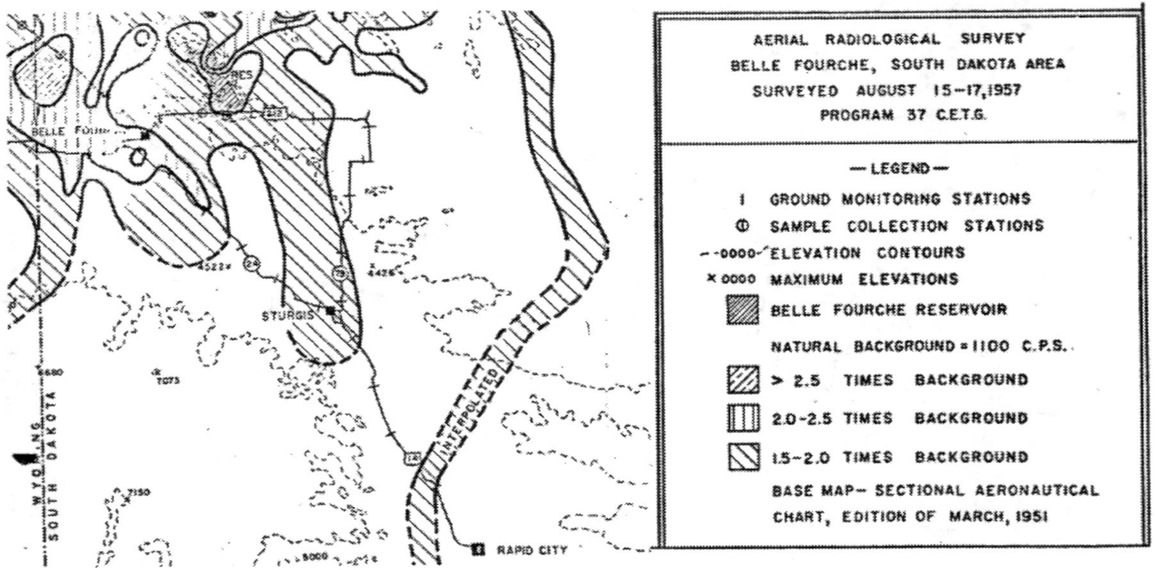

BELLE FOURCHE , S.D. RADIATION SURVEY MAP: AEC 1957

SHOT 8: PLUMBBOB SHASTA (PB8)

Plumbbob: Shasta

Detonation Date: 18 Aug 1957 • *Detonation Time:* 5:00 A.M. • *Area:* 2a • *Sponsor:* Livermore Laboratory • *Yield:* 17 kt • *Radiation Level at Ground Zero:* 500 R/hr • *Height of Burst:* 500 ft • *Cloud Top Height:* 32,000 ft

20,000-Foot Trajectory: NEVADA: (Belted Range), (Reveille Range), (Railroad Valley), (Pancake Range), (Pancake Summit), Cherry Creek, (Becky Peak), (Sugarloaf Peak) • UTAH: (Great Salt Lake), (Promontory Mountains), Corinne, Bear, River City, Wellsville, Deweyville, Logan, Providence, Hyrum, Round Valley, Laketown, (Bear Lake) • WYOMING: (Commissary Ridge), La Barge, Atlantic City, Gas Hills, (Pine Mountain), Bill • SOUTH DAKOTA: Edgemont, Smithwick, Manderson, Porcupine, Wounded Knee, Batesland • NEBRASKA: Merriman, Eli, Rose, Raeville, Petersburg, Creston, Leigh Ames, Fremont, Inglewood, Elk City, Bennington, Omaha • IOWA: Council Bluffs, Mineola, Malvern, Strahan, Imogene, Essex, Norwich, Yorktown, Shambaugh, Braddyville • MISSOURI: Clearmont, Pickering, Ravenwood, Conception, Gentryville, McFall, Pattenburg, Altamont, Gallatin, Breckinridge, Mooresville, Ludlow, Braymer, Coloma, Bogard, Carrollton, Bosworth, Malta Bend, Marshall, Napton, Nelson, Syracuse, Tipton, Fortuna, Latham, High Point, Etterville, Marys Home, Capps, St. Anthony, Iberia, Hancock, Dixon, Devils Elbow, Duke, Flat, Licking, Raymondville, Yukon, Eunice, Summersville, Pine Crest, Mountain View, Teresita, Thomasville, Rover, Alton, Couch, Myrtle • ARKANSAS: Dalton, Noland, Walnut Ridge, Hoxie, Sedgwick, Bono, Harrisburg, Payneway, Twist, Marion, West Memphis • MISSISSIPPI: Glower, Barks, Eudora, Arkabutle, Coldwater, Seratobia, New Town, Looxahoma, Taylor, Springdale, Water Valley, Bunner, Shepherd, Pittsboro, Calhoun City, Slate Spring, Houhenlinden, Maben, Sherwood, Reform, Bradley, Longview, Sturgis, Brooksville, Macon, Shuqualak, Paulette, Binnsville • ALABAMA: Panola, Warsaw, Gainsville, Boligee, Forkland, Prairieville, Gallion, Founsdale, Uniontown, Safford, Alberta, Boykin, Darlington, Rosebud, Oakhill, Pine Apple, Red Level, River Falls, Andalusa, Carolina, Libertyville, Florala • FLORIDA: Paxton (highest point in Florida), Glendale, Ponce de Leon, Live Oak, New Hope, Crystal Lake, Bennett, Wewahitchka, Sumatra, Carrabelle, (Gulf of Mexico), Palm Harbor, Ozona, Tampa, Center, Balm, Fort Lonesome, Fort Green, Wauchula, Zolfo Springs, Lake Placid, Brighton, (Lake Okeechobee), Palm City, Stuart, Jensen Beach, Sewall's Point.

CHARACTERISTICS: PLUMBBOB SHASTA

The device used for the SHASTA shot resembled the one used in the DIABLO test. It weighed 1435 lbs, was 16.9 inches in diameter and 69.2 inches long. Among the nuclear tests, Plumbbob SHASTA was ranked 5th in percentage of Pb203 in fallout. The deposition pattern for this shot was most similar to Upshot-Knothole NANCY (UK2) (R: 0.49, T: 30.91). SHASTA deposited fallout on 2301 U.S. counties (74 percent). SHASTA was ranked 6th in the deposition of Be7 on U.S. counties and 15th in the deposition of Mn54. Counties of highest deposition included Eureka NV, White Pine NV, Billings ND, Brookings SD, Cherry NE.. On Aug 18, 1957 Eureka County NV received an estimated maximum activity level of 1000 μCi/sq meter. Other counties receiving high estimated maximum activity levels included White Pine NV (695 μCi/sq meter) and Park, WY (274 μCi/sq meter.)

RELATIVE TOTAL FALLOUT: PB SHASTA

AUG 18, 1957 17 KT

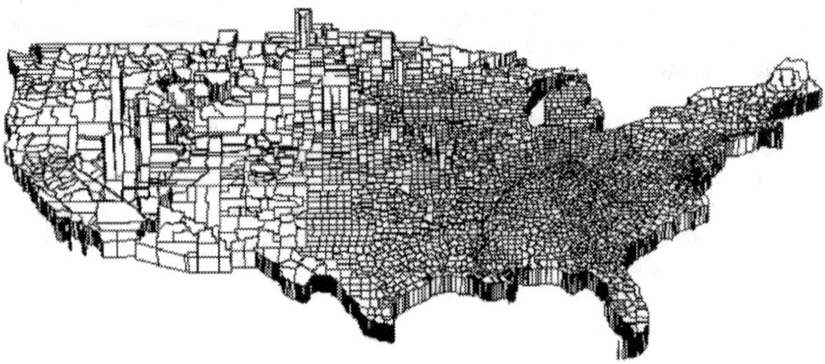

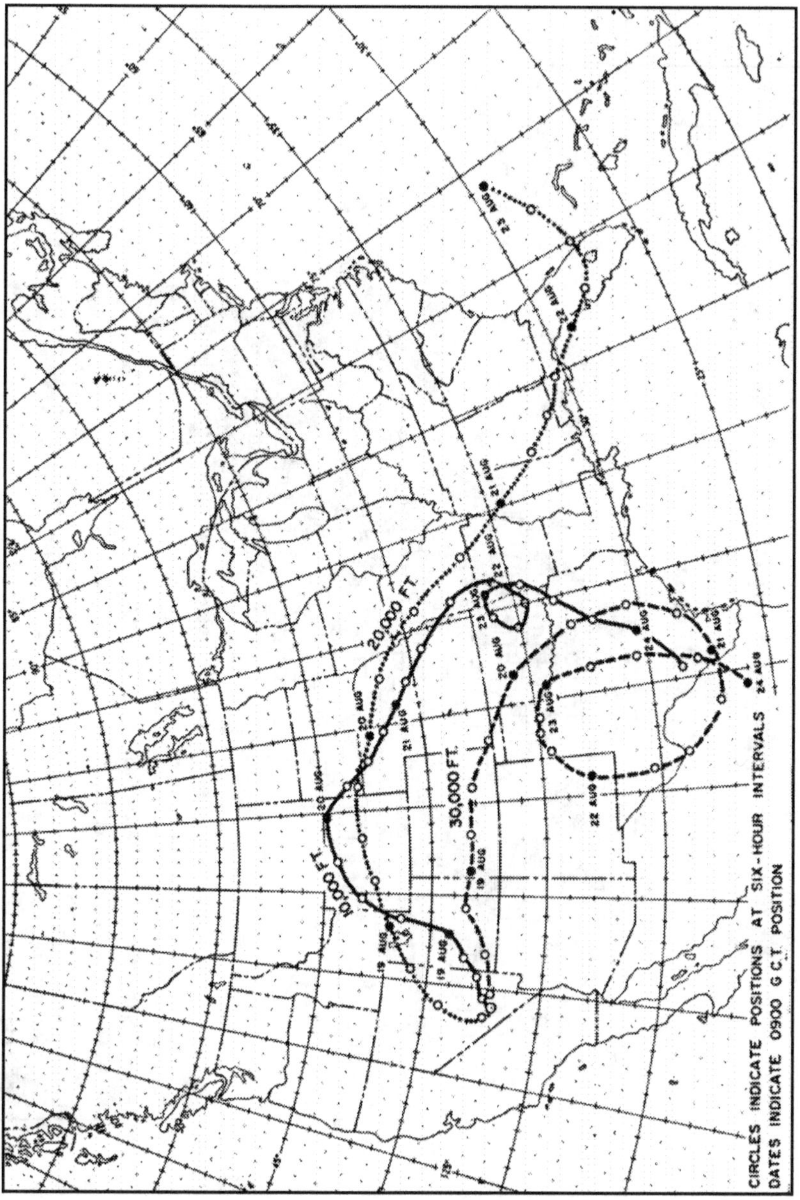

TRAJECTORY: PLUMBBOB SHASTA (PB8)

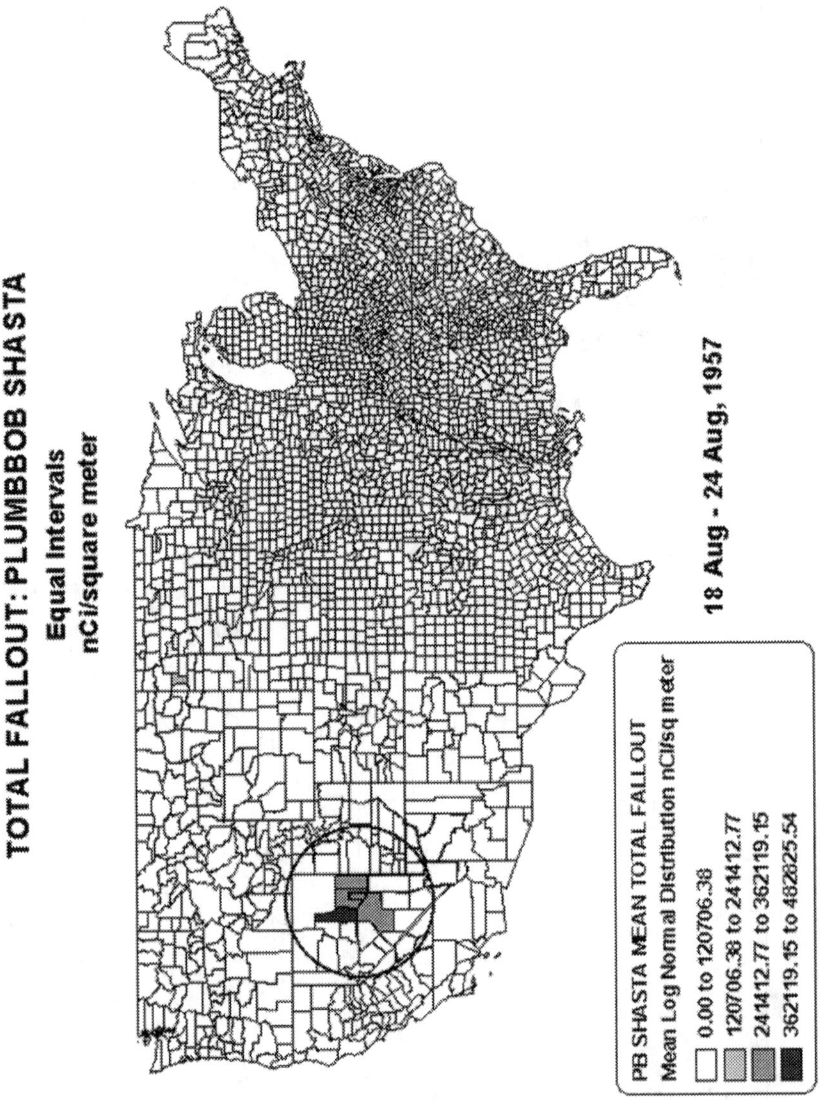
TOTAL FALLOUT: PLUMBBOB SHASTA
Equal Intervals
nCi/square meter
18 Aug - 24 Aug, 1957
PB SHASTA MEAN TOTAL FALLOUT
Mean Log Normal Distribution nCi/sq meter
0.00 to 120706.38
120706.38 to 241412.77
241412.77 to 362119.15
362119.15 to 482825.54

SHOT 9: PLUMBBOB DOPPLER (PB9)

PLUMBBOB: DOPPLER

Detonation Date: 23 Aug 1957 • *Detonation Time:* 5:30 A.M. • *Area:* 7a • *Sponsor:* Los Alamos • *Yield:* 11 kt • *Radiation Level at Ground Zero:* 50 R/hr • *Height of Burst:* 1,500 ft • *Cloud Top Height:* 38,000 ft

40,000-Foot Trajectory: NEVADA: (Papoose Range), (Pahranga Range), Hiko, (Wilson Creek Range) • UTAH: (Indian Peak), (Black Rock Desert), Flowell, Holden, Fayette, Ephraim, Manti, Hiawatha, Price, Wellington, Sunnyside, East Carbon, (Desolation Canyon), (Sweetwater Canyon) • COLORADO: (Douglas Pass), Rio Blanco, Burns, Bond, Green Mountain Camp, Geeny, (Arapahoe National Forest), Black Hawk, Central City, Golden, Arvada, Denver, Englewood, Aurora, Watkins, Bennett, Strasburg, Byers, Deer Trail, Shaw, Seibert, Vona, Stratton, Bethune, Burlington • KANSAS: Kanorado, Brownville, Monument, (Castle Rock), Brownell, McCracken, Hargrave, La Crosse, Bison, Shaffer, Otis, Olmitz, Albert, Heizer, Great Bend, Ellinwood, Silica, Raymond, Alden, Saxman, Sterling, Nickerson, Medora, Hutchinson, Halstead, Furley, Towanda, Haverhill, Leon, Latham, Elk Falls, Elk City, Bolton, Jefferson, Deating, Coffeyville • OKLAHOMA: South Coffeyville, Centralia, Vinita, Ketchum, Jay, Colcord, Siloam Springs, Watts, Ballard • ARKANSAS: Weddington, Farmington, Greenland, Prairie, Brentwood, Winslow, Ozark, Hunt, Atlas, Coal Hill, Midway, (Magazine Mountain, highest point in Arkansas: 2,753 feet), Bellevile, Plainview, Rover, Mountain Pine, Royal, Lake Hamilton, Bismark, Caddo Valley, Arkadelphia, Curtis, Amy, Chidester, Camden, Buena Vista, Louann, Smackover, Shaler, Newell, Wesson, Junction City • LOUISIANA: Summerfield, Lisbon, Vienna, Ruston, Grambling, Clay, Ausley, Quitman, Hudson, Georgetown, Fishville, Kolin, Echo, Bunkie, Whiteville, Beggs, Washington, Grand Coteau, Carencro, Lafayette, Broussard, Youngsville, Abbevile, Delcambre, Avery Island, (Shell Keys).

The 40,000-foot trajectory passed over the highest point in Arkansas.

CHARACTERISTICS: PLUMBBOB DOPPLER

The device used in the DOPPLER test was 17 inches in diameter, 26 inches long and weighed only 275 lbs. Among the nuclear tests, Plumbbob DOPPLER is ranked first for percentage of Na24 in fallout. The deposition pattern for this shot is most similar to Plumbbob NEWTON (PB15) (R: 0.41; T: 25.29). Plumbbob DOPPLER was ranked 9th in the amount of Mn54 deposited nationwide. Counties of highest deposition included the Iowa counties of Adair, Audubon, Boone, Carroll, Cass, Dallas, Davis, FRANKLIN, Jasper and Madison. On Aug 26, 1957 all received (estimated maximum activity) fallout levels of 509 μCi/sq meter.

The ratio of total fallout to I-131 for this shot was moderately high at 104:1.

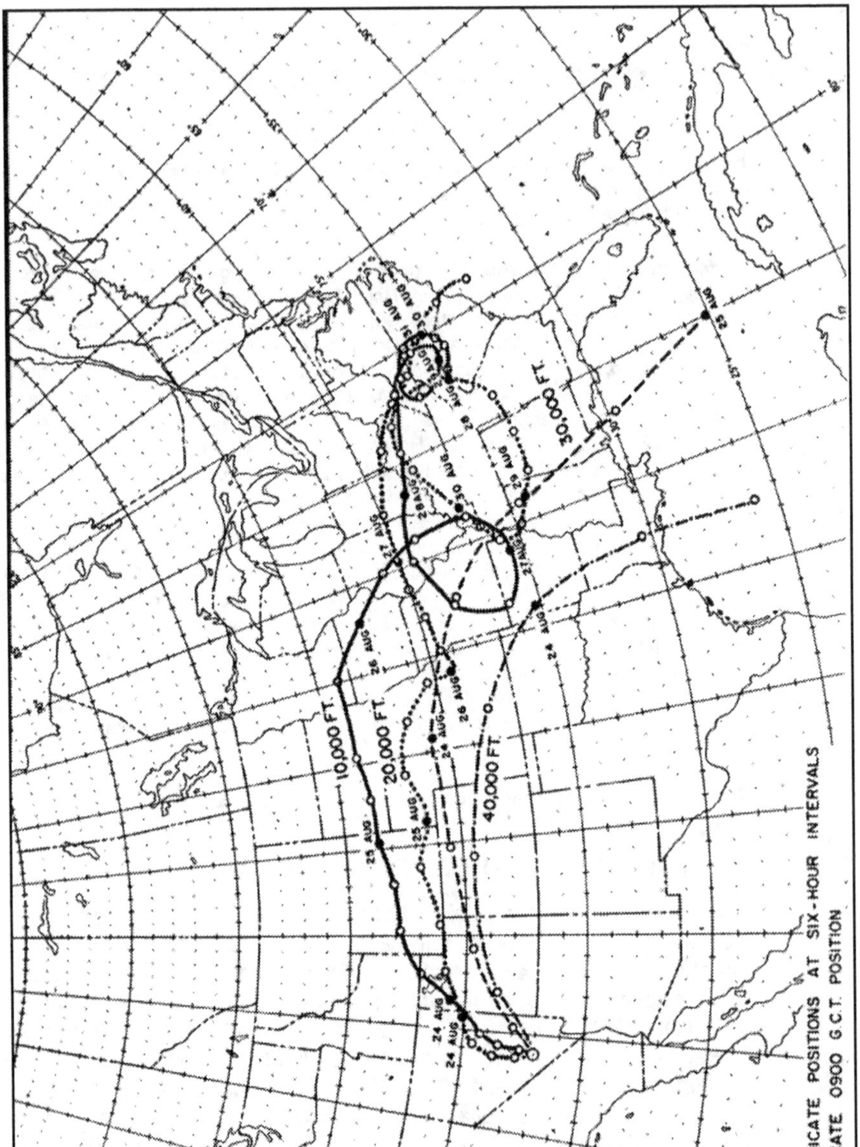

TRAJECTORY: PLUMBBOB DOPPLER (PB9)

TOTAL FALLOUT: PLUMBBOB DOPPLER

Equal Intervals

nCi/square meter

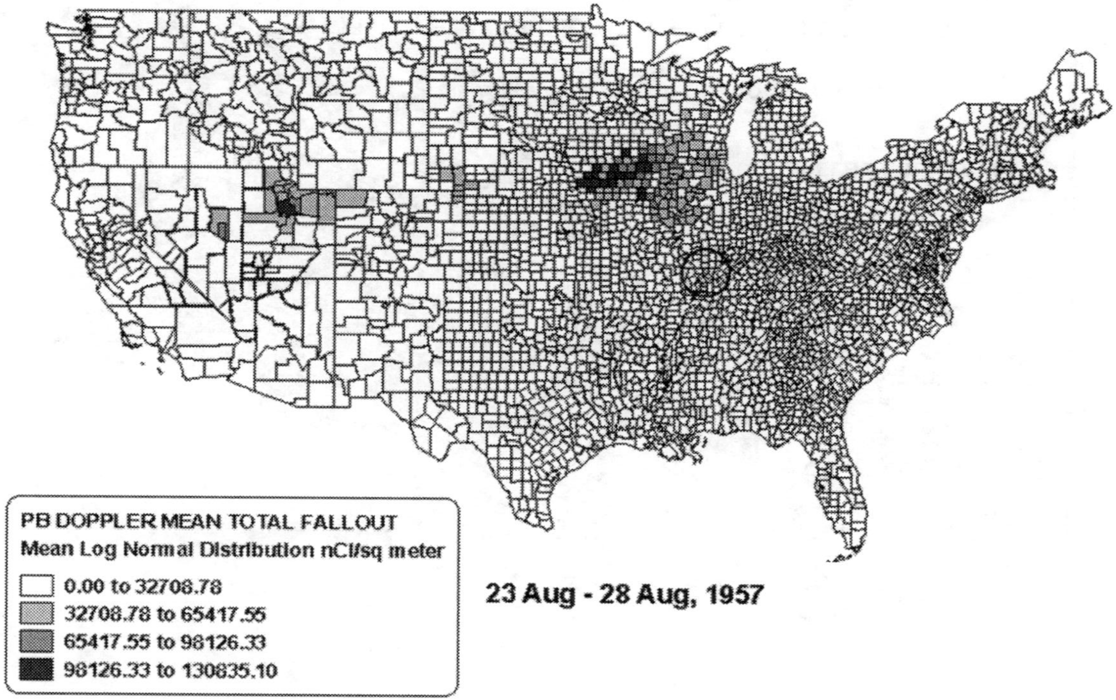

23 Aug - 28 Aug, 1957

SHELBY COUNTY, IA
DANE COUNTY, WI
RELATIVE AVERAGE TOTAL FALLOUT : PB DOPPLER
AUG 23, 1957 11 KT

SHOT 10: PLUMBBOB FRANKLIN PRIME (PB-10)

Plumbbob: Franklin Prime

Detonation Date: 30 Aug 1957 • *Detonation Time:* 5:40 A.M. • *Area:* 7b • *Sponsor:* Los Alamos • *Yield:* 4.7 kt • *Radiation Level at Ground Zero:* 100 R/hr • *Height of Burst:* 750 ft • *Cloud Top Height:* 32,000 ft

30,000-Foot Trajectory: NEVADA: Ash Springs, Alamo, Hiko, Caliento • UTAH: Deseret, Hinckley, Nephi, Provo, Orem, Hailstone, Lake Town • WYOMING: Cokeville, Bondurant, Valley, (Buffalo Bill Dam), (Bear Tooth Pass) • MONTANA: Red Lodge, Bear Creek, Joliet, Laurel, Billings, Musselshell, Sand Springs, Glasgow, Geltana • NEW YORK: Potsdam, Parishville, Saranac Lake, Whitehall • VERMONT: Poultney, (Green Mountain National Forest), Houghtonville, Cambridgeport, Bellows Falls • NEW HAMPSHIRE: Walpole, Surry, West Peterborough, Noone, Greenville, Mason • MASSACHUSETTS: Woburn, Lexington, Arlington, Medford, Boston, Beechwood, Egypt, Green Garbor, Hyannis Port.

CHARACTERISTICS: PLUMBBOB FRANKLIN PRIME

This was a retest of the fizzled FRANKLIN shot that had taken place early in the test series. This device had a diameter of 21.4 inches and was 45.8 inches long. Among the nuclear tests, Plumbbob FRANKLIN PRIME was theoretically 1st for percentages of Nb95 and Zr95 in fallout. According to the National Cancer Institute, however, there was no fallout deposition recorded from this test.

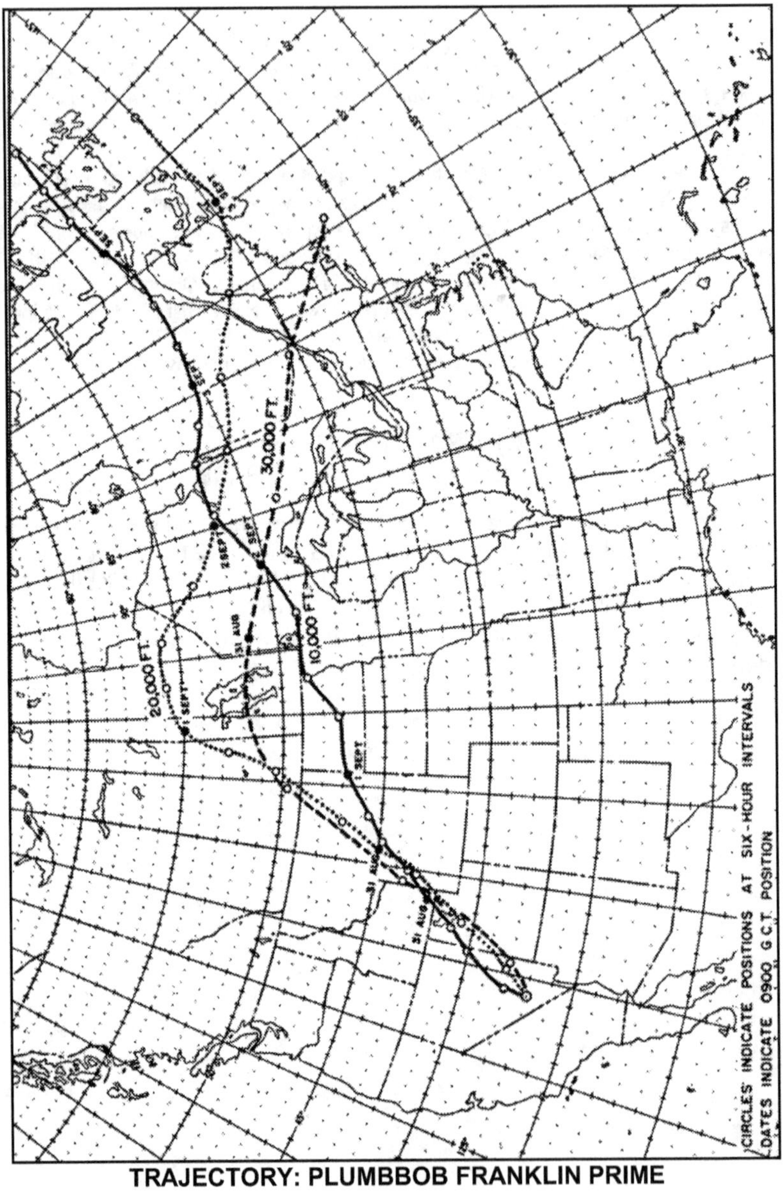

TRAJECTORY: PLUMBBOB FRANKLIN PRIME

SHOT 11: PLUMBBOB SMOKY (PB11)

Plumbbob: Smoky

Detonation Date: 31 Aug 1957 • *Detonation Time:* 5:30 A.M. • *Area:* 2c • *Sponsor:* Livermore Laboratory • *Yield:* 44 kt • *Radiation Level at Ground Zero:* 300 R/hr • *Height of Burst:* 700 ft • *Cloud Top Height:* 38,000 ft

10,000-Foot Trajectory: NEVADA: (Shoshone Peak), Beatty, (Funeral Mountains) • CALIFORNIA: (Death Valley), Baker, Kelso, Essex • ARIZONA: Wikieup, (Snow Mountain), Valentine, Truxton, Peach Springs, Lake Mead, (Hoover Dam) • NEVADA: (Sheep Range), (Timpahute Range), (Grant Range), Preston, Ely, East Ely, (Shell Creek Range), Goshute Indian Reservation • UTAH: Goshute Indian Reservation, (Great Salt Lake Desert), (Cedar Mountains), Grantsville, (Great Salt Lake), (Antelope Island), Brigham City, Wellsville, Logan, Smithfield, Richmond • IDAHO: Weston, Franklin, Whitney, Preston, Benida, Mink Creek, Thatcher, Bench, Grace, Soda Springs, Conda, Freedom • WYOMING: Etna, Freedom, Merna, (Wind River Range), (Wolverine Peak), Fort Washakie, Ethete, Kinnear, Riverton, Moneta, Hiland, Waltman, Powder River, Natrona, Bill • SOUTH DAKOTA: Rockford, Nemo, Piedmont, Elm Springs, Milesville, Hayes, Fort Pierre, Pierre, Crow Creek Indian Reservation, Gann Valley, Crow Lake, Storia, Mitchell, Ethan, Dimrock, Milltown, Menno, Mayfields, Volin, Vermillion, Elk Point, Jefferson, North Sioux City, Sioux City, South Sioux City • IOWA: Bronson, Climbing Hill, Holly Springs, Smithland, Ticonic, Ute, Dunlap, Earling, Westphalia, Marne, Atlantic, Cumberland, Mount Etna, Prescott, Stringtown, Diagonal, Mount Ayr • MISSOURI: Hatfield, Blue Ridge, Gilmar City, Bedford, Avalon, Hale, (Swan Lake), Triplett, Brunswick, Glasgow, Fayette, New Franklin, Rocheport, Wooldridge, Huntsdale, Columbia, Ashland, Holts Summit, Wainwright, Loose Creek, Bland, Bele, Cuba, Steelville, Cherryville, Courtois, Edgehill, (Taum Sauk Mountain, highest point in Missouri), Glover, Sabula, Coldwater, Gipsy, Zalma, Sturdivant, Advance, Painton, Morley, Blodgett, Diehlstadt, Charleston, Wyatt, Wilson City • KENTUCKY: Bardwell, Kirbyton, Fancy Farm, Mayfield, Kirksey, Murray, New Concord • TENNESSEE: Faxon, Waverly, Hurricane Mills, Bold Spring, Spot, Shipps Bend, Centerville, Twomey, Chapel Hill, Mount Pleasant, Culleoka, Mooresville, Lewisburg, Bellfast, Richmond, Booneville, Lynchburg, Winchester, Estill Springs, Decherd, Alto, Sewanee, Monteagle, Sequatchie, Victoria, (Signal Mountain), Ridgeside, McDonald, Waterville, Parksville, Benton, Reliance, Servilla, Cokercreek, (Great Smoky Mountains National Park) • NORTH CAROLINA: Webster, Waynesville, Asheville, Little Switzerland, Celo, Collettsville, Valmead, Kings Creek, Hamptonville, Flint Hill, Rural Hall, Dennis, Reidsville, Ruffin, Pelham, Providence, Blanch, Milton • VIRGINIA: Danville, Turbeville, South Boston, Red Oak, Fairview, Chase City, Dundas, McKenney, DeWitt, Burrowsville, Brandon, Williamsburg, Gloucester, Ware Neck, Gwynn, (Chesapeake Bay), Onancock, Tasley, Mappsville, Chincoteague.

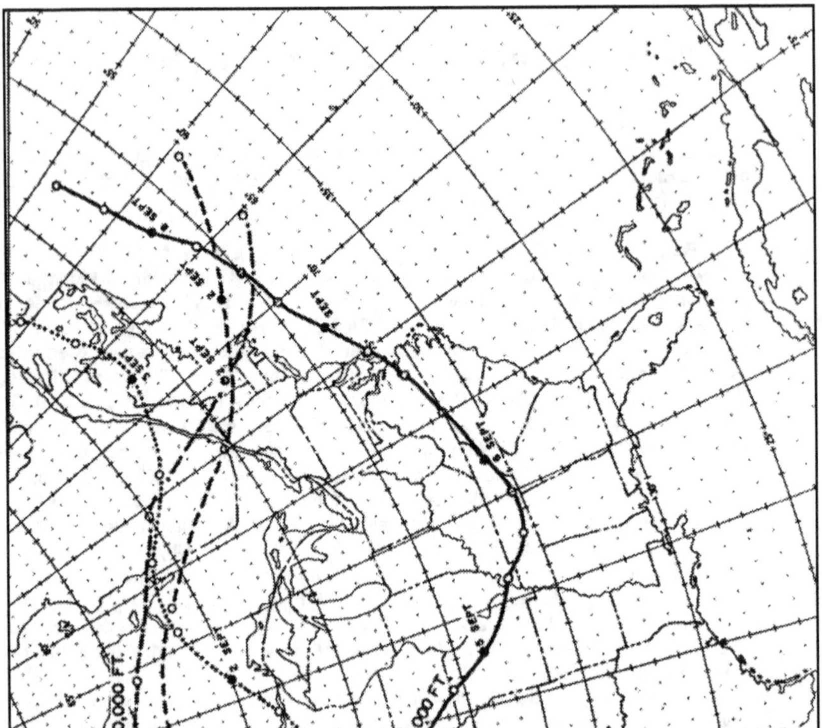

TRAJECTORY: PLUMBBOB SMOKY (PB11)

TOTAL FALLOUT: PLUMBBOB SMOKY

Equal Intervals

nCi/square meter

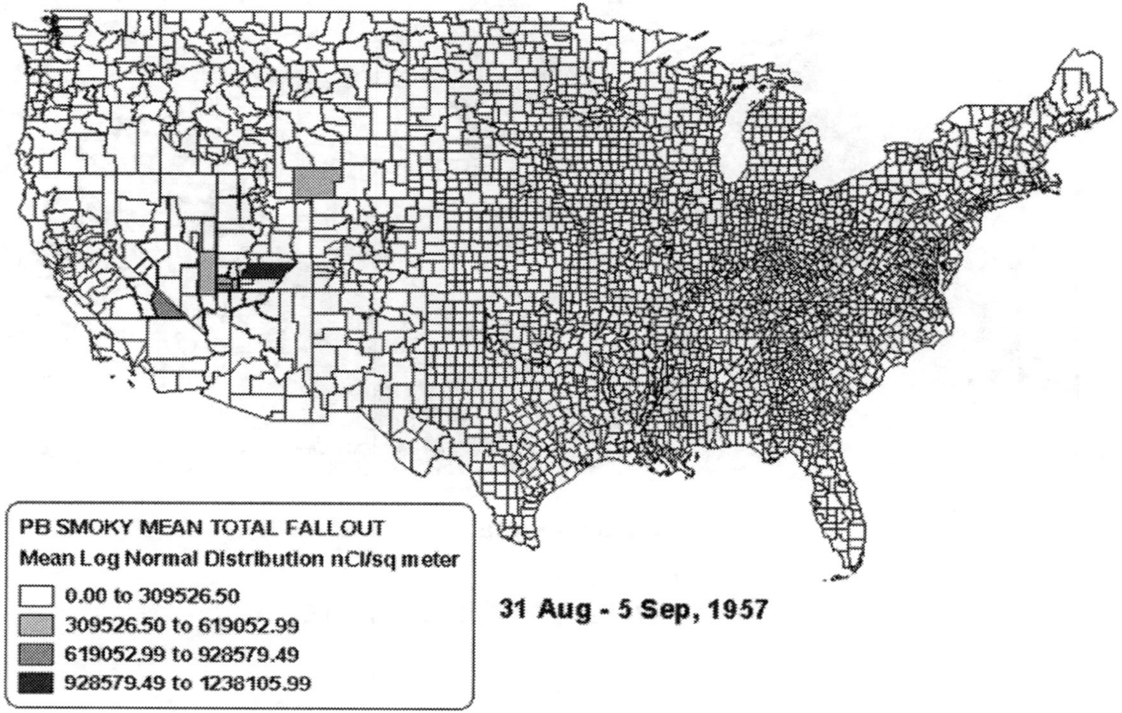

31 Aug - 5 Sep, 1957

TOTAL FALLOUT: PLUMBBOB SMOKY

Equal Number

nCi/square meter

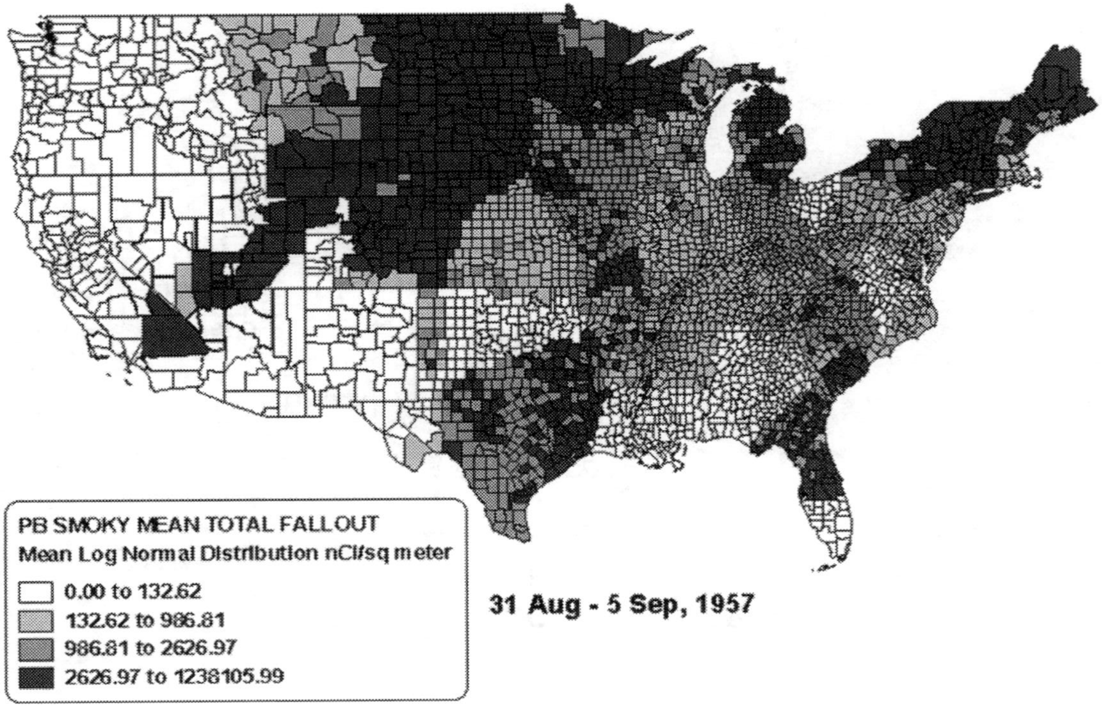

31 Aug - 5 Sep, 1957

CHARACTERISTICS: PLUMBBOB SMOKY

The device for the SMOKY shot was 50 inches in diameter and was 10.5 feet long, and weighed nearly five tons. Among the nuclear tests, Plumbbob SMOKY was 4th in the percentage of Be7 in fallout and 7th in the percentage of Cd115, In115m and In117 in fallout. The deposition pattern for this shot was most similar to Plumbbob NEWTON (PB15) (R: 0.41; T: 25.29). Plumbbob SMOKY was ranked 5th for deposition of Be7, 11th for deposition of Mn54, 15th for deposition of Sr90 and 28th for deposition of Ru106 nationwide.

Counties of highest deposition included Garfield UT, Inyo CA, Washington UT, Sweetwater WY and Lincoln NV. Estimated maximum activities in μCi/sq meter resulting from this shot included the following: Inyo CA—2638; Garfield UT—2566; Washington UT—1046; Sweetwater WY—965. The hottest average activity level (total fallout) resulting from this test occurred at Garfield County, UT: 1536 μCi/sq meter (1.536 milliCuries per square meter activity).

The ratio of total fallout to I-131 for shot SMOKY was relatively low at 87:1.

RELATIVE TOTAL FALLOUT: PB SMOKY

AUG 31, 1957 44 KT

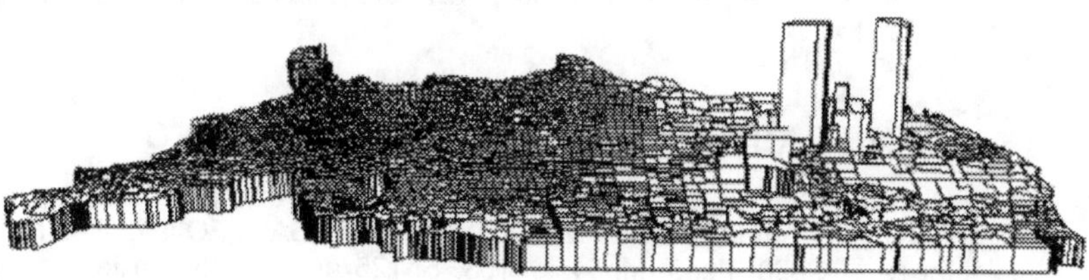

SHOT 12: PLUMBBOB GALILEO (PB12)

1

PLUMBBOB: GALILEO

Detonation Date: 2 Sept 1957 • *Detonation Time:* 5:40 A.M. • *Area:* 1 • *Sponsor:* Los Alamos • *Yield:* 11 kt • *Radiation Level at Ground Zero:* 200 R/hr • *Height of Burst:* 500 ft • *Cloud Top Height:* 37,000 ft

40,000-Foot Trajectory: NEVADA: (Pahute Mesa), Ash Springs • UTAH: (Dixie National Forest), Central, New Harmony, Kanarraville, Hatch, Tropic, (Natural Bridges National Monument) • COLORADO: Northdale, Dove Creek, Ophir, Red Mountain, (Spring Creek Pass), (Mesa Peak), Timpas, Toonerville • KANSAS: Dodge City, Windthorst, Preston, Varner, Waterloo, Garden Plain, Wichita, Latham, Howard, Neodesha, Pittsburg • MISSOURI: Mindonmines, Lockwood, Springfield, Cabool, Pine Crest, Garwod, Bell City • KENTUCKY: Cairo, Paducah • ILLINOIS: Unionville, Hamletsburg, New Liberty • KENTUCKY: Smithland, Salem, Shady Grove, Manitou, Livermore, Bedo, Sulphur Springs, Hardin Springs, Stephensburg, Elizabethtown, Frederickstown, Mooresville, Willisburg, Duncan, McAffee, Mayo, Troy, Lexington, Mount Sterling, Morehead, Soldier, Jacobs, Lawton, Grahn, Grayson, Kilgore, Rush • WEST VIRGINIA: Huntington, Milton, Hometown, Goldtown, Kentuck, Speed, Milestone, Glenville, Troy, Westan, Lost Creek, Quiet Dell, Bridgeport, Grafton, McGee, Arthurdale, Morgantown, Hazelton • VIRGINIA: Markleysburg, Addison, Listonburg, St. Paul, Glencoe, Fairhope, Rainsburg, Clearville, Crystal Spring, Harrisonville, Metal, Shippenburg, Walnut Bottom, Mount Holly Springs, Lisburn, Harrisburg, High Springs, Lebanon, Mount Aetna, Trexler, Wanamakers, Jacksonville, Palmerton, Effort, Reeders, Henryville, Milford, Millrift • NEW YORK: Rio, Port Jervis, Slate Hill, Johnson, Ridgebury, New Hampton, Goshen, Blooming Grove, Salisbury Mills, Mountainville, Cornell-on-Hudson, Nelsonville, Kent Cliffs, Lake Carmel, Putnam Lake • CONNECTICUT: New Fairfield, Hawleyville, Southbury, Waterbury, Marion, Plantsville, Kensington, Berlin, Wethersfield, Rocky Hill, Glastonbury, Manchester, South Willington, Westford, North Ashford, Quinebaug • MASSACHUSETTS: Dudley Hill, Dudley, Webster, Manchaug, West Sutton, Sutton, Riverdale, Northbridge, Upton, Milford, Holliston, Sherborn, Harding, Dover, Needham, Dedham, Brookline, Boston, Winthrop.

CHARACTERISTICS: PLUMBBOB GALILEO

The device for the GALILEO test was 12.4 inches in diameter, 5 ft. long and weighed 848 lbs. Among the nuclear tests, Plumbbob GALILEO is ranked first for percentages of Co60, Fe59, Mn54, Sn123, Sb125 and Te125m in fallout. The deposition pattern for this shot is most similar to that of Plumbbob SHASTA (PB8) (R: 0.44; T: 27.28). Plumbbob GALILEO deposited fallout over 2975 counties, 96 percent of the total U.S. counties. Counties of highest deposition included LaFourche LA, St. Charles LA, Nye NV, Sumner KS, El Paso CO, McPherson KS, Sedgwick KS. Overall, activity levels for this shot were rather low:, On Sep 7, 1957 the highest estimated maximum estimated deposition was for LaFourche, LA for a level of 286 μCi/sq meter. The total fallout to I-131 ratio for this shot was moderately low at 92:1.

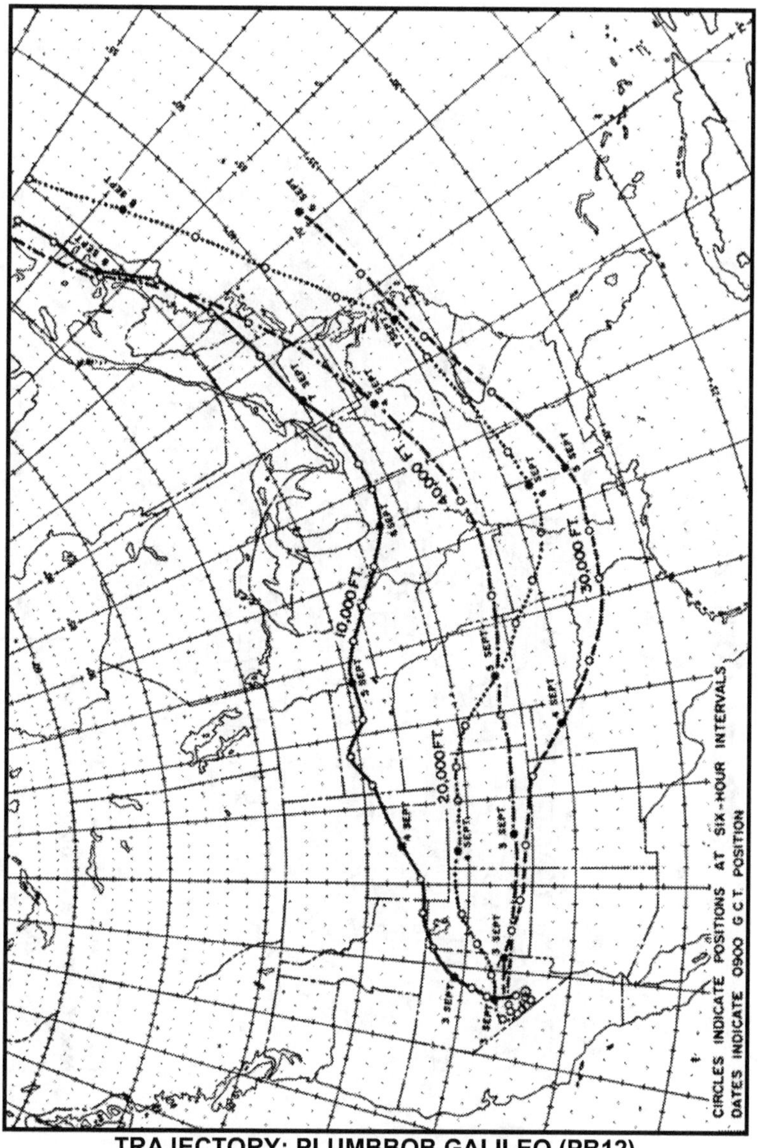

TRAJECTORY: PLUMBBOB GALILEO (PB12)

TOTAL FALLOUT: PLUMBBOB GALILEO

Equal Intervals

nCi/square meter

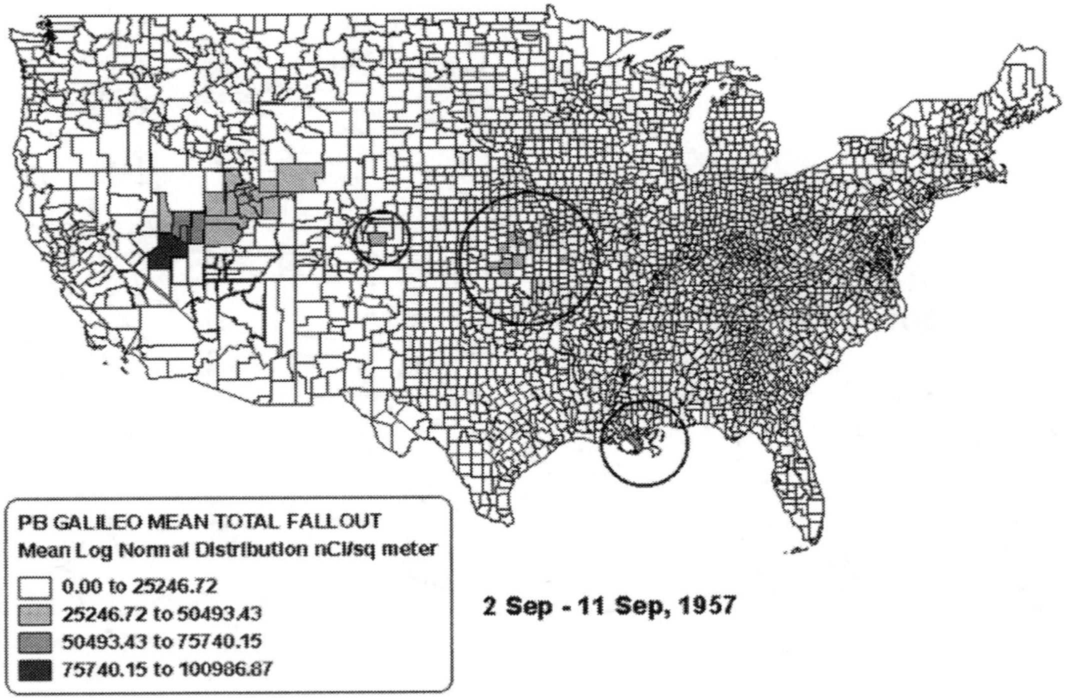

2 Sep - 11 Sep, 1957

SHOT13a,b,c: PLUMBBOB WCL (PB13)

SHOTS WHEELER, COULOMB B AND LAPLACE

Plumbbob: Wheeler

Detonation Date: 6 Sept 1957 • *Detonation Time:* 5:45 A.M. • *Area:* 9a • *Sponsor:* Livermore Laboratory • *Yield:* 197 tons • *Radiation Level at Ground Zero:* 10 R/hr • *Height of Burst:* 500 ft • *Cloud Top Height:* 17,000 ft

10,000-Foot Trajectory: NEVADA: Goldfield, Hawthorne, Smith, Weed Heights, Reno, Sparks • CALIFORNIA: Twin Bridges, Dardanelle, Mammoth Lakes, Death Valley, Needles • ARIZONA: Drake, Sedonia, Winslow, Holbrook • NEW MEXICO: Fence Lake, Grants, Seyboyeta, Los Alamos, Santa Fe, San Ignacio, Mosquero • TEXAS: Dalhart, Morse (30 miles north of Wheeler, Texas), Higgins • OKLAHOMA: Arnet, Cestos, Longdale, Lacey, Bison, Douglas, Covington, Red Rock, Bowring • KANSAS: Dearing, Mound Valley, Fort Scott • MISSOURI: Nevada, Walker, Schell City, Rockville, Deepwater, Clinton, Windsor, Green Ridge, Sedalia, Boonville, Clark, Paris, Hannibal • ILLINOIS: Hall, Beverly, Fishhook, Beardstown, Kilbourne, Delavan, Danvers, El Paso, Streator, Lisbon, Plattville, Plainfield, Cicero, Chicago • MICHIGAN: Beechwood, Lamont, Sparta, Pierson, Lakeview, Clare, Gladwin, Selkirk, Lincoln.

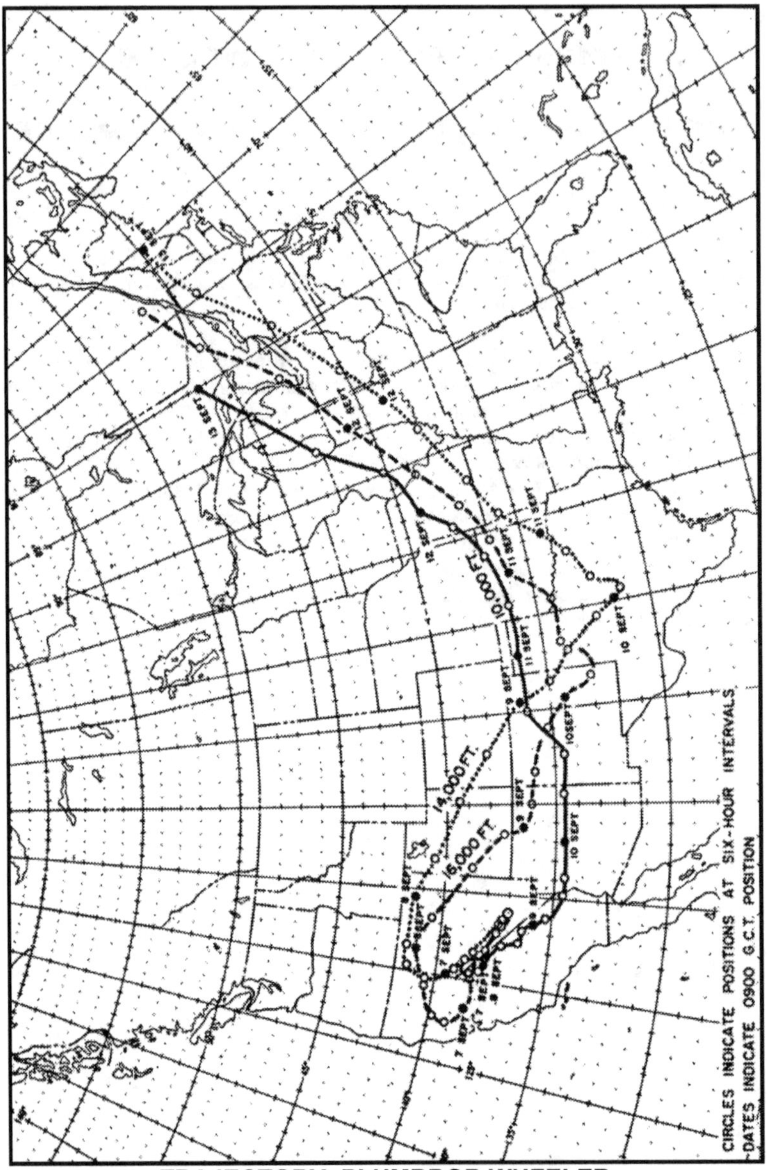

TRAJECTORY: PLUMBBOB WHEELER

Plumbbob: Coulomb B

Detonation Date: 6 Sept 1957 • *Detonation Time:* 1:05 P.M. • *Area:* 3g • *Sponsor:* Los Alamos • *Yield:* 0.3 kt • *Radiation Level at Ground Zero:* 100 R/hr • *Height of Burst:* 3 ft • *Cloud Top Height:* 18,000 ft

10,000-Foot Trajectory: NEVADA: (Shoshone Peak), (Sarcobatus Flat) • CALIFORNIA: (Scotty's Castle), Bishop Indian Reservation, Bishop, Chalfant, Benton • NEVADA: (Boundary Peak, highest point in Nevada), Mount Montgomery, Basalt, (Monte Christo Range), Tonopah, (Mud Lake), (Reveille Range), Tempiute, Caliente • ARIZONA: Littlefield, (Mount Delenbaugh), Red Lake, (Humphreys Peak, highest point in Arizona), Winona, Leupp Corner, Winslow, (Painted Desert), (Petrified Forest National Park), Zuni Indian Reservation • NEW MEXICO: Zuni Indian Reservation, Ramah Navajo Indian Reservation, Laguna Indian Reservation, Los Chavez, Dalles, Bosque Farms, Isleta, Tijeras, Edgewood, Moriarty, Clines Corners, Santa Rosa, Puerto De Luna, McAlister, House, Field, Pleasant Hill • TEXAS: Bovina, Rhea, Friona, Black, Summerfield, Hereford, Umbarger, Canyon, Lake Tanglewood, Pullman, Washburn, Panhandle, White Deer, Kings Mill, Pampa, Hoover, Miami, Canadian, Glazier, Higgins • OKLAHOMA: Shattuck, Gage, Fargo, Fort Supply, Camp Houston • KANSAS: Aetna, Medicine Lodge, Zenda, Murdock, Midway, Andale, Bentley, Sedgwick, Elbing, Florence, Cedar Point, Clements, Elmdale, Dunlap, Bushong, Eskridge, Dover, Topeka, Kiro, Meriden, Rock Creek, Valley Falls, Nortonville, Cummings, Atchison, Doniphan, Wathena, Blair • MISSOURI: St. Joseph, Avenue City, Rochester, Union, Star, King City, Gentryville, New Hampton, Bethany, Ridgeway, Blythedale • IOWA: Allerton, Corydon, Bethlehem, Confidence, Melrose, Georgetown, Hiteman, Eddyville, Rose Hill, What Cheer, Keswick, North English, Parnell, Oxford, Homestead, Amana, Swisher, Shueyville, Ely, Martelle, Anamosa, Amber, Scotch Grove, Bernard, St. Donatus • ILLINOIS: Galena, Charles Mound (highest point in Illinois) • WISCONSIN: South Waynbe, Browntown, Monroe, Brodhead, Footville, Janesville, Lima Center, Whitewater, Mukwonago, Big Bend, Franklin, Muskego, Hales Corners, Greenfield, Cudahy, St. Francis, Milwaukee • MICHIGAN: Montague, Hesperia, Aetna, Ramona, White Cloud, Big Rapids, Rodney, Barryton, Sherman City, Brinton, Farwell, Vernon City, Clare, Beaverton, Rhodes, Bentley, Sterling, Maple Ridge, Twining, Turner, National City, Tawas City, East Tawas, Au Sable, Oscoda.

CHARACTERISTICS: PB WHEELER, COULOMB B AND LAPLACE

Among the nuclear tests, Plumbbob WHEELER was first in La141, while COULOMB B is first for I-132, I-133 and I-135. Plumbbob LAPLACE is 2nd for La141. The composite deposition pattern for these three shots is most similar to Teapot APPLE-2 (TP10) (R: 0.58, T: 39.53). Counties of highest deposition: Buffalo NE, Hall NE, Mitchell KS, Kearney NE, Pawnee KS.

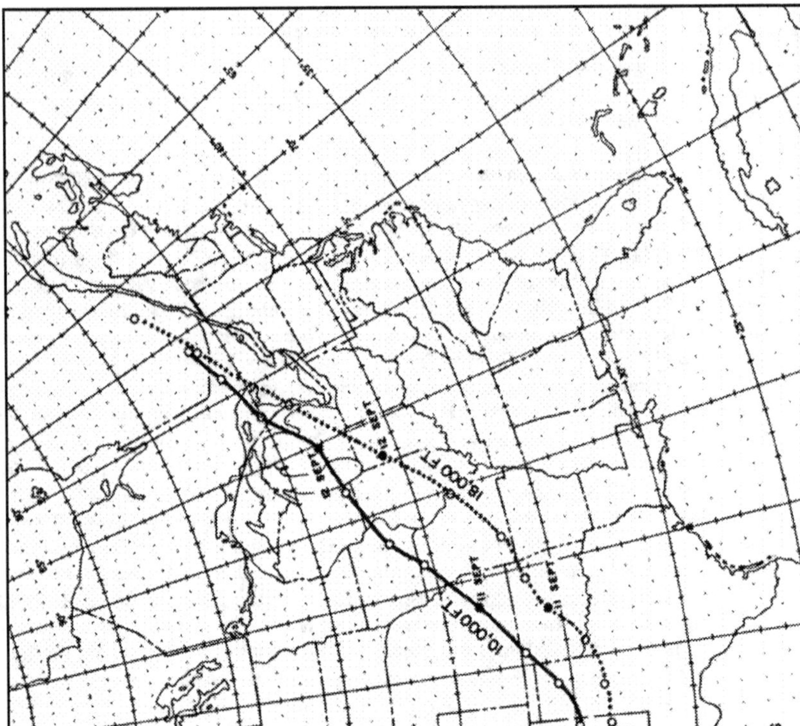

TRAJECTORY: PLUMBBOB COULOMB B

Plumbbob: La Place

Detonation Date: 8 Sept 1957 • *Detonation Time:* 6:00 A.M. • *Area:* 7b • *Sponsor:* Los Alamos • *Yield:* 1 kt • *Radiation Level at Ground Zero:* 50 R/hr • *Height of Burst:* 750 ft • *Cloud Top Height:* 20,000 ft

10,000-Foot Trajectory: NEVADA: (Spotted Range), (Sheep Range), (Valley of Fire State Park), Logandale, Overton • ARIZONA: (Virgin Mountains), (Grand Wash Cliffs), (Grand Canyon National Park), Hualiai Indian Reservation, (Trinity Mountain), Ash Fork, (Casner Mountain), Sedonia, (Mogollon Plateau), Holbrook, Woodruff, Zuni Indian Reservation • NEW MEXICO: Zuni Indian Reservation, Ramah Navajo Indian Reservation, Belen, Los Chavez, (Manzano Mountains), (Mosca Peak), Escabosa, Edgewood, Clines Corners, Pastura, Sumner Lake, House, McAlester, Forrest, Grady, Cameron • TEXAS: Vega, Ady, Exell, Masterson, Pringle, Morse, McKibben, Spearman, Waka, Farnsworth, Perryton, Twitchell, Booker • OKLAHOMA: Clear Lake, Knowles, Gate • KANSAS: Englewood, Sitka, Protection, Coldwater, Wilmore, Belvidere, Croft, Coats, Pratt, Natrona, Preston, Neola, Plevna, Abbyville, Willowbrook, Medora, Hutchinson, Buhler, Elyria, Lehigh, Durham, Pilsen, Lost Springs, Burdick, Delavan, Wilsey, Hessdale, Keene, Williard, Kiro, Topeka, Silver Lake, Elmont, Valley Falls, Half Mound, Nortonville, Cummings, Atchinson, Doniphan, Wathena, Elwood, Blair • MISSOURI: St. Joseph, Nodaway, Amazonia, Savannah, King City, Gentryville, Albany, Martinsville, Brooklyn, Eagleville, Blythdale • IOWA: Davis City, Pleasanton, Woodland, Corydon, Millerton, Melrose, Georgetown, Albia, Avery, Eddyville, Cedar, Wright, Fremont, Delta, Sigourney, South English, Iowa City, Morse, Cedar Bluff, Buchanan, Clarence, Massillon, Lost Nation, Elwood, Maquoketa, Springbrook • ILLINOIS: Savanna Army Depot, Loran, Kent, Pearl City, Freeport, Eleroy, Dakota, Davis, Shirland, Harrison, Rockton, South Beloit • WISCONSIN: Beloit, Clinton, Allen, Darien, Lyons, Springfield, Honey Creek, Rochester, Burlington, Franksville, Wind Point, North Bay • MICHIGAN: Muskegon, Cloverville, Ravenna, Baily, Kent City, Sand Lake, Trufant, Langston, Stanton, Elm Hall, Sumner, Alma, Ithaca, Nelson, Saginaw, Frankentrost, Reese, Richville, Gifford, Akron, Fairgrove, Colling, Cass City, New Greenleaf, Tyro, Parisville, Ruth.

RELATIVE TOTAL FALLOUT: PB WHEELER, COULOMB-B, LAPLACE

SEP 6, 1957 WHEELER: 197 TONS, COULOMB B 300 TONS, LAPLACE: 1 KT

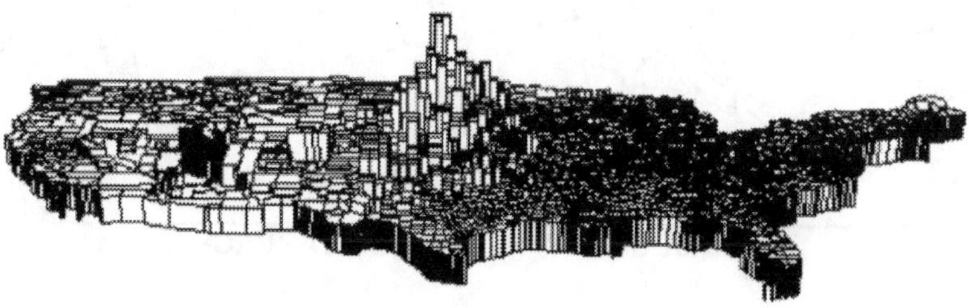

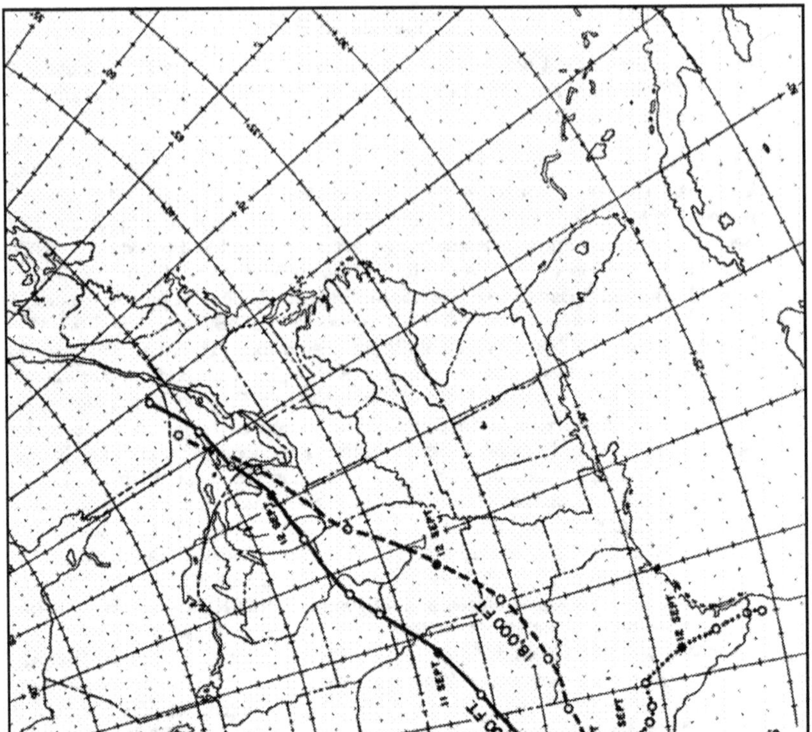

TRAJECTORY: PLUMBBOB LAPLACE

TOTAL FALLOUT: PB WHEELER, COULOMB B, LAPLACE

Equal Intervals

nCi/square meter

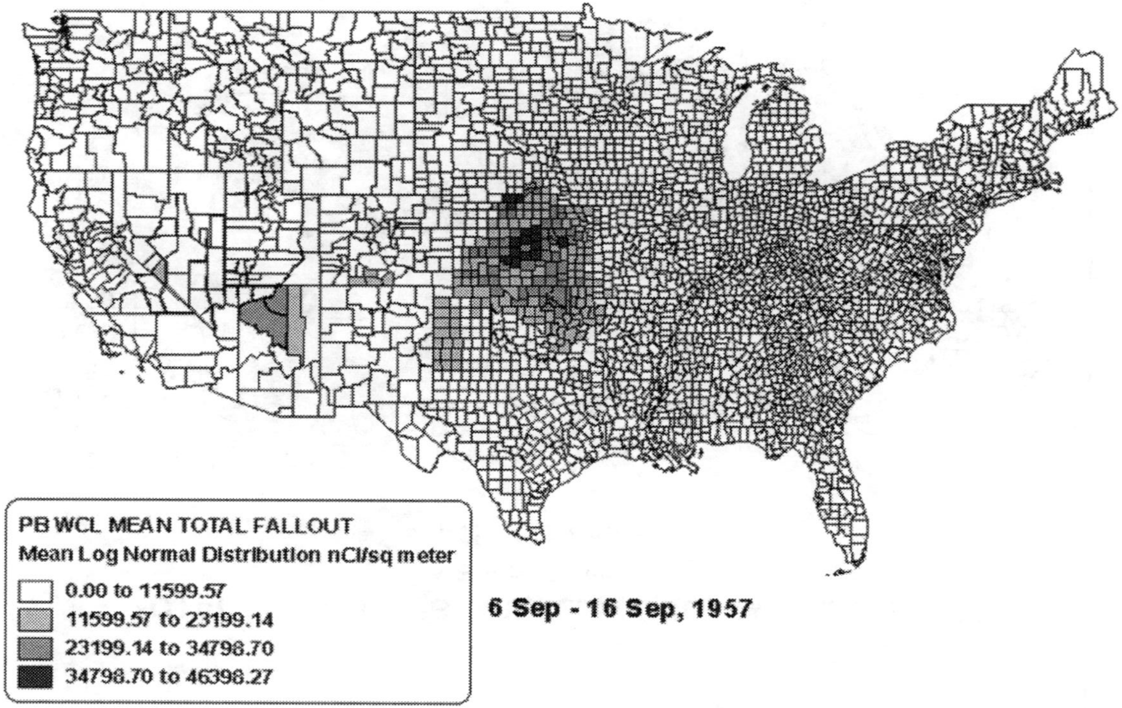

TOTAL FALLOUT: PB WHEELER, COULOMB B, LAPLACE

Equal Number

nCi/square meter

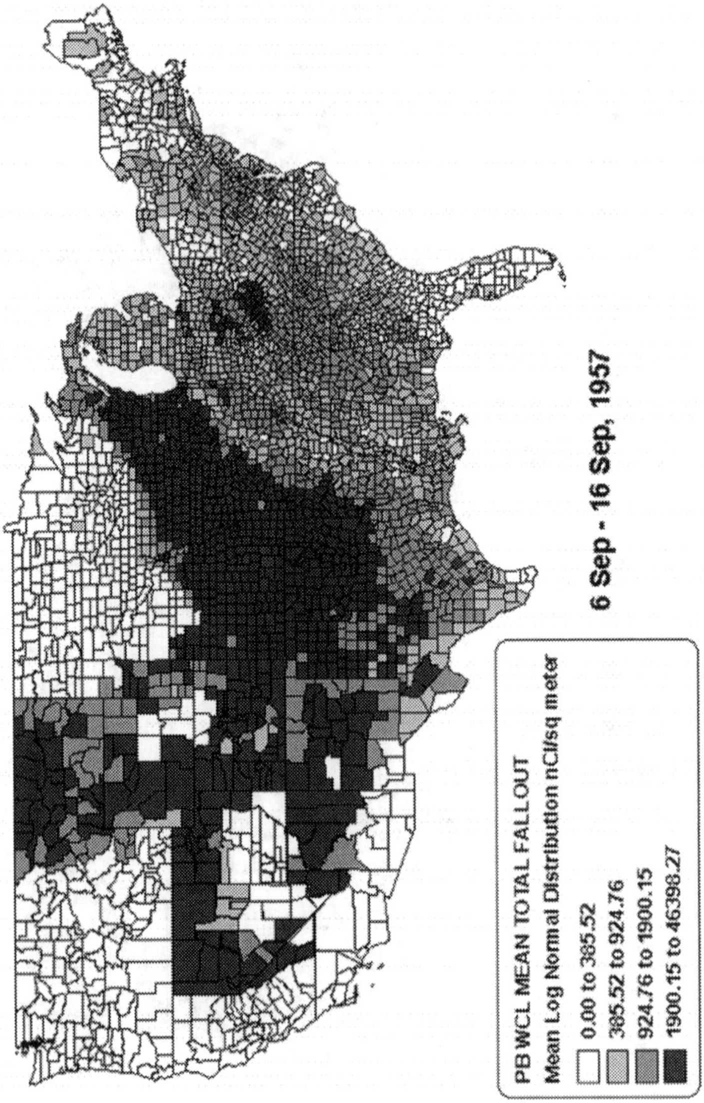

6 Sep - 16 Sep, 1957

SHOT 14: PLUMBBOB FIZEAU (PB14)

Plumbbob: Fizeau

Detonation Date: 14 Sept 1957 • *Detonation Time:* 9:45 A.M. • *Area:* 3b • *Sponsor:* Los Alamos • *Yield:* 11 kt • *Radiation Level at Ground Zero:* 100 R/hr • *Height of Burst:* 500 ft • *Cloud Top Height:* 40,000 ft

40,000-Foot Trajectory: NEVADA: (Pahute Mesa), Nellis Air Force Range, Tonopah, Luning, Shurz, Yerington, Wabuska, Carson City, Silver City, Glenbrook • CALIFORNIA: Tahoe City, Baxter, Emigrant Gap, Dutch Flat, Grass Valley, Nevada City, Loma Rica, Live Oak, Pennington, Colusa Indian Reservation, Maxwell, Ladoga, Ukiah, Calpella, Navarro, Elk, San Francisco, Alameda, Oakland, Berkeley, Orinda, Danville, Brentwood, Middle River, Byron, Stockton, Holt, French Camp, Farmington, Copperopolis, Tuttletown, Tuolumne, Sonora, (Yosemite National Park), Lee Vining • NEVADA: (Boundary Peak, highest point in Nevada), Montgomery, Coaldale, Tonopah, Warm Springs, Ursine, Pioche • UTAH: Beryl, Land, Parowan, Tropic, Cannonville, Henrieville, Mexican Hat • COLORADO: Ute Mountain Indian Reservation, Marvel, Oxford, Chimney Rock, Pagosa Springs, Capulin, La Jara, Bountiful, Sanford, Lasauses, San Acacio, Viejo San Acacio, San Luis, Guinare, Ludlow, Hoehne, Villegreen, Kim, Stonington • KANSAS: Richfield, Moscow, Fowler, Minneola, Croft, Coats, St. Leo, Murdock, Cheney, Goddard, Schulte, Wichita, Andover, Augusta, Haverhill, Leon, Climax, Toronto, Batesville, Yates Center, Rose, Piqua, Iola, Gas, LaHarpe, Moran, Bayar, Xenia, Mapleton, Mantey, Harding, Fulton • MISSOURI: Hume, Rich Hill, Prairie City, Ohio, Mount Zion, Warsaw, Whitakerville, Gravois Mills, Eldon, Etterville, Henley, St. Thomas, Westphalia, Mount Sterling, Drake, Stony Hill, Washington, Clover Bottom, Labadie, Pond, Ballwin, Manchester, St. Louis • ILLINOIS: East St. Louis, Collinsville, Edwardsville, Marine, Old Ripley, Pocahontas, Vandalia, Haferstown, Bluff City, Browstown, Altamont, Effingham, Teutopolis, Montrose, Jewett, Hidalgo, Greenup, Casey, Martinsville, Marshall • INDIANA: Prairieton, Terre Haute, Cory, Prairie City, Poland, Cunot, Quincy, Lewisville, Centerton, Brooklyn, Bargersville, Whitelaud, Needham, Shelbyville, Rays Crossing, Manilla, Rushville, Glenwood, Connersville, Alquina, Brownsville, Liberty, Kitchell • OHIO: Sugar Valley, Glenwood, Little Richmond, Trotwood, Northridge, Dayton, Sulpher Grove, Fairborn, Medway, Snyderville, Springfield, Catawba, Tradersvile, Plumwood, Hayden, Linworth, Minerva Park, Croton, Lock, Homer, Millwood, Tiverton, Killbuck, Berlin, Trail, Winesburg, Beech City, Boliver, Howenstein, North Industry, Mapleton, Robertsville, Bayard, New Alexander, New Garden, Guilford, East Fairfield, Signal, East Palestine, Negley • PENNSYLVANIA: New Galilee, Ellwood City, Mount Chestnut, Windward Heights, Chicora, Frogtown, Wattersonville, Tidal, Climax, New Bethlehem, South Bethlehem, Ringgold, Panic, Prescottville, Sandy, Oklahoma, Home Camp, (Moshannon State Forest), Keating, Cooks Run, West Renovo, Haneyville, Waterville, Quiggleville, Warrensville, Muncy Valley, Sonestown, Nordmont, Beaumont, Orange, Upper Exeter, Ransom, Scranton, Mount Cobb, Hamlin, Lakeville, Tafton, Greeley, Shohola, Pond Eddy, Millrift • NEW JERSEY: (High Point, highest point in New Jersey: 1,803 feet) • NEW YORK: Port Jervis, Huguenot, Mechanicstown, Campbell Hall, New Windsor, Beacon, Farmers Mills, Ludingtonville, Patterson, Towners • CONNECTICUT: New Fairfield, Brookfield, Brookfield Center, Roxbury Falls, Woodbury, Middleburg, Waterbury, Woodtick, Southington, Kensington, New Britain, Berlin, East Berlin, Rocky Hill, East Glastonbury, Diamond Lake, Willimantic, North Windham, South Killingly • RHODE ISLAND: Foster, North Scituate, Saundersville, Providence, East Providence • MASSACHUSETTS: Taunton, Raynham Center, North Middleboro, Warrentown, Silver Lake, Kingston, Tinkertown, South Duxbury.

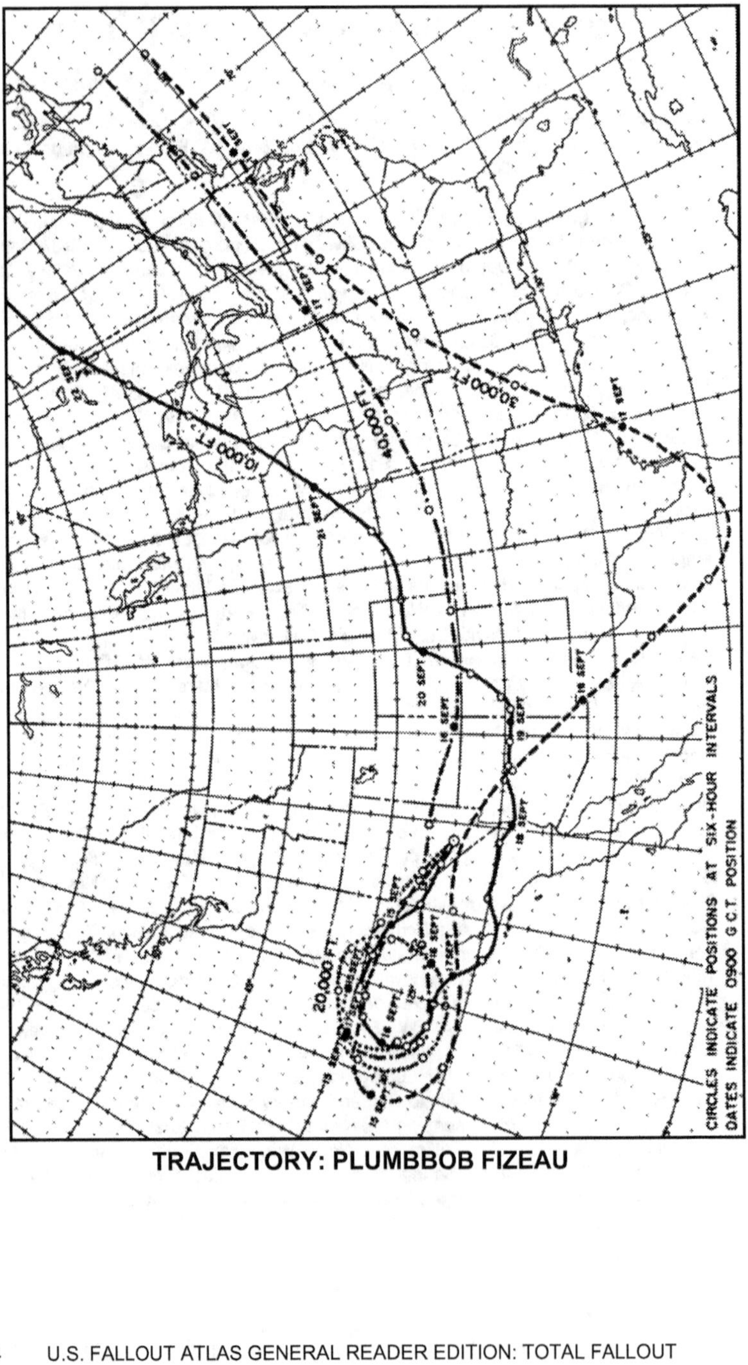

TRAJECTORY: PLUMBBOB FIZEAU

CHARACTERISTICS: PLUMBBOB FIZEAU

The device used for the FIZEAU shot was 16 inches in diameter, 31.75 inches long, and weighed just 131.3 lbs. Among the nuclear tests, Plumbbob FIZEAU (PB14) is ranked 5th for percentages of Cd115, In115m and In117 in fallout. The deposition pattern for this shot was most similar to the composite pattern for the Plumbbob shots WHEELER, COULOMB B and LAPLACE (R: 0.29, T: 16.76). Fallout from this test was deposited on only 651 counties (21 percent of the total). Counties of highest deposition included Mineral NV, Carson City NV, Douglas NV, Lyon NV, Storey NV. Fallout levels were relatively light. The estimated maximum activity for the series, occurring at Mineral, NV on Sep 14, 1957 was only 119 Ci/sq meter.

The total fallout to I-131 ratio for this shot was relatively low at 75:1.

RELATIVE TOTAL FALLOUT: PB FIZEAU

SEP 14, 1957 11 KT

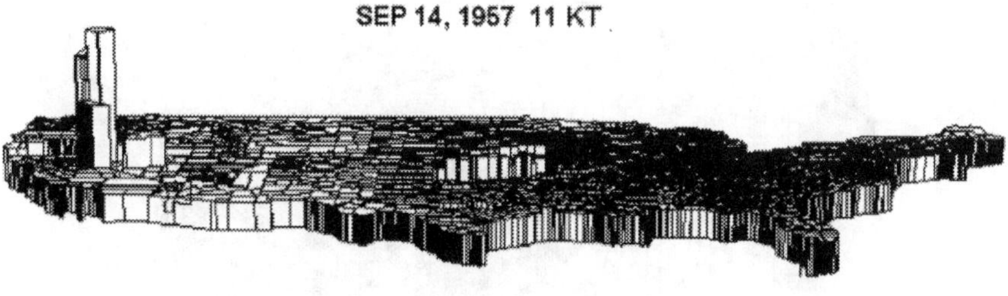

TRAJECTORY: PLUMBBOB FIZEAU (PB14)

TOTAL FALLOUT: PLUMBBOB FIZEAU

Equal Intervals

nCi/square meter

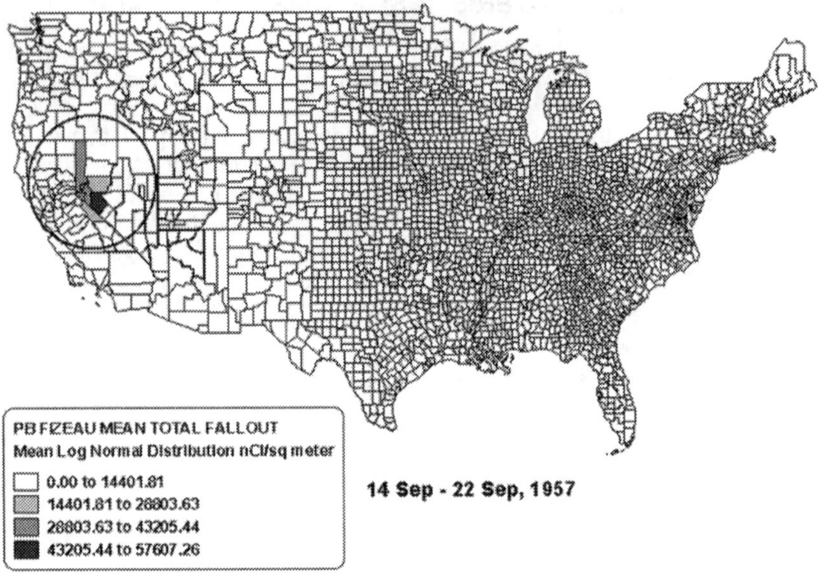

TOTAL FALLOUT: PLUMBBOB FIZEAU

Equal Number

nCi/square meter

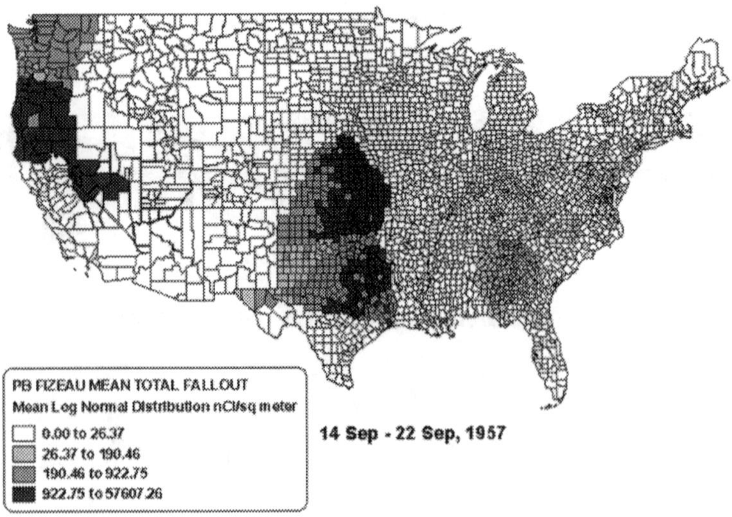

SHOT 15: PLUMBBOB NEWTON (PB15)

Plumbbob: Newton

Detonation Date: 16 Sept 1957 • *Detonation Time:* 5:50 A.M. • *Area:* 7b • *Sponsor:* Los Alamos • *Yield:* 12 kt • *Radiation Level at Ground Zero:* 50 R/hr • *Height of Burst:* 1,500 ft • *Cloud Top Height:* 32,000 ft

30,000-Foot Trajectory: NEVADA: (Papoose Lake), (Papoose Range), (Pahranagat Range), Ash Springs, Caliente, Panaca • UTAH: Uvada, Modena, Lund, Minersville, Beaver, (Delano Peak), Loa, Bicknel, Teasdale, Torrey, Cainesville, Hanksville, (Canyonlands National Park), (Lisbon Valley) • COLORADO: Norwood, Ridgeway, (Sheep Mountain), Powderhorn, Doyleville, Sargents, (Ouray Peak), Poncha Pass, Wellsville, Howard, Parkdale, (Royal Gorge), Canon City, Fort Carson, Fountain, Widefield, Ellicott, Yoder, Rush, Hugo, Arriba, Flagler, Hale • KANSAS: St. Francis • NEBRASKA: Benkelman, Danbury, Naponee, Red Cloud, Superior, Hardy, Byron, Chester, Hubbell, Thompson, Steele City, Barneston, DuBois, Falls City, Preston, Rulo • KANSAS: Cedar Bluffs, Lyle, Long Island, Woodruff, Thornburg, Womer, Northbranch, Burr Oak, Webber, Republic, Narka, Mahaska, Hollenberg, Oketo, Summerfield, Bern, Berwick, Reserve, Iowa Sac and Fox Indian Reservation, White Cloud • MISSOURI: Fortescue, Mound City, Maitland, Barnard, Guilford, Stanberry, Conception Junction, Albany, Bethany, Mount Moriah, Modena, Mill Grove, Newtown, Pollock, Lemons, Graysville, Worthington, Downing, Memphis, Arbela, Farmington, Athens • IOWA: Argyle, Summitville, Montrose, New Boston • ILLINOIS: Nauvoo, Fort Madison, Niota, Raritan, Swan Creek, St. Augustine, Rapatee, Yates City, Elmwood, Kickapoo, Alta, Rome, Chillicothe, Washburn, Rutland, Long Point, Manville, Blackstone, Gardner, Braceville, Symerton, Wilton Center, Monee, Chicago Heights • INDIANA: Highland, Griffith, Gary, Lake Station, Ogden Dunes, Burns Harbor, Town of Pines, Trail Creek • MICHIGAN: Bertrand, Eagle Lake, Adamsville, Constantine, Centreville, Nottawa, Colon, Fairfax, Sherwood, Hodunk, Union City, Tekonsha, Homer, Concord, Parma, Jackson, Waterloo, Hamburg, New Hudson, Farmington Hills, Pontiac, Birmingham, Rochester Hills, New Baltimore, Marine City • NEW YORK: Sandy Pond, Ellisburg, Pierrepont Manor, Adams, Lorraine, Worth, Copenhagen, West Carthage, Carthage, Indian River, (Alder Bed Mountain), Sabattis, (Seymour Mountain), Keene Valley, Elizabethtown, Wadhams, Whallonsburg • VERMONT: Westport, Vergennes, Irasville, Waitsfield, Montpelier, Berlin, Barre, Groton, Ryegate • NEW HAMPSHIRE: Lyman, Lisbon, Franconia, Bethlehem, Twin Mountain, Bretton Woods, North Chatham • MAINE: Center, Louell, Waterford, South Paris, Hebron, North Monmouth, Monmouth, Winthrop, Hallowell, Gardinier, Randolph, Coopers Mills, Washington, Northport, Lincolnville, Harborside, Islesboro, Brooksville, Blue Falls, North Sedgwick, Bar Harbor, Corea, Winter Harbor, Petit Manan Pt.

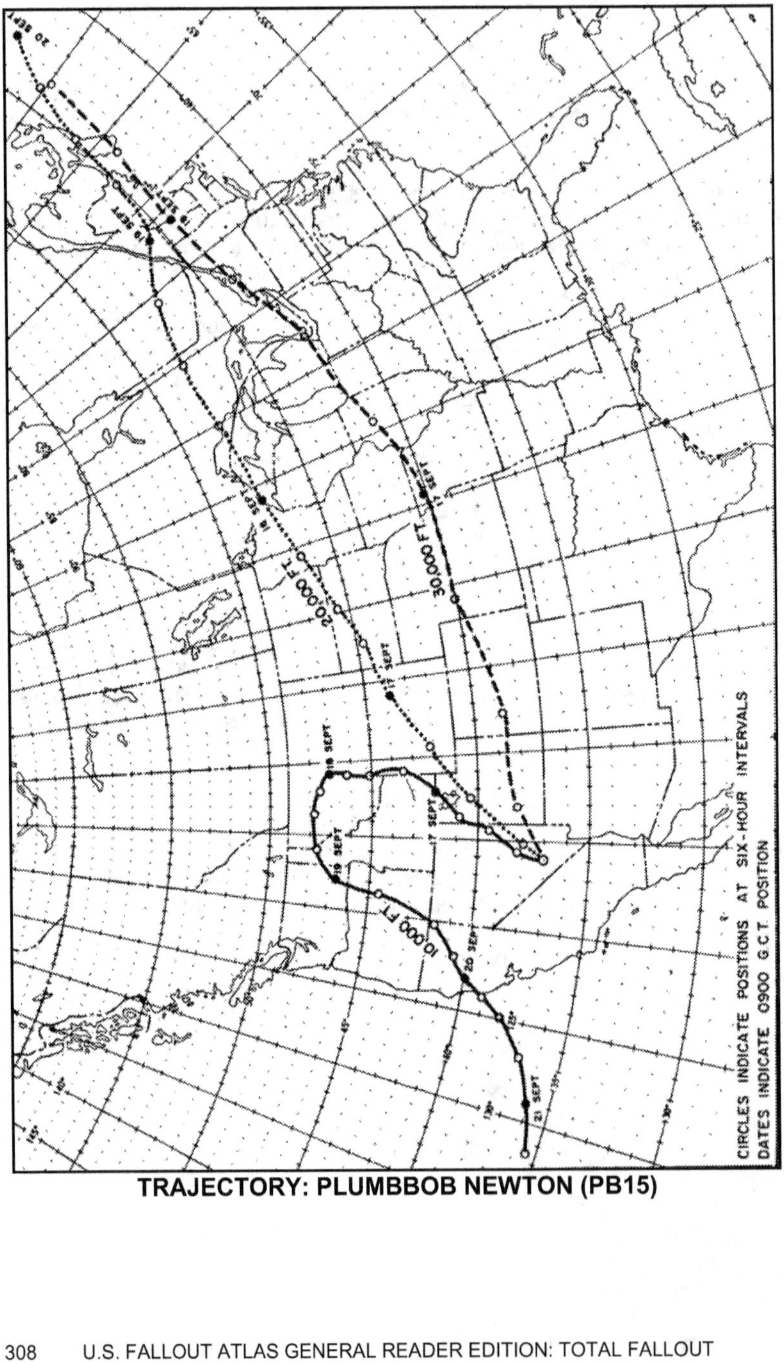

TRAJECTORY: PLUMBBOB NEWTON (PB15)

CHARACTERISTICS: PLUMBBOB NEWTON

The device used for the NEWTON shot was 28 inches in diameter, 39 inches long and weighed 1346 lbs. Among the nuclear tests, Plumbbob NEWTON was first for percentage of Cm242 in fallout. The deposition pattern for this shot is most similar to that of Tumbler-Snapper HOW (TS8) (R: 0.53, T: 34.85). Fallout from this test was deposited on 1531 counties (49 percent of the total.) Counties of highest deposition included Garfield NE, Codington SD, Kingsbury SD, Hamlin SD, Miner SD and Clark SD. Fallout levels were relatively light. The estimated maximum activity at Codington, SD, occurring on Sep 17, 1957, was only 204 μCi/sq meter.

TOTAL FALLOUT: PLUMBBOB NEWTON
Equal Intervals
nCi/square meter

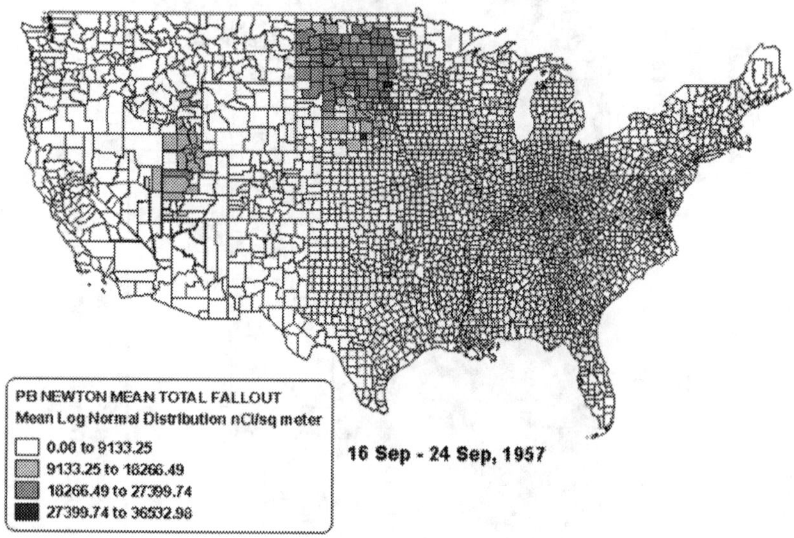

TOTAL FALLOUT: PLUMBBOB NEWTON

Equal Number

nCi/square meter

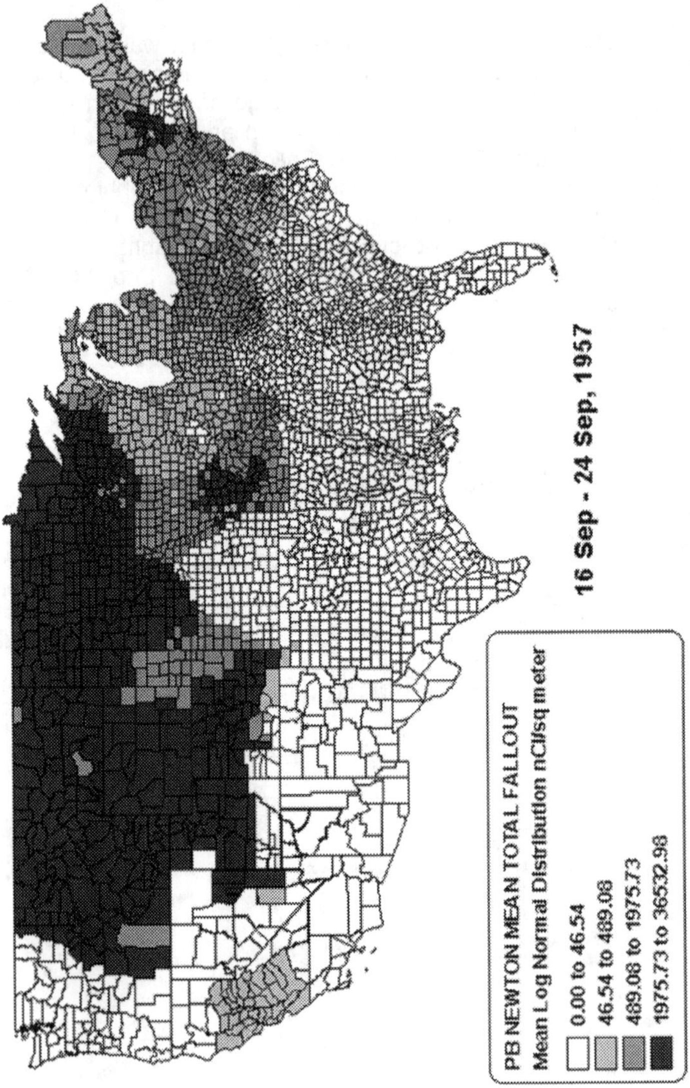

RELATIVE TOTAL FALLOUT: PB NEWTON

SEP 14, 1957 12 KT

SHOT 16: PLUMBBOB WHITNEY (PB16)

PLUMBBOB: WHITNEY

Detonation Date: 23 Sept 1957 • *Detonation Time:* 5:30 A.M. • *Area:* 2 • *Sponsor:* Livermore Laboratory • *Yield:* 19 kt • *Radiation Level at Ground Zero:* 200 R/hr • *Height of Burst:* 500 ft • *Cloud Top Height:* 30,000 ft

10,000-Foot Trajectory: NEVADA: Warm Springs, Crescent Valley, Carlin, Metropolis, Pequop, Montello • UTAH: (Grouse Creek Mountains), Park Valley • IDAHO: Holbrook, Buist, Arbon, Pocatello, Chubbuck, Riverside, Terreton, Dubois, Spencer, Humphrey • MONTANA: Monida, Sheridan, Twin Bridges, Silver Star, Butte, McQueen, Basin, Rimini, Blossburg, Marysville, Canyon Creek, Choteau, Miller Colony, Bynum, Pendroy, Valier, Ethridge, Sweetgrass.

RELATIVE TOTAL FALLOUT: PB WHITNEY

SEP 23, 1957 19 KT

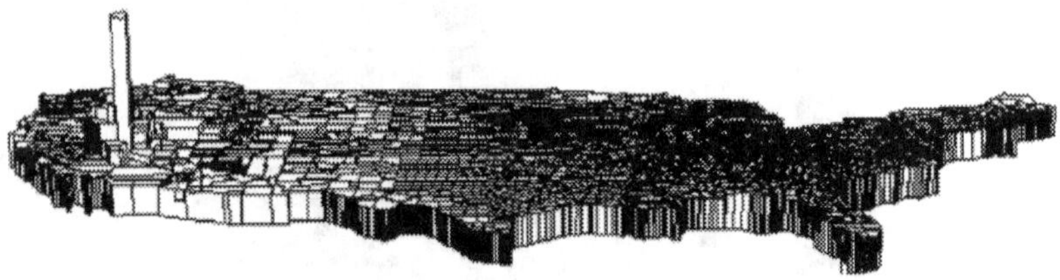

CHARACTERISTICS: PLUMBBOB WHITNEY

Among the nuclear tests, Plumbbob WHITNEY is ranked 2nd for percentages of Am241 and Fe59 and 3rd for percentage of Pb203 in fallout. The deposition pattern for this shot is most similar to that of Buster-Jangle CHARLIE (R: 0.30; T: 17.78). WHITNEY deposited fallout on 2982 counties (96 percent). Counties of highest deposition included Esmeralda NV, Churchill NV, Mineral NV, Lander NV, Mono CA. Fallout from this test had low to moderate activity. Esmeralda, NV received an estimated maximum activity of 569 μCi/sq meter, while Mineral, NV received an estimated maximum activity of only 179 μCi/sq meter, both on Sep 23, 1957.

The total fallout to I-131 ratio for Plumbbob WHITNEY was relatively low at 91:1.

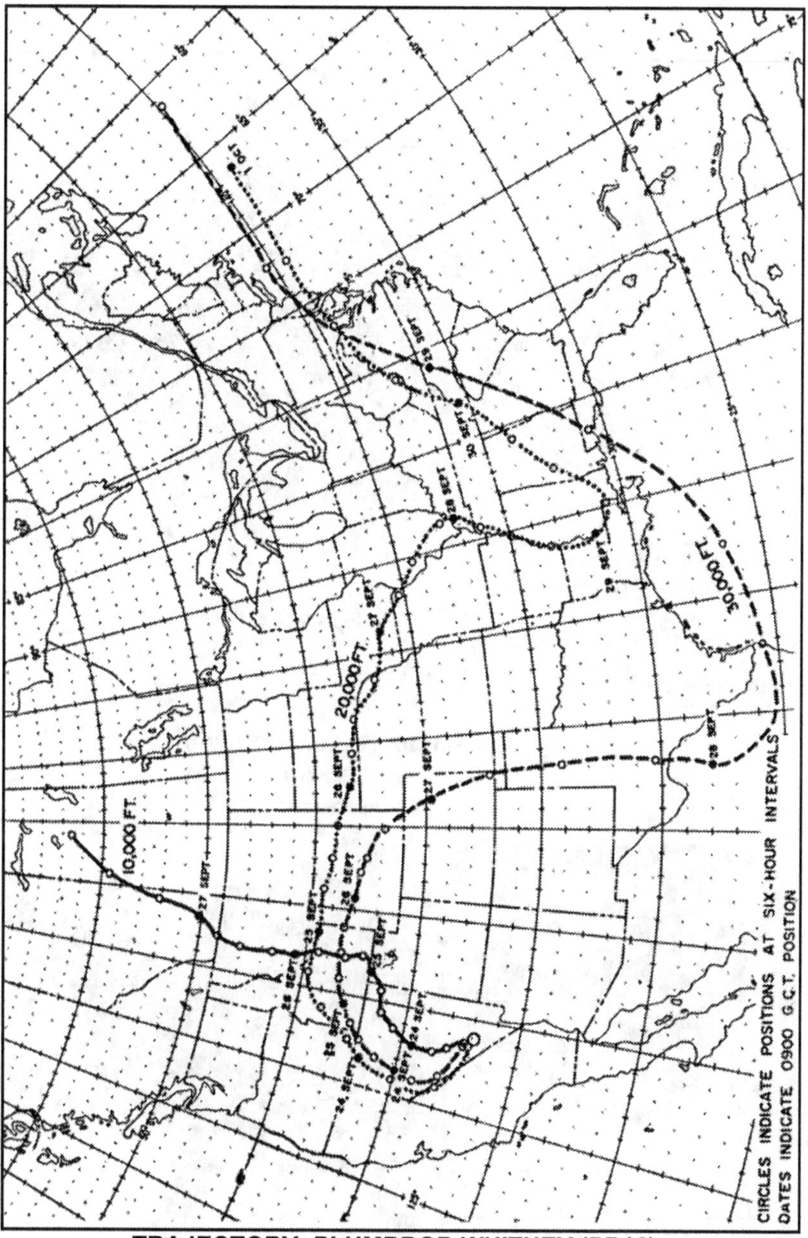

TRAJECTORY: PLUMBBOB WHITNEY (PB16)

TOTAL FALLOUT: PLUMBBOB WHITNEY

Equal Intervals

nCi/square meter

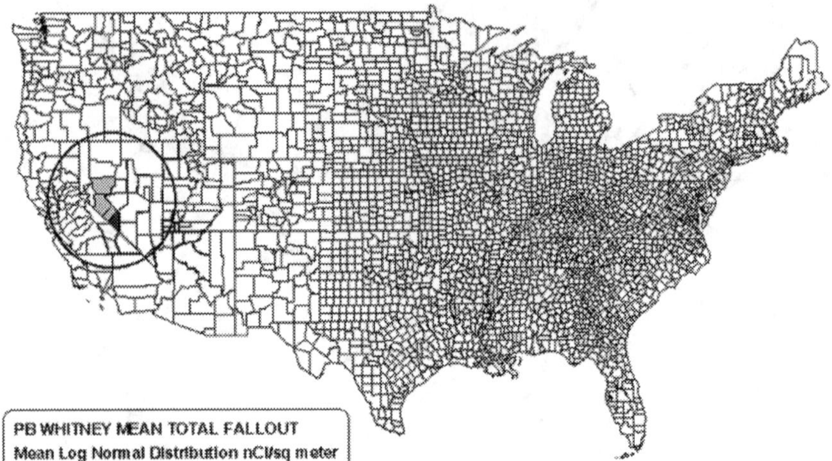

PB WHITNEY MEAN TOTAL FALLOUT
Mean Log Normal Distribution nCi/sq meter

- 0.00 to 68588.42
- 68588.42 to 137176.83
- 137176.83 to 205765.25
- 205765.25 to 274353.67

23 Sep - 5 Oct, 1957

TOTAL FALLOUT: PLUMBBOB WHITNEY

Equal Number

nCi/square meter

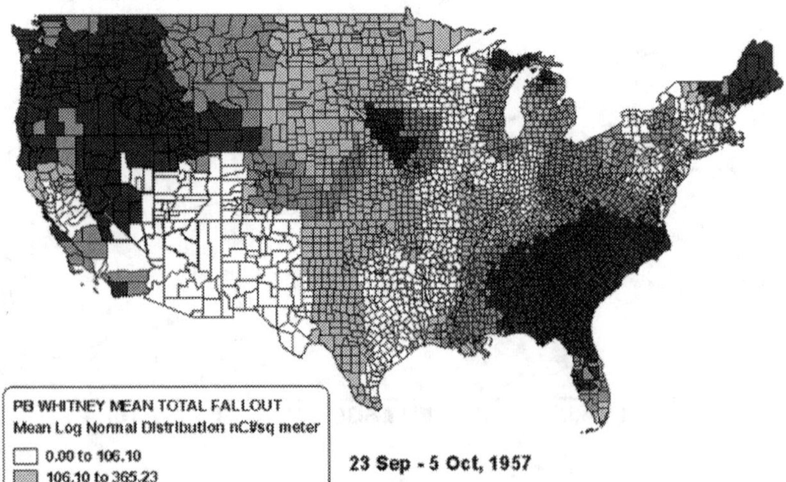

PB WHITNEY MEAN TOTAL FALLOUT
Mean Log Normal Distribution nCi/sq meter

- 0.00 to 106.10
- 106.10 to 365.23
- 365.23 to 1120.33
- 1120.33 to 274353.67

23 Sep - 5 Oct, 1957

SHOT 17: PLUMBBOB CHARLESTON (PB17)

PLUMBBOB: CHARLESTON

Detonation Date: 28 Sept 1957 • *Detonation Time:* 6:00 A.M. • *Area:* 9a • *Sponsor:* Livermore Laboratory • *Yield:* 12 kt • *Radiation Level at Ground Zero:* 50 R/hr • *Height of Burst:* 1,500 ft • *Cloud Top Height:* 32,000 ft

20,000-Foot Trajectory: NEVADA: (Belted Range), (Revielle Range), (Pancake Range), (Diamond Mountains), Ruby Valley, Wells, (Delano Peak) • IDAHO: Oakley, Marion, Burley, Rupert, Minidoka, Atomic City, Taber, Roberts, Terneton, Hamer, St. Anthony, Island Park • WYOMING: (Yellowstone National Park: Old Faithful), Canyon, Silver Gate • MONTANA: (Granite Peak; highest point in Montana), Alpine, Red Lodge, Luther, Roberts, Franberg, Edgar, Pryor, Crow Indian Reservation, Hardin, Brandenberg, Valborg, Powderville, Ekalaka, Mill Iron • SOUTH DAKOTA: Buffalo, Reva, Sorum, Usta, Red Elm, Dupree, Eagle Butte, Mission Ridge, Blunt, Stephen, Wessington Springs, Letcher, Farmer, Emery, Monroe, Lennox, Worthing, Canton • IOWA: Beloit, Lebanon, Struble, West LeMars, LeMars, Moville, Climbing Hill, Holly Springs, Rodney, Castana, Moorhead, Logan, Neola, McClelland, Treynor, Hastings, Malvern, Randolph, Anderson, Riverton, Hamburg • MISSOURI: Watson • NEBRASKA: Brownville, Nemaha, Verdun, Dawson • KANSAS: Berwick, Sabetha, Kelley, Goff, Corning, Havensville, Onaga, Louisville, Wamego, Wabaunsee, Alta Vista, Council Grove, Strong City, Elmdale, Matfield Green, El Dorado, Haverhill, Douglass, Rock, Akron, Udall, Oxford, Geuda Springs, Ashton • OKLAHOMA: Braman, Billings, Covington, Hayward, Lovell, Crescent, Cashion, Piedmont, El Reno, Union City, Cogar, Dutton, Anadarko, Cyril, Apache, Fletcher, Geronimo, Devol • TEXAS: Burkburnett, Mankins, Dundee, Padgett, Crystal Falls, Breckenridge, Eastland, Rising Star, Brownwood, Trickham, Whon, Fife, Lohn, Pear Valley, London, Junction, Segovia, Barksdale, Camp Wood, (Military Mountain), (Kelly Peak), (Boiling Mountain), (Salmon Peak), (Turkey Mountain), Spofford, (Cline Mountain), Blewett, Dabney, Normandy, Eagle Pass.

CHARACTERISTICS: PLUMBBOB CHARLESTON

The device used for the CHARLESTON shot was 22.4 inches in diameter, 46.6 inches long and weighed 1225 lbs. Among the nuclear tests, Plumbbob CHARLESTON was ranked 2nd for percentage of Ce141 in fallout. The deposition pattern for this shot is most similar to Plumbbob WILSON (PB2) (R: 0.35; T: 20.94). Fallout from this test covered 3067 counties, ranking its fallout pattern the 3rd most dispersed in the above-ground test series (after Tumbler-Snapper FOX (1) and Upshot-Knothole SIMON (2). Fallout activity, however, was light. Counties with highest fallout amounts included Coffee TN Oconee SC and Clay NC, each with an estimated maximum activity: 137 μCi/sq meter on Sep 30, 1957. The counties of Grundy TN, Sequatchie TN, Bedford TN, and Towns GA each received slightly lower estimated maximums on the same day. The total fallout to I-131 ratio for Plumbbob CHARLESTON for day 1 was average at 106:1.

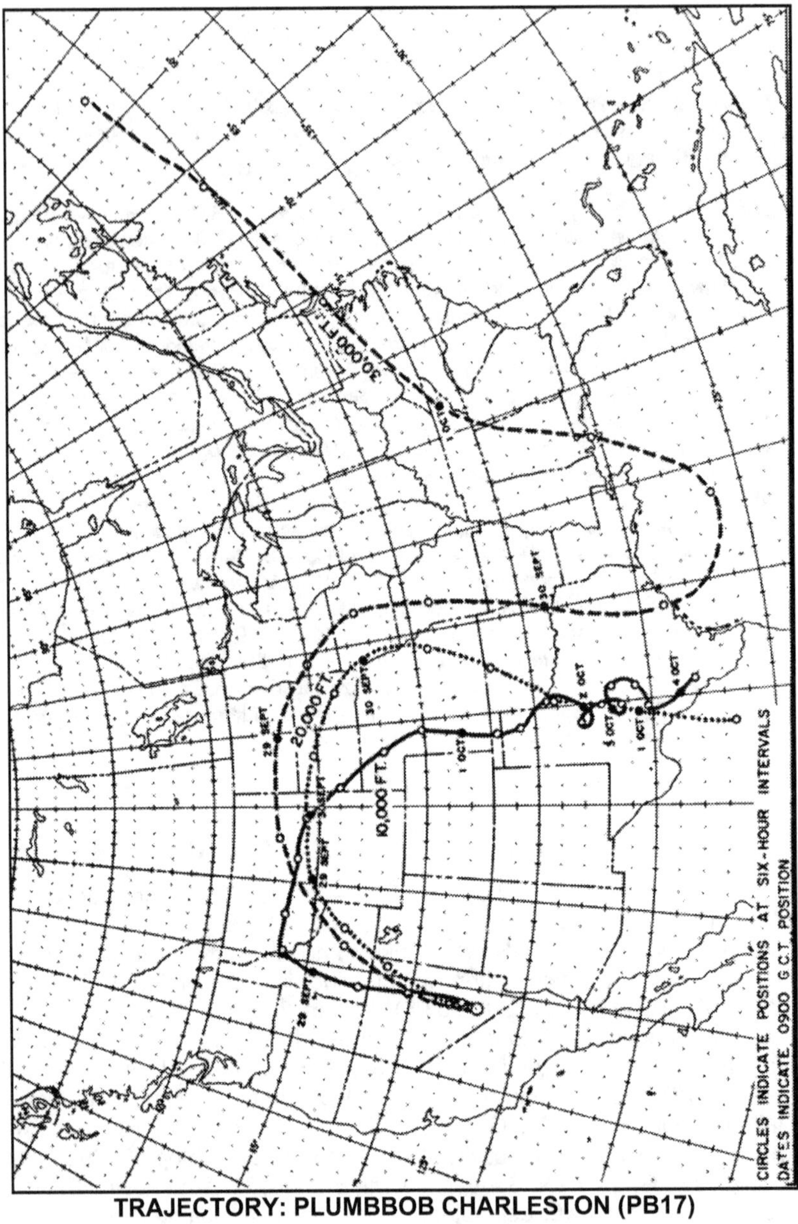

TRAJECTORY: PLUMBBOB CHARLESTON (PB17)

TOTAL FALLOUT: PLUMBBOB CHARLESTON

Equal Intervals
nCi/square meter

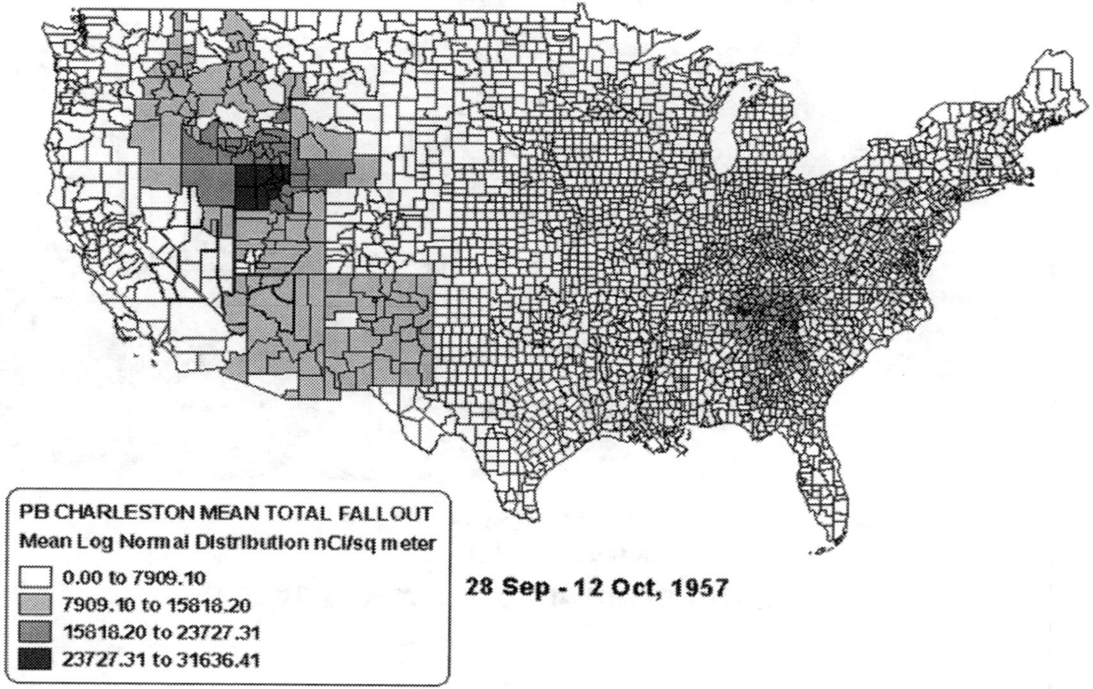

28 Sep - 12 Oct, 1957

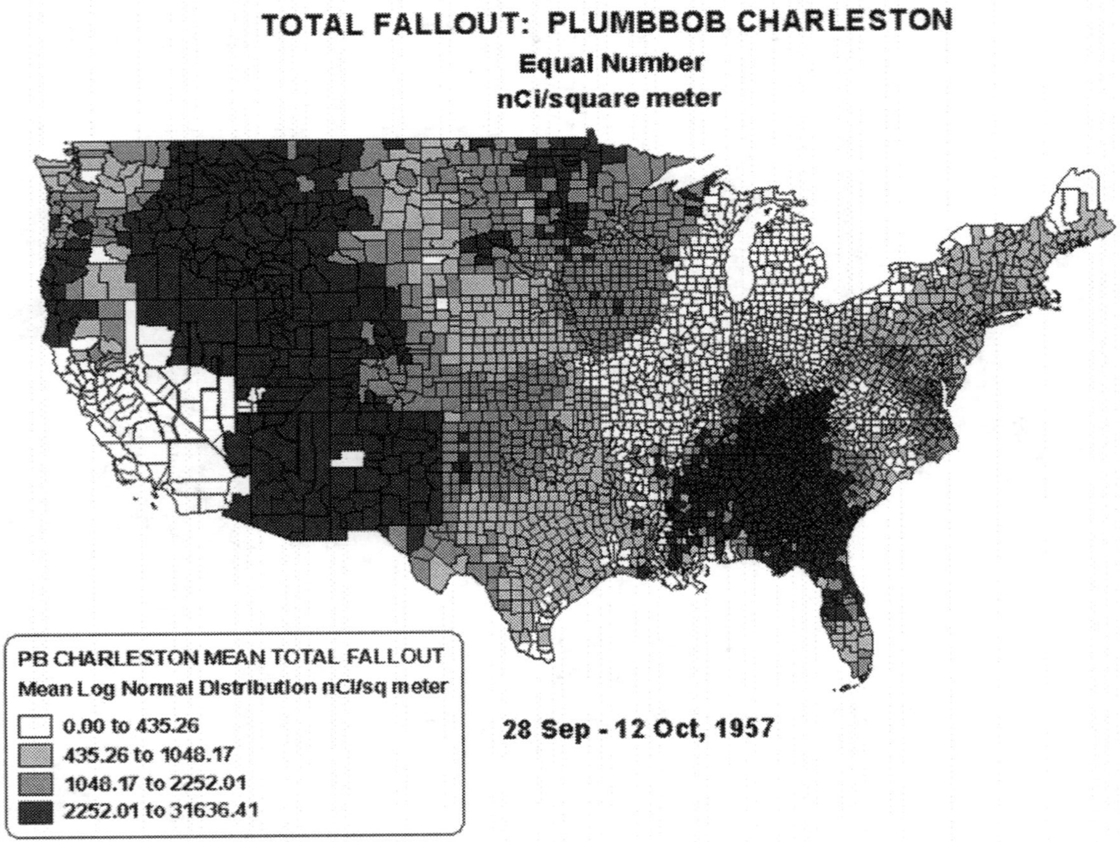
TOTAL FALLOUT: PLUMBBOB CHARLESTON
Equal Number
nCi/square meter
28 Sep - 12 Oct, 1957
PB CHARLESTON MEAN TOTAL FALLOUT
Mean Log Normal Distribution nCi/sq meter
0.00 to 435.26
435.26 to 1048.17
1048.17 to 2252.01
2252.01 to 31636.41

RELATIVE TOTAL FALLOUT: PB CHARLESTON

SEP 23, 1957 12 KT

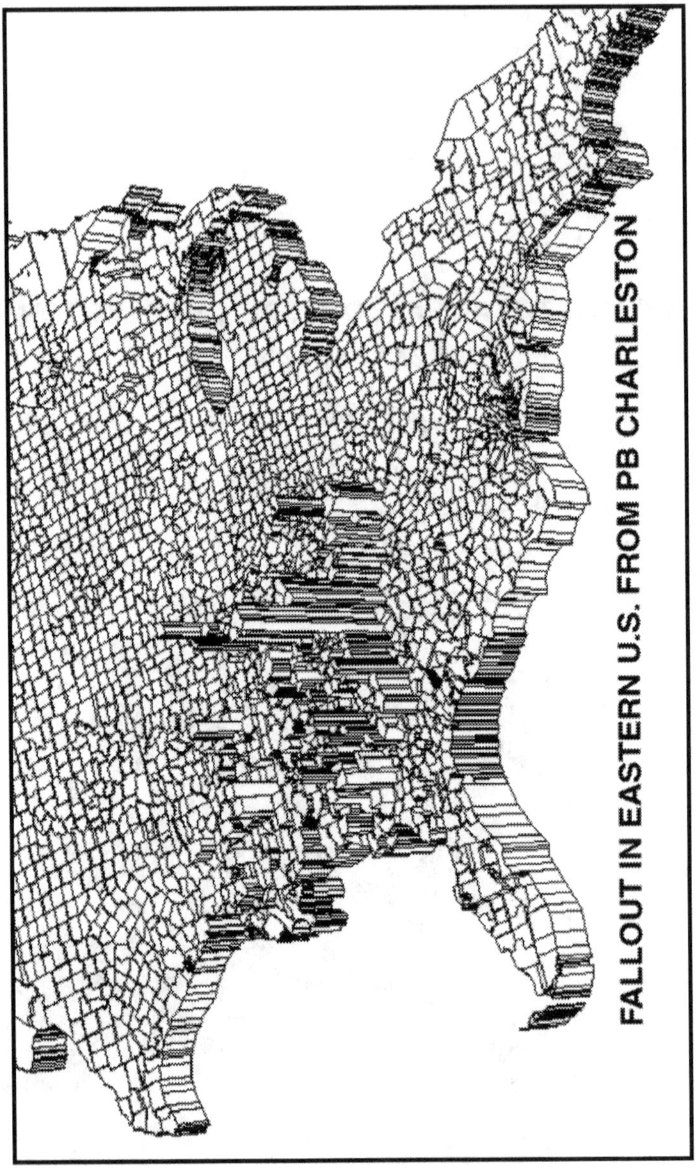
FALLOUT IN EASTERN U.S. FROM PB CHARLESTON

SHOT 18: PLUMBBOB MORGAN (PB18)

PLUMBBOB: MORGAN

Detonation Date: 7 Oct 1957 • *Detonation Time:* 5:00 A.M. • *Area:* 9a • *Sponsor:* Livermore Laboratory • *Yield:* 8 kt • *Radiation Level at Ground Zero:* 100 R/hr • *Height of Burst:* 500 ft • *Cloud Top Height:* 40,000 ft

20,000-Foot Trajectory: NEVADA: (Spotted Range), (Pintwater Range), (Sheep Range), (Mormon Peak) • ARIZONA: Colorado City, Cane Beds, Moccasin, Kaibab Indian Reservation, Fredonia, Page, (Monument Valley), Mexican Water, Teec Nos Pos • UTAH: St. George, Hildale, Kanab, Glen Canyon, (Navajo Mountain), (Rainbow Bridge National Monument), (Monument Valley), Mexican Hat, Ute Mountain Indian Reservation • COLORADO: Ute Mountain Indian Reservation, (Mesa Verde National Park), Hesperus, Durango, Trimble, Hermosa, (Sunlight Peak), Spar City, Wagon Wheel Gap, (Rio Grande National Forest), (Sangre De Cristo Mountains), Lincoln Park, Rockvale, Florence, Canon City, Penrose, Portland, Fort Carson, Fountain, Yoder, Rush, Hugo, Flagler, Seibert, Hale • KANSAS: St. Francis • NEBRASKA: Benkelman, Stratton, Trenton, Stockville, Eustis, Lexington, Miller, Amherst, Pleasanton, Poole, Ravenna, Boelus, Cairo, Chapman, Hordville, Marquette, Polk, Arborville, Gresham, Ulysses, Bee, Dwight, Valparaiso, Agnew, Ceresco, Ashland, Memphis, Gretna, Springfield, Cedar Creek, LaPlatte, Bellevue • IOWA: Mineola, Silver City, Henderson, Macedonia, Griswold, Lyman, Cumberland, Berea, Fontanelle, How, Arbor Hill, Cumming, Norwalk, West Des Moines, Colfax, Prairie City, Newton, Oakland Acres, Grinnell, Belle Plaine, Luzerne, Blairstown, Norway, Fairfax, Covington, Cedar Rapids, Puralta, Springville, Fairview, Anamosa, Amber, Center Junction, Onslow, Fulton, Andrew, Springbrook, Green Island • ILLINOIS: Savanna, Pearl City, German Valley, Seward, Winnebago, Rockford, Belvidere, Marengo, Union, Crystal Lake, Libertyville, Wauconda, Lake Bluff, Lake Forest, North Chicago • MICHIGAN: Paw Paw Lake, Glendale, Lawrence, Kalamazoo, Westwood, Galesburg, Springfield, Battle Creek, Brownlee Park, Mershall, Devereaux, Jackson, Munith, Stockbridge, Unadilla, Pinckney, Hell, Hamburg, Lakeland, New Hudson, Wixom, Walled Lake, Orchard Lake, Sylvan Lake, Pontiac, Birmingham, Troy, Rochester Hills, Sterling, New Baltimore, Anchorville, Fair Haven, Pearl Beach • NEW YORK: (Lake Ontario), Dexter, Brownville, Watertown, Evans Mills, Philadelphia, Antwerp, Oxbow, Wegatchie, Somerville, Natural Dam, Gouverneur, Richville, DeKalb, Rensselaer Falls, Canton, Morley, Madrid, Chamberlain Corners, Chase Mills, Louisville, Massena, Massena Center, Rooseveltown • VERMONT: Rouses Point, Alburg Springs, Highgate Springs, East Franklin, Rickford, Stevens Mills, Missisquoi, North Troy, Beebe Plain, Rock Island, Derby Line, Norton, Canaan, Beecher Falls • NEW HAMPSHIRE: Pittsburg, (Shatney Mountain), Happy Corner, The Glen, (Deer Mountain), (Salmon Mountain) • MAINE: Coburn Gore, (Kirby Mountain), (Spencer Mountain), (Coburn Mountain), (Indian Pond), (Big Squaw Mountain), (Whitecap Mountain), (Jo-Mary Mountain), Millinocket, East Millinocket, Reed, (Wytopilock Lake), Haynesville, Orient, North Amity.

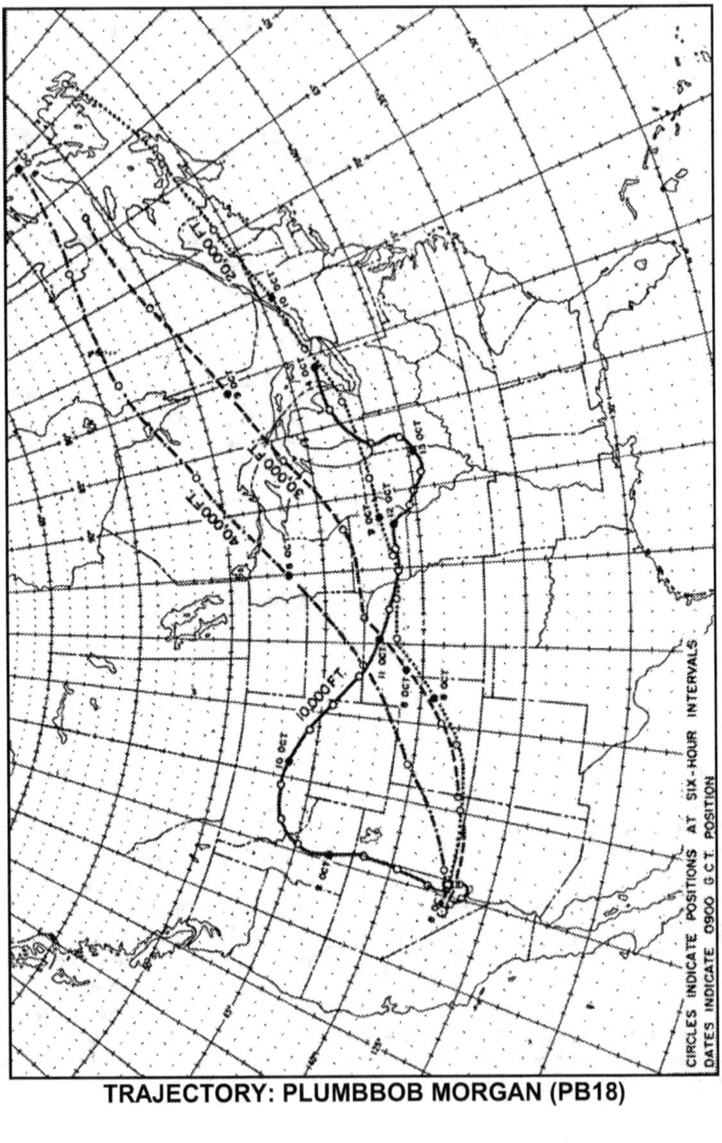

TRAJECTORY: PLUMBBOB MORGAN (PB18)

CHARACTERISTICS: PLUMBBOB MORGAN

Among the nuclear tests, Plumbbob MORGAN was 3rd in percentage of Cm242 in fallout. (Plumbbob WILSON was 2nd in Cm242, while Plumbbob NEWTON was 1st for percentage of Cm242.). The deposition pattern for this shot is most similar to Upshot-Knothole SIMON (UK7) (R: 0.51; T: 32.67). Fallout from this test covered 2328 counties (75 percent of the total). Counties of highest deposition included El Paso CO (estimated maximum activity: 56.27 μCi/sq meter on Oct 8, 1957); Cheyenne CO, Kit Carson CO, Yuma CO, Boulder CO and Washington CO. Interestingly, one of the highest estimated maximum activities occurred on the eighth day postshot (Oct 14, 1957) for the following Missouri counties of Adair, Marion, Shelby and Monroe—the last being the author's home county. Shot MORGAN is ranked 7th in the amount of Mn54 deposited in fallout. The estimated maximum fallout activity for these counties equaled 56 μCi/sq meter. Total fallout to I-131 ratio for MORGAN was 102:1.

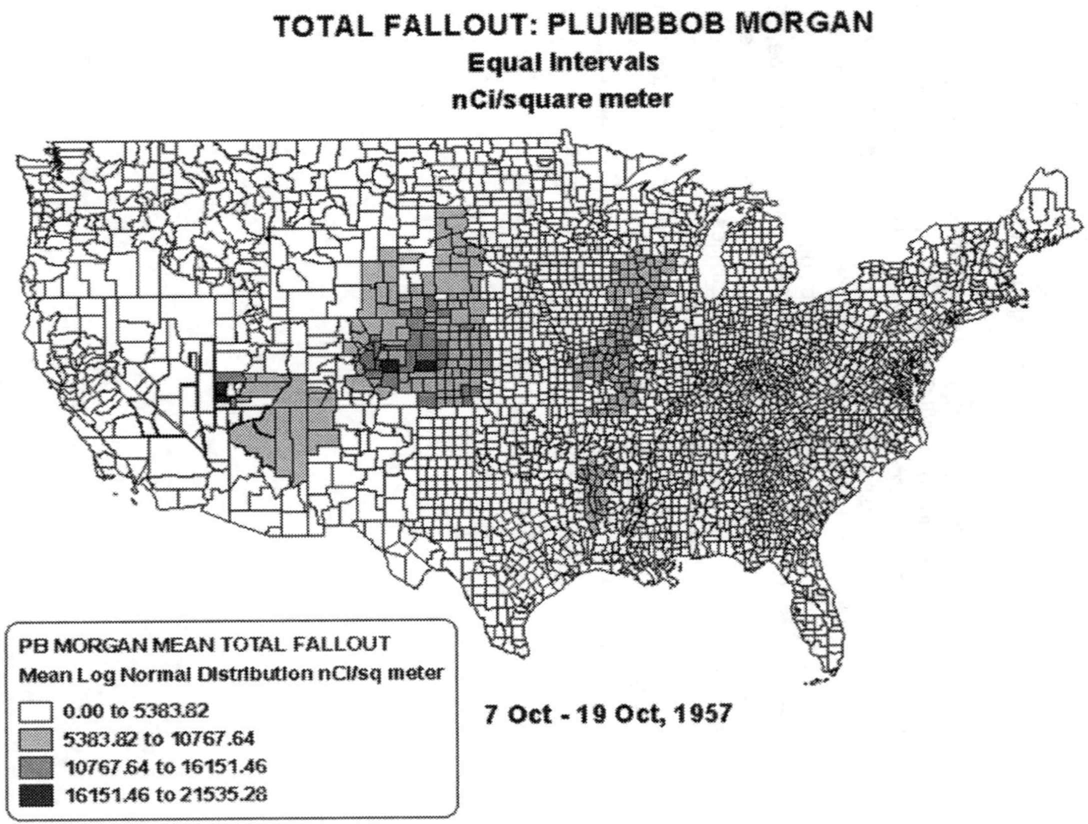
TOTAL FALLOUT: PLUMBBOB MORGAN
Equal Intervals
nCi/square meter
PB MORGAN MEAN TOTAL FALLOUT
Mean Log Normal Distribution nCi/sq meter
0.00 to 5383.82
5383.82 to 10767.64
10767.64 to 16151.46
16151.46 to 21535.28
7 Oct - 19 Oct, 1957

TOTAL FALLOUT: PLUMBBOB MORGAN

Equal Number

nCi/square meter

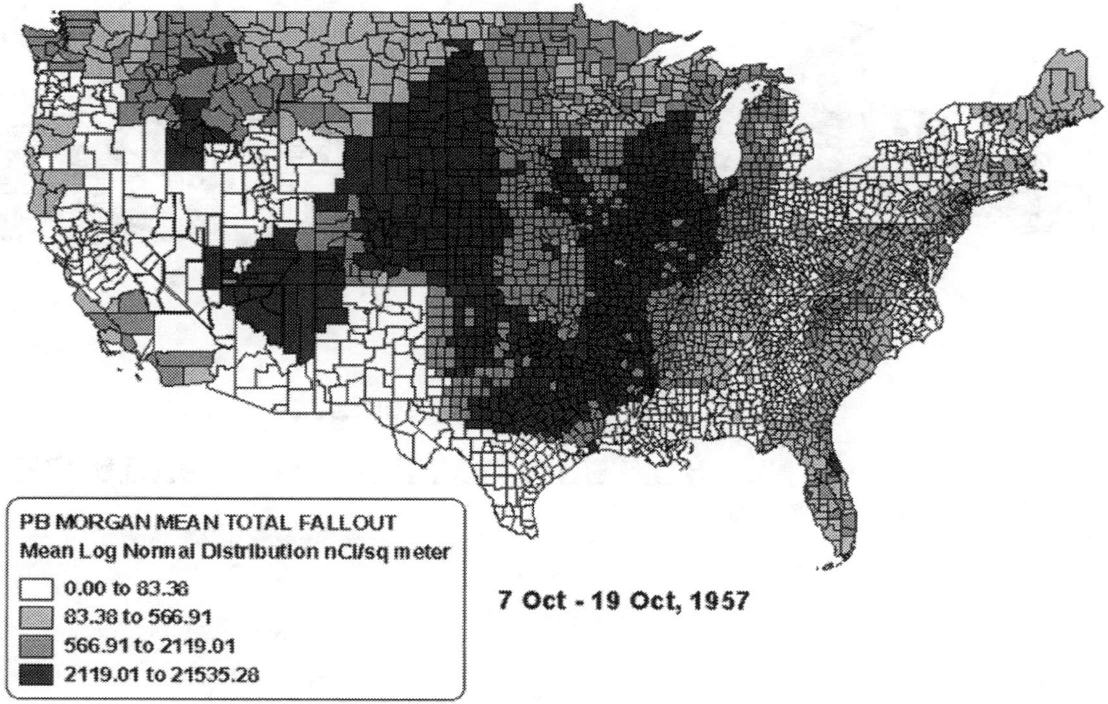

7 Oct - 19 Oct, 1957

RELATIVE TOTAL FALLOUT: PB MORGAN

OCT 7, 1957 8 KT

SECTION 7

HARDTACK II

Sep-Oct 1958

The Hardtack II Series

The massive Plumbbob series of 1957 was followed in 1958 by the much smaller Hardtack II series. This series involved 37 nuclear tests, most after counties in New Mexico. The amount of radiation released to the atmosphere from the Hardtack II series was relatively small and was limited to counties near the Nuclear Test Site.

Of the 37 tests associated with Hardtack II, only 9 were evaluated for I-131 deposition by the National Cancer Institute: Eddy, Hidalgo, Quay, Lea, Vesta, Rio Arriba, Socorro, Wrangell and Sanford. They are described briefly here:

EDDY: Detonated September 19, 1958 from a balloon suspended 500 feet above Area 7b at the Nevada Test Site, the original device weighed 146 lbs. Yield 83 tons. The bottom of the mushroom cloud reached 7,500 ft MSL, while the top of the cloud reached 11,000 ft MSL. The maximum radiation at ground zero, measured 1 hour after detonation, was 1 Roentgen/hr. The debris cloud headed first northeast, then north, but crossed the Utah border 85 miles north of the Utah-Arizona border. According to the 1997 NCI report, fallout from shot Eddy was deposited only on Lincoln 2 NV (estimated activity: 0.58 uCi/sq meter) , Nye 2 (estimated activity: 0.27 uCi/sq meter) and Lincoln 1 (estimated activity: 0.26 uCi/sq meter).

HIDALGO: This shot was a safety experiment. It was detonated October 5, 1958 from a balloon suspended 377 feet above Area 7b. Yield 77 tons. The bottom of the mushroom cloud reached 8,000 ft MSL, while the top of the cloud reached 12,000 MSL. Maximum radiation at ground zero reached 3 Roentgens/hr measured 1 hour after detonation. According to the NCI 1997 report, the only county recording deposition from this test was Lincoln 2 NV. Estimated total fallout activity: 2.09 uCi/sq meter.

QUAY: Detonated October 10, 1958 from a 100-ft tower. Yield 79 tons. Despite the small yield, the bottom of the mushroom cloud reached 7,500 ft. MSL and the top of the cloud reached 10,000 ft MSL. The debris cloud traveled southwest, and crossed the California border. At the California-Nevada border the surface radiation was measured at 0.002 Roentgens/hr. According to the 1997 NCI report, only Nye 3 NV received fallout from this nuclear test. Estimated activity was 7.03 uCi/sq meter total fallout.

LEA: Detonated October 13, 1958 from a balloon suspended 1500 feet above area 7b at the Nevada Test Site, the original device was only 11.7 inches in diameter, 15 inches in length and weighed only 63 lbs, but yielded 1.4 kilotons. The cloud bottom reached 12,000 MSL while the cloud top reached 17, 000 ft MSL. The debris cloud drifted due north. Radiation was detected 180 miles north of Area 7b at a level of 0.001 Roentgen/hr. According to the 1997 NCI report, only Nye 2 NV received fallout from this test. Estimated activity: 6.35 uCi/sq meter total fallout.

VESTA: Some Detonated October 17, 1958 from Area 9c, this device was one of the few surface shots, and was detonated in a wooden building covered with

20 ft of gravel. The yield was only 24 tons, but the cloud top reached 10,000 ft MSL. Radiation at the point of detonation was measured at 2,000 Roentgens per hour. The cloud drifted almost due north. Surface radiation was detected from shot Vesta 35 miles due north of Area 9c. According to the 1997 NCI report, only Nye 2 NV received fallout. Estimated activity: 0.08 uCi/sq meter.

RIO ARRIBA: Detonated October 18, 1958 from a 72.5-ft wooden tower at area 3s. Yield: 90 tons. The debris cloud drifted north-northeast at first, but then headed due east, crossing the Utah border 150 miles north of the Arizona border. The bottom of the mushroom cloud reached 11,000 ft. MSL, while the top reached 13, 500 ft. MSL. According to the 1997 NCI report, the only country receiving fallout from this nuclear test was Nye 2 NV, with an estimated total fallout activity of 5.12 uCi/sq meter.

SOCORRO: Detonated October 22, 1958 from a balloon suspended 1450 feet above Area 7b, the Socorro device yielded 6 Kilotons. The debris cloud drifted toward the northeast. The cloud bottom height was 20,000 ft, while the cloud top height was 26,000 ft. According to the 1997 NCI report, the only county receiving fallout from this test was Clark 2, NV with an estimated total fallout activity of 1.39 uCi/Sq meter.

WRANGELL: Detonated October 22, 1958 from a balloon suspended 1500 feet over Area 5, the Wrangell test was considered a fizzle and yielded only 115 tons. The cloud formed by shot Wrangell first drifted west, then turned to the northeast. The rate of rise of the debris cloud was 5000 ft/hour. Cloud bottom height: 7,000 ft MSL; cloud top: 10,000 ft MSL. No crater was formed. According to the 1997 NCI report, only two counties received fallout from this nuclear test: Lincoln 2 NV (estimated total fallout activity: 0.12 uCi/Sq meter) and Nye 2, NV (estimated total fallout activity: 0.12 uCi/sq meter.)

SANFORD: Detonated October 26, 1958 from a balloon suspended 1500 feet over area 5, the Sanford device yielded 4.9 kilotons. The bottom of the mushroom cloud reached 12,500 feet MSL, with the top of the cloud reaching a height of 26,000 ft MSL. No crater was formed. According to the 1997 NCI report, only two counties received fallout from this nuclear test: Esmeralda 2 NV (estimated total fallout activity: 6.7 uCi/sq meter) and Clark 2 NV (estimated total fallout activity: 1.41 uCi/sq meter.)

SECTION 8

THE UNDERGROUND SERIES

1961-1970

THE UNDERGROUND SERIES
1961-1970

After the above-ground tests of 1958, nuclear detonations at the Nevada Test Site went underground. That is not to say that there was less fallout. In fact, many of the underground tests, such as Danny Boy, Sedan, Small Boy and others resulted in significant fallout deposition in the Midwestern United States. Sedan, for example, resulted in a greater estimated deposition of beryllium than all other Nevada nuclear tests combined. In preparing their 1997 report, *Estimated Exposures and Thyroid Doses Received by the American People from Iodine-131 in Fallout Following Nevada Atmospheric Nuclear Bomb Tests*, National Cancer Institute researchers included only 16 underground tests: *Antler, Danny Boy, Platte, Eel, Des Moines, Johnie Boy, Sedan, Small Boy, Bandicoot, Pike, Sulky, Palanquin, Pin Stripe, Cabriolet, Buggy, Schooner* and *Baneberry.* Though Sedan was technically an underground test--the device was buried 635 feet below the surface—the resultant debris cloud reached an altitude of 16,000 ft. MSL before traveling northeast toward South Dakota, Iowa and Missouri.

The Baneberry test of Dec 18, 1970 resulted in a venting of spectacular proportions: the debris cloud reached 8,000 ft MSL before traveling northeast. Nuclear fallout from this event reached counties near the Eastern seaboard.

Between 1951 and 1957, he U.S. Atomic Energy Commission carefully tracked debris clouds across the country. However, the trajectory maps that exist for the underground series appear to be drawn with broad strokes. The fallout trajectory for Eel and Des Moines are drawn as hundred-mile wide arrows snaking across the country toward the Atlantic. Sedan's arrow seems to stop near Monroe County in Northeast Missouri---even though radioactive fallout was detected at sites hundreds of miles to the east.

ANTLER (UE1) Detonated on September 15, 1961 from a depth of 1319 ft. The yield was 2.4 Kt. No crater was formed.

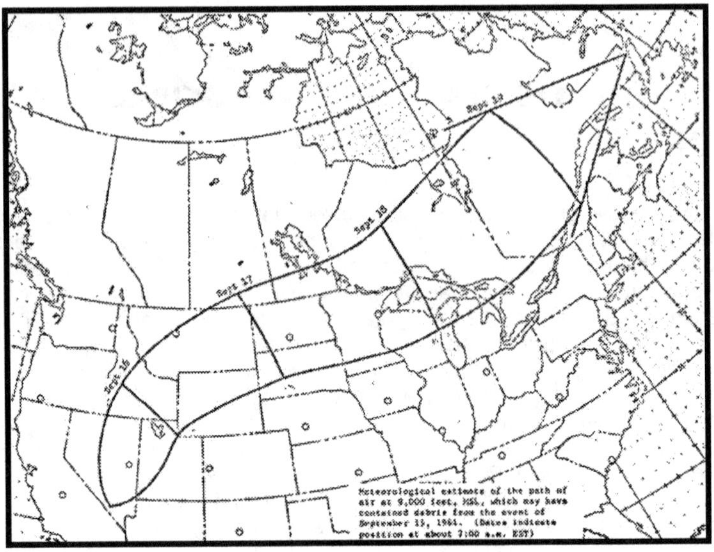

TRAJECTORY: ANTLER

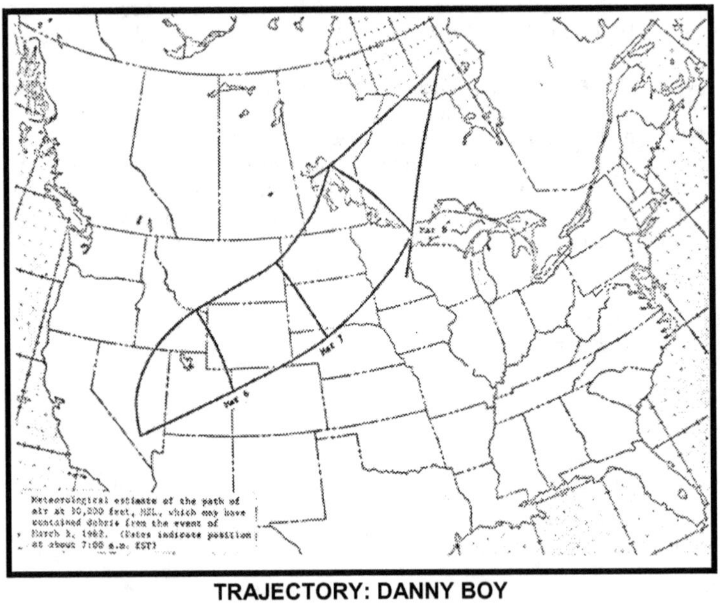

TRAJECTORY: DANNY BOY

DANNY BOY (UE2) Detonated on March 5, 1962 from a depth of 110 ft. The total yield was 0.42 Kt. The event formed a crater with a diameter of 214 ft and a depth of 62 ft.

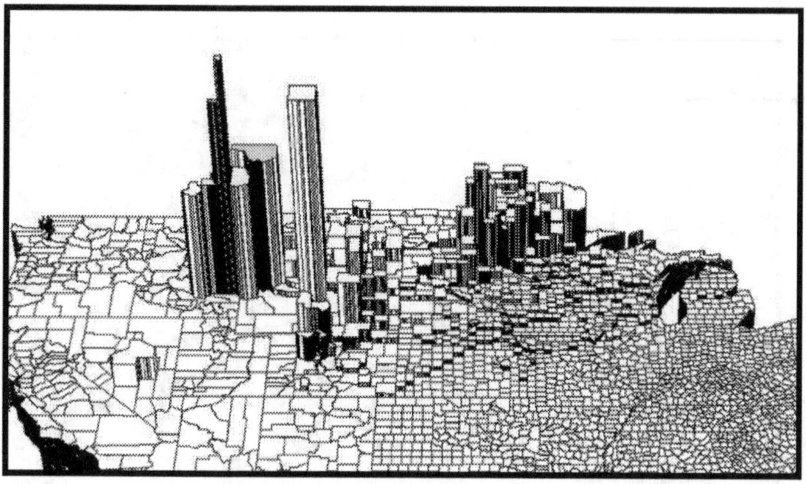

DANNY BOY FALLOUT PATTERN

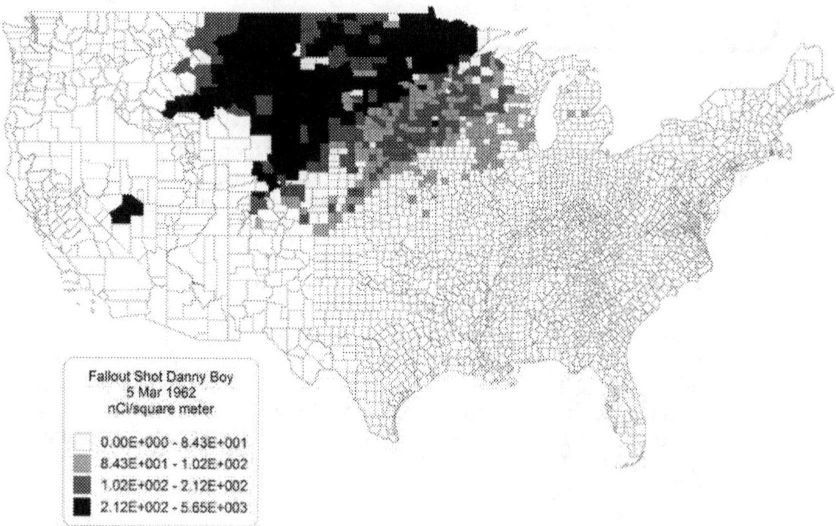

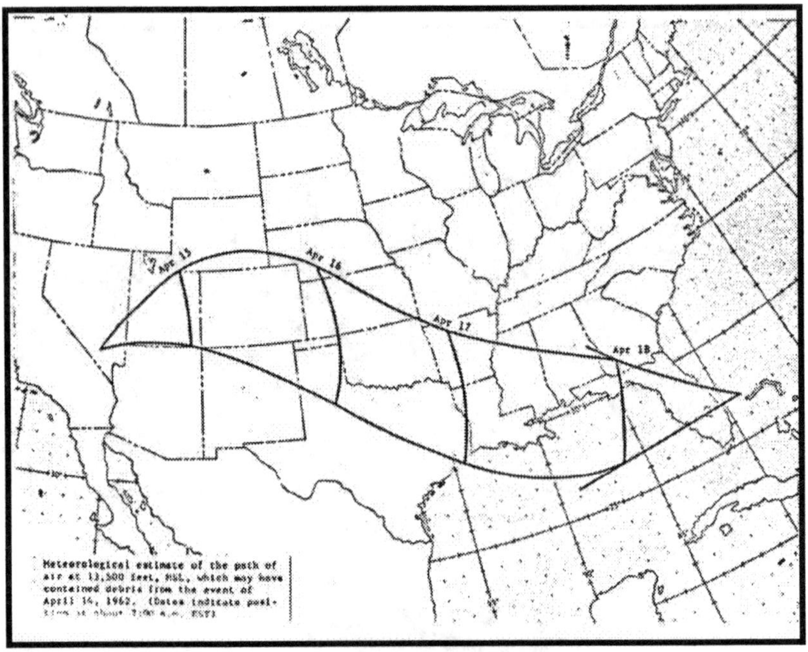

TRAJECTORY: PLATTE

PLATTE (UE3) Detonated on April 14, 1962 inside a tunnel 628 ft below the surface and 560 ft inside a hill. The yield, according to DASA-1251, was 1.7 Kt. The detonation produced a cloud that rose to 8000 ft altitude, MSL. The DASA-1251 account:

> "Venting occurred at the tunnel portal, through fissures, and at a vent hole at H+1.5 seconds. The fissures were created on the side of the hill, and radial cracks formed on top of the hill. A persistent cloud was produced containing appreciable quantities of radioactivity associated with particles (redacted). The estimated dose rate at the tunnel portal, normalized to H+1 hour, was 20 Roentgens/hour and the estimated total release, normalized to H+1 minute, was 5×10^7 Curies. The release products contained the following known isotopes: Ru^{103}, Ru^{105}, $Zr\text{-}Nb^{95}$, Ca^{141}, Ca^{144}, K^{40}, I^{131}, I^{133}, I^{135}, Te^{132}.
>
> "The cloud drifted in a northerly direction. The radiation area at H+4.5 hours extended upwind approximately one mile from GZ and was monitored at 10 milli-Roentgen/hr. Some radioactivity was detected in off-site areas. No radiation was detected at the worksite or any other location from releases of gaseous radioactivity during post-shot drilling or tunnel re-entry operations."

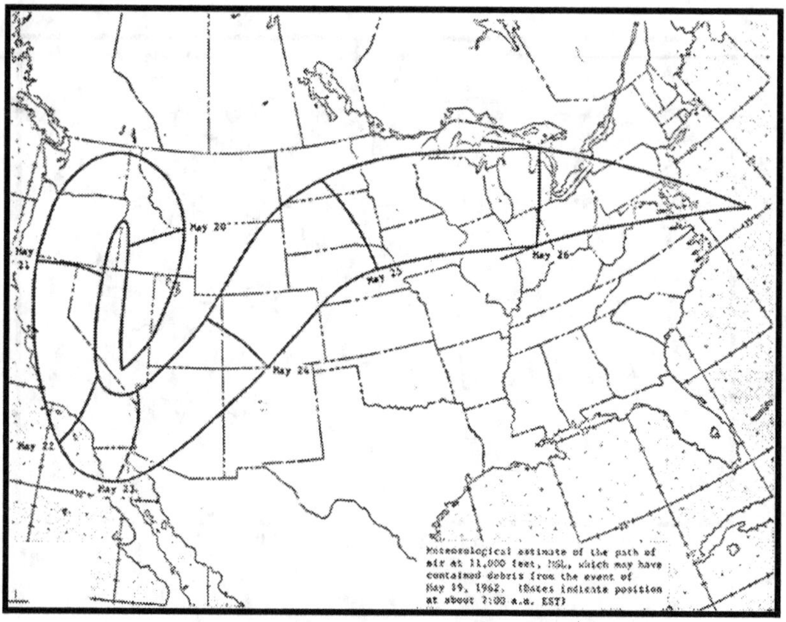

TRAJECTORY: EEL

EEL (UE4) Detonated on May 19, 1962from a depth of 714 ft. From the account in DASA 1251:

> "Venting in the form of a geyser, occurred at H+10 seconds at vent hole U9m-2 and continued steadily until H+19 minutes 42 seconds. A similar venting occurred at H+15 seconds at vent hole U9-m3 and lasted until H+21 minutes."

TRAJECTORY: DES MOINES

DES MOINES (UE5) Detonated on June 13, 1962 from within a tunnel with a depth of 660 ft and a slant depth of 610 ft. From remarks (DASA 1251):

> "The shot vented out of the tunnel mouth with sufficient pressure and flow rate that radioactive debris was projected entirely across the canyon and deposited on the slope behind a trailer shelter. This shelter was not shielded from fallout. The entrance to the shelter faced away from the tunnel but the door was open."

Note: The above map was originally published in *Under The Cloud: The Decades of Nuclear Testing* (Miller: 1986) and is a schematic of a Department of Energy map that is no longer available.

STORAX SEDAN

Storax: Sedan

Detonation Date: 6 Jul 1962 • *Detonation Time:* 9:00 A.M. • *Area:* U10h • *Sponsor:* Livermore Laboratory • *Yield:* 110 kt • *Radiation Level at Ground Zero:* 500 R/hr • *Height of Burst:* –635 ft • *Cloud Top Height:* 16,000 ft

16,000-Foot Trajectory: NEVADA: (Belted Range), (Reveille Range), (Railroad Valley), Duckwater, (Mount Hamilton), (Pancake Summit), (Newark Lake), (Long Valley), (Ruby Lake), (Spruce Mountain), Shafter, (Pilot Peak) • UTAH: (Great Salt Lake Desert), (Hogup Mountains), (Hansels Mountains), (Dolphin Island in Great Salt), Snowville, Portage • IDAHO: Stone, Samaria, Woodruff, Malad City, Oxford, Clifton, Dayton, Banida, Mink Creek, Swan Lake, Liberty, Bern, Bennington, Ovid, Montpelier, Geneva • WYOMING: (Salt River Range), (Commissary Ridge), (Bridger-Teton National Forest), Marbleton, Boulder, Big Sandy, (Wind River Peak), Lander, Hudson, Hiland, Waltman, Powder River, Natrona, (Teapot Dome), Bill, Redbird • SOUTH DAKOTA: Provo, Ardmore, Pine Ridge • NEBRASKA: Eli, Cody, Kilgore, Crookston, Valentine, Sparks, Springview, Winnetoon, Creighton, Marquet, McLean, Randolph, Sholes, Wakefield, Emerso, Winnebago Indian Reservation, Homer, Winnebago • IOWA: Sloan, Rodney, Grant Center, Ticonic, Castana, Ute, Dow City, Earling, Defiance, Kirkman, Red Line, Jacksonville, Kimballton, Elk Horn, Brayton, Lorah, Wiota, Cumberland, Massena, Bridgewater, Williamson, Nevinville, Spaulding, Creston, Afton, Hopeville, Westerville, Leon, Woodland, Lineville • MISSOURI: South Lineville, Ravenna, Lucerne, Newtown, Harris, Milan, New Boston, Ethel, Callao, New Cambria, Bevier, Number Eight, College Mound, Cairo, Moberly, Middle Grove, Centralia, Thompson, Mexico, Auxvasse, Shamrock, Williamsburg, Readsville, Americus, Rhineland, Hermann, Gasconade, New Haven, Stony, Beaufort, Stanton, Richwoods, Fertile, Blackwell, Cadet, Bonne Terre, Desloge, Leadwood, Elvins, Flat River, Farmington, Junction City, Fredericktown, Patton, Scopus, Millersville, Burfordville, Whitewater, Dutchtown, Chaffee, New Hamburg, Benton, Diehlstadt, Charleston • KENTUCKY: Bardwell, Arlington, Milburn, Beulah, Fulgham, Wingo, Cuba, Lynnville, Fairbanks • TENNESSEE: Jones Mill, Paris, India, Springville, Manlyville, Big Sandy, Black Center, Eva, Denver, New Johnsonville, Hustburg, Bakerville, Sycamore Landing, (Tennessee National Wildlife Refuge), Only, Coble.

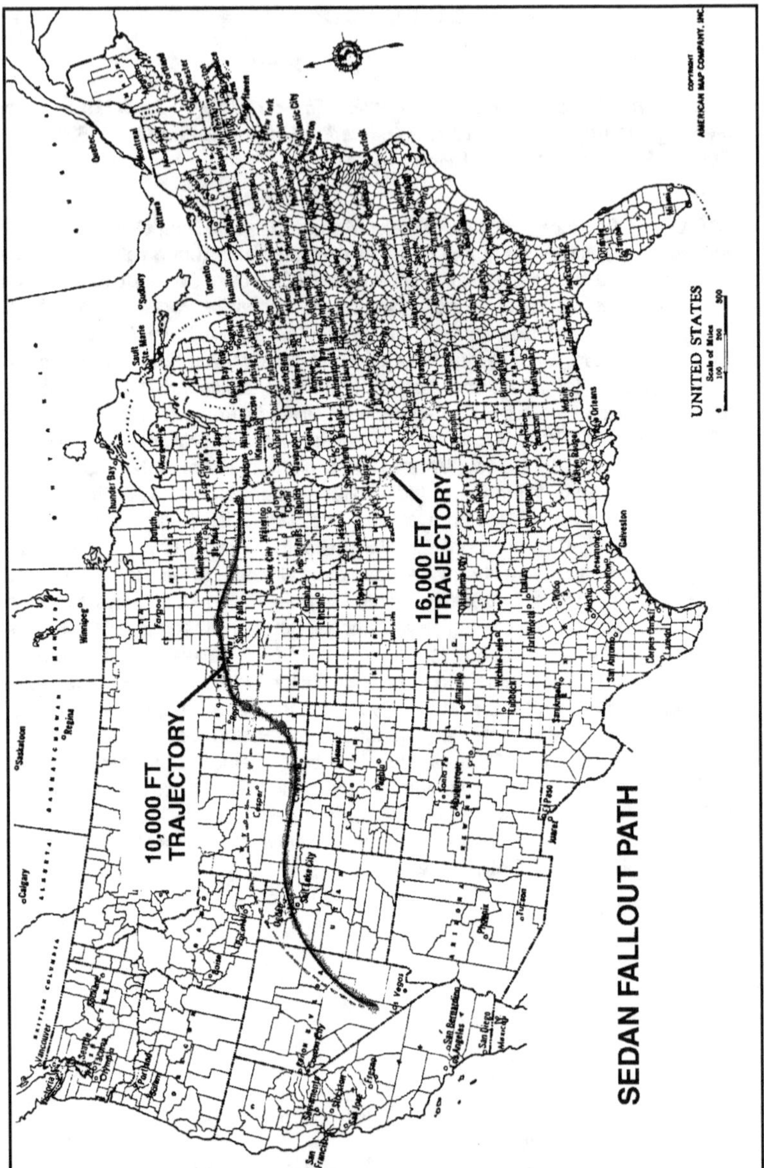

ESTIMATED SEDAN TRAJECTORIES

MAP NO. 761 REPRODUCED WITH PERMISSION FROM THE AMERICAN MAP CORPORATION

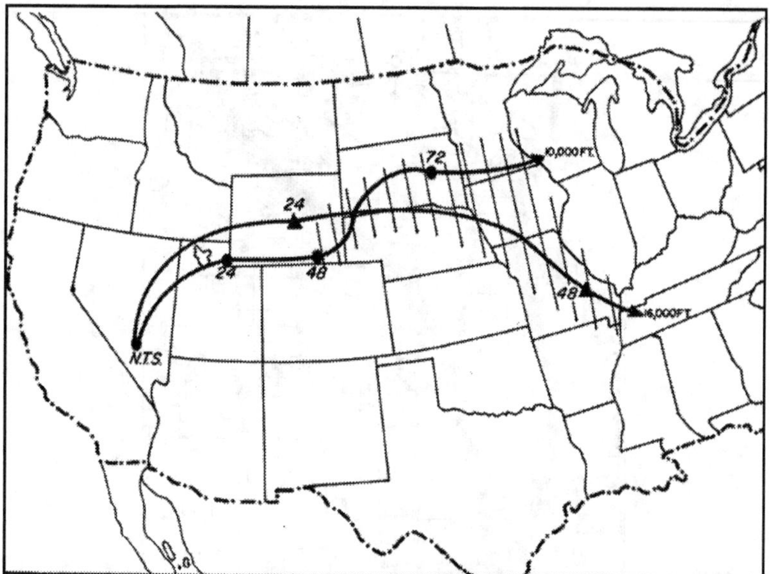

An Estimate of the 24, 48, and 72 Hour 10,000 and 16,000 Ft. MSL Wind Trajectory

TRAJECTORY: SEDAN (SOURCE: DEFENSE DEPARTMENT)

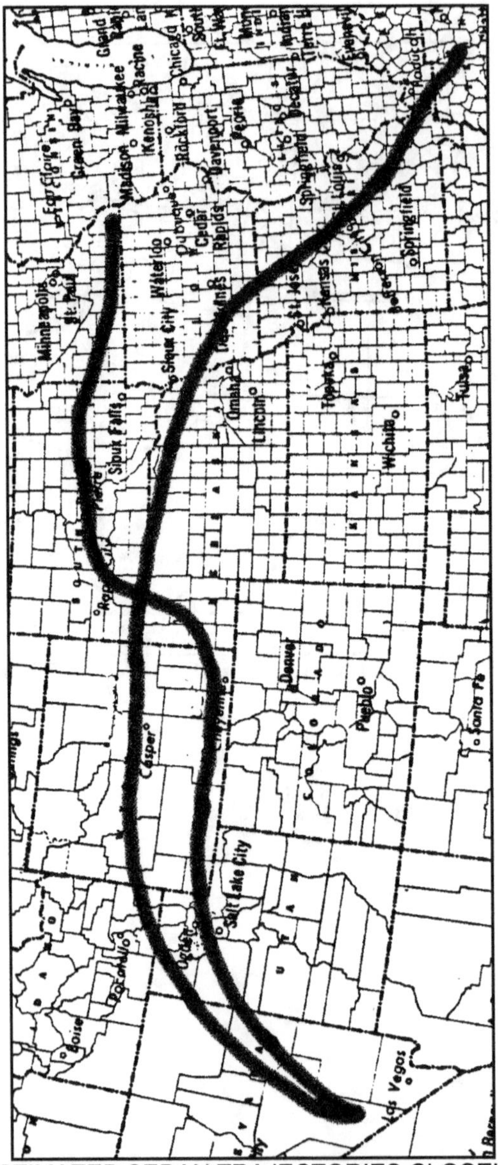

ESTIMATED SEDAN TRAJECTORIES CLOSEUP
MAP NO. 761 REPRODUCED WITH PERMISSION FROM THE AMERICAN MAP CORPORATION

TOTAL FALLOUT: STORAX SEDAN

Equal Intervals

nCi/square meter

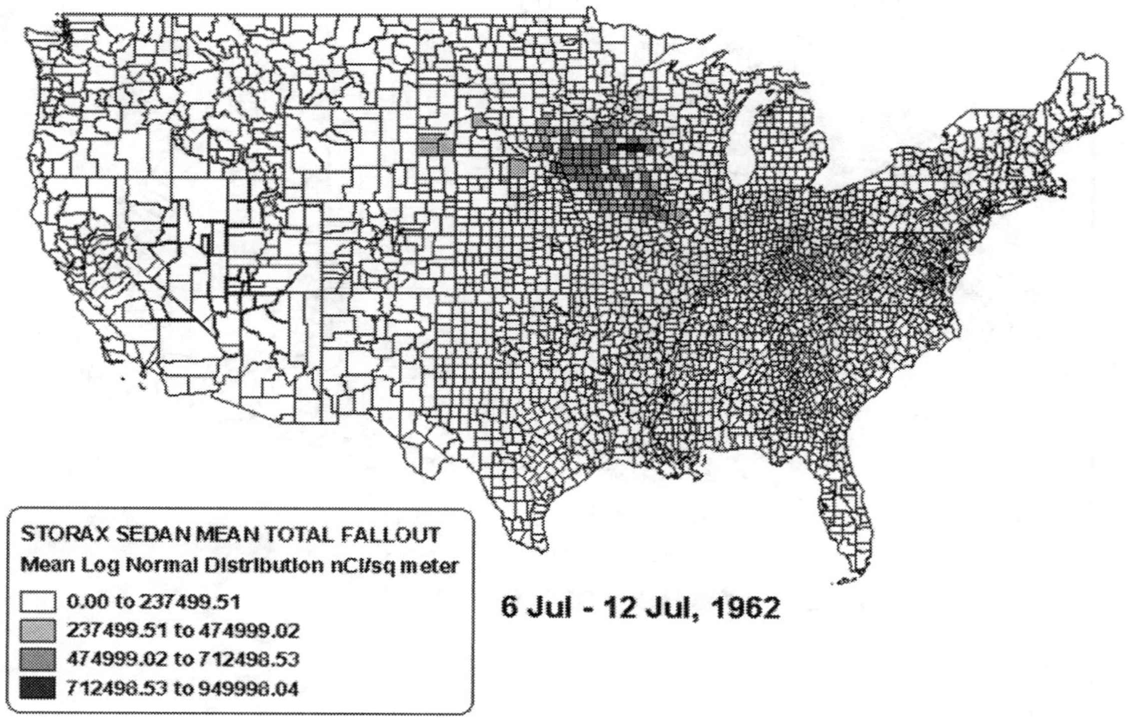

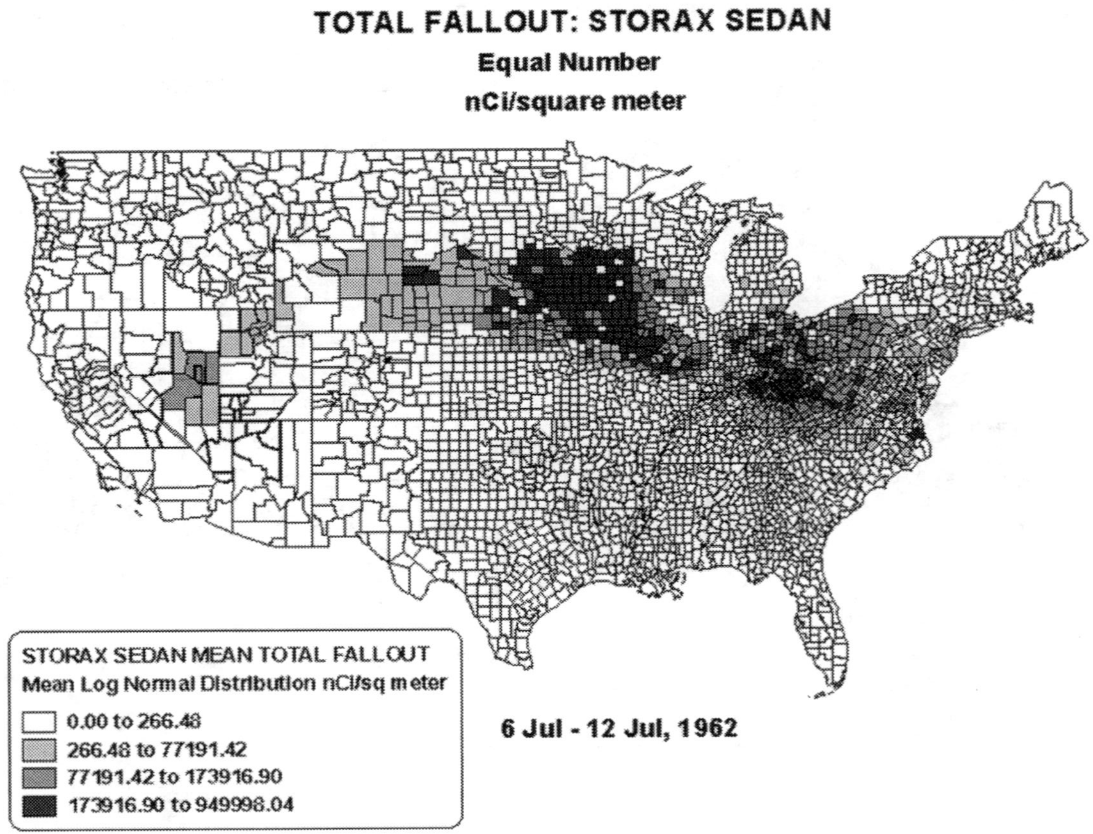
TOTAL FALLOUT: STORAX SEDAN
Equal Number
nCi/square meter
STORAX SEDAN MEAN TOTAL FALLOUT
Mean Log Normal Distribution nCi/sq meter
0.00 to 266.48
266.48 to 77191.42
77191.42 to 173916.90
173916.90 to 949998.04
6 Jul - 12 Jul, 1962

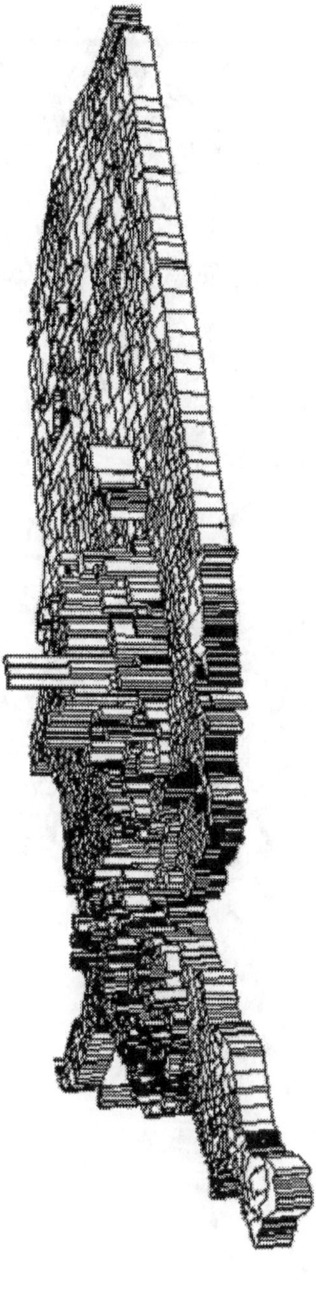

The Beryllium-7 deposition follows the SEDAN fallout pattern.

(Estimated Be7 relative deposition calculated at 1 yr postshot)

FALLOUT PATH: SHOT SEDAN JULY, 1962

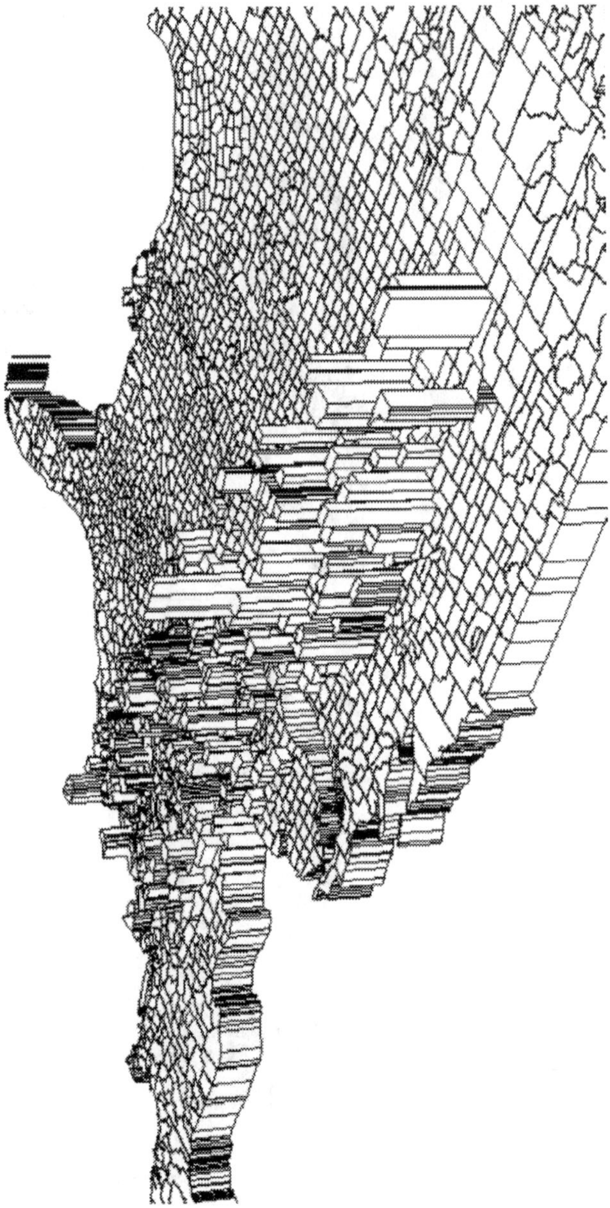

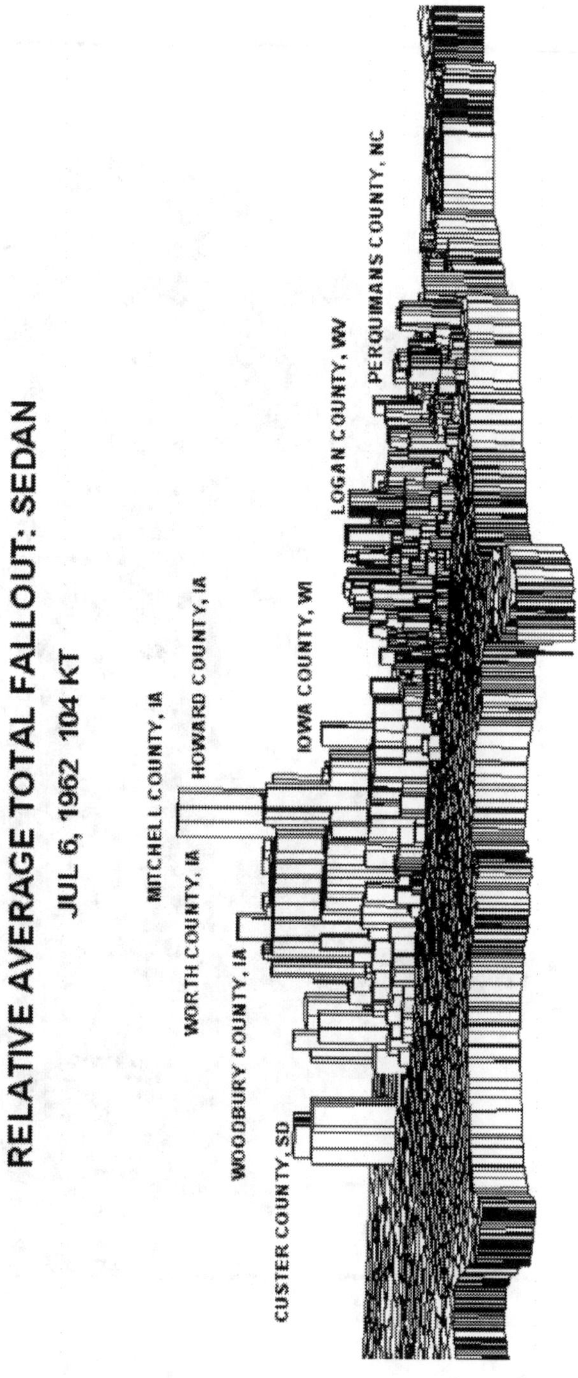
RELATIVE AVERAGE TOTAL FALLOUT: SEDAN
JUL 6, 1962 104 KT
CUSTER COUNTY, SD
WOODBURY COUNTY, IA
WORTH COUNTY, IA
MITCHELL COUNTY, IA
HOWARD COUNTY, IA
IOWA COUNTY, WI
LOGAN COUNTY, WV
PERQUIMANS COUNTY, NC

Fallout Pattern
shot Sedan
Jul 6-13, 1962

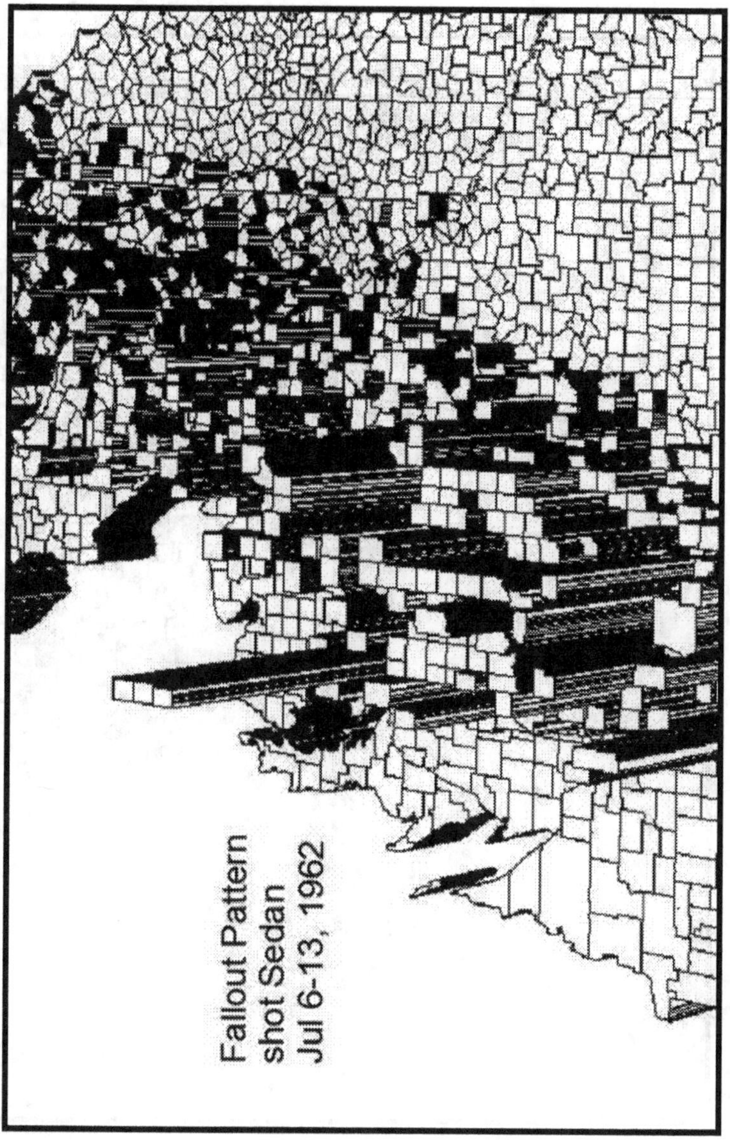
Fallout Pattern
shot Sedan
Jul 6-13, 1962

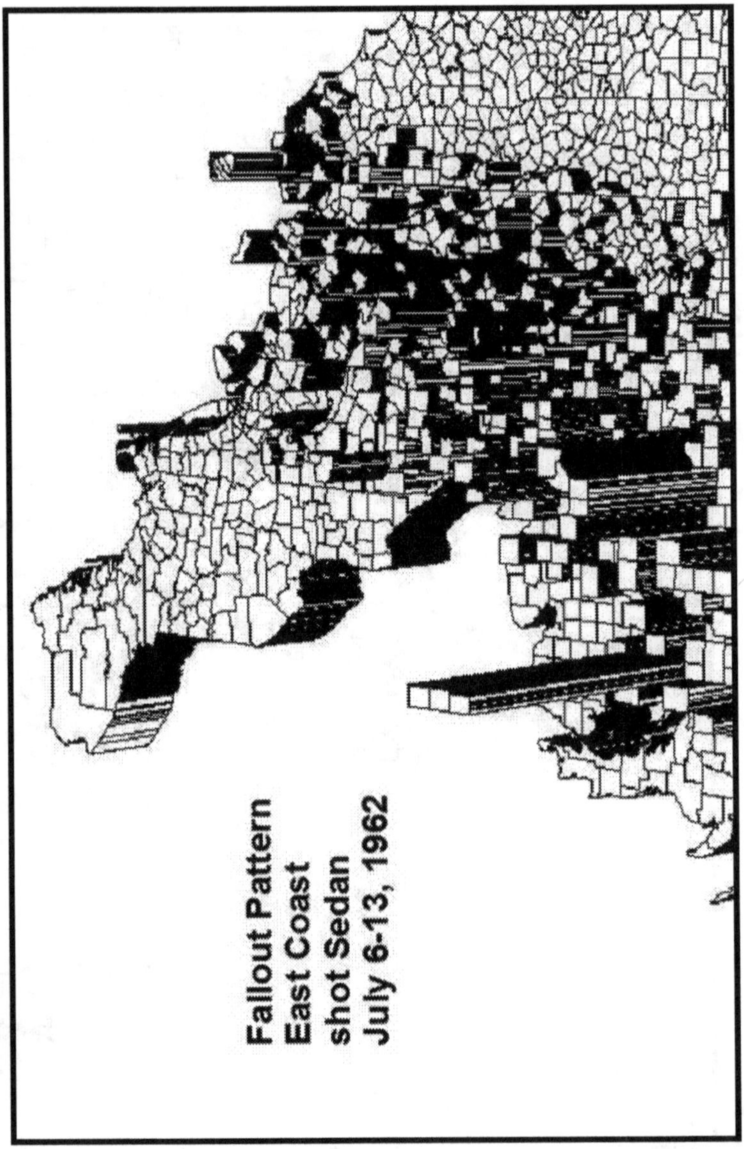
Fallout Pattern
East Coast
shot Sedan
July 6-13, 1962

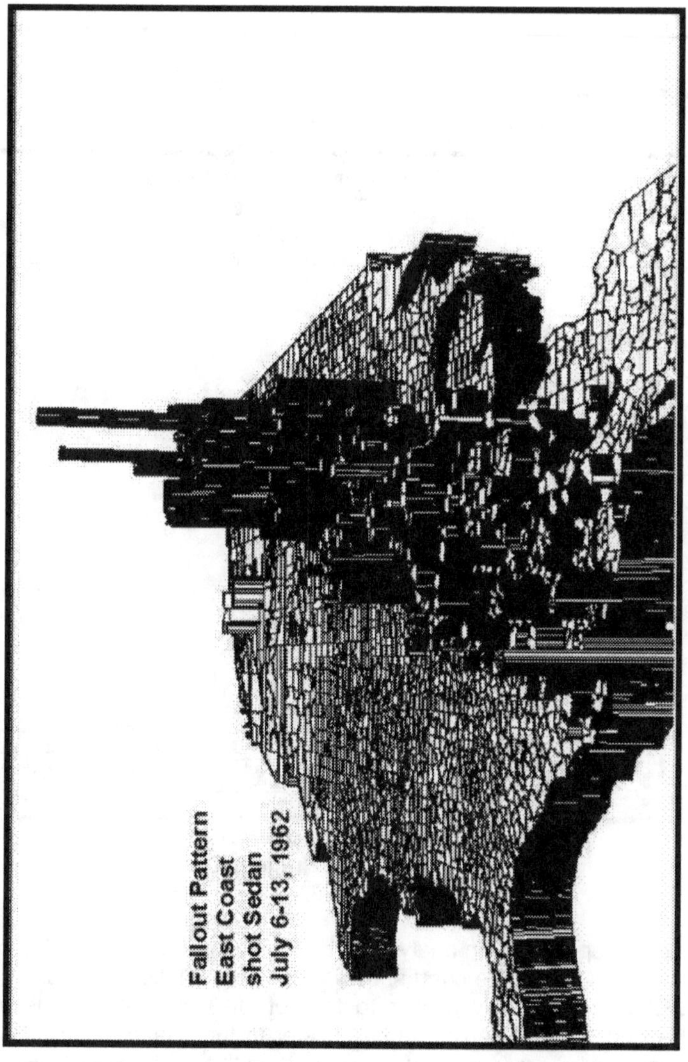
Fallout Pattern
East Coast
shot Sedan
July 6-13, 1962

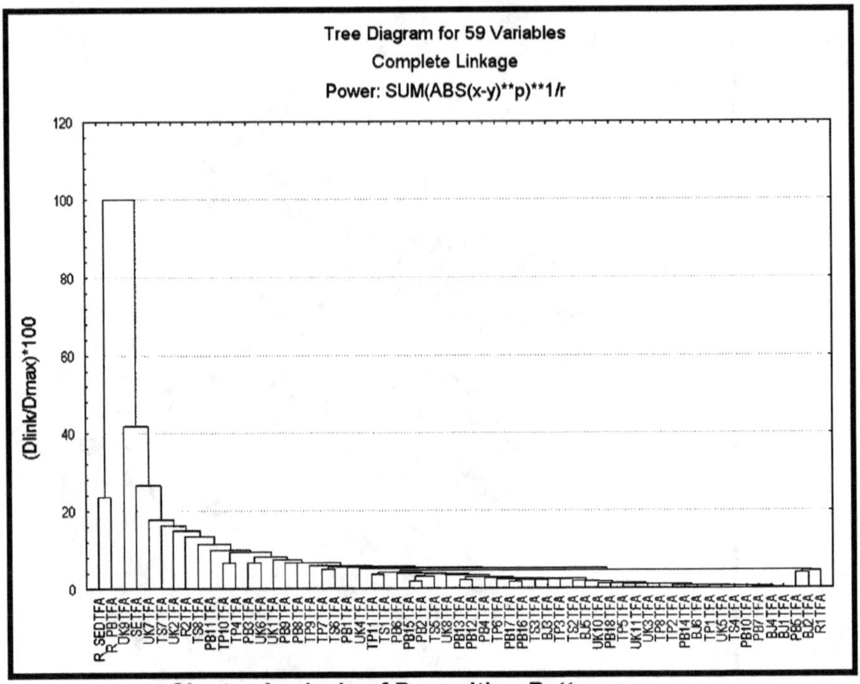

**Cluster Analysis of Deposition Patterns:
All Aboveground Nuclear Tests
1951-62**

This figure represents an analysis of deposition pattern similarities between nuclear tests. Here, the chart shows that the total fallout deposition pattern for SEDAN is relatively similar to that of UK9 and UK7. The terms R PBTFA and R-SEDTFA on the chart refer to the total fallout from 1951-57 and from 1951-62 respectively. The chart on the next page uses a different algorithm to evaluate differences and similarities between the SEDAN fallout and shots primarily from the Plumbbob series.

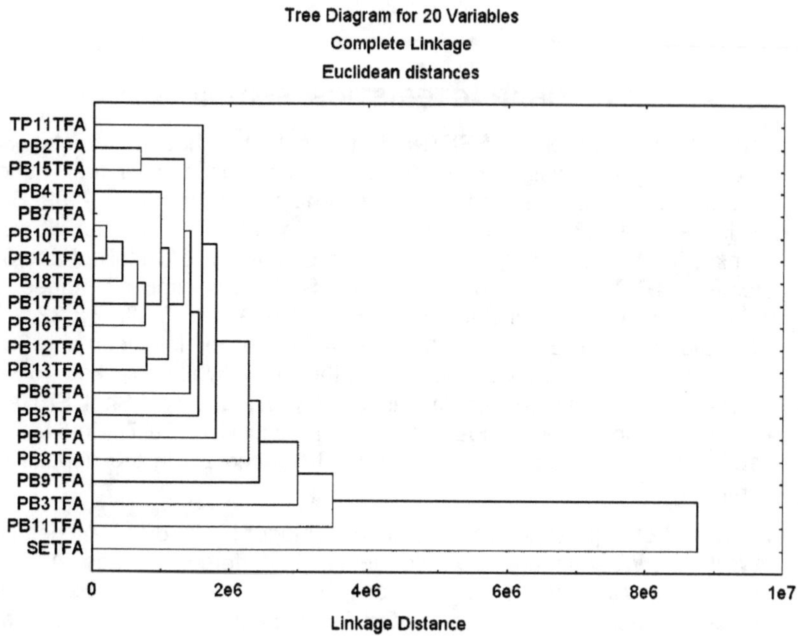
Tree Diagram for 20 Variables
Complete Linkage
Euclidean distances
TP11TFA
PB2TFA
PB15TFA
PB4TFA
PB7TFA
PB10TFA
PB14TFA
PB18TFA
PB17TFA
PB16TFA
PB12TFA
PB13TFA
PB6TFA
PB5TFA
PB1TFA
PB8TFA
PB9TFA
PB3TFA
PB11TFA
SETFA
0
2e6
4e6
6e6
8e6
1e7
Linkage Distance

CHARACTERISTICS: SEDAN

The device used for the SEDAN shot was 17.1 inches in diameter, 38 inches long and weighed 467.9 lbs. Among the nuclear tests, shot Storax SEDAN was ranked 1ST for percentages of Au-198, Au-199, Be-7, Mo-99, Nd-147, Pb-203, W-181, W-185 and W-188 in fallout. It was ranked 2nd for percentages of Co-57, Co-60 and Mn-54 in fallout and 3rd for Na-24. The deposition pattern for SEDAN is similar to those of many earlier shots, including Teapot TESLA (R: 0.38, T: 22.67); Tumbler-Snapper Easy (R: 0.36); SEDAN deposited fallout on only 752 counties, approximately 24 percent of the total. While it ranked 45th in dispersion, SEDAN ranked first in activity per county. Shot SEDAN ranked first in overall deposition of the radionuclides Be-7, Mn-54, Ru-106 and Cm-242 across the country. It ranked 2nd in the deposition amount of Te-127m.

Counties of highest fallout deposition included: Howard IA, Mitchell IA, Worth IA, Washabaugh SD, Woodbury IA, Jefferson IA, Lee IA, Van Buren IA, Washington IA, Hancock IL and Dixon NE. Estimated maximum activities were calculated for a number of Midwestern and Eastern counties. Estimated *maximum* activities in μCi/sq meter were calculated for the following counties: Washabaugh SD: 6343 μCi/sq meter on July 7, 1962 (Washabaugh county, SD no longer exists*.); Howard, Mitchell and Worth counties in Iowa :6599 μCi/sq meter (each) on Jul 8, 1962; Hancock IL: 4314 μCi/sq meter on Jul 8, 1962; Faribault, MN: 4061 μCi/sq meter on Jul 8, 1962; Greenup, KY: 2497 μCi/sq meter on Jul 9, 1962; Boone WV: 2497 μCi/sq meter on Jul 9, 1962 and Logan WV: 2497 μCi/sq meter on July 9, 1962. The day one ratio of total fallout to I-131 for SEDAN was relatively low at 93 to 1.

***Since the time of the SEDAN shot, Washabaugh county, SD was absorbed into Jackson county, SD. Washabaugh was entirely within the present boundaries of the Pine Ridge Reservation.**

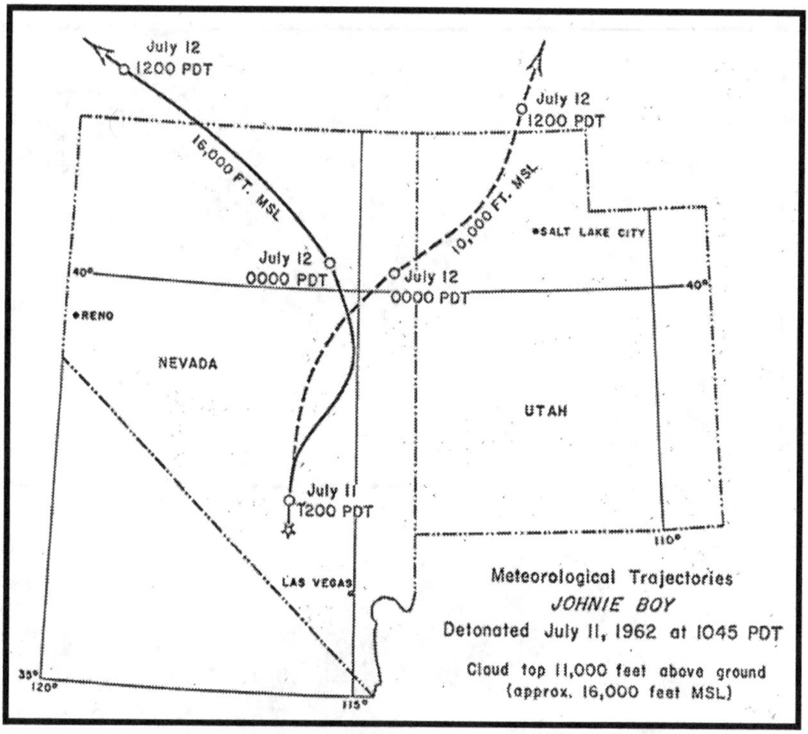

TRAJECTORY: JOHNIE BOY

JOHNIE BOY (UE7) Detonated on July 11, 1962 from a depth of 23 inches. Technically, while this would be classified as an underground test, it was actually a surface burst. The event formed a crater 30.6 ft deep and 122 ft in diameter. The bottom of the cloud reached 12,500 ft MSL and the top of the cloud reached 17,000 ft MSL. Residual gamma radiation was detected as far as 100 miles downwind.

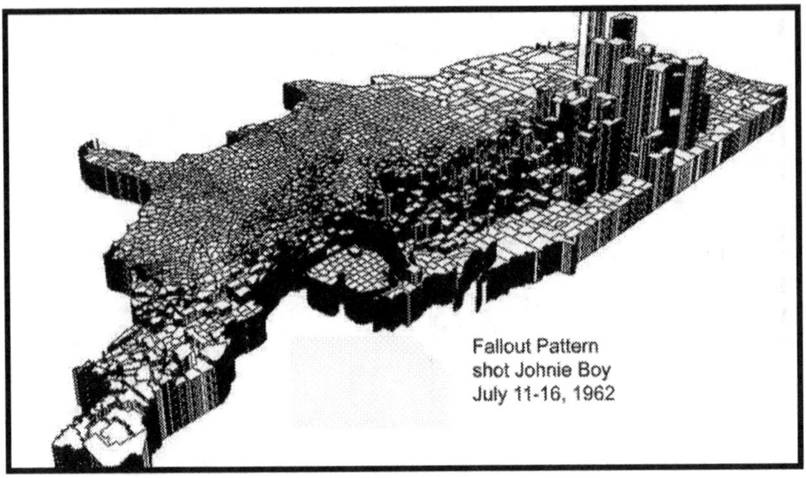
Fallout Pattern
shot Johnie Boy
July 11-16, 1962

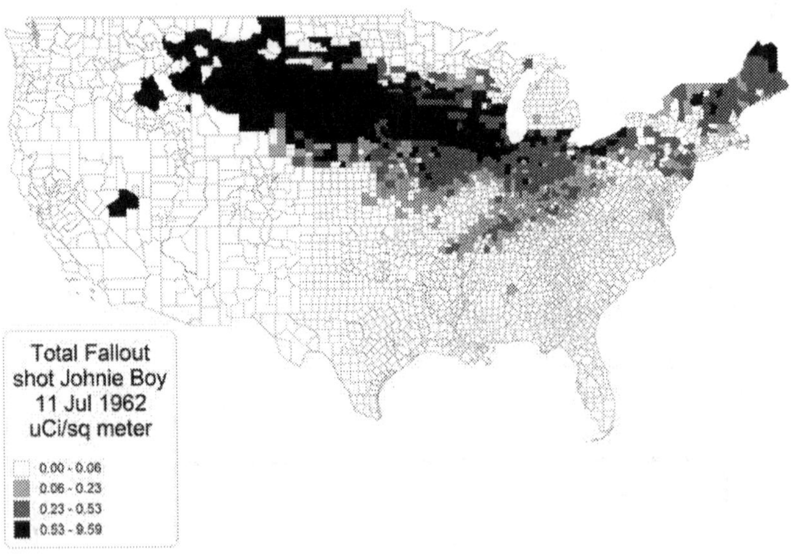
Total Fallout
shot Johnie Boy
11 Jul 1962
uCi/sq meter
0.00 - 0.06
0.06 - 0.23
0.23 - 0.53
0.53 - 9.59

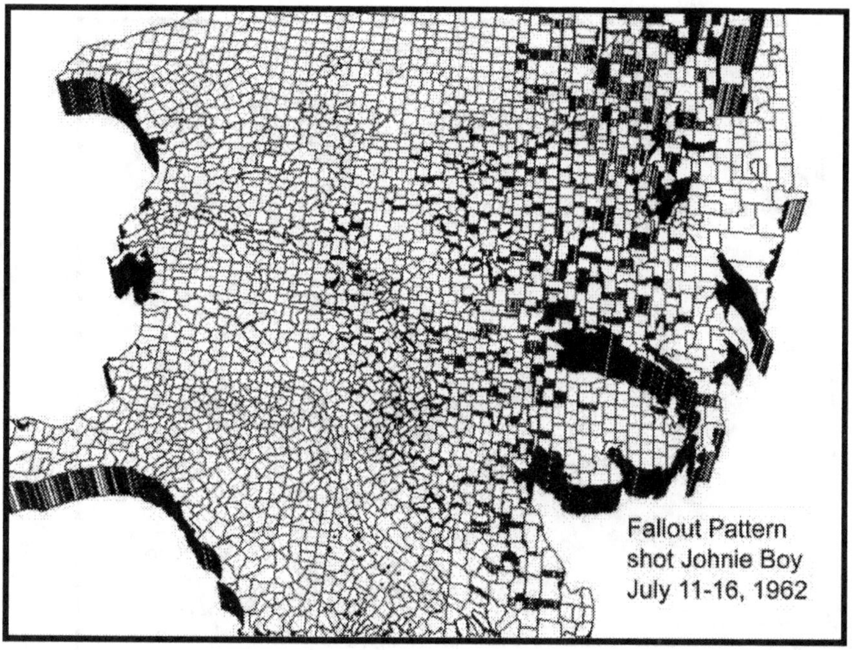
Fallout Pattern
shot Johnie Boy
July 11-16, 1962

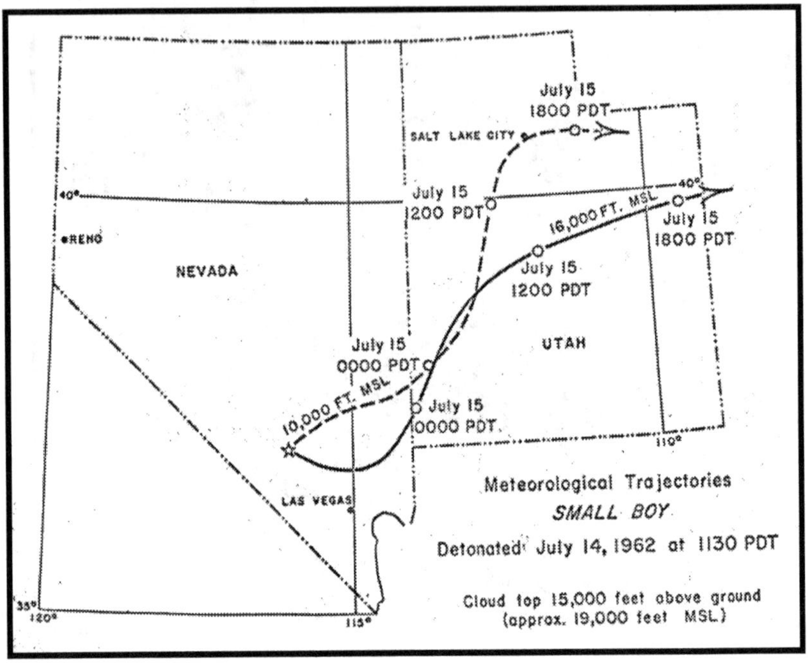

SMALL BOY (UE8) Small Boy was not an underground test. It was detonated from a tower (height not specified) on July 14, 1962. The top of the cloud reached 19,000 ft MSL. The DASA 1251 account notes that the probable path of the Small Boy cloud took it as far east as western Nebraska. The radiation level at ground zero was measured at 2000 Roentgens/hr. One official Department of Energy map (see below) shows residual gamma radiation at 100 Roentgens/hr nearly 300 miles downwind of this test.

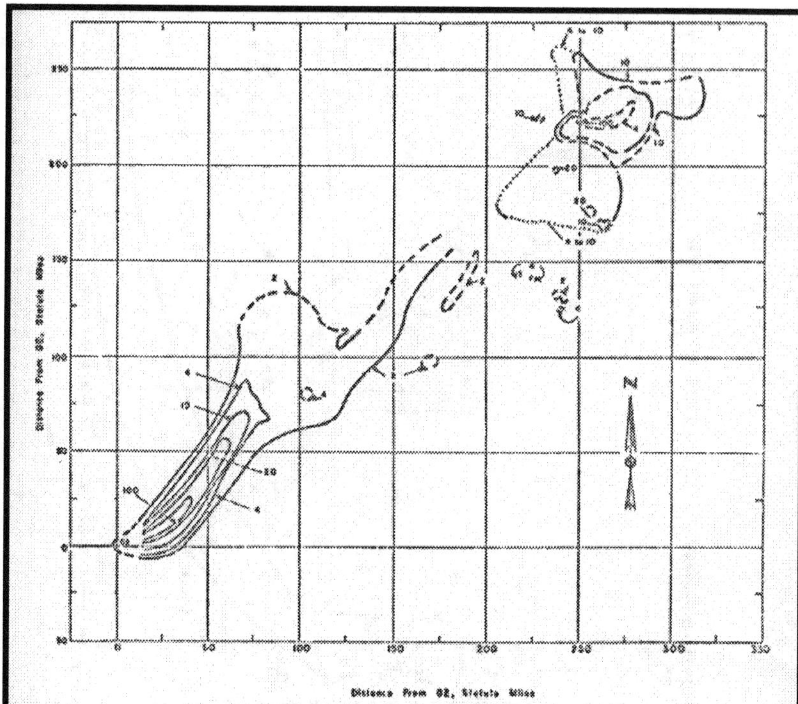

Figure 332. OPERATION SUNBEAM - Small Boy contours of residual gamma radiation in R/hr at H+1 hour to 300 miles downwind

Note the radiation contours in the upper right corner of the map. When magnified, this design shows a crescent-shaped area of 100 Roentgens/hr.

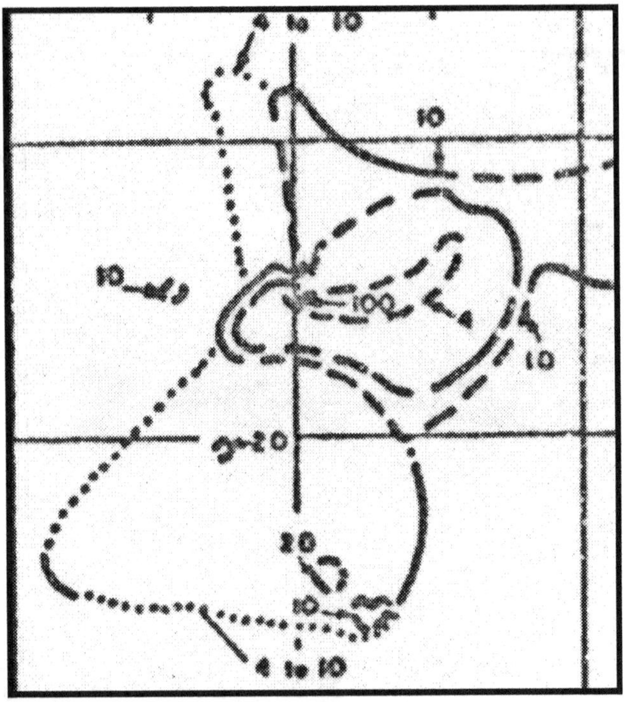

CLOSE-UP OF OFFICIAL RADIATION CONTOUR MAP: SMALL BOY

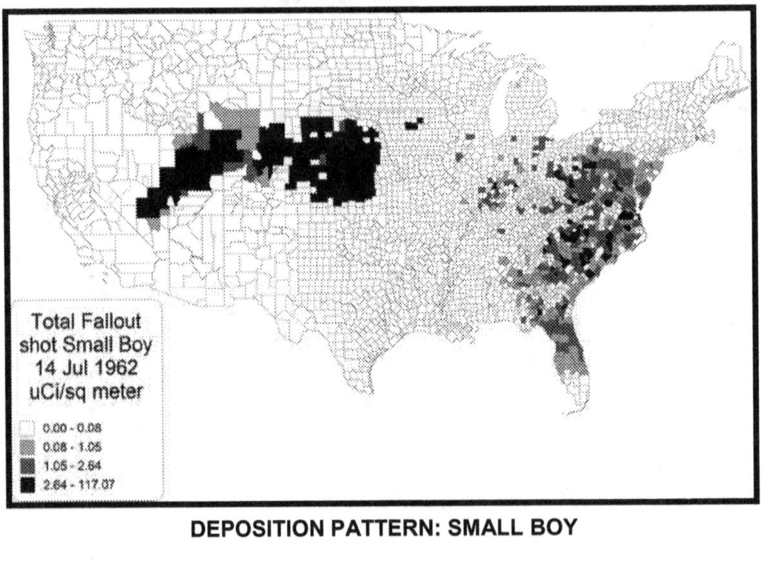

DEPOSITION PATTERN: SMALL BOY

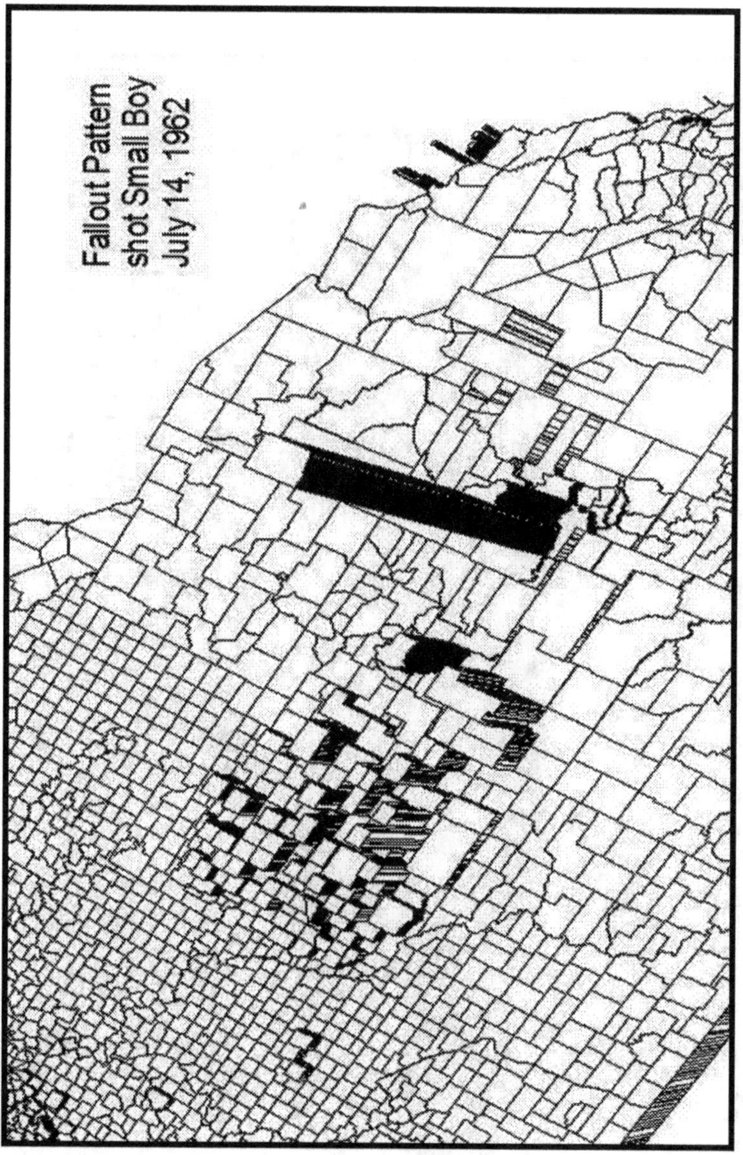
Fallout Pattern
shot Small Boy
July 14, 1962

Fallout Pattern
shot Small Boy
Eastern U.S.
July 14-19, 1962

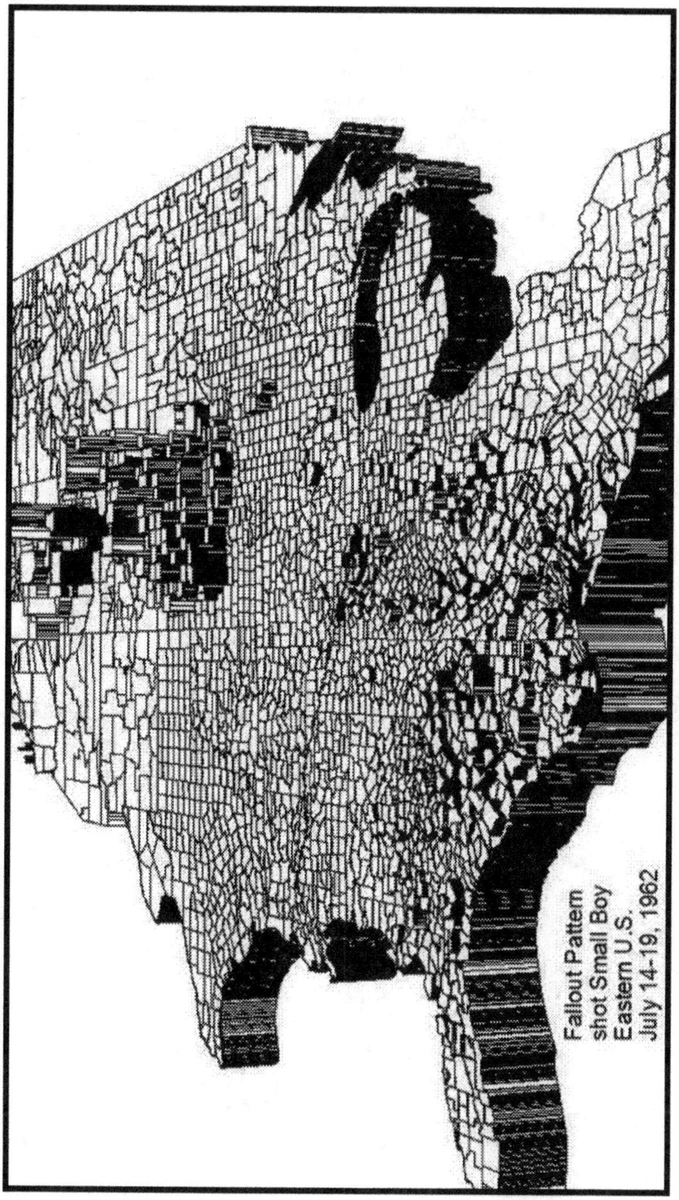
Fallout Pattern
shot Small Boy
Eastern U.S.
July 14-19, 1962

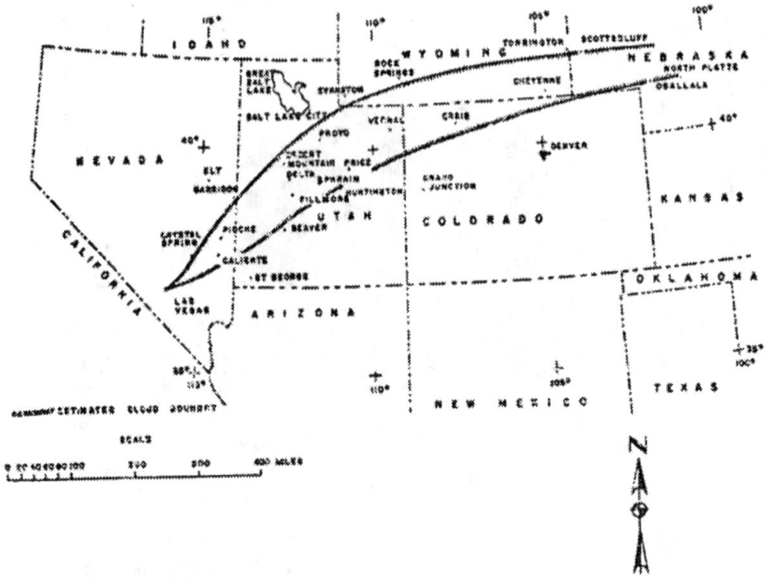

Small Boy cloud path

SMALL BOY CLOUD PATH

BANDICOOT (UE9) Detonated on October 19, 1962, from a depth of 792 ft, Bandicoot produced a yield of 12.5 Kt. The detonation produced a cloud that reached an altitude of 10,500 ft. MSL. The radioactivity at ground zero at the time of the detonation was reported to be greater than 7500 Roentgens/hour. Five hours later, the radioactivity had decreased to 150 Roentgens/hr. According to the account in DASA 1251,

> "The radioactive cloud split into two portions. The lower portion of the cloud traveled in a NNE direction to Area 9 where it remained stagnant, then went slowly across Flat Top Mesa and north to the Area 12 compound (redacted). The cloud dispersed in the valleys north of the test site and no exposures to people were detected."

Another portion of the cloud went south:

> "The upper portion of the cloud traveled in a southern direction and traversed a course over the CP Compound, Camp Mercury, Cactus Springs, Indian Springs, Lathrop Wells and Highway 95. (redacted). The cloud was first detected over Highway 95 at H+2 hours (redacted). Upon crossing Highway 95, the cloud was 9 miles wide, was diffusing rapidly, and was proceeding southwest. The intensity at ground level was approximately twice background. A maximum intensity of 50 mR/hr was detected at 4 miles west of the Mercury junction on Highway 95 (redacted). A 20-mR/hr dose rate was recorded 7 miles west of the Mercury junction at H+3.5 hours. Maximum intensities by portable instruments (3 feet underground) were recorded for Johnie, 12 mR/hr; Ash Meadows, 16 mR/hr; Death Valley Junction 3 mR/hr and Camp Mercury, 5 mR/hr. No radiation was detected off-site at Area 51, Indian Springs and Pahrump. The highest reading at ground level at Cactus Springs (36 miles from SZ) was approximately 0.6 mR/hr."

PIKE (UE10) Detonated on March 13, 1964, this device had a yield less than 20 kilotons. Not much is known of shot Pike, since many details remain classified at the time of this writing (Feb 2006.) The test did involve, however, one of the worst ventings prior to shot Baneberry in 1970. The venting occurred about 15 seconds after detonation and continued for 69 seconds. At that time, rock overlying the shot cavity collapsed and prevented further release of radioactive material. However, the materials that had been released during that 69 second interval rose to about 3,000 ft MSL. As a result of the Pike venting, a model was developed that was used to predict worst-case venting scenarios for underground nuclear tests. It became known as the Pike Model. (*The Containment of Underground Nuclear Explosions*: NTIS Order # PB90-156183, Oct. 1989.)

SULKY (UE11) Detonated from a depth of 88 ft in granite on December 18, 1964, Sulky's yield was only 0.092 Kt.

PALANQUIN (UE12) Detonated from a depth of 278 ft on April 14, 1965, the yield was estimated at 4.3 Kt.

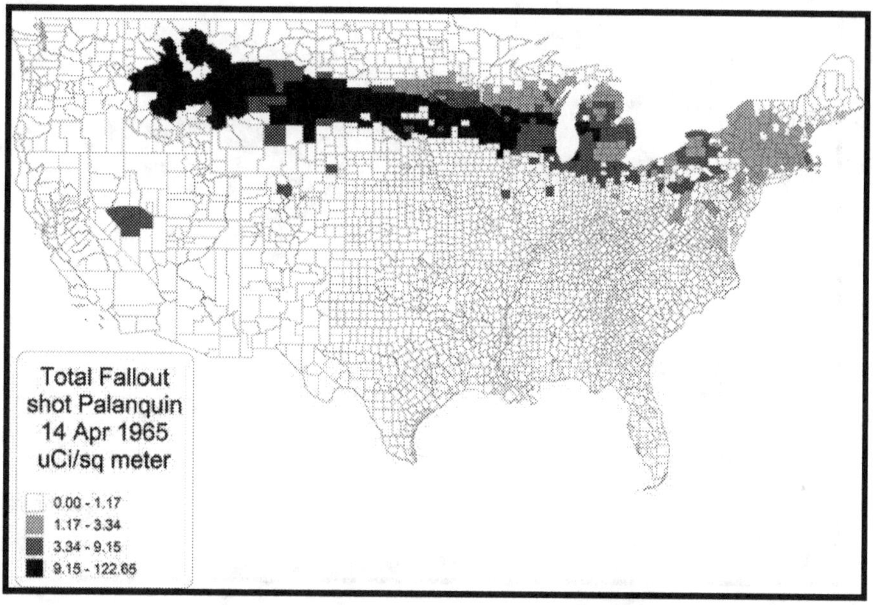
Total Fallout
shot Palanquin
14 Apr 1965
uCi/sq meter
0.00 - 1.17
1.17 - 3.34
3.34 - 9.15
9.15 - 122.65

Fallout Pattern
shot Palanquin
Apr 14, 1965

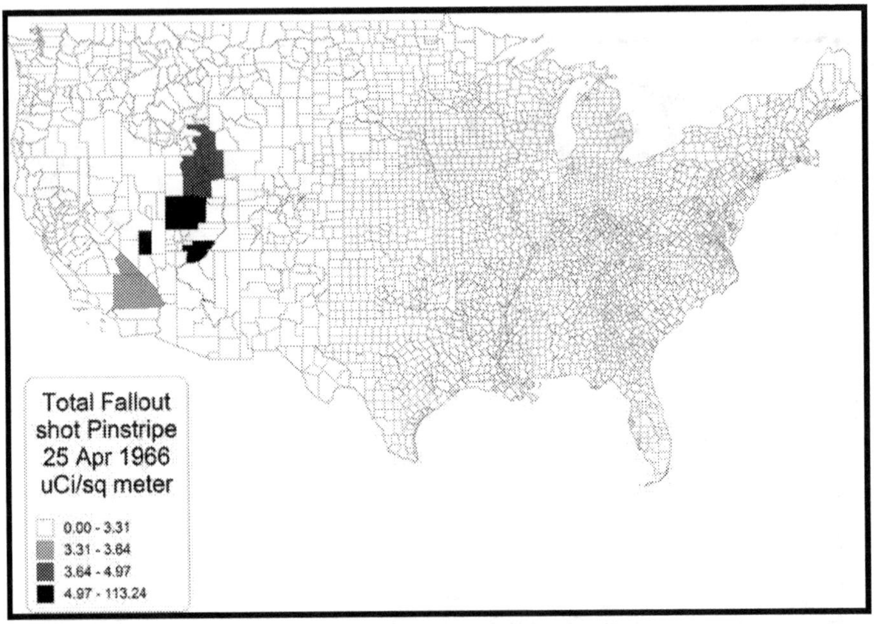
Total Fallout
shot Pinstripe
25 Apr 1966
uCi/sq meter
0.00 - 3.31
3.31 - 3.64
3.64 - 4.97
4.97 - 113.24

PIN STRIPE (UE13) Pin Stripe was detonated on April 25, 1966 with a yield of about 4.0 Kt.

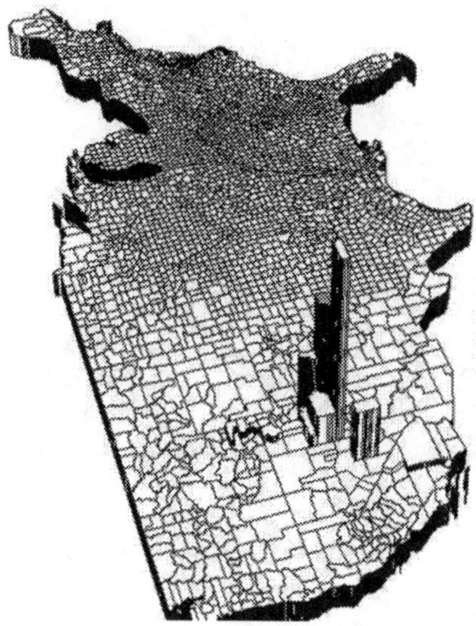

Fallout Pattern
shot Pin Stripe
Apr 25, 1966

CABRIOLET (UE14) Detonated from a depth of 65 ft on January 1st 1968, Cabriolet's yield was estimated at 2.3 Kt.

BUGGY (UE15) March 12, 1968. Buggy was actually a series of five separate nuclear detonations spaced 150 feet apart and fired from a depth of 130 ft. The resultant composite crater formed a trench 850 ft long, 250 ft wide and 65 ft deep.

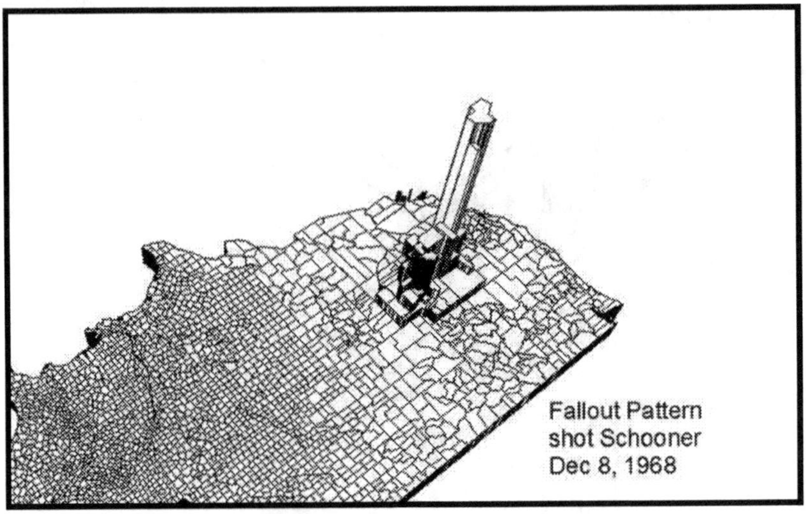

SCHOONER (UE16) Detonated on December 8, 1968 with a yield of 30 Kt.

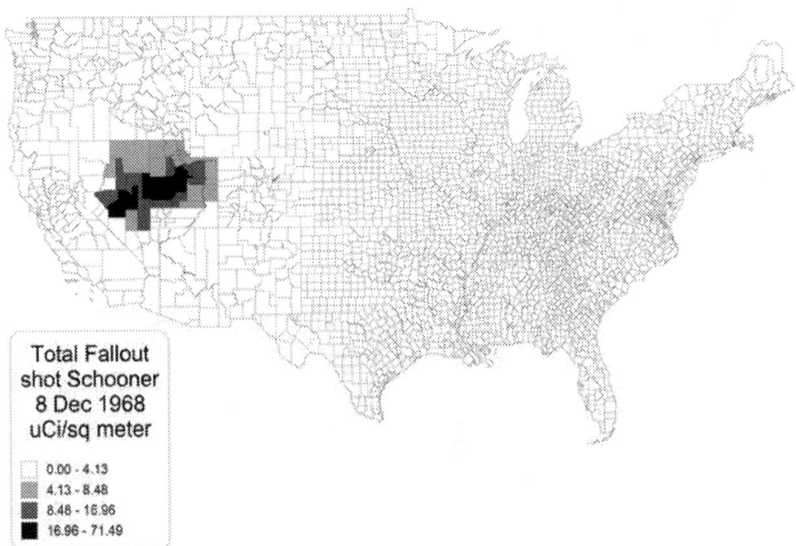

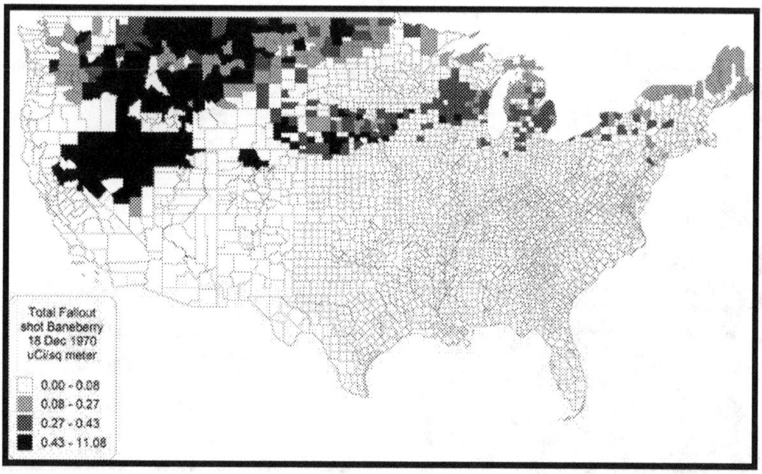

BANEBERRY (UE17) Detonated from a depth of 910 ft on December 18, 1970, the shot vented, producing a cloud that reached an altitude of 8000 ft MSL.

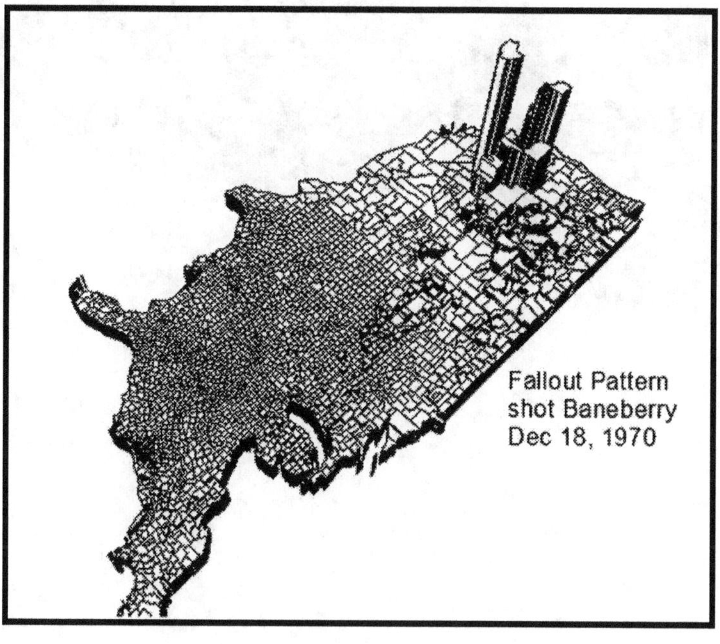

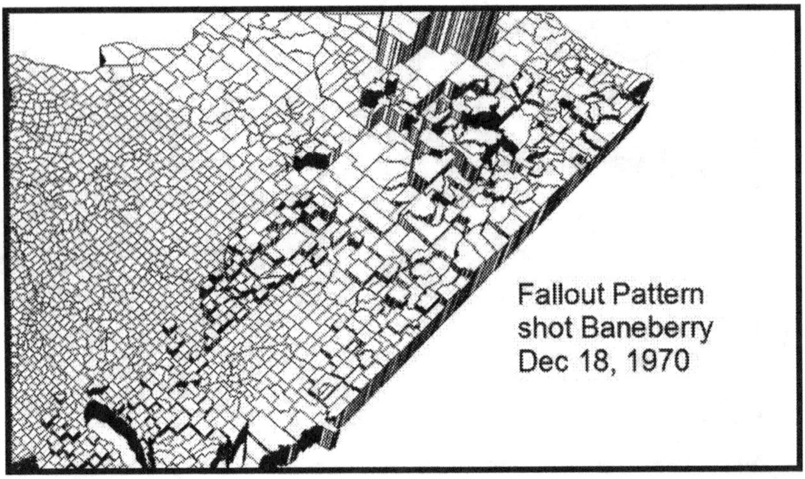

Fallout Pattern
shot Baneberry
Dec 18, 1970

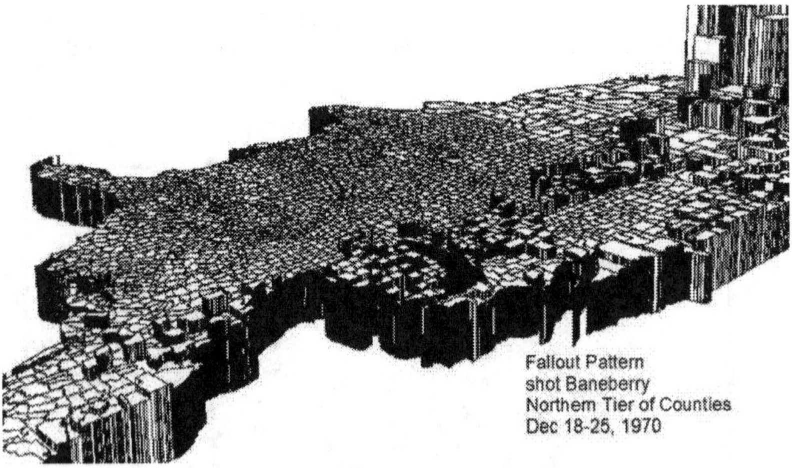

Fallout Pattern
shot Baneberry
Northern Tier of Counties
Dec 18-25, 1970

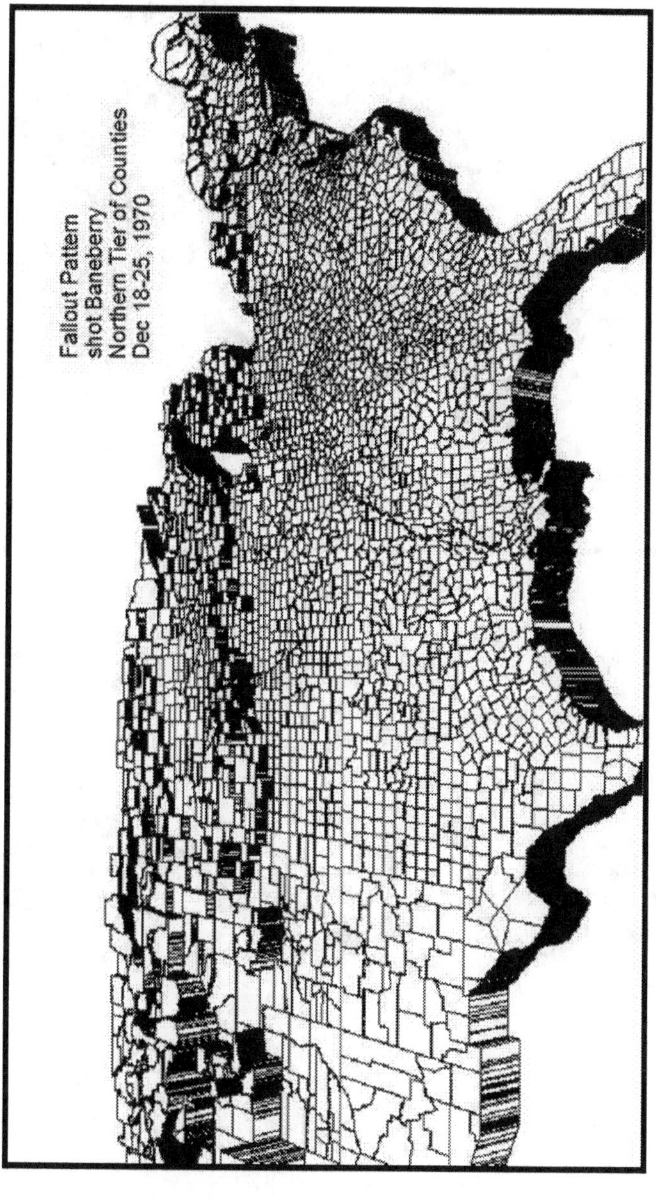
Fallout Pattern
shot Baneberry
Northern Tier of Counties
Dec 18-25, 1970

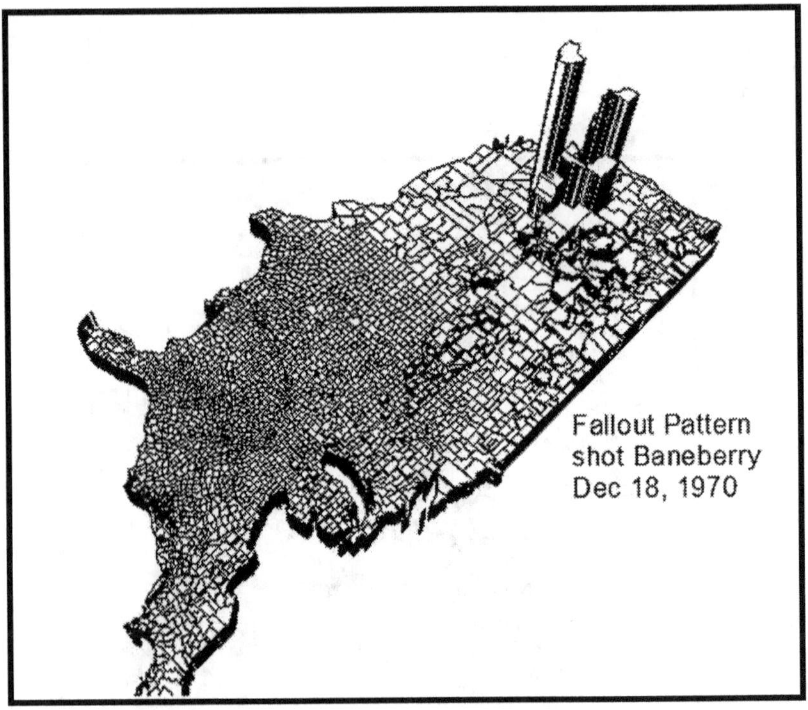
Fallout Pattern
shot Baneberry
Dec 18, 1970

PHOTOS FROM THE NEVADA TEST SITE: 1951-1962

BUSTER CHARLIE

BUSTER DOG

TUMBLER-SNAPPER ABLE

TUMBLER-SNAPPER GEORGE

UPSHOT-KNOTHOLE NANCY

UPSHOT-KNOTHOLE GRABLE

UPSHOT-KNOTHOLE HARRY

TEAPOT ESS

TEAPOT MET

PLUMBBOB HOOD

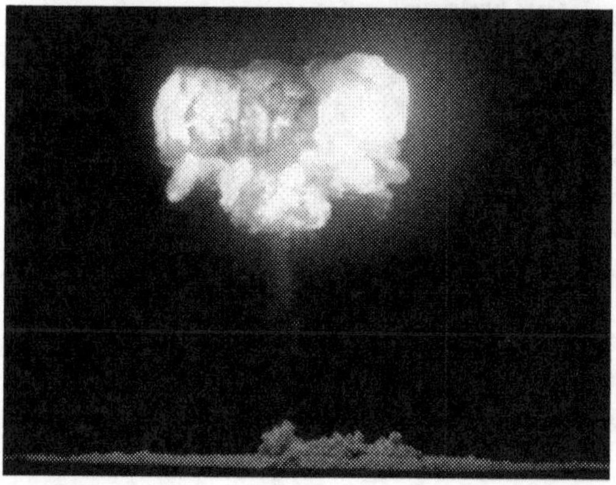

PLUMBBOB
CHARLESTON

PLUMBBOB STOKES

PLUMBBOB PRISCILLA

PLUMBBOB FIZEAU

PLUMBBOB FRANKLIN

STORAX SEDAN

SEDAN CRATER

SECTION 9

U.S. COUNTIES RANKED FOR TOTAL FALLOUT

1951-1970

COUNTIES OF HIGHEST AVERAGE FALLOUT FROM SHOT:

RANGER BAKER-1 (R-1) Jan 28, 1951 8 Kt

(microCuries per square meter)

Each Row Includes:
County Rank (most fallout = 1), State, County, Fallout Level in microCuries/square meter

1	CO	LA PLATA	190.05
2	VT	WINDHAM	48.70
3	OH	BELMONT	48.61
4	OH	GUERNSEY	48.19
5	WV	OHIO	48.19
6	TX	DALLAM	46.74
7	WV	PLEASANTS	41.49
8	MO	AUDRAIN	39.50
9	MO	PIKE	39.50
10	OH	ATHENS	38.66
11	MA	ESSEX	38.35
12	MA	MIDDLESEX	38.01
13	NH	CARROLL	37.92
14	MA	NORFOLK	37.60
15	MA	SUFFOLK	37.60
16	NH	ROCKINGHAM	37.60
17	NH	STRAFFORD	37.60
18	MO	GASCONADE	32.70
19	PA	WAYNE	32.31
20	NY	RENSSELAER	31.23
21	CT	LITCHFIELD	30.19
22	CT	TOLLAND	30.19
23	MA	HAMPDEN	30.19
24	NY	GREENE	30.19
25	NY	SULLIVAN	30.19
26	NY	ULSTER	30.19
27	MO	BENTON	29.96
28	MO	CALLAWAY	27.53
29	MO	RALLS	26.72
30	MO	DENT	25.01
31	MO	PHELPS	25.01
32	CO	ARCHULETA	24.57
33	CO	CONEJOS	24.57
34	PA	CAMERON	23.52
35	PA	ELK	23.52
36	PA	WASHINGTON	23.52
37	PA	ALLEGHENY	23.29
38	PA	INDIANA	23.29
39	PA	WESTMORELAND	23.29
40	OH	COLUMBIANA	23.23
41	OH	GEAUGA	23.23
42	OH	MORROW	23.23
43	OH	RICHLAND	23.23
44	OH	SENECA	23.23
45	PA	BLAIR	23.18
46	PA	CAMBRIA	23.18
47	OH	MONROE	22.77
48	OH	MORGAN	22.77
49	OH	WASHINGTON	22.77
50	PA	ARMSTRONG	22.77
51	PA	BEAVER	22.77
52	PA	CENTRE	22.77
53	PA	CLEARFIELD	22.77
54	PA	FAYETTE	22.77
55	PA	GREENE	22.77
56	PA	HUNTINGDON	22.77
57	PA	MONTOUR	22.77
58	PA	SNYDER	22.77
59	PA	SULLIVAN	22.77
60	PA	SUSQUEHANNA	22.77
61	PA	TIOGA	22.77
62	PA	UNION	22.77
63	WV	MARSHALL	22.77
64	WV	TYLER	22.77
65	OH	ALLEN	22.15
66	OH	MONTGOMERY	21.66
67	NY	OTSEGO	21.62
68	VT	BENNINGTON	21.62
69	NY	DELAWARE	21.17
70	OH	ASHLAND	20.92
71	OH	CUYAHOGA	20.92
72	OH	CLARK	20.91
73	OH	DELAWARE	20.91
74	OH	MARION	20.91
75	OH	MIAMI	20.91
76	OH	KNOX	20.40
77	WV	WOOD	20.12
78	IN	FRANKLIN	19.88
79	OH	COSHOCTON	19.88
80	OH	LICKING	19.88

COUNTIES OF HIGHEST AVERAGE FALLOUT FROM SHOT:

RANGER BAKER-2 (R-2) Feb 2, 1951 8 Kt

(microCuries per square meter)
Each Row Includes:
County Rank (most fallout = 1), State, County, Fallout Level in microCuries/square meter

1	TX REFUGIO	1066.98
2	FL DADE	3.76
3	FL PALM BEACH	1.82
4	FL COLLIER	1.63
5	FL BROWARD	1.44
6	FL LAKE	1.18
7	FL HILLSBOROUGH	1.18
8	FL FRANKLIN	0.95
9	FL LIBERTY	0.95
10	MD SOMERSET	0.90
11	CT TOLLAND	0.78
12	MA NANTUCKET	0.78
13	FL VOLUSIA	0.72
14	FL OKEECHOBEE	0.71
15	GA LOWNDES	0.71
16	GA DOOLY	0.71
17	FL POLK	0.57
18	FL CITRUS	0.57
19	FL SARASOTA	0.57
20	FL LAFAYETTE	0.57
21	FL JEFFERSON	0.57
22	FL MADISON	0.57
23	FL DE SOTO	0.57
24	ME PISCATAQUIS	0.56
25	NJ OCEAN	0.56
26	NJ CAPE MAY	0.56
27	MD WORCESTER	0.56
28	VA ACCOMACK	0.56
29	VA MATHEWS	0.56
30	VA NORFOLK/CHES	0.56
31	VA HAMPTON	0.56
32	FL BREVARD	0.54
33	FL MANATEE	0.54
34	FL DUVAL	0.54
35	FL LEVY	0.54
36	FL DIXIE	0.54
37	FL GILCHRIST	0.54
38	FL PINELLAS	0.54
39	GA ECHOLS	0.54
40	FL HERNANDO	0.52
41	FL ORANGE	0.52
42	FL COLUMBIA	0.52
43	FL SUWANNEE	0.52
44	FL TAYLOR	0.52
45	FL SEMINOLE	0.52
46	FL MARTIN	0.52
47	GA CHARLTON	0.52
48	GA CAMDEN	0.52
49	FL HAMILTON	0.52
50	GA GLYNN	0.52
51	GA EVANS	0.52
52	GA CHATHAM	0.52
53	GA COLUMBIA	0.47
54	GA JOHNSON	0.47
55	AL GENEVA	0.47
56	FL OKALOOSA	0.47
57	PA BUCKS	0.45
58	NC HYDE	0.45
59	NC CHOWAN	0.45
60	NC CARTERET	0.45
61	ME LINCOLN	0.41
62	NH ROCKINGHAM	0.41
63	ME KNOX	0.41
64	MA MIDDLESEX	0.41
65	CT WINDHAM	0.37
66	ME CUMBERLAND	0.37
67	ME SAGADAHOC	0.37
68	CT NEW LONDON	0.37
69	RI PROVIDENCE	0.37
70	ME YORK	0.37
71	MA SUFFOLK	0.37
72	RI KENT	0.37
73	RI NEWPORT	0.37
74	ME PENOBSCOT	0.37

COUNTIES OF HIGHEST AVERAGE FALLOUT FROM SHOT:

BUSTER BAKER (BJ1) Oct 28, 1951 3.5 Kt

(microCuries per square meter)

Each row includes:
County Rank (most fallout = 1), State, County, Fallout Level in microCuries/square meter

Rank	State	County	Fallout
1	CA	FRESNO	12.11
2	CA	KERN	12.11
3	CA	KINGS	12.11
4	CA	MONTEREY	12.11
5	CA	SAN BENITO	12.11
6	CA	SAN LUIS OBI	12.11
7	CA	TULARE	12.11
8	CA	VENTURA	12.11
9	CA	SANTA BARBAR	5.03
10	GA	MERIWETHER	2.15
11	GA	PIKE	2.15
12	VA	GRAYSON	1.72
13	WV	PUTNAM	1.72
14	WV	RALEIGH	1.72
15	WV	SUMMERS	1.72
16	MS	MONROE	1.60
17	AL	RANDOLPH	1.29
18	GA	BALDWIN	1.29
19	GA	CLARKE	1.29
20	GA	DAWSON	1.29
21	GA	GREENE	1.29
22	GA	GWINNETT	1.29
23	GA	HANCOCK	1.29
24	GA	JACKSON	1.29
25	GA	MORGAN	1.29
26	GA	OCONEE	1.29
27	GA	PAULDING	1.29
28	GA	TALIAFERRO	1.29
29	GA	TROUP	1.29
30	GA	WARREN	1.29
31	GA	WILKINSON	1.29
32	NC	MACON	1.29
33	VA	BEDFORD	1.18
34	VA	PULASKI	1.18
35	VA	ROANOKE	1.18
36	VA	TAZEWELL	1.18
37	WV	BRAXTON	1.18
38	WV	KANAWHA	1.18
39	WV	MERCER	1.18
40	WV	TAYLOR	1.18
41	MS	ATTALA	0.91
42	MS	CARROLL	0.91
43	MS	HOLMES	0.91
44	MS	HUMPHREYS	0.91
45	MS	SHARKEY	0.91
46	MS	TATE	0.91
47	MS	WINSTON	0.91
48	GA	BARTOW	0.86
49	GA	BUTTS	0.86
50	GA	CHEROKEE	0.86
51	GA	DE KALB	0.86
52	GA	DODGE	0.86
53	GA	DOUGLAS	0.86
54	GA	FRANKLIN	0.86
55	GA	HALL	0.86
56	GA	HEARD	0.86
57	GA	HENRY	0.86
58	GA	JOHNSON	0.86
59	GA	LAMAR	0.86
60	GA	LAURENS	0.86
61	GA	MADISON	0.86
62	GA	NEWTON	0.86
63	GA	OGLETHORPE	0.86
64	GA	PICKENS	0.86
65	GA	RABUN	0.86
66	GA	ROCKDALE	0.86
67	GA	SPALDING	0.86
68	GA	STEPHENS	0.86
69	GA	UNION	0.86
70	GA	WALTON	0.86
71	GA	WASHINGTON	0.86
72	NC	HAYWOOD	0.86
73	SC	OCONEE	0.86
74	WV	RANDOLPH	0.71
75	AR	ARKANSAS	0.69
76	AR	GREENE	0.69
77	AR	JACKSON	0.69
78	AR	VAN BUREN	0.69
79	AR	WOODRUFF	0.69
80	CA	RIVERSIDE	0.69

COUNTIES OF HIGHEST AVERAGE FALLOUT FROM SHOT:

BUSTER CHARLIE (BJ-2) Oct 30, 1951 14 Kt

(microCuries per square meter)

Each Row Includes:
County Rank (most fallout = 1), State, County, Fallout Level in microCuries/square meter

Rank	State	County	Fallout
1	ME	AROOSTOOK	86.45
2	CA	INYO3	72.80
3	CA	LOS ANGELES	71.77
4	NY	OSWEGO	68.30
5	NY	ERIE	59.25
6	NY	GENESEE	56.61
7	NY	TOMPKINS	56.52
8	ME	PENOBSCOT	53.60
9	ME	PISCATAQUIS	50.86
10	NY	MONROE	50.14
11	NY	WAYNE	48.50
12	NY	JEFFERSON	48.48
13	NY	LIVINGSTON	48.31
14	NY	WYOMING	48.15
15	ME	WASHINGTON	47.73
16	NY	NIAGARA	46.00
17	NY	ORLEANS	45.42
18	NY	OTSEGO	44.89
19	NY	CHEMUNG	43.83
20	NY	ONTARIO	43.12
21	NY	HAMILTON	40.53
22	NY	DELAWARE	40.13
23	NY	SENECA	38.19
24	NY	CAYUGA	37.72
25	NY	LEWIS	37.57
26	ME	HANCOCK	37.54
27	NH	CARROLL	36.08
28	NY	ONONDAGA	35.72
29	NY	MADISON	35.24
30	NY	YATES	35.01
31	NY	CHENANGO	34.37
32	NY	CORTLAND	34.35
33	NY	FRANKLIN	34.31
34	NY	ONEIDA	33.25
35	NY	ALLEGANY	32.93
36	NC	CRAVEN	32.32
37	NY	SCHUYLER	31.72
38	ME	SAGADAHOC	31.50
39	NY	STEUBEN	31.13
40	ME	CUMBERLAND	30.70
41	ME	FRANKLIN	30.62
42	PA	WAYNE	30.49
43	NC	PITT	30.26
44	ME	OXFORD	30.03
45	ME	WALDO	30.02
46	NY	SULLIVAN	29.91
47	NC	ALAMANCE	29.84
48	NY	ESSEX	29.72
49	NY	ST LAWRENCE	29.54
50	NY	HERKIMER	29.49
51	VA	ARLINGTON	28.68
52	PA	WYOMING	28.05
53	NC	DUPLIN	27.72
54	DC	WASHINGTON	27.68
55	PA	LACKAWANNA	27.61
56	MD	PRINCE GEORG	27.08
57	NY	SARATOGA	26.98
58	ME	KENNEBEC	26.77
59	NC	HARNETT	26.75
60	NY	BROOME	26.33
61	VT	WINDSOR	26.24
62	NC	COLUMBUS	26.19
63	SC	MARION	26.06
64	ME	KNOX	25.84
65	NC	LEE	25.78
66	NC	CHATHAM	25.74
67	NY	TIOGA	25.63
68	VA	MECKLENBURG	25.51
69	NH	BELKNAP	25.46
70	NH	GRAFTON	25.33
71	NC	PENDER	25.22
72	NY	GREENE	25.03
73	SC	HORRY	24.98
74	NY	FULTON	24.93
75	ME	YORK	24.87
76	NY	CATTARAUGUS	24.73
77	NC	BLADEN	24.65
78	VA	FAIRFAX	24.57
79	MD	CHARLES	24.47
80	SC	FLORENCE	24.23

COUNTIES OF HIGHEST AVERAGE FALLOUT FROM SHOT:

BUSTER DOG (BJ3) Nov 1, 1951, 21 Kt

(microCuries per square meter)

Each Row Includes: County Rank (most fallout = 1), State, County, Fallout Level in microCuries/square meter

Rank	State	County	Fallout
1	NY	NIAGARA	43.70
2	NY	CHAUTAUQUA	40.67
3	NY	ERIE	36.48
4	NY	ORLEANS	35.31
5	NY	GENESEE	34.26
6	NY	JEFFERSON	30.38
7	NY	ALLEGANY	29.56
8	NY	ONTARIO	29.56
9	PA	POTTER	29.56
10	PA	WARREN	29.56
11	NY	STEUBEN	28.98
12	NY	WAYNE	28.39
13	NY	WYOMING	28.39
14	NY	CATTARAUGUS	25.36
15	PA	ELK	25.36
16	PA	FOREST	25.36
17	PA	MCKEAN	25.36
18	NY	YATES	24.20
19	NY	LIVINGSTON	23.15
20	NY	SENECA	22.84
21	NJ	MORRIS	19.38
22	NY	PUTNAM	19.38
23	NY	ULSTER	19.38
24	VA	CHARLES CITY	14.15
25	VA	PRINCE GEORG	14.15
26	NC	CAMDEN	13.38
27	NC	CURRITUCK	13.38
28	NC	DARE	13.38
29	NC	PASQUOTANK	13.38
30	NC	PERQUIMANS	13.38
31	VA	NORFOLK/CHES	13.38
32	VA	LANCASTER	12.94
33	VA	MIDDLESEX	12.94
34	VA	NORTHUMBERLA	12.94
35	TN	ANDERSON	12.70
36	TN	SEQUATCHIE	12.70
37	NC	NORTHAMPTON	12.62
38	VA	ISLE OF WIGH	12.62
39	VA	SUFFOLK/NANS	12.62
40	VA	GREENSVILLE	12.45
41	MD	SOMERSET	12.24
42	NC	BEAUFORT	12.24
43	NC	BERTIE	12.24
44	NC	CARTERET	12.24
45	NC	CHOWAN	12.24
46	NC	GATES	12.24
47	NC	HERTFORD	12.24
48	NC	HYDE	12.24
49	NC	MARTIN	12.24
50	NC	WASHINGTON	12.24
51	VA	ACCOMACK	12.24
52	VA	GLOUCESTER	12.24
53	VA	HAMPTON	12.24
54	VA	JAMES CITY	12.24
55	VA	MATHEWS	12.24
56	VA	NEWPORT NEWS	12.24
57	VA	NORTHAMPTON	12.24
58	VA	SOUTHAMPTON	12.24
59	VA	SURRY	12.24
60	VA	SUSSEX	12.24
61	VA	YORK	12.24
62	CT	FAIRFIELD	11.68
63	NJ	ESSEX	11.68
64	NJ	PASSAIC	11.68
65	NJ	WARREN	11.68
66	NY	BRONX	11.68
67	NY	NASSAU	11.68
68	NY	WESTCHESTER	11.68
69	NC	PAMLICO	11.05
70	OH	ERIE	10.53
71	GA	CATOOSA	9.63
72	GA	WALKER	9.63
73	GA	WHITFIELD	9.63
74	NC	CHEROKEE	9.63
75	TN	LOUDON	9.63
76	VA	CHESTERFIELD	9.52
77	VA	HENRICO	9.52
78	GA	FLOYD	9.46
79	TN	GRUNDY	9.46

COUNTIES OF HIGHEST AVERAGE FALLOUT FROM SHOT:

BUSTER EASY (BJ4) Nov 5, 1951, 31 Kt

(microCuries per square meter)

Each Row Includes:
County Rank (most fallout = 1), State, County, Fallout Level in microCuries/square meter

Rank	State	County	Fallout
1	DE	KENT	2.92
2	DE	NEW CASTLE	2.92
3	DE	SUSSEX	2.92
4	MD	WORCESTER	2.92
5	NJ	BURLINGTON	2.92
6	NJ	CAMDEN	2.92
7	NJ	CAPE MAY	2.92
8	NJ	CUMBERLAND	2.92
9	NJ	GLOUCESTER	2.92
10	NJ	MERCER	2.92
11	NJ	OCEAN	2.92
12	NJ	SALEM	2.92
13	PA	BUCKS	2.92
14	PA	DELAWARE	2.92
15	PA	MONTGOMERY	2.92
16	PA	PHILADELPHIA	2.92
17	MI	WAYNE	2.75
18	MD	SOMERSET	2.70
19	NC	BEAUFORT	2.70
20	NC	BERTIE	2.70
21	NC	CAMDEN	2.70
22	NC	CARTERET	2.70
23	NC	CHOWAN	2.70
24	NC	CURRITUCK	2.70
25	NC	DARE	2.70
26	NC	GATES	2.70
27	NC	HERTFORD	2.70
28	NC	HYDE	2.70
29	NC	MARTIN	2.70
30	NC	NORTHAMPTON	2.70
31	NC	PAMLICO	2.70
32	NC	PASQUOTANK	2.70
33	NC	PERQUIMANS	2.70
34	NC	TYRRELL	2.70
35	NC	WASHINGTON	2.70
36	VA	ACCOMACK	2.70
37	VA	CHARLES CITY	2.70
38	VA	CHESTERFIELD	2.70
39	VA	DINWIDDIE	2.70
40	VA	GLOUCESTER	2.70
41	VA	GREENSVILLE	2.70
42	VA	HAMPTON	2.70
43	VA	HENRICO	2.70
44	VA	ISLE OF WIGH	2.70
45	VA	JAMES CITY	2.70
46	VA	KING AND QUE	2.70
47	VA	KING WILLIAM	2.70
48	VA	LANCASTER	2.70
49	VA	MATHEWS	2.70
50	VA	MIDDLESEX	2.70
51	VA	NEW KENT	2.70
52	VA	NEWPORT NEWS	2.70
53	VA	NORFOLK/CHES	2.70
54	VA	NORTHAMPTON	2.70
55	VA	NORTHUMBERLA	2.70
56	VA	PRINCE GEORG	2.70
57	VA	SOUTHAMPTON	2.70
58	VA	SUFFOLK/NANS	2.70
59	VA	SURRY	2.70
60	VA	SUSSEX	2.70
61	VA	YORK	2.70
62	MI	HURON	2.35
63	OH	ERIE	2.21
64	OH	LAKE	2.21
65	OH	OTTAWA	2.21
66	OH	HURON	1.94
67	OH	LORAIN	1.94
68	OH	RICHLAND	1.94
69	OH	SANDUSKY	1.94
70	OH	WYANDOT	1.94
71	MI	MACOMB	1.81
72	OH	SUMMIT	1.81
73	MI	OAKLAND	1.54
74	MI	SANILAC	1.54
75	MI	ST CLAIR	1.54
76	OH	ASHTABULA	1.54
77	OH	DELAWARE	1.54
78	OH	GEAUGA	1.54
79	OH	MARION	1.54
80	OH	MEDINA	1.54

COUNTIES OF HIGHEST AVERAGE FALLOUT FROM SHOT:

JANGLE SUGAR (BJ5) Nov 19, 1951 1.2 Kt

(microCuries per square meter)

Each Row Includes:
County Rank (most fallout = 1), State, County, Fallout Level in microCuries/square meter

Rank	State	County	Fallout
1	NV		152.96
2	NV	EUREKA	106.82
3	NV	ELKO	105.23
4	ID	OWYHEE	91.04
5	ID	ADA	59.69
6	NV	LANDER1	56.41
7	ID	CLEARWATER	52.92
8	ID	ADAMS	50.30
9	ID	VALLEY	49.13
10	ID	IDAHO	48.01
11	ID	GEM	47.97
12	ID	ELMORE	45.35
13	ID	CANYON	45.24
14	ID	BOISE	44.79
15	NV	WHITE PINE2	40.22
16	NV	WHITE PINE1	40.01
17	NV	WHITE PINE3	37.79
18	ID	PAYETTE	36.45
19	ID	WASHINGTON	34.26
20	ID	CUSTER	33.20
21	ID	CAMAS	29.37
22	MT	MINERAL	28.25
23	NV	HUMBOLDT	27.53
24	UT	JUAB	18.73
25	NV	PERSHING	18.62
26	OR	HARNEY	18.28
27	OR	MALHEUR	18.28
28	ID	GOODING	18.16
29	ID	TWIN FALLS	18.16
30	ID	LINCOLN	9.14
31	ID	MINIDOKA	9.14
32	UT	BOX ELDER1	9.14
33	ID	JEROME	9.03
34	UT	TOOELE1	8.72
35	ID	CASSIA	8.62
36	NV	LANDER2	7.56
37	OK	GARFIELD	6.33
38	OK	MAJOR	6.23
39	UT	BOX ELDER2	4.63
40	KS	COWLEY	3.85
41	KS	SUMNER	3.85
42	OK	ALFALFA	3.85
43	OK	BLAINE	3.85
44	OK	BRYAN	3.85
45	OK	CANADIAN	3.85
46	OK	CHEROKEE	3.85
47	OK	CREEK	3.85
48	OK	DEWEY	3.85
49	OK	GRANT	3.85
50	OK	LINCOLN	3.85
51	OK	LOGAN	3.85
52	OK	MAYES	3.85
53	OK	MUSKOGEE	3.85
54	OK	NOBLE	3.85
55	OK	OKMULGEE	3.85
56	OK	OSAGE	3.85
57	OK	PAWNEE	3.85
58	OK	PAYNE	3.85
59	OK	ROGER MILLS	3.85
60	OK	ROGERS	3.85
61	OK	WAGONER	3.85
62	OK	WOODWARD	3.85
63	TX	GRAYSON	3.85
64	TX	WHEELER	3.85
65	NV	LINCOLN2	3.84
66	OK	GRADY	3.75
67	OK	KAY	3.75
68	OK	WASHINGTON	3.75
69	NE	ROCK	3.18
70	NV	CHURCHILL	2.82
71	UT	TOOELE2	2.82
72	OK	JACKSON	2.63
73	OK	WOODS	2.63
74	KS	CLARK	2.58
75	KS	EDWARDS	2.58
76	KS	HARPER	2.58
77	KS	KINGMAN	2.58
78	KS	KIOWA	2.58
79	KS	PRATT	2.58
80	KS	RUSH	2.58

COUNTIES OF HIGHEST AVERAGE FALLOUT FROM SHOT:

JANGLE UNCLE (BJ6) Nov 29, 1951 1.2 Kt

(microCuries per square meter)

Each Row Includes:
County Rank (most fallout = 1), State, County, Fallout Level in microCuries/square meter

Rank	State	County	Fallout
1	NV	WHITE PINE2	88.82
2	NV	LINCOLN2	85.19
3	NV	NYE2	79.78
4	NV	WHITE PINE1	46.98
5	NV	WHITE PINE3	28.14
6	NV	ELKO	8.56
7	UT	TOOELE1	5.89
8	NV	EUREKA	4.65
9	UT	JUAB	3.63
10	UT	BOX ELDER1	2.72
11	UT	TOOELE2	2.72
12	WY	UINTA	2.61
13	WY	SWEETWATER	2.46
14	ID	OWYHEE	1.82
15	ID	TWIN FALLS	1.82
16	UT	BOX ELDER2	1.82
17	WY	CARBON	1.82
18	UT	MILLARD	1.71
19	ID	CASSIA	1.70
20	ID	BEAR LAKE	0.91
21	ID	BINGHAM	0.91
22	ID	BONNEVILLE	0.91
23	ID	CANYON	0.91
24	ID	CARIBOU	0.91
25	ID	ELMORE	0.91
26	ID	FRANKLIN	0.91
27	ID	GOODING	0.91
28	ID	JEROME	0.91
29	ID	LINCOLN	0.91
30	ID	MINIDOKA	0.91
31	ID	ONEIDA	0.91
32	ID	POWER	0.91
33	OR	MALHEUR	0.91
34	UT	CACHE	0.91
35	UT	DAGGETT	0.91
36	UT	DUCHESNE	0.91
37	UT	RICH	0.91
38	UT	SUMMIT	0.91
39	UT	UINTAH	0.91
40	UT	WASATCH	0.91
41	WY	FREMONT	0.91
42	WY	LINCOLN	0.91
43	WY	SUBLETTE	0.91
44	ID	ADA	0.86
45	ID	BANNOCK	0.86
46	UT	DAVIS	0.86
47	UT	MORGAN	0.86
48	UT	SALT LAKE	0.86
49	UT	UTAH	0.86
50	UT	WEBER	0.86
51	NV	LINCOLN1	0.42

COUNTIES OF HIGHEST AVERAGE FALLOUT FROM SHOT:

TUMBLER-SNAPPER ABLE (TS1) Apr 1, 1952 1 Kt

(microCuries per square meter)

Each Row Includes:
County Rank (most fallout = 1), State, County, Fallout Level in microCuries/square meter

Rank	State	County	Fallout
1	NE	DAWES	140.77
2	NE	BANNER	140.57
3	NE	SIOUX	140.57
4	MO	CLAY	120.74
5	CO	PITKIN	98.21
6	KS	GRANT	80.31
7	KS	HODGEMAN	80.31
8	MO	MARIES	76.23
9	MO	VERNON	76.23
10	KS	WYANDOTTE	74.62
11	MO	BATES	74.62
12	MO	CEDAR	74.62
13	MO	GREENE	74.62
14	MO	HOWARD	74.62
15	MO	RAY	74.62
16	MO	SHELBY	74.62
17	MO	ST CLAIR	74.62
18	CO	ADAMS	71.13
19	CO	DENVER	71.13
20	CO	LARIMER	71.13
21	CO	LOGAN	71.13
22	CO	SEDGWICK	71.13
23	NE	CHEYENNE	71.13
24	NE	KIMBALL	71.13
25	NE	MORRILL	71.13
26	NE	SHERIDAN	71.13
27	WY	GOSHEN	71.13
28	WY	LARAMIE	71.13
29	CO	GILPIN	70.93
30	NE	BOX BUTTE	70.93
31	NE	HOOKER	70.93
32	WY	ALBANY	70.93
33	WY	CONVERSE	70.93
34	WY	NIOBRARA	70.93
35	CO	LAKE	65.97
36	KS	SHAWNEE	65.42
37	NE	RICHARDSON	65.42
38	NE	JOHNSON	63.82
39	KS	LYON	63.41
40	KS	NEOSHO	63.41
41	OK	GRANT	63.01
42	OK	LINCOLN	62.61
43	UT	BEAVER	56.63
44	MO	CAMDEN	51.73
45	MO	HOWELL	51.73
46	MO	MORGAN	51.73
47	MO	ST FRANCOIS	51.73
48	TX	RAINS	50.87
49	IL	CALHOUN	50.12
50	MO	ADAIR	50.12
51	MO	LAFAYETTE	50.12
52	MO	LINN	50.12
53	MO	POLK	50.12
54	MO	RANDOLPH	50.12
55	MO	SALINE	50.12
56	KS	LEAVENWORTH	46.39
57	KS	WABAUNSEE	46.39
58	KS	SALINE	44.79
59	NE	PAWNEE	44.79
60	KS	ALLEN	43.58
61	KS	BOURBON	43.58
62	KS	LINN	43.58
63	KS	WOODSON	43.58
64	AR	MONTGOMERY	43.07
65	OK	CHOCTAW	43.07
66	OK	MCCURTAIN	43.07
67	TX	CAMP	43.07
68	TX	CASS	43.07
69	TX	HOPKINS	43.07
70	TX	LAMAR	43.07
71	KS	CHASE	42.78
72	OK	KAY	42.78
73	MO	CHRISTIAN	42.76
74	MO	STONE	42.76
75	OK	ALFALFA	42.38
76	AR	NEWTON	42.13
77	KS	CRAWFORD	42.13
78	MO	BARRY	42.13
79	MO	DADE	42.13
80	MO	MCDONALD	42.13

COUNTIES OF HIGHEST AVERAGE FALLOUT FROM SHOT:

TUMBLER-SNAPPER BAKER (TS2) Apr 15, 1952 1 Kt

(microCuries per square meter)

Each Row Includes:
County Rank (most fallout = 1), State, County, Fallout Level in microCuries/square meter

Rank	State	County	Fallout
3	CA	INYO3	193.83
2	AZ	MOHAVE3	193.08
3	AZ	MOHAVE4	193.08
4	CA	SAN BERNADIN	91.14
5	AZ	YAVAPAI	75.75
6	NV	WHITE PINE2	33.67
7	CA	LOS ANGELES	24.28
8	NV	EUREKA	22.96
9	CA	ORANGE	18.93
10	CA	SAN DIEGO	18.82
11	CA	VENTURA	17.54
12	NV	WHITE PINE1	16.83
13	NV	WHITE PINE3	16.83
14	CA	KERN	15.89
15	UT	UTAH	15.35
16	AZ	SANTA CRUZ	15.31
17	AZ	PIMA	14.44
18	AZ	MARICOPA	14.38
19	AZ	PINAL	14.38
20	UT	IRON1	14.38
21	UT	JUAB	14.38
22	UT	MILLARD	14.38
23	CA	SANTA BARBAR	14.24
24	ID	OWYHEE	13.92
25	ID	TWIN FALLS	13.92
26	AZ	YUMA	13.56
27	ID	CASSIA	13.22
28	CA	IMPERIAL	12.09
29	ID	GOODING	10.90
30	ID	IDAHO	10.73
31	ID	JEROME	10.20
32	ID	ELMORE	9.97
33	AZ	COCONINO1	9.41
34	AZ	MOHAVE1	9.41
35	AZ	MOHAVE2	9.41
36	ID	ADA	9.41
37	ID	LINCOLN	9.28
38	ID	CANYON	8.81
39	NV	ELKO	8.31
40	NV	LANDER2	7.89
41	UT	BEAVER	7.19
42	UT	SANPETE	7.19
43	UT	BOX ELDER1	6.96
44	UT	TOOELE1	6.96
45	CA	RIVERSIDE	6.89
46	UT	KANE1	6.68
47	ID	MINIDOKA	6.49
48	AZ	COCHISE	6.26
49	AZ	GILA	6.26
50	AZ	GRAHAM	6.26
51	AZ	GREENLEE	6.26
52	UT	IRON2	6.26
53	UT	KANE2	6.26
54	AZ	COCONINO3	5.91
55	ID	ADAMS	5.63
56	ID	BOISE	5.63
57	ID	VALLEY	5.63
58	OR	GRANT	5.63
59	AZ	COCONINO2	5.33
60	CA	INYO1	5.10
61	CA	INYO2	5.10
62	NV	LANDER1	4.41
63	OR	BAKER	4.27
64	NV	CHURCHILL	4.17
65	UT	IRON3	4.17
66	UT	SEVIER	4.17
67	OR	MALHEUR	3.94
68	UT	TOOELE2	3.94
69	NV	DOUGLAS	3.25
70	NV	LYON	3.25
71	NV	PERSHING	3.25
72	NV	STOREY	3.25
73	NV	CARSON CITY	3.01
74	UT	BOX ELDER2	3.01
75	UT	PIUTE	3.01
76	AZ	APACHE	2.97
77	AZ	NAVAJO	2.97
78	ID	CAMAS	2.92
79	ID	CUSTER	2.92
80	ID	GEM	2.92

COUNTIES OF HIGHEST AVERAGE FALLOUT FROM SHOT:

TUMBLER-SNAPPER CHARLIE (TS3) Apr 22, 1952 31 Kt

(microCuries per square meter)

Each Row Includes:
County Rank (most fallout = 1), State, County, Fallout Level in microCuries/square meter

Rank	State	County	Fallout
1	CA	GLENN	37.19
2	AL	LOWNDES	27.19
3	AL	GENEVA	27.16
4	AL	WILCOX	27.03
5	AL	CONECUH	26.87
6	AL	GREENE	26.87
7	FL	HOLMES	26.87
8	CA	TEHAMA	25.20
9	SC	LEE	25.04
10	CA	SONOMA	24.93
11	FL	HILLSBOROUGH	24.93
12	AZ	MOHAVE2	23.11
13	GA	MONTGOMERY	22.64
14	GA	WHEELER	22.64
15	SC	CLARENDON	22.08
16	SC	FLORENCE	20.66
17	CA	CALAVERAS	20.64
18	CA	BUTTE	20.52
19	CA	YUBA	20.52
20	CA	NEVADA	20.51
21	CA	AMADOR	20.47
22	CA	ALPINE	20.43
23	AZ	COCONINO3	19.47
24	SC	SUMTER	18.92
25	AL	MARENGO	18.46
26	AL	CHAMBERS	18.35
27	AL	PIKE	18.30
28	AL	COVINGTON	18.19
29	AL	DALLAS	18.19
30	AL	CLARKE	17.97
31	AL	DALE	17.97
32	AL	HALE	17.97
33	AL	SUMTER	17.97
34	FL	OKALOOSA	17.97
35	VA	HANOVER	17.59
36	VA	CHARLES CITY	17.50
37	VA	PRINCE GEORG	17.29
38	VA	GOOCHLAND	17.18
39	VA	CULPEPER	17.17
40	VA	HENRICO	17.06
41	VA	DINWIDDIE	17.06
42	VA	BRUNSWICK	16.86
43	VA	KING GEORGE	16.46
44	VA	FAUQUIER	16.35
45	VA	KING WILLIAM	16.11
46	VA	KING AND QUE	15.84
47	VA	NEW KENT	15.70
48	VA	GREENSVILLE	15.62
49	GA	LONG	15.49
50	GA	LIBERTY	15.47
51	GA	BULLOCH	15.41
52	GA	SCREVEN	15.41
53	GA	BRYAN	15.39
54	CA	INYO1	14.64
55	CA	INYO2	14.64
56	CA	INYO3	14.64
57	CA	MONO	14.64
58	SC	DORCHESTER	14.63
59	SC	CHARLESTON	14.55
60	SC	MARION	14.51
61	SC	DILLON	14.44
62	CA	SAN DIEGO	14.42
63	GA	EFFINGHAM	14.19
64	GA	WAYNE	14.12
65	SC	JASPER	14.12
66	AZ	COCONINO2	14.11
67	GA	MCINTOSH	14.05
68	CA	PLACER	13.84
69	CA	COLUSA	13.80
70	PA	MONROE	13.73
71	PA	PIKE	13.72
72	CA	SACRAMENTO	13.64
73	PA	LUZERNE	13.33
74	VA	ORANGE	13.31
75	PA	WAYNE	13.23
76	VA	RICHMOND	13.16
77	VA	STAFFORD	13.12
78	VA	ESSEX	13.05
79	VA	WESTMORELAND	13.05
80	PA	LACKAWANNA	12.98

COUNTIES OF HIGHEST AVERAGE FALLOUT FROM SHOT:

TUMBLER-SNAPPER DOG (TS4) May 1, 1952 19 Kt

(microCuries per square meter)

Each Row Includes:
County Rank (most fallout = 1), State, County, Fallout Level in microCuries/square meter

Rank	State	County	Fallout
1	UT	IRON1	13.73
2	UT	IRON2	8.39
3	AZ	MOHAVE3	6.41
4	AZ	MOHAVE4	6.41
5	AZ	MOHAVE1	6.26
6	AZ	MOHAVE2	6.26
7	AZ	COCONINO2	5.86
8	UT	BEAVER	5.86
9	AZ	COCONINO1	4.86
10	UT	IRON3	4.86
11	UT	MILLARD	4.86
12	OK	CIMARRON	4.81
13	TX	MOORE	4.81
14	TX	SHERMAN	4.81
15	AZ	NAVAJO	4.57
16	AZ	APACHE	4.43
17	NM	MCKINLEY	4.43
18	NM	SAN JUAN	4.43
19	AZ	COCONINO3	4.31
20	NM	VALENCIA	4.29
21	NV	LINCOLN2	4.24
22	UT	SANPETE	3.86
23	UT	SEVIER	3.86
24	IN	CARROLL	3.70
25	IN	CLINTON	3.70
26	IN	TIPPECANOE	3.70
27	AZ	YAVAPAI	3.57
28	CO	LA PLATA	3.57
29	CO	MONTEZUMA	3.57
30	UT	PIUTE	3.57
31	CO	DELTA	3.43
32	CO	DOLORES	3.43
33	CO	MONTROSE	3.43
34	CO	OURAY	3.43
35	CO	SAN JUAN	3.43
36	CO	SAN MIGUEL	3.43
37	NM	CATRON	3.43
38	NM	SANDOVAL	3.43
39	UT	EMERY	3.43
40	UT	GRAND	3.43
41	UT	WAYNE	3.43
42	AL	LIMESTONE	3.23
43	AL	MADISON	3.23
44	NM	DE BACA	3.14
45	NM	LINCOLN	3.14
46	NM	ROOSEVELT	3.14
47	KS	MEADE	3.07
48	KS	SEWARD	3.07
49	OK	BEAVER	3.07
50	OK	TEXAS	3.07
51	TX	DEAF SMITH	3.07
52	TX	HANSFORD	3.07
53	TX	HARTLEY	3.07
54	TX	LIPSCOMB	3.07
55	TX	OCHILTREE	3.07
56	NM	CHAVES	2.97
57	NM	EDDY	2.86
58	NM	LEA	2.86
59	UT	KANE2	2.86
60	UT	GARFIELD	2.72
61	UT	KANE1	2.72
62	UT	SAN JUAN	2.72
63	IN	HOWARD	2.66
64	IN	TIPTON	2.66
65	CO	LAS ANIMAS	2.64
66	OK	ELLIS	2.64
67	AZ	GILA	2.57
68	CO	GARFIELD	2.57
69	CO	MOFFAT	2.57
70	CO	RIO BLANCO	2.57
71	NM	LOS ALAMOS	2.57
72	NM	RIO ARRIBA	2.57
73	MO	BUCHANAN	2.54
74	NM	GUADALUPE	2.43
75	NM	SANTA FE	2.43
76	NM	SOCORRO	2.43
77	NM	TORRANCE	2.43
78	CO	MESA	2.43
79	IN	RIPLEY	2.37
80	IA	MONONA	2.30

COUNTIES OF HIGHEST AVERAGE FALLOUT FROM SHOT:

TUMBLER-SNAPPER EASY (TS5) May 7, 1952 12 Kt

(microCuries per square meter)
Each Row Includes:
County Rank (most fallout = 1), State, County, Fallout Level in microCuries/square meter

Rank	State/County	Fallout
1	UT TOOELE1	868.7
2	UT TOOELE2	418.7
3	UT BOX ELDER2	417.2
4	UT WEBER	417.2
5	UT RICH	417.2
6	UT DAVIS	375.5
7	UT MORGAN	375.5
8	NV WHITE PINE1	365.9
9	NV WHITE PINE3	337.4
10	NV WHITE PINE2	335.3
11	UT UTAH	335.0
12	UT CACHE	333.8
13	UT SUMMIT	333.8
14	UT SALT LAKE	315.9
15	UT WASATCH	250.8
16	WY UINTA	249.4
17	WY SUBLETTE	248.5
18	WY FREMONT	247.6
19	UT MILLARD	179.4
20	NV NYE2	171.4
21	UT JUAB	167.5
22	ID ONEIDA	166.9
23	ID FRANKLIN	166.9
24	ID BEAR LAKE	166.9
25	WY LINCOLN	166.3
26	UT SANPETE	88.7
27	UT BOX ELDER1	83.5
28	ID CARIBOU	83.5
29	NV LINCOLN2	77.4
30	ID POWER	50.1
31	ID BANNOCK	47.2
32	NE SIOUX	42.6
33	WY HOT SPRINGS	42.4
34	ID BONNEVILLE	41.7
35	WY NATRONA	40.5
36	WY NIOBRARA	38.9
37	NE DAWES	38.0
38	SD FALL RIVER	35.9
39	WY TETON	34.4
40	UT IRON1	33.9
41	UT BEAVER	33.9
42	WY CARBON	33.0
43	SD SHANNON	32.0
44	WY GOSHEN	28.6
45	ID TETON	28.5
46	SD CUSTER	28.0
47	ID CAMAS	27.1
48	NE SHERIDAN	26.9
49	WY PLATTE	26.6
50	ID CUSTER	26.2
51	ID BOISE	25.6
52	ID MINIDOKA	24.9
53	ID CASSIA	24.9
54	WY JOHNSON	24.8
55	SD WASHABAUGH	24.6
56	WY CONVERSE	23.7
57	NE BANNER	23.6
58	NV ELKO	23.6
59	WY SWEETWATER	23.5
60	ID IDAHO	23.5
61	ID GEM	23.0
62	WY WASHAKIE	22.3
63	OR BAKER	21.8
64	OR GRANT	20.8
65	ID BLAINE	20.5
66	ID FREMONT	20.4
67	SD BENNETT	19.8
68	SD MELLETTE	19.6
69	WY PARK	19.3
70	WY WESTON	17.6
71	SD LAWRENCE	17.3
72	NE BOX BUTTE	17.0
73	NC DAVIDSON	16.8
74	NV EUREKA	16.8

COUNTIES OF HIGHEST AVERAGE FALLOUT FROM SHOT:

TUMBLER-SNAPPER FOX (TS6) May 25, 1952 11 Kt

(microCuries per square meter)
Each Row Includes:
County Rank (most fallout = 1), State, County, Fallout Level in microCuries/square meter

Rank	County	Fallout
1	NV LINCOLN2	521.62
2	UT IRON1	509.71
3	NV LINCOLN1	302.36
4	UT IRON2	130.89
5	UT MILLARD	120.23
6	UT BEAVER	119.89
7	CO HINSDALE	99.67
8	CO PHILLIPS	99.10
9	UT SEVIER	92.16
10	UT PIUTE	92.16
11	UT EMERY	91.50
12	UT WAYNE	82.99
13	UT GRAND	82.10
14	UT CARBON	72.49
15	CO GARFIELD	72.49
16	UT GARFIELD	69.51
17	CO MESA	68.57
18	UT SANPETE	64.31
19	CO DELTA	63.65
20	CO RIO BLANCO	63.43
21	UT UINTAH	63.21
22	CO MOFFAT	62.88
23	WY CARBON	62.21
24	NE CHASE	57.34
25	CO SAGUACHE	56.48
26	UT JUAB	55.69
27	UT KANE2	55.69
28	UT KANE1	55.69
29	CO LOGAN	55.61
30	UT SAN JUAN	54.48
31	CO MONTROSE	54.48
32	KS CHEYENNE	52.73
33	UT IRON3	52.32
34	NE DEUEL	48.79
35	NE PERKINS	47.60
36	UT UTAH	46.41
37	AZ MOHAVE2	46.41
38	AZ MOHAVE1	46.41
39	AZ COCONINO1	46.41
40	CO SAN MIGUEL	45.42
41	CO SAN JUAN	45.42
42	CO OURAY	45.42
43	CO MONTEZUMA	45.42
44	CO LA PLATA	45.42
45	CO DOLORES	45.42
46	UT DUCHESNE	45.31
47	NM SIERRA	45.20
48	AZ COCONINO2	45.08
49	AZ NAVAJO	44.75
50	NM SAN JUAN	44.42
51	AZ APACHE	44.42
52	CO YUMA	43.66
53	NM VALENCIA	43.54
54	NM SANDOVAL	43.54
55	NM MCKINLEY	43.54
56	NM LOS ALAMOS	43.54
57	CO PUEBLO	43.13
58	AZ COCONINO3	42.52
59	NE HAYES	41.59
60	CO HUERFANO	40.81
61	CO GUNNISON	40.49
62	NE PHELPS	39.71
63	NE KEITH	38.83
64	CO DOUGLAS	37.96
65	CO SEDGWICK	36.86
66	NE DUNDY	36.37
67	UT SUMMIT	36.35
68	UT WASATCH	36.35
69	UT DAGGETT	36.13
70	NM BERNALILLO	35.64
71	NM RIO ARRIBA	34.81
72	CO WELD	34.74
73	CO ARCHULETA	34.49
74	CO GRAND	34.31

COUNTIES OF HIGHEST AVERAGE FALLOUT FROM SHOT:

TUMBLER-SNAPPER GEORGE (TS7) Jun 1, 1952 15 Kt

(microCuries per square meter)

Each Row Includes:
County Rank (most fallout = 1), State, County, Fallout Level in microCuries/square meter

Rank	State	County	Fallout
1	MO	KNOX	351.71
2	MO	LEWIS	324.08
3	WY	NIOBRARA	310.25
4	IA	APPANOOSE	277.35
5	MO	SCOTLAND	274.93
6	SD	CUSTER	246.94
7	IA	MONROE	240.64
8	MO	RALLS	211.81
9	MO	AUDRAIN	208.57
10	MO	MACON	201.62
11	MO	MARION	198.74
12	MO	PIKE	198.74
13	IL	FORD	189.28
14	IL	PIATT	176.23
15	IL	CHRISTIAN	174.20
16	IL	MOULTRIE	174.17
17	IA	FREMONT	174.10
18	MO	SULLIVAN	173.28
19	IL	SHELBY	166.46
20	IL	MACON	164.91
21	IL	DOUGLAS	161.50
22	IL	ADAMS	161.46
23	MO	PUTNAM	155.34
24	IL	EFFINGHAM	151.62
25	IL	JASPER	150.31
26	MT	MEAGHER	149.80
27	WY	CONVERSE	141.80
28	MO	SCHUYLER	140.42
29	IA	WAYNE	132.25
30	MI	INGHAM	131.62
31	MI	LIVINGSTON	131.51
32	NE	SIOUX	131.27
33	IL	LAWRENCE	127.34
34	IL	RICHLAND	126.35
35	MI	CALHOUN	124.44
36	IN	KNOX	123.66
37	IA	RINGGOLD	117.91
38	IA	DAVIS	117.73
39	MI	KALAMAZOO	110.52
40	MI	GENESEE	110.29
41	OH	PUTNAM	109.19
42	MO	ADAIR	106.04
43	IA	TAMA	105.16
44	MI	VAN BUREN	104.90
45	MI	SHIAWASSEE	102.07
46	CO	MINERAL	102.04
47	IL	IROQUOIS	102.02
48	IN	WARREN	101.12
49	MI	EATON	100.47
50	MI	OTTAWA	98.95
51	TX	EL PASO	98.02
52	MI	CLINTON	96.75
53	IL	KANKAKEE	95.57
54	MO	LINCOLN	95.23
55	IA	MARION	94.54
56	NY	ST LAWRENCE	91.91
57	MI	LAPEER	91.85
58	OH	HANCOCK	91.44
59	MO	FRANKLIN	90.38
60	MO	LINN	89.49
61	MI	BARRY	87.74
62	IL	WILL	86.19
63	MO	SHELBY	85.60
64	IA	SHELBY	81.66
65	IN	DUBOIS	81.60
66	IN	CASS	81.30
67	IA	CASS	80.60
68	IA	ADAMS	80.22
69	SD	TURNER	79.84
70	SD	HAAKON	79.67
71	IL	CHAMPAIGN	78.41
72	MI	HILLSDALE	78.41
73	IA	MAHASKA	77.89
74	SD	HUTCHINSON	76.95
75	IA	UNION	76.47
76	MI	SAGINAW	76.26
77	IN	PULASKI	75.97
78	SD	DOUGLAS	75.74
79	OK	TULSA	75.66
80	IN	CARROLL	73.42

COUNTIES OF HIGHEST AVERAGE FALLOUT FROM SHOT:
TUMBLER-SNAPPER HOW (TS8) Jun 5, 1952 14 Kt

(microCuries per square meter)
Each Row Includes:
County Rank (most fallout = 1), State, County, Fallout Level in microCuries/square meter

Rank	State County	Fallout
1	ID GEM	878.1
2	ID CUSTER	843.6
3	ID BLAINE	795.2
4	ID LEMHI	578.2
5	MT DEER LODGE	517.4
6	ID VALLEY	399.0
7	ID IDAHO	361.9
8	ID CAMAS	358.2
9	MT CHOUTEAU	349.3
10	MT BEAVERHEAD	335.4
11	MT JUDITH BASIN	320.2
12	MT POWELL	313.6
13	ID GOODING	311.6
14	MT SILVER BOW	311.5
15	ID BOISE	309.4
16	MT JEFFERSON	306.6
17	MT BROADWATER	282.7
18	ID WASHINGTON	267.2
19	OR WALLOWA	257.5
20	ID LEWIS	255.8
21	MT MEAGHER	244.6
22	MT MISSOULA	243.4
23	MT MADISON	238.3
24	NV LANDER1	236.2
25	ID OWYHEE	233.7
26	ID LINCOLN	233.7
27	ID ELMORE	233.7
28	ID BUTTE	227.1
29	ID ADA	222.0
30	NV NYE2	209.5
31	NV ELKO	207.2
32	MT WHEATLAND	202.5
33	NV LANDER2	199.0
34	ID PAYETTE	194.2
35	ID JEFFERSON	192.3
36	MT GALLATIN	190.5
37	OR BAKER	184.9
38	ID CLEARWATER	182.9
39	MT LEWIS AND CL	180.1
40	MT MINERAL	178.7
41	MT TETON	158.5
42	MT HILL	157.1
43	ID SHOSHONE	157.0
44	ID TWIN FALLS	155.8
45	ID MINIDOKA	155.8
46	ID JEROME	155.8
47	ID CASSIA	155.8
48	NV HUMBOLDT	154.8
49	MT FLATHEAD	147.8
50	WA GARFIELD	133.4
51	MT PARK	127.7
52	MT RAVALLI	126.0
53	MT CASCADE	122.9
54	MT SANDERS	122.1
55	MT FERGUS	120.7
56	MT TOOLE	120.1
57	MT LIBERTY	120.1
58	UT BOX ELDER1	117.3
59	ID NEZ PERCE	115.8
60	MT GRANITE	107.9
61	MT LAKE	107.5
62	ID MADISON	106.5
63	MT PONDERA	92.5
64	MT BLAINE	90.9
65	WA ASOTIN	88.7
66	OR UNION	84.5
67	ID ADAMS	80.3
68	ID LATAH	79.4
69	ID CANYON	78.2
70	NV EUREKA	75.3
71	ID FREMONT	68.6
72	MT LINCOLN	68.2
73	MT GLACIER	66.9
74	OR GRANT	66.7

COUNTIES OF HIGHEST AVERAGE FALLOUT FROM SHOT:

UPSHOT-KNOTHOLE ANNIE (UK1) Mar 17, 1953 16 Kt

(microCuries per square meter)

Each Row Includes:
County Rank (most fallout = 1), State, County, Fallout Level in microCuries/square meter

Rank	State	County	Fallout
1	UT	WASHINGTON2	1303.21
2	UT	WASHINGTON3	596.14
3	UT	KANE2	213.70
4	NV	LINCOLN1	120.42
5	UT	SAN JUAN	65.17
6	NJ	MONMOUTH	63.97
7	NJ	OCEAN	62.43
8	NY	KINGS	59.87
9	NY	QUEENS	55.75
10	NJ	ATLANTIC	54.88
11	NY	NASSAU	52.39
12	NJ	BURLINGTON	52.29
13	NV	CLARK1	51.94
14	NY	NEW YORK	50.45
15	NJ	MIDDLESEX	48.50
16	CO	MONTEZUMA	47.94
17	NY	RICHMOND	47.85
18	CO	LA PLATA	46.06
19	NJ	HUDSON	45.35
20	NJ	UNION	44.15
21	NJ	MERCER	42.87
22	NJ	ESSEX	41.47
23	NJ	CAPE MAY	41.42
24	NY	BRONX	41.27
25	NJ	BERGEN	40.02
26	NJ	CAMDEN	37.38
27	NY	WESTCHESTER	34.98
28	NJ	SOMERSET	33.76
29	NJ	CUMBERLAND	31.71
30	NJ	GLOUCESTER	30.93
31	NY	ROCKLAND	30.18
32	CT	FAIRFIELD	28.57
33	PA	PHILADELPHIA	28.44
34	AZ	APACHE	27.92
35	NY	SUFFOLK	27.80
36	NM	SAN JUAN	27.70
37	NM	RIO ARRIBA	27.48
38	NM	TAOS	27.48
39	NJ	PASSAIC	26.63
40	CO	CONEJOS	25.51
41	NJ	MORRIS	24.14
42	NJ	SALEM	23.63
43	NJ	HUNTERDON	22.81
44	NM	COLFAX	22.33
45	CO	LAS ANIMAS	22.22
46	TN	KNOX	21.73
47	PA	BUCKS	21.22
48	PA	DELAWARE	21.21
49	DE	SUSSEX	20.72
50	CO	COSTILLA	20.64
51	TN	ANDERSON	20.05
52	NY	PUTNAM	19.87
53	DE	KENT	18.91
54	AR	POLK	18.78
55	TX	DALLAS	18.76
56	TN	SEVIER	18.65
57	TX	GRAYSON	18.57
58	AR	LOGAN	18.44
59	NM	LOS ALAMOS	18.36
60	CO	BACA	18.28
61	NM	CHAVES	18.25
62	NM	DE BACA	18.25
63	NM	GUADALUPE	18.25
64	NM	MORA	18.25
65	NM	SAN MIGUEL	18.25
66	NM	CURRY	18.14
67	NM	HARDING	18.14
68	NM	QUAY	18.14
69	NM	ROOSEVELT	18.14
70	NM	UNION	18.14
71	OK	PUSHMATAHA	18.12
72	DE	NEW CASTLE	17.95
73	TX	ROCKWALL	17.90
74	AZ	MOHAVE2	17.83
75	TN	COCKE	17.77
76	TX	FANNIN	17.72
77	MO	TANEY	17.58
78	PA	MONTGOMERY	17.29
79	TN	CAMPBELL	17.29
80	AR	SCOTT	17.29

COUNTIES OF HIGHEST AVERAGE FALLOUT FROM SHOT:

UPSHOT-KNOTHOLE NANCY (UK2) Mar 24, 1953 24 Kt

(microCuries per square meter)

Each Row Includes:
County Rank (most fallout = 1), State, County, Fallout Level in microCuries/square meter

1	UT	MORGAN	725.74
2	UT	DAVIS	723.10
3	UT	WEBER	679.47
4	UT	TOOELE2	635.02
5	UT	RICH	623.63
6	UT	SALT LAKE	598.89
7	WY	LINCOLN	568.59
8	WY	FREMONT	564.43
9	WY	SUBLETTE	564.43
10	UT	CACHE	536.50
11	NV	WHITE PINE3	492.93
12	UT	SUMMIT	451.90
13	ID	BEAR LAKE	443.88
14	UT	BOX ELDER2	443.88
15	WY	UINTA	286.32
16	ID	FRANKLIN	263.50
17	NV	NYE2	243.64
18	WY	CONVERSE	195.15
19	WY	CAMPBELL	194.99
20	WY	JOHNSON	194.72
21	MT	POWDER RIVER	194.67
22	WY	WASHAKIE	194.44
23	ID	CARIBOU	176.27
24	NV	LINCOLN2	168.92
25	MT	DAWSON	122.97
26	MT	CUSTER	122.96
27	ND	BOTTINEAU	122.86
28	ND	MCHENRY	122.86
29	NV	WHITE PINE2	122.36
30	UT	TOOELE1	119.78
31	UT	UTAH	96.49
32	ND	DUNN	92.54
33	ND	WARD	92.41
34	ND	BURKE	92.40
35	ND	MCKENZIE	92.36
36	UT	BOX ELDER1	88.82
37	ID	ONEIDA	87.55
38	NV	WHITE PINE1	87.15
39	WY	NATRONA	81.65
40	ND	MCLEAN	63.71
41	ND	SLOPE	62.14
42	MT	FALLON	62.09
43	ND	GOLDEN VALLE	62.09
44	ND	HETTINGER	62.07
45	MT	WIBAUX	62.04
46	ND	STARK	62.04
47	ND	BILLINGS	62.00
48	ND	MERCER	61.97
49	ND	GRANT	61.95
50	ND	MOUNTRAIL	61.91
51	ND	MORTON	61.90
52	ND	OLIVER	61.90
53	MT	SHERIDAN	61.87
54	MT	RICHLAND	61.85
55	MT	PRAIRIE	61.84
56	MT	ROOSEVELT	61.84
57	ND	RENVILLE	61.81
58	ND	DIVIDE	61.80
59	ND	BURLEIGH	61.79
60	ND	SHERIDAN	61.79
61	MT	DANIELS	61.78
62	MT	MCCONE	61.69
63	MT	VALLEY	61.57
64	AR	CRAIGHEAD	58.07
65	AR	FULTON	57.49
66	AR	PRAIRIE	49.98
67	UT	WASATCH	47.74
68	AR	CRITTENDEN	47.56
69	SD	LAWRENCE	47.17
70	SD	BUTTE	46.83
71	WY	CROOK	46.82
72	SD	HARDING	46.78
73	SD	MEADE	46.78
74	WY	WESTON	46.75
75	SD	SHANNON	46.72
76	MT	CARTER	46.71
77	SD	PERKINS	46.68
78	SD	CUSTER	46.68
79	ND	ADAMS	46.64
80	ND	BOWMAN	46.63

COUNTIES OF HIGHEST AVERAGE FALLOUT FROM SHOT:

UPSHOT-KNOTHOLE RUTH (UK3) Mar 31, 1953 0.2 Kt

(microCuries per square meter)

Each Row Includes:
County Rank (most fallout = 1), State, County, Fallout Level in microCuries/square meter

1	AZ	MARICOPA	25.62
2	AZ	GILA	18.78
3	AZ	MOHAVE2	15.93
4	AZ	YAVAPAI	14.51
5	OK	BECKHAM	14.12
6	OK	GREER	14.12
7	OK	HARMON	14.12
8	OK	ROGER MILLS	14.12
9	OK	WASHITA	14.12
10	TX	ARMSTRONG	14.12
11	TX	CARSON	14.12
12	TX	COLLINGSWORT	14.12
13	TX	DONLEY	14.12
14	TX	PARMER	14.12
15	TX	WHEELER	14.12
16	AZ	PINAL	14.08
17	AZ	MOHAVE3	11.23
18	AZ	MOHAVE4	11.23
19	CA	SAN BERNADIN	11.23
20	AZ	COCONINO2	10.24
21	AZ	COCONINO3	9.52
22	AR	MADISON	9.41
23	AR	WASHINGTON	9.41
24	MO	NEWTON	9.41
25	OK	NOWATA	9.41
26	AR	BENTON	9.27
27	OK	DELAWARE	9.27
28	OK	MAYES	9.27
29	TX	HALL	9.13
30	AZ	GRAHAM	9.10
31	OK	ROGERS	9.07
32	OK	TULSA	9.07
33	OK	BEAVER	8.66
34	OK	ELLIS	8.66
35	OK	TEXAS	8.66
36	TX	BRISCOE	8.66
37	TX	CASTRO	8.66
38	TX	CHILDRESS	8.66
39	TX	DEAF SMITH	8.66
40	TX	GRAY	8.66
41	TX	HEMPHILL	8.66
42	TX	HOCKLEY	8.66
43	TX	LIPSCOMB	8.66
44	TX	LYNN	8.66
45	TX	ROBERTS	8.66
46	TX	TERRY	8.66
47	OK	CRAIG	8.57
48	OK	MCINTOSH	6.71
49	AR	CRAWFORD	6.68
50	MO	BARRY	6.68
51	OK	ADAIR	6.68
52	OK	CREEK	6.68
53	OK	HUGHES	6.68
54	OK	MUSKOGEE	6.68
55	OK	PITTSBURG	6.68
56	OK	CHEROKEE	6.54
57	OK	OKFUSKEE	6.54
58	OK	OKMULGEE	6.54
59	OK	OTTAWA	6.54
60	OK	SEMINOLE	6.54
61	OK	WAGONER	6.34
62	AR	FAULKNER	6.31
63	MO	MCDONALD	6.27
64	AR	FRANKLIN	5.97
65	KS	MEADE	5.93
66	KS	SEWARD	5.93
67	TX	HALE	5.93
68	TX	HANSFORD	5.93
69	TX	HUTCHINSON	5.93
70	TX	LUBBOCK	5.93
71	TX	MOTLEY	5.93
72	TX	OCHILTREE	5.93
73	TX	OLDHAM	5.93
74	TX	POTTER	5.93
75	TX	SHERMAN	5.93
76	AR	BOONE	5.84
77	MO	LAWRENCE	5.84
78	AR	LOGAN	5.61
79	AZ	NAVAJO	5.40
80	KS	CHEROKEE	5.31

COUNTIES OF HIGHEST AVERAGE FALLOUT FROM SHOT:

UPSHOT-KNOTHOLE DIXIE (UK4) Apr 6, 1953 11 Kt

(microCuries per square meter)

Each Row Includes:
County Rank (most fallout = 1), State, County, Fallout Level in microCuries/square meter

1	MA	PLYMOUTH	213.71
2	MA	NORFOLK	191.58
3	MA	BRISTOL	156.46
4	MA	MIDDLESEX	136.12
5	MA	SUFFOLK	126.99
6	RI	BRISTOL	101.13
7	RI	WASHINGTON	100.81
8	MA	DUKES	96.78
9	MA	BARNS	95.69
10	CT	MIDDLESEX	95.31
11	MA	ESSEX	94.31
12	RI	KENT	84.09
13	RI	PROVIDENCE	74.30
14	RI	NEWPORT	67.23
15	CT	NEW LONDON	60.66
16	CT	WINDHAM	56.64
17	CT	NEW HAVEN	55.07
18	MA	WORCESTER	53.19
19	NH	ROCKINGHAM	52.93
20	MA	NANTUCKET	49.57
21	CT	TOLLAND	44.32
22	NY	SUFFOLK	34.02
23	NH	HILLSBOROUGH	33.09
24	CT	HARTFORD	29.69
25	NH	CHESHIRE	23.57
26	NH	MERRIMACK	23.40
27	MA	HAMPDEN	21.48
28	NH	BELKNAP	20.05
29	NH	STRAFFORD	18.77
30	FL	LEON	15.57
31	NH	SULLIVAN	15.54
32	MA	HAMPSHIRE	14.87
33	ME	YORK	13.96
34	CT	FAIRFIELD	11.47
35	MA	FRANKLIN	11.35
36	ME	CUMBERLAND	10.77
37	NH	CARROLL	10.30
38	NH	GRAFTON	10.12
39	NH	COOS	9.48
40	CT	LITCHFIELD	9.36
41	ME	SAGADAHOC	8.39
42	ME	KENNEBEC	7.91
43	ME	ANDROSCOGGIN	7.87
44	FL	DIXIE	7.78
45	FL	GILCHRIST	7.78
46	FL	HAMILTON	7.78
47	FL	MADISON	7.78
48	GA	BACON	7.78
49	GA	JEFF DAVIS	7.78
50	ME	OXFORD	7.73
51	VT	CALEDONIA	7.33
52	VT	ESSEX	7.30
53	ME	LINCOLN	6.09
54	ME	FRANKLIN	5.65
55	VT	WINDHAM	5.18
56	NY	PUTNAM	4.84
57	FL	ALACHUA	3.89
58	FL	BAKER	3.89
59	FL	BRADFORD	3.89
60	FL	BREVARD	3.89
61	FL	CITRUS	3.89
62	FL	CLAY	3.89
63	FL	COLUMBIA	3.89
64	FL	FLAGLER	3.89
65	FL	HERNANDO	3.89
66	FL	HILLSBOROUGH	3.89
67	FL	JEFFERSON	3.89
68	FL	LAFAYETTE	3.89
69	FL	LAKE	3.89
70	FL	LEVY	3.89
71	FL	MARION	3.89
72	FL	NASSAU	3.89
73	FL	ORANGE	3.89
74	FL	OSCEOLA	3.89
75	FL	PASCO	3.89
76	FL	PINELLAS	3.89
77	FL	POLK	3.89
78	FL	PUTNAM	3.89
79	FL	SEMINOLE	3.89
80	FL	ST JOHNS	3.89

COUNTIES OF HIGHEST AVERAGE FALLOUT FROM SHOT:

UPSHOT-KNOTHOLE RAY (UK5) Apr 11, 1953 0.2 Kt

(microCuries per square meter)

Each Row Includes:
County Rank (most fallout = 1), State, County, Fallout Level in microCuries/square meter

1	AZ	YUMA	46.04
2	CA	SAN BERNADIN	22.71
3	AZ	MOHAVE3	10.24
4	AZ	MOHAVE4	10.24
5	UT	IRON1	6.53
6	UT	IRON3	6.23
7	AZ	YAVAPAI	5.56
8	UT	BEAVER	5.51
9	CA	INYO3	5.42
10	UT	IRON2	5.20
11	AZ	COCONINO3	5.12
12	AZ	COCONINO1	5.05
13	AZ	MOHAVE2	5.05
14	UT	GARFIELD	5.05
15	UT	KANE1	5.05
16	UT	KANE2	5.05
17	UT	PIUTE	5.05
18	UT	SEVIER	5.05
19	UT	WAYNE	5.05
20	AZ	COCONINO2	4.90
21	AZ	MOHAVE1	4.73
22	CO	ROUTT	4.27
23	AR	HOT SPRING	4.17
24	AR	PULASKI	4.17
25	AR	BRADLEY	4.11
26	AR	CLEVELAND	4.11
27	AR	DALLAS	4.11
28	AR	GARLAND	4.11
29	AR	GRANT	4.11
30	AR	SALINE	4.11
31	AR	CALHOUN	4.05
32	AR	CLARK	4.05
33	AR	COLUMBIA	4.05
34	AR	HEMPSTEAD	4.05
35	AR	HOWARD	4.05
36	AR	LAFAYETTE	4.05
37	AR	LITTLE RIVER	4.05
38	AR	MONTGOMERY	4.05
39	AR	NEVADA	4.05
40	AR	PIKE	4.05
41	AR	SEVIER	4.05
42	LA	BOSSIER	4.05
43	LA	LINCOLN	4.05
44	LA	UNION	4.05
45	LA	WEBSTER	4.05
46	OK	CHOCTAW	4.05
47	OK	MCCURTAIN	4.05
48	TX	BOWIE	4.05
49	TX	RED RIVER	4.05
50	TX	UPSHUR	4.05
51	UT	EMERY	4.01
52	CO	PITKIN	3.98
53	UT	GRAND	3.86
54	AZ	NAVAJO	3.62
55	TN	COCKE	3.03
56	ME	CUMBERLAND	3.01
57	NH	ROCKINGHAM	2.99
58	MA	NORFOLK	2.95
59	NC	CLAY	2.92
60	TN	SEVIER	2.92
61	MA	MIDDLESEX	2.89
62	ME	SAGADAHOC	2.85
63	MA	ESSEX	2.84
64	MA	NANTUCKET	2.81
65	ME	OXFORD	2.78
66	NC	GRAHAM	2.78
67	NH	HILLSBOROUGH	2.78
68	ME	ANDROSCOGGIN	2.73
69	AR	UNION	2.72
70	LA	CADDO	2.72
71	LA	NATCHITOCHES	2.72
72	LA	RED RIVER	2.72
73	TX	CAMP	2.72
74	TX	CASS	2.72
75	TX	FRANKLIN	2.72
76	TX	HARRISON	2.72
77	TX	MARION	2.72
78	TX	MORRIS	2.72
79	TX	TITUS	2.72
80	AZ	APACHE	2.64

COUNTIES OF HIGHEST AVERAGE FALLOUT FROM SHOT:

UPSHOT-KNOTHOLE BADGER (UK6) Apr 18, 1953 23 Kt

(microCuries per square meter)

Each Row Includes:
County Rank (most fallout = 1), State, County, Fallout Level in microCuries/square meter

1	AZ	COCONINO2	328.54
2	NV	CLARK1	324.13
3	AZ	NAVAJO	321.49
4	AZ	APACHE	319.19
5	NV	CLARK2	255.63
6	CO	ARCHULETA	217.12
7	NM	MCKINLEY	211.30
8	NM	VALENCIA	211.30
9	NM	SANDOVAL	156.71
10	NM	BERNALILLO	108.40
11	NM	CATRON	104.94
12	NM	SOCORRO	103.41
13	NM	SANTA FE	82.78
14	NM	TORRANCE	82.78
15	TX	HARRIS	82.03
16	TX	JASPER	81.94
17	TX	WALKER	81.89
18	TX	WALLER	81.80
19	LA	ACADIA	81.78
20	LA	ALLEN	81.78
21	LA	EVANGELINE	81.78
22	LA	RAPIDES	81.78
23	LA	VERMILION	81.78
24	LA	CAMERON	81.75
25	LA	CALCASIEU	81.73
26	TX	NEWTON	81.73
27	TX	FORT BEND	81.69
28	TX	CHAMBERS	81.66
29	LA	SABINE	81.65
30	TX	MONTGOMERY	81.64
31	TX	POLK	81.64
32	TX	SAN AUGUSTIN	81.62
33	LA	VERNON	81.59
34	TX	HARDIN	81.59
35	TX	ORANGE	81.59
36	TX	ANGELINA	81.55
37	TX	GRIMES	81.53
38	TX	LIBERTY	81.50
39	NM	LOS ALAMOS	81.49
40	TX	AUSTIN	81.46
41	TX	TYLER	81.40
42	TX	BRAZORIA	81.37
43	LA	LAFAYETTE	81.34
44	TX	GALVESTON	81.31
45	TX	TRINITY	81.30
46	TX	SABINE	81.30
47	TX	WASHINGTON	81.28
48	TX	SAN JACINTO	81.27
49	LA	GRANT	81.24
50	LA	BEAUREGARD	81.12
51	LA	JEFFERSON DA	81.08
52	NV	CLARK3	57.72
53	UT	IRON1	55.53
54	NM	GUADALUPE	40.75
55	NM	LINCOLN	40.75
56	NM	SIERRA	40.75
57	NM	SAN MIGUEL	40.49
58	NM	MORA	39.59
59	LA	ST HELENA	39.08
60	MS	PEARL RIVER	39.01
61	LA	ST BERNARD	38.99
62	MS	WALTHALL	38.97
63	AL	MOBILE	38.97
64	LA	IBERIA	38.94
65	LA	EAST FELICIA	38.87
66	LA	JEFFERSON	38.87
67	MS	JACKSON	38.82
68	MS	FORREST	38.80
69	MS	GEORGE	38.80
70	MS	GREENE	38.80
71	MS	PERRY	38.80
72	LA	PLAQUEMINES	38.77
73	LA	ST CHARLES	38.77
74	LA	ST TAMMANY	38.77
75	MS	HANCOCK	38.77
76	MS	HARRISON	38.77
77	LA	TERREBONNE	38.76
78	LA	LIVINGSTON	38.75
79	LA	ST JOHN THE	38.75
80	LA	WASHINGTON	38.75

COUNTIES OF HIGHEST AVERAGE FALLOUT FROM SHOT:

UPSHOT-KNOTHOLE SIMON Apr 25, 1953 43 Kt

(microCuries per square meter)

Each Row Includes:
County Rank (most fallout = 1), State, County, Fallout Level in microCuries/square meter

1	NV	CLARK1	2185.05
2	AZ	COCONINO2	912.14
3	AZ	APACHE	537.54
4	NY	WASHINGTON	533.92
5	NY	WARREN	507.23
6	NY	RENSSELAER	488.57
7	NM	MCKINLEY	436.89
8	NM	CHAVES	393.20
9	AZ	NAVAJO	363.55
10	NM	SANDOVAL	338.39
11	NM	SANTA FE	335.90
12	NM	LINCOLN	333.53
13	NY	SARATOGA	302.04
14	NY	FULTON	294.53
15	NM	VALENCIA	262.11
16	VT	BENNINGTON	254.90
17	NM	TORRANCE	251.96
18	NM	CURRY	250.15
19	NM	DE BACA	250.15
20	NM	GUADALUPE	250.15
21	NM	ROOSEVELT	250.15
22	NY	SCHENECTADY	230.23
23	NM	QUAY	208.52
24	NY	ALBANY	186.57
25	NY	COLUMBIA	176.93
26	VT	ADDISON	176.61
27	AZ	COCONINO3	172.07
28	NM	LOS ALAMOS	166.77
29	NM	RIO ARRIBA	166.77
30	NM	SAN JUAN	166.77
31	NM	SAN MIGUEL	166.77
32	CO	DOLORES	165.53
33	AZ	MOHAVE1	156.16
34	AZ	COCONINO1	155.03
35	AZ	MOHAVE3	146.09
36	AZ	MOHAVE4	146.09
37	NM	BERNALILLO	145.15
38	AZ	MOHAVE2	139.35
39	UT	SAN JUAN	127.39
40	CO	MONTEZUMA	125.13
41	CO	LA PLATA	124.23
42	NV	LINCOLN1	117.10
43	VT	CHITTENDEN	102.45
44	CO	MESA	102.26
45	NY	MONTGOMERY	96.58
46	TX	COCHRAN	94.34
47	UT	BEAVER	93.64
48	TX	YOAKUM	90.45
49	UT	KANE1	90.27
50	UT	KANE2	90.27
51	CO	ARCHULETA	88.85
52	UT	IRON1	85.77
53	TX	BAILEY	84.16
54	VT	RUTLAND	83.89
55	CO	DELTA	83.38
56	CO	MONTROSE	83.38
57	CO	OURAY	83.38
58	CO	SAN JUAN	83.38
59	CO	SAN MIGUEL	83.38
60	NM	EDDY	83.38
61	NM	HARDING	83.38
62	NM	LEA	83.38
63	NM	MORA	83.38
64	NM	TAOS	83.38
65	NY	ESSEX	78.90
66	TX	GAINES	73.11
67	NY	SCHOHARIE	72.54
68	TX	HOCKLEY	71.59
69	CO	KIOWA	71.01
70	TX	TERRY	69.79
71	TX	PARMER	69.49
72	UT	WASHINGTON2	69.39
73	TX	LAMB	67.49
74	UT	GRAND	66.68
75	NY	GREENE	66.02
76	UT	WASHINGTON1	64.00
77	TX	GRAYSON	61.80
78	TX	FANNIN	60.25
79	NY	HAMILTON	58.38
80	CO	GARFIELD	58.33

COUNTIES OF HIGHEST AVERAGE FALLOUT FROM SHOT:

UPSHOT-KNOTHOLE ENCORE (UK8) May 8, 1953 27 Kt

(microCuries per square meter)

Each Row Includes:
County Rank (most fallout = 1), State, County, Fallout Level in microCuries/square meter

1	MT	CUSTER	95.13
2	MT	RICHLAND	95.02
3	MT	MCCONE	62.76
4	MT	PRAIRIE	59.22
5	ND	BILLINGS	59.22
6	ND	BURKE	59.22
7	MT	DAWSON	59.12
8	MT	FALLON	59.12
9	MT	ROOSEVELT	59.12
10	MT	SHERIDAN	59.12
11	MT	WIBAUX	59.12
12	ND	DIVIDE	59.12
13	ND	GOLDEN VALLE	59.12
14	ND	MCKENZIE	59.12
15	ND	RENVILLE	59.12
16	ND	WARD	59.12
17	ND	MOUNTRAIL	57.30
18	ND	STARK	57.30
19	ND	MERCER	54.67
20	ND	OLIVER	54.67
21	MT	GARFIELD	53.15
22	MT	BIG HORN	42.04
23	ND	BOTTINEAU	39.65
24	ND	DUNN	39.54
25	ND	SLOPE	39.54
26	WY	SHERIDAN	39.50
27	MT	PETROLEUM	39.41
28	MT	ROSEBUD	38.89
29	MT	FERGUS	38.80
30	MT	MUSSELSHELL	38.80
31	MT	PHILLIPS	36.78
32	MT	BLAINE	36.47
33	ND	WILLIAMS	26.11
34	MT	TREASURE	25.66
35	MT	CARBON	24.78
36	MT	DANIELS	23.21
37	MT	GOLDEN VALLE	23.03
38	MT	STILLWATER	22.51
39	MT	VALLEY	21.50
40	ND	HETTINGER	21.39
41	ND	MCLEAN	21.39
42	WY	BIG HORN	20.58
43	MT	WHEATLAND	20.40
44	ND	GRANT	19.57
45	ND	MCHENRY	19.57
46	ND	MORTON	19.57
47	WY	HOT SPRINGS	19.09
48	WY	PARK	16.54
49	MT	SWEET GRASS	12.31
50	ND	SHERIDAN	10.49
51	ND	BURLEIGH	10.19
52	AR	CROSS	9.74
53	MT	YELLOWSTONE	9.48
54	AR	CRITTENDEN	9.11
55	AR	MONROE	9.11
56	AR	ARKANSAS	8.89
57	AR	JEFFERSON	8.89
58	AR	PHILLIPS	8.89
59	AR	PRAIRIE	8.89
60	AR	ST FRANCIS	8.89
61	AR	WHITE	8.89
62	AR	CLEBURNE	8.82
63	MS	DE SOTO	8.82
64	AR	LEE	8.50
65	OK	ROGER MILLS	8.46
66	MS	COAHOMA	8.40
67	AR	JACKSON	8.28
68	AR	POINSETT	8.28
69	AR	WOODRUFF	8.28
70	AR	INDEPENDENCE	8.21
71	TN	TIPTON	8.06
72	MS	PANOLA	8.04
73	MS	TATE	8.04
74	AR	CRAIGHEAD	8.01
75	AR	MISSISSIPPI	8.01
76	AR	LAWRENCE	7.94
77	AR	LONOKE	7.94
78	MS	QUITMAN	7.79
79	OK	CIMARRON	7.77
80	OK	WASHITA	7.77

COUNTIES OF HIGHEST AVERAGE FALLOUT FROM SHOT:

UPSHOT-KNOTHOLE HARRY (UK9) May 19, 1953 32 Kt

(microCuries per square meter)

Each Row Includes:
County Rank (most fallout = 1), State, County, Fallout Level in microCuries/square meter

1	UT	WASHINGTON2	4392.12
2	UT	WASHINGTON3	3642.40
3	UT	WASHINGTON1	2758.95
4	UT	KANE1	2334.06
5	UT	KANE2	2334.06
6	AZ	MOHAVE2	1986.84
7	NV	LINCOLN1	1716.20
8	CO	GUNNISON	1586.86
9	AZ	COCONINO1	1522.26
10	CO	CONEJOS	1452.03
11	AZ	MOHAVE1	1045.83
12	CO	ARCHULETA	926.47
13	CO	MINERAL	771.98
14	CO	HINSDALE	765.75
15	CO	RIO GRANDE	658.57
16	CO	PITKIN	627.41
17	UT	GARFIELD	568.41
18	UT	IRON3	502.13
19	UT	SAN JUAN	470.30
20	CO	DOLORES	466.90
21	CO	LA PLATA	466.90
22	CO	OURAY	466.90
23	CO	SAN JUAN	466.90
24	CO	SAN MIGUEL	466.90
25	CO	MONTROSE	463.51
26	CO	ALAMOSA	453.95
27	CO	SAGUACHE	406.29
28	UT	MILLARD	376.15
29	UT	SANPETE	376.15
30	CO	COSTILLA	376.07
31	CO	MONTEZUMA	374.79
32	UT	CARBON	374.79
33	CO	MESA	365.84
34	CO	LAKE	355.51
35	NM	BERNALILLO	351.01
36	CO	DELTA	323.30
37	NV	LINCOLN2	321.44
38	UT	IRON1	296.90
39	UT	WAYNE	282.11
40	AZ	APACHE	281.09
41	UT	EMERY	281.09
42	UT	JUAB	281.09
43	UT	GRAND	280.07
44	UT	UTAH	280.07
45	UT	WASATCH	280.07
46	NM	SANDOVAL	279.05
47	CO	GARFIELD	278.15
48	NM	MCKINLEY	278.15
49	NM	TORRANCE	275.09
50	UT	IRON2	266.15
51	NM	SAN JUAN	264.14
52	CO	EAGLE	238.74
53	UT	BEAVER	189.43
54	UT	PIUTE	188.75
55	UT	SEVIER	188.07
56	NM	LOS ALAMOS	186.72
57	NM	RIO ARRIBA	186.72
58	NM	SANTA FE	186.04
59	NM	VALENCIA	186.04
60	NM	TAOS	185.36
61	CO	CUSTER	164.71
62	CO	CHAFFEE	156.16
63	CO	SUMMIT	151.91
64	CO	HUERFANO	138.34
65	CO	FREMONT	118.59
66	NM	COLFAX	112.06
67	CO	LAS ANIMAS	107.60
68	CO	ROUTT	101.78
69	TX	HARTLEY	99.29
70	TX	DALLAM	99.03
71	NV	CLARK1	95.70
72	NM	MORA	92.00
73	NM	SAN MIGUEL	92.00
74	NM	GUADALUPE	91.66
75	NM	HARDING	91.32
76	NM	SOCORRO	91.32
77	TX	OLDHAM	90.89
78	NM	UNION	90.64
79	OK	CIMARRON	88.62
80	WI	SAWYER	76.50

COUNTIES OF HIGHEST AVERAGE FALLOUT FROM SHOT:

UPSHOT-KNOTHOLE GRABLE (UK10) May 25, 1953 15 Kt

(microCuries per square meter)

Each Row Includes:
County Rank (most fallout = 1), State, County, Fallout Level in microCuries/square meter

1	NV	WHITE PINE3	35.05
2	WA	FERRY	35.04
3	UT	MILLARD	33.91
4	UT	SANPETE	33.63
5	UT	SEVIER	33.63
6	UT	JUAB	33.20
7	UT	UTAH	33.20
8	CO	DELTA	32.92
9	UT	DUCHESNE	32.92
10	UT	CARBON	32.77
11	WY	CARBON	31.06
12	CO	MESA	28.89
13	WA	PEND OREILLE	28.73
14	NV	PERSHING	27.36
15	NV	NYE2	26.05
16	WA	STEVENS	24.81
17	WY	SWEETWATER	24.06
18	WA	OKANOGAN	23.39
19	UT	EMERY	22.66
20	UT	GRAND	22.37
21	UT	WASATCH	22.37
22	CO	GARFIELD	22.23
23	CO	MONTROSE	22.23
24	CO	RIO BLANCO	22.09
25	CO	MOFFAT	21.94
26	UT	DAGGETT	21.94
27	UT	UINTAH	21.94
28	WY	UINTA	21.94
29	UT	SALT LAKE	20.96
30	WY	FREMONT	20.95
31	ID	BONNER	20.36
32	NV	LANDER2	20.09
33	NV	CARSON CITY	18.52
34	NV	CHURCHILL	18.52
35	NV	STOREY	18.52
36	NV	HUMBOLDT	17.47
37	NV	WASHOE	17.47
38	UT	BEAVER	17.07
39	UT	SUMMIT	16.39
40	MT	GARFIELD	14.08
41	WA	DOUGLAS	14.03
42	ID	KOOTENAI	13.69
43	NV	LYON	13.68
44	ID	LATAH	13.26
45	WA	GARFIELD	13.03
46	ID	BOUNDARY	12.71
47	MT	PETROLEUM	12.49
48	UT	GARFIELD	10.97
49	UT	PIUTE	10.97
50	UT	WAYNE	10.97
51	MT	WHEATLAND	10.94
52	UT	BOX ELDER2	10.69
53	UT	DAVIS	10.69
54	UT	MORGAN	10.69
55	UT	TOOELE2	10.69
56	UT	WEBER	10.69
57	ID	CASSIA	10.54
58	ID	ELMORE	10.54
59	ID	GOODING	10.54
60	ID	OWYHEE	10.54
61	ID	TWIN FALLS	10.54
62	UT	CACHE	10.54
63	UT	RICH	10.54
64	ID	JEROME	10.40
65	ID	LINCOLN	10.40
66	ID	MINIDOKA	10.40
67	ID	ONEIDA	10.40
68	ID	POWER	10.40
69	WA	CHELAN	10.39
70	ID	BEAR LAKE	10.26
71	ID	CARIBOU	10.26
72	ID	FRANKLIN	10.26
73	WY	LINCOLN	10.26
74	WY	SUBLETTE	10.26
75	ID	ADA	9.94
76	CO	PITKIN	9.91
77	VA	NEWPORT NEWS	9.90
78	WA	LINCOLN	9.85
79	VA	HAMPTON	9.80
80	NV	LANDER1	9.69

COUNTIES OF HIGHEST AVERAGE FALLOUT FROM SHOT:

UPSHOT-KNOTHOLE CLIMAX (UK11) Jun 4, 1953 61 Kt

(microCuries per square meter)

Each Row Includes:
County Rank (most fallout = 1), State, County, Fallout Level in microCuries/square meter

1	AZ	COCONINO2	115.24
2	NV	CLARK1	87.97
3	UT	KANE1	47.69
4	AZ	COCONINO1	44.82
5	AZ	NAVAJO	38.55
6	UT	KANE2	36.40
7	AZ	MOHAVE2	34.93
8	UT	GARFIELD	33.82
9	AZ	APACHE	33.12
10	UT	SAN JUAN	32.99
11	NM	SAN JUAN	25.20
12	AZ	MOHAVE1	22.83
13	AZ	MOHAVE3	22.41
14	AZ	MOHAVE4	22.41
15	UT	EMERY	22.13
16	UT	GRAND	21.85
17	CO	DELTA	21.71
18	CO	DOLORES	21.71
19	CO	GARFIELD	21.71
20	CO	MONTROSE	21.71
21	CO	SAN MIGUEL	21.71
22	CO	LA PLATA	21.57
23	CO	OURAY	21.57
24	CO	SAN JUAN	21.57
25	NM	MCKINLEY	21.43
26	CO	MONTEZUMA	21.13
27	WY	CARBON	21.02
28	UT	WAYNE	20.87
29	CO	MESA	20.61
30	CO	ROUTT	19.67
31	NV	CLARK2	19.07
32	CO	RIO BLANCO	16.42
33	NM	LOS ALAMOS	16.28
34	NM	RIO ARRIBA	16.28
35	NM	SANDOVAL	16.28
36	WY	SWEETWATER	14.44
37	AZ	COCONINO3	13.39
38	UT	IRON3	11.83
39	UT	CARBON	11.55
40	CO	JACKSON	11.23
41	NM	BERNALILLO	11.16
42	CO	MOFFAT	11.13
43	NM	VALENCIA	11.13
44	WY	FREMONT	10.86
45	KS	WYANDOTTE	10.75
46	CO	LARIMER	10.60
47	WY	ALBANY	10.05
48	MO	BUCHANAN	9.35
49	KS	MIAMI	8.82
50	KS	ANDERSON	8.79
51	CO	GRAND	8.25
52	KS	JEFFERSON	8.15
53	KS	DOUGLAS	8.02
54	KS	DONIPHAN	7.94
55	WY	JOHNSON	7.45
56	KS	ATCHISON	7.43
57	KS	LEAVENWORTH	7.30
58	UT	PIUTE	7.10
59	CO	JEFFERSON	7.08
60	KS	LINN	6.98
61	WY	PLATTE	6.90
62	NM	TAOS	6.82
63	KS	BOURBON	6.72
64	CO	PITKIN	6.59
65	KS	ALLEN	6.49
66	CO	PHILLIPS	6.45
67	WY	LARAMIE	6.41
68	KS	WOODSON	6.39
69	MO	ANDREW	6.37
70	CO	SUMMIT	6.36
71	CO	GUNNISON	6.25
72	CO	YUMA	6.16
73	WY	SHERIDAN	6.03
74	WY	NATRONA	6.00
75	CO	SEDGWICK	5.95
76	MO	CASS	5.94
77	CO	LAKE	5.89
78	CO	BOULDER	5.88
79	CO	EAGLE	5.83
80	UT	IRON2	5.79

COUNTIES OF HIGHEST AVERAGE FALLOUT FROM SHOT:

TEAPOT WASP (TP1) Feb 18, 1955 1 Kt

(microCuries per square meter)

Each Row Includes:
County Rank (most fallout = 1), State, County, Fallout Level in microCuries/square meter

1	CA	SAN BERNADIN	107.56
2	AZ	YUMA	68.08
3	AZ	SANTA CRUZ	0.90
4	AZ	PIMA	0.85

COUNTIES OF HIGHEST AVERAGE FALLOUT FROM SHOT:

TEAPOT MOTH (TP2) Feb 22, 1955 2 Kt

(microCuries per square meter)

Each Row Includes:
County Rank (most fallout = 1), State, County, Fallout Level in microCuries/square meter

1	NV	CLARK2	85.77
2	NV	CLARK1	1.99

COUNTIES OF HIGHEST AVERAGE FALLOUT FROM SHOT:

TEAPOT TESLA (TP3) Mar 1, 1955 7 Kt

(microCuries per square meter)

Each Row Includes:
County Rank (most fallout = 1), State, County, Fallout Level in microCuries/square meter

Rank	State	County	Fallout
1	UT	WASHINGTON2	254.31
2	UT	KANE2	91.68
3	UT	WASHINGTON3	68.79
4	UT	KANE1	64.42
5	NV	LINCOLN2	57.19
6	NV	LINCOLN1	49.78
7	CO	ADAMS	49.41
8	CO	BOULDER	45.13
9	CO	SUMMIT	45.13
10	CO	ARAPAHOE	44.96
11	CO	CLEAR CREEK	44.96
12	CO	DOUGLAS	44.96
13	CO	EAGLE	44.96
14	CO	GRAND	44.96
15	CO	JEFFERSON	44.96
16	CO	LAKE	44.96
17	CO	GILPIN	44.90
18	CO	MORGAN	44.90
19	UT	GARFIELD	38.02
20	UT	IRON3	36.62
21	AZ	MOHAVE2	30.06
22	AZ	MOHAVE1	29.87
23	AZ	COCONINO1	29.68
24	UT	IRON1	29.48
25	UT	IRON2	29.00
26	UT	SAN JUAN	28.13
27	UT	PIUTE	19.01
28	CO	DENVER	18.69
29	CO	SAN MIGUEL	18.52
30	CO	DOLORES	18.43
31	CO	OURAY	18.43
32	CO	SAN JUAN	18.43
33	UT	WASHINGTON1	16.54
34	UT	BEAVER	15.25
35	UT	SEVIER	14.26
36	CO	ROUTT	9.81
37	NE	LINCOLN	9.49
38	UT	EMERY	9.41
39	UT	WAYNE	9.41
40	UT	GRAND	9.31
41	CO	LA PLATA	9.21
42	CO	MONTEZUMA	9.21
43	CO	GUNNISON	9.00
44	CO	PITKIN	9.00
45	CO	BACA	8.92
46	CO	CHEYENNE	8.92
47	CO	KIOWA	8.92
48	CO	KIT CARSON	8.92
49	CO	PHILLIPS	8.92
50	CO	PROWERS	8.92
51	CO	WASHINGTON	8.92
52	CO	YUMA	8.92
53	KS	CHEYENNE	8.92
54	KS	DECATUR	8.92
55	KS	FINNEY	8.92
56	KS	FORD	8.92
57	KS	GOVE	8.92
58	KS	GRAHAM	8.92
59	KS	GRANT	8.92
60	KS	GRAY	8.92
61	KS	GREELEY	8.92
62	KS	HAMILTON	8.92
63	KS	HASKELL	8.92
64	KS	HODGEMAN	8.92
65	KS	KEARNY	8.92
66	KS	LANE	8.92
67	KS	LOGAN	8.92
68	KS	NESS	8.92
69	KS	NORTON	8.92
70	KS	RAWLINS	8.92
71	KS	SCOTT	8.92
72	KS	SHERIDAN	8.92
73	KS	STANTON	8.92
74	KS	THOMAS	8.92
75	KS	TREGO	8.92
76	KS	WALLACE	8.92
77	KS	WICHITA	8.92
78	NE	CHASE	8.92
79	NE	DAWSON	8.92
80	NE	DUNDY	8.92

COUNTIES OF HIGHEST AVERAGE FALLOUT FROM SHOT:

TEAPOT TURK (TP4) Mar 7, 1955 43 Kt

(microCuries per square meter)

Each Row Includes:
County Rank (most fallout = 1), State, County, Fallout Level in microCuries/square meter

Rank	State	County	Fallout
1	CA	INYO3	341.80
2	NV	NYE2	322.23
3	NV	WHITE PINE2	205.41
4	CO	LAKE	168.12
5	CO	SUMMIT	164.80
6	CO	GRAND	163.68
7	CO	GILPIN	160.53
8	CO	CLEAR CREEK	160.48
9	CO	BOULDER	159.56
10	CO	EAGLE	159.07
11	CO	JEFFERSON	158.98
12	CO	ARAPAHOE	158.85
13	CO	DOUGLAS	158.80
14	CO	ADAMS	158.56
15	CO	MORGAN	157.69
16	UT	UTAH	151.68
17	NV	WHITE PINE3	138.40
18	UT	MILLARD	115.01
19	UT	JUAB	114.19
20	UT	SALT LAKE	114.06
21	UT	WASATCH	113.01
22	UT	SUMMIT	112.55
23	UT	MORGAN	97.59
24	NV	WHITE PINE1	95.05
25	CO	PITKIN	91.50
26	CO	GUNNISON	90.05
27	CO	ROUTT	83.98
28	CO	HINSDALE	82.97
29	CO	MINERAL	82.84
30	UT	DAVIS	75.34
31	UT	UINTAH	73.97
32	UT	DAGGETT	73.70
33	WY	UINTA	73.42
34	CO	MESA	69.50
35	CO	DENVER	67.84
36	CO	GARFIELD	66.12
37	UT	TOOELE1	60.65
38	UT	TOOELE2	60.65
39	UT	WEBER	60.29
40	UT	GRAND	59.56
41	CO	RIO BLANCO	58.74
42	WY	SWEETWATER	58.37
43	KS	THOMAS	53.49
44	KS	SHERIDAN	53.49
45	CO	KIT CARSON	53.48
46	KS	GOVE	53.42
47	KS	LOGAN	53.39
48	KS	NORTON	53.34
49	KS	DECATUR	53.30
50	CO	CHEYENNE	53.29
51	KS	GRAHAM	53.27
52	CO	WASHINGTON	53.26
53	KS	WICHITA	53.26
54	KS	WALLACE	53.26
55	KS	RAWLINS	53.25
56	CO	KIOWA	53.25
57	KS	SCOTT	53.20
58	KS	NESS	53.19
59	KS	TREGO	53.18
60	NE	RED WILLOW	53.13
61	NE	HITCHCOCK	53.11
62	KS	CHEYENNE	53.11
63	KS	GREELEY	53.05
64	KS	LANE	53.05
65	CO	YUMA	53.03
66	NE	FURNAS	53.01
67	KS	HODGEMAN	52.94
68	NE	DUNDY	52.93
69	KS	FINNEY	52.88
70	NE	GOSPER	52.79
71	KS	HAMILTON	52.74
72	NE	CHASE	52.74
73	NE	FRONTIER	52.70
74	KS	KEARNY	52.69
75	NE	HAYES	52.69
76	CO	PROWERS	52.65
77	KS	FORD	52.62
78	CO	PHILLIPS	52.57
79	KS	GRAY	52.50
80	NE	DAWSON	52.46

COUNTIES OF HIGHEST AVERAGE FALLOUT FROM SHOT:

TEAPOT HORNET (TP5) Mar 12, 1955 4 Kt

(microCuries per square meter)

Each Row Includes:
County Rank (most fallout = 1), State, County, Fallout Level in microCuries/square meter

Rank	State	County	Fallout
1	NV	CLARK1	126.37
2	AZ	COCONINO2	43.96
3	AZ	MOHAVE1	43.35
4	AZ	COCONINO1	42.85
5	AZ	COCONINO3	41.27
6	AZ	NAVAJO	34.78
7	UT	KANE1	34.28
8	NV	LINCOLN1	30.46
9	UT	IRON1	26.11
10	AZ	APACHE	26.01
11	AZ	MOHAVE2	23.87
12	UT	IRON3	17.34
13	NM	MCKINLEY	17.24
14	UT	IRON2	17.24
15	UT	KANE2	17.14
16	UT	GARFIELD	16.94
17	NM	VALENCIA	16.84
18	NM	SOCORRO	16.53
19	NV	LINCOLN2	15.69
20	NM	SAN JUAN	12.91
21	IN	LAWRENCE	12.81
22	NM	SANDOVAL	12.40
23	NM	LINCOLN	12.10
24	IN	MARTIN	11.91
25	IN	DUBOIS	11.42
26	IN	DAVIESS	11.38
27	IN	WASHINGTON	10.69
28	IN	ORANGE	10.65
29	IN	PIKE	10.51
30	NM	CHAVES	10.46
31	IN	JACKSON	10.10
32	IN	CLARK	9.92
33	IN	SCOTT	9.79
34	IN	GIBSON	9.22
35	KY	OLDHAM	8.94
36	IN	JEFFERSON	8.94
37	CO	DOLORES	8.67
38	CO	MONTEZUMA	8.67
39	IN	JENNINGS	8.57
40	UT	SAN JUAN	8.57
41	KY	TRIMBLE	8.39
42	NM	CATRON	8.37
43	NM	LOS ALAMOS	8.27
44	NM	RIO ARRIBA	8.27
45	NM	SIERRA	8.27
46	NM	TAOS	8.27
47	IL	WABASH	8.10
48	KY	JEFFERSON	8.07
49	NM	COLFAX	8.07
50	NM	MORA	8.07
51	NM	SAN MIGUEL	8.07
52	NM	SANTA FE	8.07
53	NM	TORRANCE	8.07
54	WY	LARAMIE	8.00
55	NM	CURRY	7.86
56	NM	DE BACA	7.86
57	NM	GUADALUPE	7.86
58	NM	HARDING	7.86
59	NM	QUAY	7.86
60	NM	ROOSEVELT	7.86
61	NM	UNION	7.86
62	KY	SHELBY	7.73
63	IL	EDWARDS	7.69
64	IN	MONROE	7.52
65	IN	PUTNAM	7.22
66	KY	WASHINGTON	7.20
67	IN	BROWN	7.12
68	IN	HENDRICKS	7.05
69	IN	GREENE	7.04
70	KY	GRANT	7.02
71	IN	OWEN	6.99
72	IN	CLAY	6.71
73	IL	LAKE	6.54
74	IN	HARRISON	6.48
75	IN	CRAWFORD	6.43
76	KY	HANCOCK	6.43
77	IN	SPENCER	6.36
78	IL	DU PAGE	6.32
79	IL	DE KALB	6.31
80	KY	PENDLETON	6.26

COUNTIES OF HIGHEST AVERAGE FALLOUT FROM SHOTS:

TEAPOT BEE AND ESS (TP6)
BEE: Mar 22, 1955 8 Kt
ESS: Mar 23, 1955 1 Kt
(microCuries per square meter)

Each Row Includes:
County Rank (most fallout = 1), State, County, Fallout Level in microCuries/square meter

Rank	State	County	Fallout
1	NV	CLARK3	104.23
2	AZ	MOHAVE2	102.51
3	AZ	MOHAVE1	56.78
4	NV	CLARK2	49.19
5	AZ	COCONINO1	42.35
6	UT	KANE1	28.08
7	NV	CLARK1	26.74
8	MS	MONTGOMERY	24.74
9	AZ	YAVAPAI	21.33
10	AZ	MOHAVE3	19.08
11	AZ	MOHAVE4	19.08
12	AL	TUSCALOOSA	18.18
13	MS	LEAKE	16.64
14	AL	MARENGO	16.62
15	AR	HEMPSTEAD	15.74
16	AR	HOWARD	15.74
17	AR	LAFAYETTE	15.74
18	AR	LITTLE RIVER	15.74
19	AR	MILLER	15.74
20	AR	MONTGOMERY	15.74
21	AR	PIKE	15.74
22	AR	POLK	15.74
23	AR	SCOTT	15.74
24	AR	SEVIER	15.74
25	OK	ATOKA	15.74
26	OK	BRYAN	15.74
27	OK	CARTER	15.74
28	OK	CHOCTAW	15.74
29	OK	CLEVELAND	15.74
30	OK	COAL	15.74
31	OK	COMANCHE	15.74
32	OK	COTTON	15.74
33	OK	GARVIN	15.74
34	OK	GRADY	15.74
35	OK	HASKELL	15.74
36	OK	HUGHES	15.74
37	OK	JEFFERSON	15.74
38	OK	JOHNSTON	15.74
39	OK	LATIMER	15.74
40	OK	LE FLORE	15.74
41	OK	LOVE	15.74
42	OK	MARSHALL	15.74
43	OK	MCCLAIN	15.74
44	OK	MCCURTAIN	15.74
45	OK	MCINTOSH	15.74
46	OK	MURRAY	15.74
47	OK	PITTSBURG	15.74
48	OK	PONTOTOC	15.74
49	OK	POTTAWATOMIE	15.74
50	OK	PUSHMATAHA	15.74
51	OK	SEMINOLE	15.74
52	OK	STEPHENS	15.74
53	TX	ANDERSON	15.74
54	TX	BELL	15.74
55	TX	BOSQUE	15.74
56	TX	BOWIE	15.74
57	TX	BRAZOS	15.74
58	TX	BURLESON	15.74
59	TX	CAMP	15.74
60	TX	CASS	15.74
61	TX	CHEROKEE	15.74
62	TX	CLAY	15.74
63	TX	COLLIN	15.74
64	TX	COOKE	15.74
65	TX	CORYELL	15.74
66	TX	DALLAS	15.74
67	TX	DELTA	15.74
68	TX	DENTON	15.74
69	TX	ELLIS	15.74
70	TX	ERATH	15.74
71	TX	FALLS	15.74
72	TX	FANNIN	15.74
73	TX	FRANKLIN	15.74
74	TX	FREESTONE	15.74
75	TX	GRAYSON	15.74
76	TX	GREGG	15.74
77	TX	HAMILTON	15.74
78	TX	HARRISON	15.74
79	TX	HENDERSON	15.74
80	TX	HILL	15.74

COUNTIES OF HIGHEST AVERAGE FALLOUT FROM SHOT:

TEAPOT APPLE-1 Mar 29, 1955 14 Kt

(microCuries per square meter)

Each Row Includes:
County Rank (most fallout = 1), State, County, Fallout Level in microCuries/square meter

Rank	State	County	Fallout
1	NV	LINCOLN2	333.26
2	OK	TEXAS	160.85
3	KS	STEVENS	160.49
4	TX	OLDHAM	159.97
5	TX	POTTER	159.85
6	TX	HANSFORD	107.99
7	TX	HUTCHINSON	107.39
8	TX	OCHILTREE	106.75
9	TX	DALLAM	106.44
10	UT	IRON3	92.64
11	UT	WASHINGTON1	89.86
12	UT	GARFIELD	72.85
13	NV	LINCOLN1	71.59
14	UT	IRON1	69.71
15	UT	IRON2	65.52
16	OK	CIMARRON	53.91
17	TX	BRISCOE	53.90
18	TX	LIPSCOMB	53.88
19	TX	HARTLEY	53.81
20	TX	WHEELER	53.57
21	UT	KANE2	46.14
22	UT	PIUTE	40.24
23	AZ	COCONINO1	32.62
24	AZ	MOHAVE1	32.62
25	UT	KANE1	32.62
26	KS	MORTON	29.88
27	OK	BECKHAM	29.33
28	TX	ROBERTS	28.90
29	OK	ELLIS	28.72
30	OK	GREER	28.63
31	TX	SHERMAN	28.51
32	OK	ROGER MILLS	28.48
33	KS	SEWARD	28.47
34	OK	BEAVER	28.41
35	TX	MOORE	28.39
36	TX	ARMSTRONG	28.39
37	TX	DEAF SMITH	28.35
38	TX	CARSON	28.35
39	TX	SWISHER	28.12
40	KS	MEADE	27.79
41	TX	DONLEY	27.77
42	TX	CASTRO	27.56
43	TX	HALL	27.55
44	OK	WASHITA	27.48
45	TX	COLLINGSWORT	27.20
46	TX	MOTLEY	27.19
47	TX	GRAY	27.19
48	OK	HARMON	27.17
49	TX	FLOYD	27.15
50	TX	HEMPHILL	27.15
51	TX	CHILDRESS	27.14
52	TX	PARMER	27.14
53	TX	CROSBY	27.12
54	TX	BAILEY	27.12
55	TX	HALE	27.12
56	TX	LAMB	27.12
57	TX	LUBBOCK	27.09
58	TX	LYNN	27.06
59	TX	HOCKLEY	27.05
60	TX	TERRY	27.02
61	TX	COCHRAN	24.80
62	TX	YOAKUM	24.78
63	TX	GAINES	24.71
64	TX	ANDREWS	24.68
65	TX	LOVING	24.62
66	AZ	MOHAVE2	24.61
67	TX	ECTOR	24.57
68	TX	WINKLER	24.57
69	TX	CULBERSON	24.56
70	TX	REEVES	24.53
71	TX	WARD	24.53
72	TX	HUDSPETH	24.49
73	TX	EL PASO	24.47
74	TX	JEFF DAVIS	24.44
75	UT	WAYNE	23.93
76	UT	BEAVER	23.12
77	NM	HARDING	22.68
78	NM	UNION	22.68
79	CO	ARCHULETA	20.86
80	NM	COLFAX	19.20

COUNTIES OF HIGHEST AVERAGE FALLOUT FROM SHOT:

TEAPOT POST (TP8) Apr 9, 1955 2 Kt

(microCuries per square meter)

Each Row Includes:
County Rank (most fallout = 1), State, County, Fallout Level in microCuries/square meter

Rank	State	County	Fallout
1	UT	TOOELE2	39.13
2	UT	SALT LAKE	36.91
3	UT	WASATCH	33.43
4	UT	MORGAN	33.17
5	UT	SUMMIT	31.24
6	UT	DAVIS	31.07
7	UT	UTAH	29.44
8	NV	WHITE PINE2	29.00
9	WY	UINTA	28.82
10	UT	MILLARD	26.16
11	NV	WHITE PINE1	25.62
12	CO	RIO BLANCO	25.10
13	CO	MOFFAT	24.92
14	UT	DUCHESNE	24.92
15	UT	WEBER	24.92
16	UT	UINTAH	24.74
17	WY	SWEETWATER	24.74
18	NV	NYE3	21.46
19	CA	INYO3	18.27
20	NV	WHITE PINE3	18.09
21	CO	MONTROSE	16.94
22	UT	CARBON	16.76
23	UT	BOX ELDER2	16.58
24	UT	DAGGETT	16.49
25	WY	CARBON	16.49
26	WY	FREMONT	16.49
27	CO	GARFIELD	15.79
28	CO	MESA	14.97
29	UT	JUAB	12.95
30	CO	OURAY	12.77
31	CO	DELTA	11.96
32	UT	BEAVER	11.44
33	NV	LINCOLN2	10.08
34	UT	IRON1	8.96
35	UT	IRON2	8.96
36	UT	IRON3	8.96
37	UT	GARFIELD	8.69
38	UT	PIUTE	8.69
39	UT	SEVIER	8.69
40	UT	EMERY	8.60
41	UT	WAYNE	8.60
42	CO	DOLORES	8.51
43	CO	SAN JUAN	8.51
44	CO	SAN MIGUEL	8.51
45	UT	GRAND	8.51
46	UT	SANPETE	8.51
47	UT	TOOELE1	8.51
48	WY	LINCOLN	8.34
49	WY	SUBLETTE	8.25
50	CO	PITKIN	6.53
51	ID	IDAHO	5.97
52	WY	NATRONA	5.96
53	ID	VALLEY	5.89
54	CO	JACKSON	5.83
55	AZ	COCONINO1	4.43
56	AZ	MOHAVE1	4.43
57	AZ	MOHAVE3	4.43
58	AZ	MOHAVE4	4.43
59	UT	KANE1	4.35
60	UT	KANE2	4.35
61	CO	LA PLATA	4.26
62	CO	MONTEZUMA	4.26
63	UT	SAN JUAN	4.26
64	UT	BOX ELDER1	4.17
65	KS	STANTON	4.14
66	ID	ADAMS	3.97
67	OR	GRANT	3.97
68	ID	BOISE	3.92
69	ID	WASHINGTON	3.89
70	ID	PAYETTE	3.88
71	ID	GEM	3.85
72	OR	BAKER	3.85
73	WY	CONVERSE	3.68
74	KS	CHEYENNE	3.51
75	CO	ROUTT	3.47
76	UT	CACHE	3.28
77	UT	RICH	3.28
78	CO	SUMMIT	3.27
79	NE	DUNDY	3.06
80	WY	LARAMIE	3.01

COUNTIES OF HIGHEST AVERAGE FALLOUT FROM SHOT:

TEAPOT MET (TP9) Apr 15, 1955 22 Kt

(microCuries per square meter)

Each Row Includes:
County Rank (most fallout = 1), State, County, Fallout Level in microCuries/square meter

Rank	State	County	Fallout
1	NV	LINCOLN1	433.28
2	UT	BEAVER	406.51
3	CO	MESA	382.05
4	UT	EMERY	325.27
5	UT	GRAND	325.27
6	UT	PIUTE	325.27
7	CO	DELTA	322.85
8	CO	GARFIELD	322.85
9	UT	SEVIER	243.92
10	NV	LINCOLN2	209.20
11	CO	RIO BLANCO	159.66
12	CO	MONTROSE	85.85
13	MI	WAYNE	85.84
14	UT	IRON1	84.89
15	OH	MEDINA	84.43
16	UT	WAYNE	80.94
17	OH	CUYAHOGA	78.35
18	OH	SENECA	70.70
19	MI	MACOMB	57.98
20	OH	ERIE	49.64
21	OH	GEAUGA	45.67
22	MI	HURON	44.33
23	OH	SANDUSKY	41.06
24	MI	EATON	38.03
25	OH	HURON	38.00
26	MI	SANILAC	37.66
27	NY	CHAUTAUQUA	37.51
28	OH	LORAIN	36.27
29	UT	MILLARD	32.46
30	OH	PORTAGE	32.34
31	MI	GENESEE	31.25
32	OH	LAKE	29.68
33	PA	ERIE	28.96
34	NE	GRANT	28.92
35	OH	TRUMBULL	27.92
36	NE	MORRILL	27.48
37	PA	MERCER	26.63
38	CO	GUNNISON	24.73
39	UT	JUAB	24.19
40	NY	ERIE	23.94
41	MI	ST CLAIR	21.48
42	PA	CRAWFORD	21.25
43	CO	DENVER	21.23
44	CO	ROUTT	19.81
45	PA	VENANGO	19.39
46	MI	CLINTON	19.37
47	KS	BARTON	19.22
48	KS	RUSSELL	18.48
49	PA	FOREST	18.28
50	CO	BOULDER	18.03
51	KS	STAFFORD	17.77
52	NY	NIAGARA	17.50
53	KS	RICE	16.95
54	MI	LAPEER	16.55
55	KS	NORTON	16.35
56	UT	IRON2	16.23
57	KS	ELLSWORTH	16.05
58	MI	OAKLAND	16.03
59	KS	RENO	15.88
60	WY	FREMONT	15.62
61	WY	CARBON	15.52
62	CO	GILPIN	15.24
63	KS	OSBORNE	15.20
64	KS	PAWNEE	15.06
65	PA	WARREN	15.04
66	KS	RUSH	14.87
67	OH	CHAMPAIGN	14.64
68	KS	ELLIS	14.57
69	KS	LINCOLN	14.35
70	KS	PRATT	14.35
71	UT	UTAH	14.11
72	UT	DAVIS	13.91
73	UT	MORGAN	13.91
74	UT	TOOELE2	13.91
75	OH	HANCOCK	13.36
76	ND	PEMBINA	13.32
77	UT	SALT LAKE	13.12
78	KS	ROOKS	12.98
79	OH	WOOD	12.98
80	KS	KINGMAN	12.96

COUNTIES OF HIGHEST AVERAGE FALLOUT FROM SHOT:

TEAPOT APPLE-2 (TP10) May 5, 1955 29 Kt

(microCuries per square meter)

Each Row Includes:
County Rank (most fallout = 1), State, County, Fallout Level in microCuries/square meter

Rank	State	County	Fallout
1	NV	WHITE PINE2	430.02
2	NV	NYE2	316.28
3	UT	UTAH	238.18
4	NV	WHITE PINE1	238.09
5	UT	WASATCH	237.40
6	UT	DUCHESNE	235.65
7	CO	MOFFAT	233.13
8	UT	DAGGETT	233.13
9	UT	UINTAH	233.13
10	CO	RIO BLANCO	232.25
11	NV	WHITE PINE3	221.15
12	UT	JUAB	158.85
13	WY	CARBON	154.87
14	CO	COSTILLA	127.67
15	UT	SALT LAKE	119.41
16	UT	DAVIS	118.26
17	UT	MORGAN	117.39
18	UT	WEBER	117.39
19	WY	SWEETWATER	116.61
20	CO	LARIMER	114.28
21	CO	WELD	113.91
22	CO	JACKSON	113.06
23	WY	ALBANY	112.25
24	WY	PLATTE	111.47
25	UT	SUMMIT	110.40
26	CO	CONEJOS	96.93
27	KS	STEVENS	80.60
28	UT	MILLARD	80.30
29	UT	TOOELE2	78.84
30	CO	GARFIELD	77.19
31	CO	CLEAR CREEK	70.00
32	CO	GILPIN	68.93
33	CO	BENT	68.82
34	CO	OTERO	68.53
35	CO	LINCOLN	67.84
36	CO	CROWLEY	67.13
37	CO	LAS ANIMAS	66.82
38	CO	CUSTER	66.28
39	CO	RIO GRANDE	66.10
40	CO	ALAMOSA	66.09
41	CO	GRAND	65.49
42	MO	CAMDEN	65.41
43	CO	HUERFANO	64.94
44	CO	LAKE	63.53
45	MO	MORGAN	63.33
46	MO	LACLEDE	63.18
47	MO	WEBSTER	62.65
48	CO	DOUGLAS	62.45
49	CO	EAGLE	62.12
50	MO	OSAGE	61.87
51	CO	JEFFERSON	61.51
52	CO	ADAMS	61.47
53	CO	SUMMIT	61.18
54	MO	PETTIS	60.99
55	AR	PERRY	59.87
56	AR	SALINE	59.59
57	CO	BOULDER	59.57
58	CO	ARAPAHOE	58.86
59	MO	WRIGHT	58.82
60	NM	COLFAX	57.87
61	NM	UNION	57.87
62	CO	MORGAN	57.65
63	KS	FINNEY	57.26
64	MO	DOUGLAS	57.01
65	MO	HENRY	55.89
66	MO	OZARK	54.80
67	TX	HARTLEY	54.24
68	OK	TEXAS	53.91
69	CO	TELLER	52.55
70	CO	ELBERT	52.27
71	IN	PERRY	52.12
72	CO	CHAFFEE	51.53
73	CO	PARK	51.38
74	CO	SAGUACHE	51.38
75	CO	FREMONT	50.90
76	TX	LIPSCOMB	50.72
77	NM	HARDING	50.59
78	AR	POPE	50.49
79	IN	SPENCER	48.53
80	AR	VAN BUREN	48.50

COUNTIES OF HIGHEST AVERAGE FALLOUT FROM SHOT:

TEAPOT ZUCCHINI (TP11) 15 May, 1955 28 Kt

(microCuries per square meter)

Each Row Includes:
County Rank (most fallout = 1), State, County, Fallout Level in microCuries/square meter

Rank	State	County	Fallout
1	UT	IRON2	413.43
2	NV	CLARK1	387.19
3	AZ	MOHAVE2	207.52
4	UT	IRON3	180.05
5	NM	COLFAX	130.08
6	NM	UNION	130.08
7	UT	WASHINGTON2	104.21
8	AZ	MOHAVE3	88.60
9	AZ	MOHAVE4	88.60
10	UT	MILLARD	87.66
11	UT	PIUTE	82.13
12	NM	HARDING	78.03
13	NM	MORA	78.03
14	UT	BEAVER	66.21
15	UT	WASHINGTON3	62.19
16	UT	WASHINGTON1	60.67
17	UT	SEVIER	57.32
18	CO	PITKIN	54.24
19	UT	SANPETE	53.79
20	NM	QUAY	52.05
21	NM	SAN MIGUEL	52.05
22	CO	MORGAN	51.43
23	CO	HINSDALE	49.80
24	WY	CARBON	43.75
25	CO	GUNNISON	43.63
26	UT	JUAB	43.26
27	UT	GARFIELD	41.11
28	CO	MINERAL	40.27
29	CO	ROUTT	38.07
30	CO	GRAND	37.29
31	WY	PLATTE	36.53
32	CO	BOULDER	36.05
33	UT	UTAH	35.45
34	CO	LARIMER	34.55
35	CO	CLEAR CREEK	30.89
36	CO	GILPIN	30.42
37	OK	PONTOTOC	29.90
38	CO	DENVER	29.75
39	WY	CAMPBELL	29.42
40	CO	JEFFERSON	29.02
41	NV	CLARK2	27.29
42	NM	CURRY	25.98
43	NM	GUADALUPE	25.98
44	CO	DOUGLAS	24.59
45	UT	EMERY	24.51
46	CO	CUSTER	24.47
47	UT	CARBON	24.41
48	UT	GRAND	24.32
49	CO	DELTA	24.12
50	CO	DOLORES	24.12
51	CO	GARFIELD	24.12
52	CO	MONTEZUMA	24.12
53	CO	MONTROSE	24.12
54	CO	SAN MIGUEL	24.12
55	CO	OURAY	24.02
56	CO	SAN JUAN	24.02
57	CO	LA PLATA	23.93
58	CO	RIO BLANCO	23.73
59	UT	UINTAH	23.73
60	UT	TOOELE2	23.66
61	UT	BOX ELDER2	23.24
62	CO	MOFFAT	23.14
63	CO	MESA	22.75
64	UT	DUCHESNE	22.56
65	ID	OWYHEE	22.36
66	ID	CANYON	22.17
67	ID	ELMORE	22.17
68	OR	HARNEY	21.87
69	OR	MALHEUR	21.87
70	UT	WASATCH	21.87
71	MT	POWDER RIVER	21.84
72	CO	ARAPAHOE	21.58
73	TX	FOARD	21.38
74	UT	DAGGETT	21.29
75	UT	SUMMIT	21.29
76	UT	DAVIS	21.09
77	UT	MORGAN	21.09
78	UT	WEBER	21.00
79	WY	UINTA	21.00
80	TX	HALE	20.94

COUNTIES OF HIGHEST AVERAGE FALLOUT FROM SHOT:

PLUMBBOB BOLTZMANN (PB12) May 28, 1957 12 Kt

(microCuries per square meter)

Each Row Includes:
County Rank (most fallout = 1), State, County, Fallout Level in microCuries/square meter

Rank	State	County	Fallout
1	NV	NYE2	766.45
2	NV	NYE1	519.96
3	NV	LANDER2	283.24
4	NV	MINERAL	282.27
5	NV	CHURCHILL	279.25
6	NV	WASHOE	253.86
7	NV	CARSON CITY	238.37
8	NV	DOUGLAS	238.37
9	NV	STOREY	237.50
10	NV	LYON	233.11
11	NV	PERSHING	160.00
12	NV	EUREKA	71.95
13	NV	LANDER1	40.48
14	UT	UTAH	24.87
15	UT	DAVIS	23.30
16	UT	MORGAN	23.30
17	UT	WEBER	23.30
18	UT	SALT LAKE	21.98
19	UT	DUCHESNE	18.70
20	UT	SUMMIT	18.70
21	UT	WASATCH	18.70
22	NV	ELKO	15.97
23	NV	HUMBOLDT	15.79
24	CO	LA PLATA	12.44
25	CO	MONTEZUMA	12.44
26	UT	CARBON	12.44
27	UT	EMERY	12.44
28	UT	GARFIELD	12.44
29	UT	KANE1	12.44
30	UT	KANE2	12.44
31	UT	SAN JUAN	12.44
32	UT	UINTAH	12.44
33	UT	WAYNE	12.44
34	AZ	MARICOPA	10.70
35	AZ	APACHE	10.41
36	AZ	COCHISE	10.41
37	AZ	COCONINO1	10.41
38	AZ	COCONINO2	10.41
39	AZ	COCONINO3	10.41
40	AZ	GILA	10.41
41	AZ	GRAHAM	10.41
42	AZ	GREENLEE	10.41
43	AZ	MOHAVE1	10.41
44	AZ	NAVAJO	10.41
45	AZ	PINAL	10.41
46	AZ	SANTA CRUZ	10.41
47	TX	WHARTON	9.49
48	TX	WALLER	9.36
49	TX	JACKSON	9.09
50	TX	COLORADO	9.05
51	TX	CALHOUN	8.73
52	NM	CATRON	8.66
53	NM	CHAVES	8.66
54	NM	COLFAX	8.66
55	NM	CURRY	8.66
56	NM	DE BACA	8.66
57	NM	DONA ANA	8.66
58	NM	EDDY	8.66
59	NM	GRANT	8.66
60	NM	GUADALUPE	8.66
61	NM	HARDING	8.66
62	NM	HIDALGO	8.66
63	NM	LEA	8.66
64	NM	LINCOLN	8.66
65	NM	LOS ALAMOS	8.66
66	NM	LUNA	8.66
67	NM	MCKINLEY	8.66
68	NM	MORA	8.66
69	NM	OTERO	8.66
70	NM	QUAY	8.66
71	NM	RIO ARRIBA	8.66
72	NM	ROOSEVELT	8.66
73	NM	SAN JUAN	8.66
74	NM	SAN MIGUEL	8.66
75	NM	SANDOVAL	8.66
76	NM	SANTA FE	8.66
77	NM	SIERRA	8.66
78	NM	SOCORRO	8.66
79	NM	TAOS	8.66
80	NM	TORRANCE	8.66

COUNTIES OF HIGHEST AVERAGE FALLOUT FROM SHOT:

PLUMBBOB WILSON (PB2) Jun 18, 1957 10 Kt

(microCuries per square meter)

Each Row Includes:
County Rank (most fallout = 1), State, County, Fallout Level in microCuries/square meter

Rank	State	County	Fallout
1	ND	RICHLAND	93.51
2	SD	ROBERTS	93.30
3	ND	CASS	56.62
4	ND	MERCER	56.52
5	SD	WALWORTH	56.43
6	ND	OLIVER	56.22
7	ND	FOSTER	56.18
8	SD	HYDE	56.12
9	SD	ZIEBACH	56.12
10	ND	SARGENT	55.98
11	SD	DAY	55.98
12	SD	FAULK	55.98
13	SD	DEWEY	55.93
14	SD	BEADLE	55.87
15	SD	CAMPBELL	55.87
16	SD	POTTER	55.87
17	SD	SULLY	55.87
18	ND	EMMONS	55.71
19	SD	BRULE	55.71
20	SD	CODINGTON	55.71
21	SD	LYMAN	55.71
22	SD	MELLETTE	55.71
23	SD	SANBORN	55.71
24	SD	SPINK	55.71
25	NE	BLAINE	55.65
26	NE	BROWN	55.65
27	SD	AURORA	55.55
28	SD	BUFFALO	55.55
29	SD	TODD	55.55
30	UT	IRON1	43.12
31	WY	NATRONA	39.86
32	ND	TRAILL	38.14
33	SD	CORSON	37.73
34	SD	MARSHALL	37.73
35	SD	HAAKON	37.68
36	SD	STANLEY	37.68
37	ND	LOGAN	37.63
38	ND	MORTON	37.63
39	ND	BURLEIGH	37.49
40	SD	BROWN	37.49
41	ND	MCINTOSH	37.38
42	ND	MCLEAN	37.38
43	ND	WALSH	37.38
44	SD	CLARK	37.38
45	SD	HAND	37.38
46	SD	TRIPP	37.38
47	ND	MCHENRY	37.22
48	ND	WARD	37.22
49	SD	BENNETT	37.22
50	SD	HAMLIN	37.22
51	SD	JERAULD	37.22
52	SD	JONES	37.22
53	SD	KINGSBURY	37.22
54	SD	MINER	37.22
55	SD	JACKSON	37.11
56	UT	IRON3	32.43
57	CA	INYO1	31.32
58	CA	INYO2	31.32
59	WY	CONVERSE	30.61
60	WY	NIOBRARA	26.21
61	UT	PIUTE	24.45
62	NV	LINCOLN2	21.88
63	UT	IRON2	21.00
64	UT	BEAVER	20.88
65	NV	NYE3	20.33
66	ND	GRANT	19.75
67	ND	BARNES	19.70
68	UT	WAYNE	19.44
69	ND	LA MOURE	19.34
70	SD	EDMUNDS	19.34
71	ND	STUTSMAN	19.29
72	ND	GRAND FORKS	19.24
73	NE	MADISON	19.24
74	ND	DICKEY	19.10
75	ND	KIDDER	19.00
76	KS	STANTON	18.85
77	ND	BOTTINEAU	18.84
78	ND	PEMBINA	18.84
79	ND	SHERIDAN	18.84
80	NE	GARFIELD	18.84

COUNTIES OF HIGHEST AVERAGE FALLOUT FROM SHOT:

PLUMBBOB PRISCILLA (PB3) Jun 24, 1957 37 Kt

(microCuries per square meter)

Each Row Includes:
County Rank (most fallout = 1), State, County, Fallout Level in microCuries/square meter

Rank	State	County	Fallout
1	UT	WASHINGTON1	188.87
2	MS	FRANKLIN	139.24
3	LA	AVOYELLES	139.24
4	MS	WILKINSON	139.03
5	LA	WEST FELICIA	138.91
6	LA	POINTE COUPE	138.78
7	LA	ACADIA	138.75
8	LA	LA SALLE	138.04
9	TX	ORANGE	138.01
10	LA	TENSAS	137.94
11	LA	GRANT	137.82
12	MS	ADAMS	137.81
13	LA	RAPIDES	137.74
14	LA	BEAUREGARD	137.45
15	LA	ST HELENA	137.04
16	LA	EAST BATON R	136.99
17	LA	EVANGELINE	136.99
18	LA	IBERVILLE	136.93
19	LA	WEST BATON R	136.93
20	LA	IBERIA	136.82
21	LA	TANGIPAHOA	136.82
22	LA	ST MARTIN	136.42
23	LA	ST MARY	136.42
24	LA	ASSUMPTION	136.37
25	LA	TERREBONNE	135.83
26	LA	LAFAYETTE	135.79
27	MS	WARREN	135.72
28	LA	CALCASIEU	135.70
29	LA	CAMERON	135.70
30	LA	CATAHOULA	85.39
31	LA	CONCORDIA	85.39
32	MS	PIKE	85.11
33	LA	ST LANDRY	84.94
34	LA	ALLEN	84.51
35	LA	JEFFERSON DA	84.43
36	MS	HINDS	84.10
37	MS	LAUDERDALE	84.01
38	MS	MARION	83.85
39	LA	EAST FELICIA	83.66
40	AL	MARENGO	83.52
41	LA	LIVINGSTON	82.72
42	LA	WASHINGTON	82.70
43	LA	ST TAMMANY	82.69
44	LA	ASCENSION	82.66
45	AL	MOBILE	82.59
46	MS	GEORGE	82.35
47	MS	HANCOCK	82.27
48	MS	PEARL RIVER	82.23
49	FL	OKALOOSA	81.80
50	LA	PLAQUEMINES	81.63
51	LA	ST BERNARD	81.63
52	LA	VERMILION	81.59
53	UT	KANE2	67.77
54	UT	IRON1	59.51
55	MS	COPIAH	58.17
56	MS	CLAIBORNE	58.10
57	MS	JEFFERSON	57.98
58	MS	CLARKE	56.81
59	NV	LINCOLN1	56.77
60	MS	WAYNE	56.74
61	MS	LAMAR	56.65
62	MS	FORREST	55.84
63	AL	CONECUH	55.80
64	MS	STONE	55.70
65	AL	MONROE	55.67
66	LA	ST JOHN THE	55.51
67	LA	ST JAMES	55.45
68	MS	GREENE	55.22
69	MS	JACKSON	54.66
70	MS	HARRISON	54.60
71	LA	LAFOURCHE	54.49
72	LA	ST CHARLES	54.49
73	UT	WASHINGTON3	52.16
74	NM	MCKINLEY	33.80
75	AZ	APACHE	32.07
76	MS	JASPER	31.42
77	MS	SMITH	31.37
78	MS	AMITE	30.83
79	MS	WALTHALL	30.20
80	AL	WILCOX	30.18

COUNTIES OF HIGHEST AVERAGE FALLOUT FROM SHOT:

PLUMBBOB HOOD (PB4) Jul 5, 1957 74 Kt

(microCuries per square meter)

Each Row Includes:
County Rank (most fallout = 1), State, County, Fallout Level in microCuries/square meter

1	AZ	PIMA	148.41
2	AZ	SANTA CRUZ	98.28
3	CO	GARFIELD	81.66
4	CO	RIO BLANCO	81.28
5	NM	COLFAX	73.63
6	NM	HARDING	73.63
7	NM	UNION	73.63
8	NM	MORA	73.09
9	NM	SAN MIGUEL	73.09
10	NM	SANTA FE	73.09
11	NM	TORRANCE	72.55
12	NM	BERNALILLO	67.97
13	NM	QUAY	55.22
14	NM	TAOS	55.22
15	NM	GUADALUPE	54.82
16	NM	LOS ALAMOS	54.41
17	NM	SANDOVAL	54.41
18	MO	OSAGE	53.81
19	NM	VALENCIA	53.60
20	MO	COLE	47.36
21	MO	MONITEAU	44.77
22	AZ	MARICOPA	41.34
23	AZ	PINAL	41.34
24	IL	GALLATIN	40.07
25	IL	KNOX	38.23
26	MO	MILLER	33.81
27	AZ	COCHISE	32.72
28	AZ	GILA	32.72
29	AZ	GRAHAM	32.72
30	MO	ST CHARLES	32.02
31	MO	MORGAN	29.85
32	MO	AUDRAIN	29.05
33	KY	CUMBERLAND	28.87
34	MO	CALLAWAY	28.57
35	IL	GREENE	27.68
36	NM	CURRY	27.61
37	NM	DE BACA	27.61
38	MO	CAMDEN	27.54
39	NM	ROOSEVELT	27.34
40	MO	PULASKI	27.12
41	IL	CHRISTIAN	27.10
42	MO	MARIES	26.77
43	MO	COOPER	26.23
44	MO	DOUGLAS	25.78
45	IL	STEPHENSON	25.72
46	IL	MACON	25.60
47	WI	GREEN	25.41
48	IL	WAYNE	25.37
49	IL	HENRY	25.34
50	MO	BOONE	25.12
51	AZ	GREENLEE	25.11
52	IL	WHITESIDE	25.09
53	MO	DALLAS	24.98
54	IL	PEORIA	24.78
55	MO	LINCOLN	24.64
56	IL	WINNEBAGO	24.49
57	IL	MONTGOMERY	23.82
58	IL	LEE	23.70
59	IL	OGLE	23.68
60	MO	PERRY	23.64
61	IL	MORGAN	23.39
62	IL	MCLEAN	23.38
63	IL	PUTNAM	23.36
64	IL	STARK	23.36
65	IL	JACKSON	23.30
66	IL	MARSHALL	23.26
67	MO	HENRY	23.18
68	KS	JOHNSON	23.02
69	MO	JOHNSON	23.00
70	IL	PERRY	22.98
71	MO	POLK	22.48
72	NC	HYDE	22.22
73	MO	GASCONADE	21.58
74	MO	SHELBY	21.02
75	MO	BATES	21.01
76	UT	GRAND	20.80
77	MO	BENTON	20.79
78	CO	DELTA	20.16
79	MO	CRAWFORD	20.16
80	MO	MONROE	20.03

COUNTIES OF HIGHEST AVERAGE FALLOUT FROM SHOT:

PLUMBBOB DIABLO (PB5) Jul 15, 1957 11 Kt

(microCuries per square meter)

Each Row Includes:
County Rank (most fallout = 1), State, County, Fallout Level in microCuries/square meter

Rank	State	County	Fallout
1	NV	NYE2	576.56
2	NV	WHITE PINE2	299.97
3	NV	LINCOLN2	222.21
4	NV	WHITE PINE3	218.27
5	NV	WHITE PINE1	111.52
6	UT	UTAH	89.61
7	UT	JUAB	73.24
8	UT	WASATCH	70.14
9	ND	STARK	48.88
10	ID	TETON	47.59
11	UT	MILLARD	45.80
12	UT	SALT LAKE	44.81
13	UT	SANPETE	44.81
14	UT	TOOELE1	44.81
15	UT	TOOELE2	44.81
16	UT	SUMMIT	43.81
17	WY	UINTA	43.81
18	WY	CARBON	42.92
19	WY	FREMONT	42.92
20	WY	SWEETWATER	42.92
21	ND	DUNN	37.91
22	ND	GOLDEN VALLE	36.51
23	MT	WIBAUX	35.11
24	ND	MOUNTRAIL	35.11
25	ID	CLARK	34.66
26	ID	JEFFERSON	34.66
27	ID	MADISON	34.66
28	UT	DUCHESNE	34.30
29	ND	ADAMS	33.39
30	NV	LINCOLN1	31.66
31	WY	WESTON	29.52
32	SD	BUTTE	28.90
33	ND	BILLINGS	27.39
34	CO	MOFFAT	25.78
35	UT	UINTAH	25.78
36	WY	SUBLETTE	25.78
37	VT	ESSEX	25.41
38	MT	DAWSON	24.74
39	MT	RICHLAND	24.59
40	ND	WILLIAMS	24.59
41	SD	LAWRENCE	23.70
42	SD	HARDING	23.39
43	WY	CROOK	23.14
44	ND	HETTINGER	22.87
45	NE	CHEYENNE	22.73
46	MT	FALLON	22.53
47	ND	BOWMAN	22.31
48	NE	KIMBALL	21.97
49	NY	CORTLAND	21.47
50	NY	STEUBEN	21.47
51	NE	HOOKER	21.00
52	WY	GOSHEN	20.99
53	WY	NIOBRARA	20.53
54	WV	MARION	20.35
55	NE	BUTLER	19.57
56	ND	BARNES	19.43
57	ND	LOGAN	19.43
58	MD	WORCESTER	19.36
59	NY	GENESEE	19.25
60	NY	ALLEGANY	18.94
61	SD	MELLETTE	18.93
62	NE	HITCHCOCK	18.89
63	VA	POWHATAN	18.89
64	CO	LOGAN	18.87
65	NY	TOMPKINS	18.45
66	NE	BANNER	18.34
67	OH	WASHINGTON	18.16
68	NE	GRANT	18.08
69	WY	CAMPBELL	18.02
70	UT	DAVIS	17.92
71	SD	PERKINS	17.82
72	NE	MORRILL	17.64
73	UT	CARBON	17.59
74	UT	MORGAN	17.59
75	UT	DAGGETT	17.37
76	NC	CHOWAN	17.34
77	OH	ATHENS	17.31
78	WV	TYLER	17.31
79	OH	COSHOCTON	17.21
80	OH	NOBLE	17.21

COUNTIES OF HIGHEST AVERAGE FALLOUT FROM SHOTS:

PLUMBBOB KEPLER AND OWENS (PB6)
KEPLER: Jul 24, 1957 10 Kt
OWENS: Jul 25, 1957 9.7 Kt

(microCuries per square meter)

Each Row Includes:
County Rank (most fallout = 1), State, County, Fallout Level in microCuries/square meter

1	NV	ESMERALDA2	328.70
2	NV	ESMERALDA1	256.26
3	NV	MINERAL	184.78
4	NV	CHURCHILL	121.94
5	CA	INYO1	88.07
6	CA	INYO2	88.07
7	NV	LYON	69.69
8	NV	DOUGLAS	55.22
9	MT	BEAVERHEAD	52.07
10	MT	GRANITE	51.14
11	MT	POWELL	47.98
12	KY	MENIFEE	47.69
13	MT	CASCADE	47.67
14	IN	DEARBORN	47.15
15	IN	RIPLEY	46.92
16	CA	MONO	46.65
17	MT	LEWIS AND CL	43.88
18	NV	CARSON CITY	43.57
19	MT	BROADWATER	41.52
20	NV	PERSHING	37.35
21	NV	WASHOE	37.35
22	MT	CHOUTEAU	37.21
23	NV	STOREY	36.99
24	UT	TOOELE1	36.99
25	MT	DEER LODGE	36.57
26	ID	LEMHI	36.31
27	KY	KENTON	36.23
28	ID	LATAH	36.23
29	ID	CUSTER	36.11
30	OR	UNION	35.99
31	ID	VALLEY	35.98
32	OR	BAKER	35.73
33	OR	GRANT	35.58
34	MT	SILVER BOW	35.13
35	OH	CLERMONT	34.91
36	MT	TOOLE	34.88
37	MT	JEFFERSON	34.60
38	MT	LAKE	33.65
39	OH	LAWRENCE	32.30
40	MT	GOLDEN VALLE	32.18
41	MT	MISSOULA	32.06
42	KY	BREATHITT	32.01
43	WV	MASON	31.91
44	OH	GALLIA	31.83
45	KY	WOLFE	31.65
46	IN	FRANKLIN	31.58
47	ID	IDAHO	31.32
48	MT	RAVALLI	30.91
49	KY	POWELL	30.50
50	ID	BONNER	30.46
51	MT	HILL	29.92
52	KY	ESTILL	29.85
53	OH	HIGHLAND	29.29
54	KY	BOYD	29.23
55	WV	CABELL	29.18
56	MT	STILLWATER	28.88
57	IN	OHIO	28.82
58	IN	DECATUR	28.67
59	MT	FERGUS	28.39
60	MT	FLATHEAD	28.26
61	MT	LIBERTY	28.07
62	MT	MINERAL	27.91
63	MT	SWEET GRASS	27.45
64	MT	TETON	27.36
65	MT	GLACIER	27.22
66	MT	JUDITH BASIN	27.10
67	ID	CLEARWATER	26.77
68	MT	SANDERS	26.67
69	NV	LANDER2	26.63
70	MT	PONDERA	26.60
71	ID	SHOSHONE	26.37
72	MT	GALLATIN	26.34
73	MT	PARK	26.13
74	MT	LINCOLN	26.12
75	ID	LEWIS	25.80
76	MT	MEAGHER	25.76
77	ID	CAMAS	25.70
78	OR	WALLOWA	25.68
79	WA	ASOTIN	25.60
80	ID	BUTTE	25.58

COUNTIES OF HIGHEST AVERAGE FALLOUT FROM SHOT:

PLUMBBOB SHASTA (PB8) 18 Aug, 1957 17 Kt

(microCuries per square meter)

Each Row Includes:
County Rank (most fallout = 1), State, County, Fallout Level in microCuries/square meter

1	NV	EUREKA	482.83
2	NV	WHITE PINE2	335.38
3	NV	WHITE PINE3	263.85
4	NV	NYE2	254.89
5	NV	WHITE PINE1	249.74
6	ND	BILLINGS	136.54
7	SD	BROOKINGS	98.45
8	NE	CHERRY	97.60
9	ND	RANSOM	94.67
10	ND	DIVIDE	89.80
11	UT	TOOELE1	84.89
12	SD	KINGSBURY	64.67
13	ND	EMMONS	57.38
14	NV	ELKO	55.23
15	ND	STARK	54.45
16	ND	SLOPE	54.08
17	ND	MCKENZIE	53.87
18	SD	HARDING	53.25
19	MT	CARTER	53.09
20	ND	DICKEY	51.87
21	ND	RICHLAND	51.87
22	NE	PHELPS	51.34
23	NE	HOOKER	50.53
24	MT	FALLON	49.63
25	ND	STUTSMAN	49.56
26	NE	MADISON	49.56
27	SD	FAULK	49.56
28	ND	BURKE	49.29
29	ND	RENVILLE	48.47
30	NE	LINCOLN	48.25
31	MT	JUDITH BASIN	46.42
32	ND	BENSON	46.23
33	ND	FOSTER	46.23
34	NE	SALINE	46.21
35	SD	MCCOOK	46.20
36	WY	PARK	45.86
37	NE	NUCKOLLS	44.85
38	KS	REPUBLIC	44.58
39	NE	PLATTE	44.48
40	MO	PETTIS	44.34
41	MO	LAFAYETTE	44.01
42	IA	DECATUR	43.91
43	MO	HARRISON	43.84
44	MO	DE KALB	43.68
45	MO	GENTRY	43.68
46	MO	GRUNDY	43.68
47	MO	CALDWELL	43.68
48	MO	LIVINGSTON	43.68
49	MO	HENRY	43.66
50	SD	MINNEHAHA	43.64
51	MO	WORTH	43.64
52	IA	TAYLOR	43.53
53	MO	DAVIESS	43.52
54	MO	JOHNSON	43.52
55	MO	MERCER	43.52
56	MO	SALINE	43.52
57	MO	POLK	43.50
58	IA	RINGGOLD	43.37
59	SD	DAVISON	43.27
60	SD	HANSON	43.27
61	SD	LAKE	43.27
62	SD	MINER	43.27
63	SD	PENNINGTON	42.39
64	UT	TOOELE2	42.39
65	SD	SANBORN	41.81
66	SD	BENNETT	40.63
67	SD	JACKSON	40.63
68	UT	BOX ELDER1	40.59
69	UT	BOX ELDER2	40.59
70	UT	UTAH	40.59
71	SD	TODD	39.41
72	SD	HAAKON	39.03
73	NE	BLAINE	37.50
74	SD	STANLEY	37.31
75	SD	MELLETTE	36.49
76	ND	DUNN	35.98
77	SD	LYMAN	35.85
78	SD	AURORA	35.71
79	SD	CORSON	35.63
80	SD	TRIPP	35.02

COUNTIES OF HIGHEST AVERAGE FALLOUT FROM SHOT:

PLUMBBOB DOPPLER (PB9) Aug 23, 1957 11 Kt

(microCuries per square meter)

Each Row Includes:
County Rank (most fallout = 1), State, County, Fallout Level in microCuries/square meter

1	IA	WAPELLO	130.84
2	IA	DAVIS	121.23
3	IA	TAMA	111.83
4	IA	JASPER	110.55
5	IA	POTTAWATTAMI	107.95
6	IA	CASS	107.70
7	IA	FRANKLIN	107.10
8	IA	SHELBY	106.60
9	IA	MONTGOMERY	106.36
10	IA	MADISON	106.12
11	IA	AUDUBON	105.90
12	IA	BOONE	105.85
13	IA	POLK	105.84
14	IA	DALLAS	105.41
15	IA	ADAIR	105.28
16	IA	CARROLL	105.26
17	UT	UTAH	103.16
18	IA	BLACK HAWK	99.05
19	IA	BENTON	98.80
20	IA	BREMER	97.95
21	IA	POWESHIEK	91.49
22	IA	LINN	82.36
23	IA	HENRY	81.99
24	IA	FAYETTE	81.01
25	IA	JACKSON	81.01
26	IA	DELAWARE	80.77
27	IA	BUCHANAN	80.64
28	IA	KEOKUK	80.64
29	NV	WHITE PINE2	79.88
30	MO	CLARK	78.57
31	IA	VAN BUREN	76.74
32	UT	WASATCH	74.47
33	IA	MARION	70.46
34	IA	MARSHALL	67.10
35	IA	STORY	67.10
36	IA	HAMILTON	65.76
37	IA	GREENE	65.41
38	IA	GUTHRIE	65.17
39	IA	HARDIN	65.17
40	IA	ADAMS	64.92
41	IA	WARREN	64.92
42	IA	CLARKE	64.79
43	IA	UNION	64.79
44	IA	LUCAS	64.09
45	IA	CEDAR	62.01
46	IA	IOWA	61.76
47	IA	CLAYTON	60.67
48	WI	DANE	60.46
49	IA	DUBUQUE	60.42
50	IL	OGLE	60.42
51	IL	DE KALB	60.28
52	IL	LOGAN	60.28
53	IA	DES MOINES	60.17
54	WI	GRANT	60.17
55	IL	MCDONOUGH	60.16
56	IL	JOHNSON	60.05
57	IL	WARREN	60.05
58	WI	SAUK	59.97
59	IL	HANCOCK	59.47
60	IL	BROWN	59.40
61	IL	SCOTT	59.40
62	IL	MORGAN	58.70
63	IL	PIKE	54.80
64	UT	JUAB	54.07
65	UT	TOOELE2	53.43
66	UT	DUCHESNE	52.28
67	UT	UINTAH	51.90
68	UT	DAVIS	49.94
69	IA	MAHASKA	48.77
70	UT	SALT LAKE	47.10
71	IA	GRUNDY	45.92
72	CO	PHILLIPS	44.49
73	IA	DECATUR	44.34
74	IA	TAYLOR	44.20
75	MO	NODAWAY	44.20
76	CO	SEDGWICK	43.50
77	IA	RINGGOLD	43.50
78	MO	WORTH	43.50
79	UT	BOX ELDER2	43.10
80	UT	SANPETE	43.10

COUNTIES OF HIGHEST AVERAGE FALLOUT FROM SHOT:

PLUMBBOB SMOKY (PB11) Aug 31, 1957 11 Kt

(microCuries per square meter)

Each Row Includes:
County Rank (most fallout = 1), State, County, Fallout Level in microCuries/square meter

1	UT	GARFIELD	1238.11
2	CA	INYO3	898.04
3	UT	WASHINGTON2	641.04
4	UT	WASHINGTON3	500.51
5	NV	LINCOLN1	393.34
6	WY	SWEETWATER	384.53
7	UT	WASHINGTON1	318.50
8	UT	UINTAH	235.72
9	UT	DAGGETT	232.42
10	AZ	MOHAVE2	230.98
11	WY	FREMONT	228.21
12	UT	IRON3	162.62
13	WY	CARBON	151.11
14	NV	CLARK1	109.14
15	UT	WAYNE	85.40
16	UT	EMERY	79.41
17	WY	SUBLETTE	76.10
18	WY	LINCOLN	60.68
19	ND	GRIGGS	60.47
20	ND	RANSOM	55.29
21	ND	RICHLAND	55.29
22	ND	LA MOURE	53.12
23	ND	BARNES	52.16
24	ND	PEMBINA	52.16
25	ND	CASS	51.07
26	ND	EDDY	51.07
27	ND	GRAND FORKS	51.07
28	ND	NELSON	51.07
29	ND	RAMSEY	51.07
30	ND	TRAILL	51.07
31	ND	WALSH	51.07
32	ND	DICKEY	49.99
33	ND	STEELE	49.99
34	ND	SARGENT	49.44
35	SD	MARSHALL	49.44
36	UT	IRON1	45.43
37	CA	SAN BERNADIN	44.25
38	UT	CARBON	41.90
39	UT	PIUTE	40.06
40	UT	DUCHESNE	38.85
41	WY	UINTA	38.15
42	ND	FOSTER	34.81
43	SD	ROBERTS	34.81
44	ND	CAVALIER	34.33
45	ND	MCINTOSH	34.27
46	ND	TOWNER	33.85
47	ND	BENSON	32.76
48	ND	ROLETTE	32.76
49	SD	EDMUNDS	32.70
50	MT	WIBAUX	32.62
51	ND	GOLDEN VALLE	31.93
52	ND	STUTSMAN	31.68
53	ND	STARK	31.65
54	ND	HETTINGER	31.38
55	ND	DUNN	31.14
56	ND	EMMONS	31.14
57	ND	KIDDER	31.14
58	ND	LOGAN	31.14
59	NE	ANTELOPE	31.14
60	NE	BOONE	31.14
61	NE	MADISON	31.14
62	SD	AURORA	31.14
63	SD	BROWN	31.14
64	SD	BRULE	31.14
65	SD	DAY	31.14
66	SD	FAULK	31.14
67	SD	MCPHERSON	31.14
68	SD	POTTER	31.14
69	SD	SPINK	31.14
70	SD	WALWORTH	31.14
71	UT	IRON2	31.07
72	UT	KANE2	30.32
73	ND	ADAMS	25.12
74	ND	BOWMAN	25.12
75	ND	SLOPE	25.12
76	NE	CEDAR	25.12
77	NE	PIERCE	25.12
78	SD	HARDING	25.12
79	SC	FLORENCE	24.42
80	UT	GRAND	23.83

COUNTIES OF HIGHEST AVERAGE FALLOUT FROM SHOT:

PLUMBBOB GALILEO (PB12) Sep 2, 1957 11 Kt

(microCuries per square meter)

Each Row Includes:
County Rank (most fallout = 1), State, County, Fallout Level in microCuries/square meter

1	NV	NYE2	100.99
2	LA	LAFOURCHE	58.17
3	LA	ST CHARLES	55.11
4	NV	WHITE PINE1	52.61
5	NV	WHITE PINE2	52.61
6	NV	WHITE PINE3	52.61
7	UT	UTAH	46.17
8	UT	MILLARD	45.16
9	UT	MORGAN	36.92
10	UT	TOOELE1	36.17
11	UT	WASATCH	36.13
12	KS	SEDGWICK	35.54
13	UT	DAVIS	34.59
14	UT	TOOELE2	34.59
15	KS	SUMNER	34.04
16	UT	SUMMIT	33.85
17	UT	SALT LAKE	32.62
18	CO	EL PASO	31.83
19	NV	EUREKA	31.50
20	KS	MCPHERSON	29.10
21	UT	WEBER	27.66
22	UT	JUAB	27.08
23	KS	KINGMAN	26.96
24	UT	DAGGETT	26.53
25	UT	DUCHESNE	26.53
26	WY	SWEETWATER	26.53
27	WY	UINTA	26.53
28	KS	MORRIS	26.30
29	CO	DENVER	26.12
30	UT	BOX ELDER2	25.91
31	KS	BUTLER	25.85
32	KS	SHAWNEE	25.65
33	MO	VERNON	25.62
34	KS	DICKINSON	25.46
35	KS	LYON	25.27
36	KS	RENO	25.18
37	KS	SALINE	25.16
38	KS	WILSON	25.01
39	KS	COWLEY	24.68
40	KS	MARION	24.58
41	KS	HARVEY	24.26
42	KS	CRAWFORD	24.21
43	OK	ROGERS	24.20
44	KS	LABETTE	24.02
45	KS	GREENWOOD	23.94
46	MO	BARRY	23.62
47	MO	CEDAR	23.36
48	CO	ADAMS	23.25
49	CO	DOUGLAS	23.25
50	MS	COVINGTON	22.95
51	KS	COFFEY	22.78
52	CO	JEFFERSON	22.61
53	KS	JEFFERSON	22.46
54	KS	CHASE	22.44
55	KS	PRATT	22.03
56	KS	LEAVENWORTH	21.94
57	KS	STAFFORD	21.84
58	CO	CUSTER	21.81
59	OK	WAGONER	21.69
60	KS	LINCOLN	21.59
61	MO	CHRISTIAN	21.36
62	KS	OSAGE	21.32
63	KS	HARPER	21.06
64	KS	OTTAWA	21.05
65	KS	ANDERSON	21.00
66	CO	ARAPAHOE	20.94
67	KS	PAWNEE	20.61
68	CO	BOULDER	20.48
69	AR	CARROLL	20.26
70	KS	ATCHISON	20.26
71	KS	JOHNSON	20.12
72	MO	LAWRENCE	20.05
73	AR	BENTON	20.02
74	KS	EDWARDS	19.96
75	CO	PUEBLO	19.88
76	MO	GREENE	19.74
77	CO	LARIMER	19.55
78	OK	TULSA	19.52
79	KS	LINN	19.49
80	KS	RUSH	19.49

COUNTIES OF HIGHEST AVERAGE FALLOUT FROM SHOTS PLUMBBOB WHEELER, COULOMB B AND LaPLACE (PB13)

WHEELER: Sep 6, 1957, 0.197 Kt
COULOMB-B: Sep 6, 1957, 0.3 Kt
LaPLACE: Sep 8, 1957, 1 Kt
(microCuries per square meter)
Each Row Includes:
County Rank (most fallout = 1), State, County, Fallout Level in microCuries/square meter

Rank	State	County	Fallout
1	NE	BUFFALO	46.40
2	NE	HALL	45.83
3	NE	KEARNEY	44.53
4	KS	MITCHELL	43.62
5	KS	MCPHERSON	40.15
6	KS	PAWNEE	39.66
7	KS	BARTON	38.16
8	KS	OTTAWA	37.65
9	KS	SALINE	37.49
10	KS	RUSSELL	37.38
11	KS	CLOUD	37.29
12	KS	STAFFORD	36.46
13	KS	LINCOLN	36.23
14	KS	RICE	35.44
15	KS	WABAUNSEE	34.88
16	KS	FORD	34.77
17	NE	PHELPS	34.52
18	KS	ELLSWORTH	34.48
19	KS	DICKINSON	34.29
20	KS	CLAY	33.76
21	KS	GEARY	33.24
22	KS	EDWARDS	32.61
23	OK	GRANT	32.57
24	KS	CLARK	32.18
25	KS	RILEY	32.11
26	KS	RUSH	31.86
27	KS	NESS	31.68
28	NE	NANCE	31.28
29	NE	MERRICK	30.81
30	KS	MARION	30.34
31	AZ	COCONINO2	30.09
32	AZ	COCONINO3	30.09
33	KS	POTTAWATOMIE	29.91
34	KS	OSBORNE	29.63
35	KS	TREGO	29.60
36	KS	ELLIS	29.39
37	KS	MEADE	28.72
38	KS	RENO	28.49
39	KS	HARVEY	28.45
40	KS	LANE	28.34
41	KS	SHAWNEE	28.33
42	KS	MARSHALL	28.28
43	KS	WASHINGTON	27.93
44	KS	LABETTE	27.79
45	KS	SCOTT	27.74
46	OK	HARPER	27.51
47	NE	HARLAN	27.44
48	KS	SUMNER	27.38
49	NE	POLK	26.99
50	KS	COWLEY	26.67
51	KS	JACKSON	26.25
52	KS	JEFFERSON	26.12
53	OK	WASHINGTON	26.05
54	NE	JOHNSON	25.93
55	KS	HASKELL	25.88
56	OK	TULSA	25.79
57	KS	GRANT	25.78
58	OK	OSAGE	25.62
59	KS	BUTLER	25.49
60	OK	CREEK	25.18
61	NE	PAWNEE	24.89
62	NE	GAGE	24.87
63	KS	CHASE	24.81
64	OK	NOBLE	24.72
65	KS	GREENWOOD	24.56
66	OK	ALFALFA	24.54
67	OK	KAY	24.21
68	KS	ELK	24.01
69	OK	ROGERS	23.81
70	OK	PAWNEE	23.78
71	KS	CHAUTAUQUA	23.71
72	OK	PAYNE	23.69
73	OK	MAJOR	23.48
74	NE	FRANKLIN	23.25
75	OK	WOODS	23.12
76	KS	MORRIS	23.05
77	OK	MAYES	22.82
78	OK	GARFIELD	22.79
79	KS	LYON	22.79
80	OK	LINCOLN	22.72

COUNTIES OF HIGHEST AVERAGE FALLOUT FROM SHOT:

PLUMBBOB FIZEAU (PB14) Sep 14, 1957 11 Kt

(microCuries per square meter)

Each Row Includes:
County Rank (most fallout = 1), State, County, Fallout Level in microCuries/square meter

Rank	State	County	Fallout
1	NV	MINERAL	57.61
2	NV	CARSON CITY	39.80
3	NV	DOUGLAS	39.80
4	NV	LYON	39.80
5	NV	STOREY	39.80
6	NV	WASHOE	39.80
7	CA	MONO	23.62
8	NV	CHURCHILL	15.92
9	NV	NYE2	12.09
10	NV	NYE1	7.89
11	KS	BUTLER	6.01
12	KS	CHASE	6.01
13	KS	FRANKLIN	6.01
14	KS	LYON	6.01
15	KS	MORRIS	6.01
16	KS	OSAGE	6.01
17	KS	WOODSON	6.01
18	KS	COFFEY	5.68
19	KS	DOUGLAS	5.68
20	KS	GREENWOOD	5.68
21	KS	JOHNSON	5.68
22	KS	ANDERSON	5.53
23	KS	LINN	5.53
24	KS	MIAMI	5.53
25	KS	COWLEY	5.34
26	OK	OKLAHOMA	5.34
27	KS	ALLEN	5.30
28	KS	BOURBON	5.30
29	KS	ELK	5.30
30	KS	WILSON	5.30
31	MO	VERNON	5.30
32	OK	CADDO	5.30
33	OK	CANADIAN	5.30
34	OK	CHEROKEE	5.30
35	OK	GRANT	5.30
36	OK	LINCOLN	5.30
37	OK	MUSKOGEE	5.30
38	OK	OKFUSKEE	5.30
39	OK	WASHITA	5.30
40	KS	HARVEY	3.93
41	KS	JEFFERSON	3.93
42	KS	LEAVENWORTH	3.93
43	KS	NEOSHO	3.93
44	KS	WABAUNSEE	3.93
45	KS	HARPER	3.60
46	KS	JACKSON	3.60
47	KS	MARION	3.60
48	KS	MCPHERSON	3.60
49	KS	SUMNER	3.60
50	MO	BATES	3.60
51	OK	CUSTER	3.60
52	CA	LASSEN	3.55
53	CA	MODOC	3.55
54	CA	SISKIYOU	3.55
55	CA	TEHAMA	3.55
56	OR	CROOK	3.55
57	OR	LAKE	3.55
58	KS	CRAWFORD	3.45
59	OK	KAY	3.45
60	OK	TULSA	3.45
61	OK	ALFALFA	3.26
62	KS	CHEROKEE	3.22
63	MO	BARTON	3.22
64	MO	JASPER	3.22
65	NE	SEWARD	3.22
66	OK	BECKHAM	3.22
67	OK	BLAINE	3.22
68	OK	CRAIG	3.22
69	OK	CREEK	3.22
70	OK	DELAWARE	3.22
71	OK	KINGFISHER	3.22
72	OK	MAYES	3.22
73	OK	NOWATA	3.22
74	OK	OKMULGEE	3.22
75	OK	OTTAWA	3.22
76	OK	PAYNE	3.22
77	OK	WAGONER	3.22
78	KS	KINGMAN	2.89
79	KS	NEMAHA	2.89
80	CA	SHASTA	2.66

COUNTIES OF HIGHEST AVERAGE FALLOUT FROM SHOT:

PLUMBBOB NEWTON (PB15) Sep 14, 1957 12 Kt

(microCuries per square meter)

Each Row Includes:
County Rank (most fallout = 1), State, County, Fallout Level in microCuries/square meter

1	NE	GARFIELD	36.53
2	SD	CODINGTON	35.71
3	SD	KINGSBURY	25.12
4	SD	HAMLIN	25.06
5	SD	MINER	24.82
6	SD	CLARK	23.94
7	SD	ROBERTS	23.76
8	ND	LA MOURE	23.62
9	ND	RICHLAND	23.62
10	SD	BEADLE	23.51
11	SD	MINNEHAHA	23.51
12	ND	BARNES	23.50
13	ND	TRAILL	23.50
14	ND	LOGAN	23.42
15	SD	WALWORTH	23.40
16	NE	LOGAN	23.17
17	SD	HUTCHINSON	23.17
18	SD	LAKE	23.17
19	ND	CASS	22.96
20	ND	DICKEY	22.96
21	ND	KIDDER	22.96
22	SD	MCPHERSON	22.84
23	ND	MORTON	22.75
24	ND	STEELE	22.75
25	ND	STUTSMAN	22.75
26	SD	DAY	22.75
27	SD	SPINK	22.70
28	ND	SARGENT	22.69
29	ND	MERCER	22.65
30	SD	AURORA	22.65
31	SD	FAULK	22.65
32	SD	EDMUNDS	22.62
33	ND	GRIGGS	22.41
34	ND	EMMONS	22.31
35	SD	HYDE	22.19
36	SD	MARSHALL	22.12
37	ND	BURLEIGH	21.97
38	ND	GRAND FORKS	21.97
39	ND	SHERIDAN	21.97
40	SD	DEWEY	21.97
41	SD	HAND	21.97
42	SD	JACKSON	21.97
43	SD	POTTER	21.97
44	SD	HAAKON	21.87
45	UT	JUAB	21.81
46	ND	MCLEAN	21.71
47	SD	ZIEBACH	21.67
48	ND	NELSON	21.63
49	SD	CORSON	21.57
50	ND	EDDY	21.54
51	ND	GRANT	21.52
52	ND	BENSON	21.48
53	SD	STANLEY	21.47
54	ND	OLIVER	21.04
55	SD	CAMPBELL	21.04
56	SD	SULLY	21.04
57	ND	BOWMAN	20.54
58	ND	GOLDEN VALLE	20.54
59	ND	HETTINGER	20.54
60	ND	STARK	20.54
61	SD	HARDING	20.54
62	ND	ADAMS	20.37
63	ND	BILLINGS	20.37
64	ND	DUNN	20.37
65	ND	MCKENZIE	20.37
66	SD	MEADE	20.37
67	SD	PERKINS	20.37
68	SD	GREGORY	17.63
69	SD	CHARLES MIX	16.89
70	SD	BRULE	16.83
71	SD	JERAULD	16.83
72	SD	SANBORN	16.83
73	SD	TRIPP	16.83
74	SD	BUFFALO	16.78
75	NE	BOYD	16.49
76	SD	MCCOOK	16.49
77	ND	FOSTER	16.44
78	ND	MCINTOSH	16.07
79	NE	CHERRY	16.02
80	SD	LYMAN	15.97

COUNTIES OF HIGHEST AVERAGE FALLOUT FROM SHOT:

PLUMBBOB WHITNEY (PB16) Sep 23, 1957 19 Kt

(microCuries per square meter)

Each Row Includes:
County Rank (most fallout = 1), State, County, Fallout Level in microCuries/square meter

1	NV	ESMERALDA2	274.35
2	NV	ESMERALDA1	107.32
3	NV	CHURCHILL	91.75
4	NV	MINERAL	86.53
5	NV	LANDER2	78.71
6	NV	PERSHING	54.49
7	CA	MONO	48.97
8	NV	CARSON CITY	45.88
9	NV	DOUGLAS	45.88
10	NV	LYON	45.88
11	NV	STOREY	45.88
12	NV	WASHOE	45.88
13	NV	NYE1	39.08
14	NV	LANDER1	36.74
15	GA	FAYETTE	29.86
16	CA	INYO1	29.60
17	CA	INYO2	29.60
18	NV	HUMBOLDT	28.46
19	GA	CLAYTON	27.52
20	GA	FULTON	26.65
21	GA	DE KALB	26.61
22	GA	MONROE	24.93
23	GA	DOUGLAS	24.59
24	GA	GWINNETT	23.14
25	GA	CRAWFORD	23.13
26	GA	BIBB	22.06
27	GA	SCHLEY	22.01
28	GA	JONES	21.58
29	GA	DOOLY	21.57
30	GA	SPALDING	21.39
31	GA	PEACH	21.31
32	GA	PULASKI	21.24
33	GA	COLUMBUS	20.72
34	GA	COBB	20.31
35	GA	SUMTER	20.10
36	GA	PIKE	19.81
37	GA	HOUSTON	19.71
38	GA	LAMAR	19.70
39	GA	BUTTS	19.69
40	GA	TWIGGS	19.50
41	GA	CRISP	19.43
42	GA	MACON	19.14
43	GA	PAULDING	19.12
44	NV	ELKO	18.94
45	GA	HARRIS	18.76
46	GA	NEWTON	18.66
47	GA	HENRY	18.59
48	GA	WEBSTER	18.58
49	GA	CARROLL	18.47
50	GA	UPSON	18.24
51	GA	COWETA	18.21
52	GA	DODGE	18.19
53	GA	LEE	17.68
54	GA	MERIWETHER	17.52
55	GA	ROCKDALE	17.38
56	OR	HARNEY	17.37
57	OR	MALHEUR	17.37
58	GA	JASPER	17.27
59	GA	BLECKLEY	17.22
60	GA	TERRELL	16.88
61	ID	OWYHEE	16.70
62	WA	SPOKANE	16.50
63	WA	WALLA WALLA	16.41
64	GA	TALBOT	16.09
65	GA	WALTON	15.92
66	GA	TAYLOR	15.84
67	WA	COLUMBIA	15.81
68	ID	CANYON	15.64
69	GA	DOUGHERTY	15.56
70	AL	CLAY	15.32
71	GA	CHATTAHOOCHE	15.05
72	GA	EMANUEL	14.86
73	GA	RANDOLPH	14.86
74	GA	MORGAN	14.68
75	GA	PUTNAM	14.53
76	GA	HARALSON	14.51
77	GA	WILCOX	14.38
78	GA	BERRIEN	14.21
79	GA	TROUP	14.11
80	NV	NYE2	14.06

COUNTIES OF HIGHEST AVERAGE FALLOUT FROM SHOT:

PLUMBBOB CHARLESTON (PB17) Sep 28, 1957 12 Kt

(microCuries per square meter)

Each Row Includes:
County Rank (most fallout = 1), State, County, Fallout Level in microCuries/square meter

1	SC	OCONEE	31.64
2	NC	CLAY	31.09
3	TN	COFFEE	31.06
4	UT	SALT LAKE	24.38
5	TN	GRUNDY	24.15
6	UT	BOX ELDER1	23.90
7	UT	BOX ELDER2	23.90
8	UT	CACHE	23.90
9	UT	DAVIS	23.90
10	UT	MORGAN	23.90
11	UT	RICH	23.90
12	UT	TOOELE1	23.90
13	UT	TOOELE2	23.90
14	UT	WEBER	23.90
15	GA	TOWNS	22.94
16	WY	SWEETWATER	22.80
17	WY	UINTA	22.80
18	NC	MACON	22.76
19	GA	RABUN	22.60
20	TN	SEQUATCHIE	22.42
21	TN	BEDFORD	22.22
22	TN	WARREN	22.08
23	TN	HAMILTON	21.85
24	NC	CHEROKEE	21.82
25	TN	CUMBERLAND	21.72
26	TN	ROANE	21.67
27	NC	SWAIN	21.59
28	TN	BLOUNT	21.53
29	TN	MARSHALL	21.39
30	NC	HAYWOOD	21.24
31	TN	BLEDSOE	21.23
32	NC	TRANSYLVANIA	21.21
33	NC	JACKSON	21.20
34	TN	MCMINN	21.02
35	TN	RHEA	20.93
36	TN	POLK	20.89
37	TN	BRADLEY	20.88
38	TN	VAN BUREN	20.85
39	TN	SEVIER	20.83
40	NV	ELKO	20.12
41	ID	FREMONT	20.12
42	WY	LINCOLN	19.02
43	GA	CRAWFORD	18.99
44	GA	PEACH	18.32
45	KY	FLOYD	18.21
46	GA	TALIAFERRO	18.13
47	GA	CLARKE	18.00
48	GA	WARREN	17.53
49	ID	BANNOCK	17.32
50	ID	BEAR LAKE	17.32
51	ID	BINGHAM	17.32
52	ID	BONNEVILLE	17.32
53	ID	CARIBOU	17.32
54	ID	CASSIA	17.32
55	ID	ELMORE	17.32
56	ID	FRANKLIN	17.32
57	ID	GOODING	17.32
58	ID	JEROME	17.32
59	ID	LINCOLN	17.32
60	ID	MINIDOKA	17.32
61	ID	ONEIDA	17.32
62	ID	OWYHEE	17.32
63	ID	POWER	17.32
64	ID	TWIN FALLS	17.32
65	GA	MCDUFFIE	17.23
66	GA	CLAY	17.18
67	GA	OGLETHORPE	16.79
68	GA	QUITMAN	16.73
69	MS	CHICKASAW	16.00
70	UT	SUMMIT	15.85
71	UT	UTAH	15.85
72	UT	WASATCH	15.85
73	ID	BLAINE	15.71
74	GA	DOUGLAS	15.58
75	GA	FAYETTE	15.55
76	GA	PAULDING	15.52
77	GA	GLASCOCK	15.39
78	WY	CARBON	15.24
79	WY	FREMONT	15.24
80	WY	SUBLETTE	15.24

COUNTIES OF HIGHEST AVERAGE FALLOUT FROM SHOT:

PLUMBBOB MORGAN (PB18) Oct 7m 1957 8 Kt

(microCuries per square meter)

Each Row Includes:
County Rank (most fallout = 1), State, County, Fallout Level in microCuries/square meter

Rank	State	County	Fallout
1	UT	IRON1	21.54
2	UT	WASHINGTON1	20.12
3	CO	EL PASO	19.98
4	CO	CHEYENNE	16.28
5	UT	IRON2	15.50
6	UT	IRON3	15.50
7	CO	KIT CARSON	15.37
8	UT	WASHINGTON3	15.36
9	CO	YUMA	15.10
10	UT	BEAVER	14.62
11	CO	BOULDER	13.25
12	CO	WASHINGTON	13.25
13	CO	BACA	13.10
14	CO	LINCOLN	12.87
15	KS	FORD	12.79
16	KS	GRAY	12.79
17	KS	MEADE	12.79
18	MO	SHELBY	12.71
19	MO	MARION	12.41
20	MO	MONROE	12.37
21	NE	KIMBALL	12.23
22	CO	LOGAN	12.17
23	CO	FREMONT	12.14
24	MO	ADAIR	12.11
25	MO	CARROLL	12.07
26	NE	GARDEN	12.07
27	CO	PUEBLO	11.95
28	CO	PHILLIPS	11.59
29	NE	CHEYENNE	11.51
30	CO	CLEAR CREEK	11.27
31	CO	PARK	11.27
32	CO	TELLER	11.27
33	KS	WALLACE	11.21
34	IL	HANCOCK	11.13
35	NE	KEITH	11.11
36	CO	ARAPAHOE	11.09
37	KS	CHEYENNE	11.09
38	KS	SHERMAN	11.09
39	CO	GILPIN	10.96
40	CO	JEFFERSON	10.94
41	NE	FRONTIER	10.81
42	NE	LINCOLN	10.72
43	KS	HASKELL	10.49
44	KS	SEWARD	10.49
45	CO	CUSTER	10.48
46	CO	HUERFANO	10.48
47	CO	SEDGWICK	10.48
48	KS	STEVENS	10.37
49	CO	ADAMS	10.32
50	NE	CUSTER	10.32
51	KS	ELLIS	10.30
52	KS	ROOKS	10.30
53	UT	GARFIELD	10.18
54	UT	PIUTE	10.18
55	UT	WAYNE	10.18
56	KS	MORTON	10.17
57	IA	CEDAR	10.11
58	CO	LA PLATA	10.06
59	CO	MONTEZUMA	10.06
60	AZ	APACHE	9.94
61	AZ	NAVAJO	9.94
62	UT	SAN JUAN	9.94
63	NE	DUNDY	9.91
64	IA	LOUISA	9.84
65	IA	VAN BUREN	9.84
66	NM	MCKINLEY	9.82
67	NM	SAN JUAN	9.82
68	IA	LEE	9.78
69	MO	CLARK	9.78
70	KS	LOGAN	9.75
71	KS	THOMAS	9.75
72	IA	DES MOINES	9.52
73	UT	KANE2	9.49
74	KS	GRAHAM	9.44
75	KS	NESS	9.44
76	KS	NORTON	9.44
77	KS	TREGO	9.44
78	NE	HAYES	9.43
79	AR	MILLER	9.35
80	KS	WICHITA	9.33

TOTAL FALLOUT FROM THE HARDTACK II NUCLEAR TEST SERIES SEP-OCT 1958

Estimated Average Values in MicroCuries/Square Meter

NUCLEAR TEST	DATE	Yield	COUNTIES RECEIVING FALLOUT	Micro-Curies/Sq Meter
EDDY (HT1)	Sep 19 1958	83 tons	NV Lincoln 2	0.58
			NV Nye 2	0.27
			Lincoln	0.26
HIDALGO (HT2)	Oct 5, 1958	77 tons	NV Lincoln 2	2.09
			NV Lincoln 1	0.37
QUAY (HT3)	Oct 10, 1958	79 tons	NV Nye 3	7.03
LEA (HT4)	Oct 13, 1958	1.4 Kt	NV Nye 2	6.35
VESTA (HT5)	Oct 17, 1958	24 tons	NV Nye 2	0.08
RIO ARRIBA (HT6)	Oct 18, 1958	90 tons	NV Nye 2	5.12
			NV Lincoln 2	1.45
SOCORRO (HT7)	Oct 22, 1958	6 Kt	NV Clark 2	1.39
WRANGELL (HT8)	Oct 22, 1958	115 tons	NV Lincoln 2	0.12
			NV Nye 2	0.12
SANFORD (HT9)	Oct 26, 1958	4.9 Kt	NV Esmeralda 2	6.7
			NV Clark 2	1.41

NUCLEAR FALLOUT FROM MINOR TESTS IN THE UNDERGROUND NUCLEAR TEST SERIES Estimated Average Values in MicroCuries/Square Meter

NUCLEAR TEST	DATE	Yield	COUNTIES RECEIVING FALLOUT	Total Fallout (TF) or I-131 values	Micro-Curies/Sq Meter
ANTLER (UE-1)	Sep 15, 1961	2.4 Kt	NV Nye 2	I-131	2.51
PLATTE (UE-3)	Apr 14, 1962	1.7 Kt	NV Nye 2	I-131	55.9
EEL (UE-4)	May 19, 1962	4.9 Kt	NV Nye 2	I-131	6.63
DES MOINES (UE-5)	Jun 13, 1962	2.9 Kt	NV Nye 2	TF	104.73
BANDICOOT (UE-9)	Oct 19, 1962	12.5 Kt	NV Nye 3	I-131	46.78
			NV Lincoln 2	I-131	3.28
			NV Clark 2	I-131	2.98
			NV Clark 1	I-131	1.39
			NV Nye 2	I-131	0.46
PIKE (UE-10)	Mar 13, 1964	<20Kt	NV Clark 2	I-131	0.99
SULKY (UE-11)	Dec 18, 1964	0.092 Kt	NV Nye 2	TF	0.17
CABRIOLET (UE-14)	Jan 26, 1968	2.3 Kt	NV Nye 2	TF	3.66
BUGGY (UE-15)	Mar 12, 1968	5.4 Kt	NV Nye 2	TF	0.99

COUNTIES OF HIGHEST AVERAGE FALLOUT FROM SHOT:

DANNY BOY (UE-2) Mar 5, 1962 0.42 Kt

(microCuries per square meter)
Each Row Includes:
County Rank (most fallout = 1), State, County, Fallout Level in microCuries/square meter

1	ID TETON	5.65
2	WY NATRONA	5.12
3	ID MADISON	4.66
4	WY PARK	3.23
5	ID FREMONT	2.89
6	WY TETON	2.82
7	ID JEFFERSON	2.73
8	ID BUTTE	2.59
9	SD DAY	2.58
10	ID CLARK	2.07
11	WY GOSHEN	1.84
12	WY NIOBRARA	1.83
13	ND GRIGGS	1.52
14	ND STEELE	1.52
15	MN DOUGLAS	1.50
16	ND TRAILL	1.49
17	MN HUBBARD	1.47
18	SD CODINGTON	1.40
19	NE DAWES	1.40
20	SD BROWN	1.39
21	MN ROSEAU	1.37
22	MN WADENA	1.37
23	WY LARAMIE	1.31
24	ND DICKEY	1.29
25	SD MARSHALL	1.29
26	SD ROBERTS	1.28
27	MN CROW WING	1.20
28	MN BECKER	1.19
29	MN KITTSON	1.18
30	MN MARSHALL	1.18
31	MN TRAVERSE	1.15
32	NE SCOTTS BLUFF	1.15
33	MN BIG STONE	1.08
34	MN CLAY	1.08
35	MN NORMAN	1.08
36	MN POLK	1.08
37	CO GRAND	1.08
38	CO ROUTT	1.08
39	MN GRANT	1.08
40	MN OTTER TAIL	1.08
41	NE BOX BUTTE	1.07
42	MN KOOCHICHING	1.06
43	MN ST LOUIS	1.04
44	MN CASS	1.03
45	MN PENNINGTON	1.03
46	MN BELTRAMI	0.93
47	MN CLEARWATER	0.93
48	MN MAHNOMEN	0.93
49	MN RED LAKE	0.93
50	SD SPINK	0.92
51	MT BIG HORN	0.90
52	SD BEADLE	0.80
53	WY CONVERSE	0.77
54	NE BANNER	0.71
55	CO PITKIN	0.69
56	NE MORRILL	0.69
57	SD JERAULD	0.69
58	MT CUSTER	0.68
59	MT POWDER RIVER	0.68
60	MT ROSEBUD	0.68
61	SD PERKINS	0.66
62	WY BIG HORN	0.65
63	MN ITASCA	0.65
64	MN AITKIN	0.62
65	MN LAKE OF THE	0.62
66	MN MORRISON	0.61
67	MT TREASURE	0.58
68	WY ALBANY	0.55
69	NV NYE2	0.55
70	CO SUMMIT	0.55
71	WY CROOK	0.53
72	WY SHERIDAN	0.52
73	MT CARBON	0.48
74	NE KIMBALL	0.47

COUNTIES OF HIGHEST AVERAGE FALLOUT FROM SHOT:

STORAX SEDAN Jul 6, 1962 110 Kt

(microCuries per square meter)

Each Row Includes:
County Rank (most fallout = 1), State, County, Fallout Level in microCuries/square meter

1	**IA**	**HOWARD**	**950.00**
2	IA	MITCHELL	950.00
3	IA	WORTH	950.00
4	SD	WASHABAUGH	860.52
5	IA	WOODBURY	665.68
6	IA	JEFFERSON	621.15
7	IA	LEE	621.15
8	IA	VAN BUREN	621.15
9	IA	WASHINGTON	621.15
10	IL	HANCOCK	621.15
11	MN	FARIBAULT	584.61
12	MN	MOWER	584.61
13	MN	WATONWAN	584.61
14	NE	DIXON	557.97
15	IA	PLYMOUTH	553.32
16	IA	OSCEOLA	551.97
17	IA	IDA	548.72
18	IA	BENTON	527.44
19	IA	BOONE	527.44
20	IA	BUENA VISTA	527.44
21	IA	CALHOUN	527.44
22	IA	CARROLL	527.44
23	IA	CHEROKEE	527.44
24	IA	CLAY	527.44
25	IA	CRAWFORD	527.44
26	IA	DICKINSON	527.44
27	IA	EMMET	527.44
28	IA	FLOYD	527.44
29	IA	GREENE	527.44
30	IA	HAMILTON	527.44
31	IA	HANCOCK	527.44
32	IA	HUMBOLDT	527.44
33	IA	IOWA	527.44
34	IA	JASPER	527.44
35	IA	KOSSUTH	527.44
36	IA	LYON	527.44
37	IA	MARSHALL	527.44
38	IA	MONONA	527.44
39	IA	O BRIEN	527.44
40	IA	PALO ALTO	527.44
41	IA	POCAHONTAS	527.44
42	IA	SAC	527.44
43	IA	SIOUX	527.44
44	IA	STORY	527.44
45	IA	WINNEBAGO	527.44
46	IA	WINNESHIEK	527.44
47	MN	ROCK	527.44
48	NE	DAKOTA	527.44
49	NE	STANTON	527.44
50	NE	THURSTON	527.44
51	SD	TURNER	527.44
52	SD	HAMLIN	451.85
53	NE	LOUP	436.77
54	SD	CUSTER	434.17
55	NE	HOLT	398.55
56	SD	KINGSBURY	361.42
57	SD	MOODY	361.42
58	KY	GREENUP	359.52
59	KY	LAWRENCE	359.52
60	KY	LEWIS	359.52
61	KY	TRIMBLE	359.52
62	OH	SCIOTO	359.52
63	WV	BOONE	359.52
64	WV	LOGAN	359.52
65	MO	CLARK	358.08
66	MO	KNOX	358.08
67	MO	SCOTLAND	358.08
68	KY	POWELL	354.59
69	SD	HUTCHINSON	346.94
70	SD	MCCOOK	346.94
71	SD	FALL RIVER	345.60
72	SD	SHANNON	345.60
73	SD	STANLEY	345.60
74	IA	ADAIR	345.56
75	IA	ADAMS	345.56
76	IA	APPANOOSE	345.56
77	IA	AUDUBON	345.56
78	IA	CASS	345.56
79	IA	CLARKE	345.56
80	IA	DAVIS	345.56

COUNTIES OF HIGHEST AVERAGE FALLOUT FROM SHOT:

JOHNIE BOY (UE-7) Jul 11, 1962 0.5 Kt

(microCuries per square meter)

Each Row Includes:

County Rank (most fallout = 1), State, County, Fallout Level in microCuries/square meter

1	NV NYE2	9.59
2	WY BIG HORN	5.30
3	MT POWDER RIVER	4.80
4	WY JOHNSON	4.37
5	WY SHERIDAN	4.37
6	MT CUSTER	4.20
7	MT TREASURE	4.20
8	MT YELLOWSTONE	4.20
9	WY HOT SPRINGS	3.96
10	MT GARFIELD	3.93
11	MT MUSSELSHELL	3.85
12	MT CHOUTEAU	3.68
13	SD JONES	3.43
14	MT WHEATLAND	3.35
15	SD BUTTE	3.23
16	SD CORSON	3.23
17	SD LAWRENCE	3.23
18	WY CROOK	3.23
19	MT BIG HORN	3.03
20	WY PARK	2.98
21	ID TETON	2.90
22	MT GALLATIN	2.90
23	SD HAAKON	2.88
24	SD MEADE	2.88
25	SD PENNINGTON	2.88
26	MT STILLWATER	2.86
27	ND MCLEAN	2.86
28	MT CARTER	2.76
29	ND BOWMAN	2.76
30	MT MEAGHER	2.75
31	MT PARK	2.75
32	SD CUSTER	2.61
33	MT ROSEBUD	2.58
34	ND KIDDER	2.48
35	ID LEMHI	2.47
36	SD DOUGLAS	2.46
37	SD HANSON	2.46
38	SD BEADLE	2.24
39	SD HYDE	2.24
40	SD JACKSON	2.24
41	SD STANLEY	2.24
42	SD MELLETTE	2.22
43	MT PETROLEUM	2.20
44	ID CUSTER	2.20
45	SD LAKE	2.11
46	IL OGLE	1.99
47	SD DEUEL	1.98
48	SD CHARLES MIX	1.95
49	SD BRULE	1.93
50	SD TRIPP	1.93
51	SD PERKINS	1.91
52	IA HANCOCK	1.87
53	MN LAC QUI PARL	1.85
54	MT FALLON	1.85
55	SD FALL RIVER	1.84
56	IA OSCEOLA	1.82
57	SD BUFFALO	1.82
58	SD GREGORY	1.82
59	SD HUTCHINSON	1.82
60	SD JERAULD	1.82
61	SD LYMAN	1.82
62	SD MCCOOK	1.82
63	SD TODD	1.82
64	IA BENTON	1.78
65	IA BUCHANAN	1.78
66	IA CALHOUN	1.78
67	IA LINN	1.78
68	IA WEBSTER	1.78
69	IA CHEROKEE	1.76
70	SD HARDING	1.76
71	WY WASHAKIE	1.75
72	MT ROOSEVELT	1.74
73	MN CHIPPEWA	1.74
74	SD DEWEY	1.74

COUNTIES OF HIGHEST AVERAGE FALLOUT FROM SHOT:

SMALL BOY (UE-8) Jul 14, 1962 1.65 Kt

(microCuries per square meter)
Each Row Includes:
County Rank (most fallout = 1), State, County, Fallout Level in microCuries/square meter

Rank	County	Fallout
1	UT DUCHESNE	117.07
2	UT WASATCH	38.78
3	NE HOOKER	31.93
4	CO GRAND	26.31
5	NV LINCOLN2	25.53
6	WY ALBANY	23.06
7	NE FRONTIER	22.72
8	CO KIT CARSON	18.17
9	CO PHILLIPS	18.17
10	CO SEDGWICK	18.17
11	NE ARTHUR	18.17
12	NE CHEYENNE	18.17
13	NE MCPHERSON	18.17
14	NE PERKINS	18.17
15	NE SHERMAN	17.74
16	KS RUSSELL	15.75
17	NE BLAINE	15.66
18	NE DAWSON	15.57
19	CO SUMMIT	14.58
20	UT MILLARD	14.49
21	KS BARTON	14.44
22	KS LINCOLN	14.44
23	KS MITCHELL	14.44
24	KS OSBORNE	14.44
25	KS PHILLIPS	14.44
26	KS ROOKS	14.44
27	KS SMITH	14.44
28	NE BUFFALO	14.44
29	NE CUSTER	14.44
30	NE KEARNEY	14.44
31	KS SALINE	13.67
32	NE BANNER	13.28
33	UT IRON1	13.11
34	KS GRAHAM	12.27
35	KS NESS	12.27
36	KS CLOUD	12.07
37	NE GREELEY	12.01
38	KS RAWLINS	11.85
39	NE LINCOLN	11.85
40	UT SANPETE	11.71
41	WY PLATTE	11.68
42	NE WHEELER	11.32
43	CO LARIMER	11.10
44	UT EMERY	10.98
45	NE VALLEY	10.88
46	KS NORTON	10.80
47	CO WELD	10.77
48	NE KIMBALL	10.77
49	WY GOSHEN	10.77
50	NE KEITH	10.75
51	KS DICKINSON	10.54
52	KS MARION	10.54
53	KS MCPHERSON	10.54
54	KS OTTAWA	10.54
55	KS REPUBLIC	10.54
56	NE CLAY	10.54
57	NE HAMILTON	10.54
58	NE MERRICK	10.54
59	NE NUCKOLLS	10.54
60	NE THAYER	10.54
61	UT WASHINGTON1	10.51
62	UT SEVIER	9.51
63	WY LARAMIE	9.26
64	CO JEFFERSON	8.62
65	NV LINCOLN1	8.40
66	CO BOULDER	8.27
67	UT JUAB	8.05
68	UT UTAH	8.05
69	KS ELLSWORTH	8.02
70	KS JEWELL	8.02
71	KS RICE	8.02
72	KS STAFFORD	8.02
73	NE ADAMS	8.02
74	NE FRANKLIN	8.02

COUNTIES OF HIGHEST AVERAGE FALLOUT FROM SHOT:
PALANQUIN (UE-12) Apr 14, 1965 4.5 Kt
(microCuries per square meter)
Each Row Includes:
County Rank (most fallout = 1), State, County, Fallout Level in microCuries/square meter

Rank	State/County	Fallout
1	MT LEWIS AND CL	122.65
2	MT CASCADE	94.25
3	MT GALLATIN	73.49
4	MT JUDITH BASIN	58.91
5	MT MUSSELSHELL	58.91
6	MT TETON	58.91
7	MT MADISON	57.83
8	WY TETON	56.70
9	MT CARBON	51.75
10	MT STILLWATER	51.09
11	ID TETON	45.80
12	MT PARK	41.56
13	MT SWEET GRASS	39.57
14	WY PARK	39.52
15	MT BEAVERHEAD	38.04
16	MT MEAGHER	38.04
17	MT SILVER BOW	36.05
18	MT BIG HORN	33.66
19	ID IDAHO	33.38
20	MT MISSOULA	33.38
21	ID LEMHI	33.29
22	MT DEER LODGE	33.29
23	MT GRANITE	33.29
24	MT YELLOWSTONE	33.00
25	MT RAVALLI	29.29
26	WY JOHNSON	29.29
27	SD MEADE	26.63
28	MT PONDERA	25.53
29	IA BREMER	22.99
30	MN DODGE	22.99
31	MN RICE	22.99
32	MN STEELE	22.99
33	MN WINONA	22.99
34	MT GOLDEN VALLE	22.53
35	MT WHEATLAND	22.53
36	SD LAWRENCE	21.54
37	SD PENNINGTON	21.54
38	WY CROOK	21.54
39	MT POWDER RIVER	20.51
40	MT LAKE	19.79
41	SD MINNEHAHA	17.16
42	SD DEWEY	15.26
43	SD POTTER	15.26
44	SD ZIEBACH	15.26
45	IL LAKE	14.56
46	IL MCHENRY	14.56
47	MI BERRIEN	14.56
48	MI IONIA	14.56
49	MI MONTCALM	14.56
50	MI MUSKEGON	14.56
51	MI NEWAYGO	14.56
52	MI OCEANA	14.56
53	MN GOODHUE	14.56
54	MN WABASHA	14.56
55	WI ADAMS	14.56
56	WI CALUMET	14.56
57	WI FOND DU LAC	14.56
58	WI GREEN LAKE	14.56
59	WI JACKSON	14.56
60	WI JUNEAU	14.56
61	WI MONROE	14.56
62	WI SHEBOYGAN	14.56
63	WI TREMPEALEAU	14.56
64	WI WAUSHARA	14.56
65	WI WINNEBAGO	14.56
66	MT TREASURE	13.52
67	IA CLAY	11.43
68	IA SIOUX	11.43
69	MN JACKSON	11.43
70	MN LINCOLN	11.43
71	MN NOBLES	11.43
72	MN PIPESTONE	11.43
73	SD BON HOMME	11.43
74	SD CHARLES MIX	11.43

COUNTIES OF HIGHEST AVERAGE FALLOUT FROM SHOT:
PALANQUIN (UE-12) Apr 14, 1965 4.5 Kt
(microCuries per square meter)
Each Row Includes:
County Rank (most fallout = 1), State, County, Fallout Level in microCuries/square meter

Rank	County	Fallout
75	SD DAVISON	11.43
76	SD DOUGLAS	11.43
77	SD JERAULD	11.43
78	SD KINGSBURY	11.43
79	SD TURNER	11.43
80	WY NIOBRARA	11.43
81	WY WESTON	11.43
82	MT CARTER	10.84
83	IA ALLAMAKEE	9.84
84	IA CERRO GORDO	9.84
85	IA CLAYTON	9.84
86	IA DELAWARE	9.84
87	IA DICKINSON	9.84
88	IA FLOYD	9.84
89	IA HOWARD	9.84
90	IA LINN	9.84
91	IA LYON	9.84
92	IA O BRIEN	9.84
93	MN BROWN	9.84
94	MN COTTONWOOD	9.84
95	MN FILLMORE	9.84
96	MN FREEBORN	9.84
97	MN LYON	9.84
98	MN MARTIN	9.84
99	MN MOWER	9.84
100	MN MURRAY	9.84

COUNTIES OF HIGHEST AVERAGE I-131 DEPOSITION FROM SHOT:

PIN STRIPE (UE-13) Apr 25, 1966 Approx. 4.0 Kt

(microCuries per square meter)

Each Row Includes:

County Rank (most fallout = 1), State, County, Fallout Level in microCuries/square meter

Rank	County	Fallout
1	UT BEAVER	113.24
2	UT SEVIER	74.83
3	UT PIUTE	40.29
4	UT SANPETE	39.58
5	NV LINCOLN2	31.85
6	UT MILLARD	18.87
7	UT JUAB	7.28
8	UT KANE1	5.30
9	AZ COCONINO1	4.97
10	ID BEAR LAKE	3.89
11	ID BONNEVILLE	3.89
12	ID CARIBOU	3.89
13	ID FRANKLIN	3.89
14	UT BOX ELDER2	3.89
15	UT CACHE	3.89
16	UT DAVIS	3.89
17	UT MORGAN	3.89
18	UT RICH	3.89
19	UT SALT LAKE	3.89
20	UT SUMMIT	3.89
21	UT TOOELE2	3.89
22	UT WASATCH	3.89
23	UT WEBER	3.89
24	WY LINCOLN	3.89
25	WY UINTA	3.89
26	UT UTAH	3.64
27	CA INYO3	3.31
28	CA SAN BERNADINO	3.31

COUNTIES OF HIGHEST AVERAGE FALLOUT FROM SHOT:

SCHOONER (UE-16) Dec 8, 1968 30 Kt

(microCuries per square meter)

Each Row Includes:

County Rank (most fallout = 1), State, County, Fallout Level in microCuries/square meter

Rank	State County	Fallout
1	NV NYE2	71.49
2	NV WHITE PINE2	63.61
3	NV WHITE PINE3	26.98
4	UT SANPETE	26.98
5	UT UTAH	26.98
6	UT MILLARD	25.45
7	UT JUAB	16.96
8	UT BEAVER	13.49
9	UT CARBON	13.49
10	UT SEVIER	13.49
11	NV WHITE PINE1	12.72
12	NV NYE1	12.40
13	UT DUCHESNE	9.60
14	UT TOOELE2	9.60
15	UT WASATCH	9.60
16	UT PIUTE	8.99
17	NV EUREKA	8.48
18	NV LINCOLN1	8.48
19	NV ELKO	4.80
20	UT BOX ELDER1	4.80
21	UT BOX ELDER2	4.80
22	UT CACHE	4.80
23	UT DAGGETT	4.80
24	UT DAVIS	4.80
25	UT EMERY	4.80
26	UT GRAND	4.80
27	UT MORGAN	4.80
28	UT SALT LAKE	4.80
29	UT SUMMIT	4.80
30	UT TOOELE1	4.80
31	UT UINTAH	4.80
32	UT WAYNE	4.80
33	UT WEBER	4.80
34	NV LANDER1	4.50
35	NV LINCOLN2	4.13

COUNTIES OF HIGHEST AVERAGE FALLOUT FROM SHOT:

BANEBERRY (UE-17) Dec 18, 1970 10 Kt

(microCuries per square meter)

Each Row Includes:

County Rank (most fallout = 1), State, County, Fallout Level in microCuries/square meter

Rank	County	Fallout
1	NV NYE2	11.08
2	CA LASSEN	7.57
3	CA SIERRA	6.98
4	CA NEVADA	4.52
5	CA PLACER	4.52
6	NV EUREKA	4.39
7	NV LANDER2	4.39
8	NV HUMBOLDT	3.88
9	NV WHITE PINE1	3.09
10	NV MINERAL	2.16
11	NV WHITE PINE2	2.04
12	NV WHITE PINE3	2.04
13	UT SALT LAKE	1.91
14	ID BONNEVILLE	1.84
15	ID BUTTE	1.81
16	ID BOISE	1.76
17	CA YUBA	1.64
18	ID CUSTER	1.63
19	WY TETON	1.50
20	MT GALLATIN	1.49
21	ID FREMONT	1.46
22	ID JEFFERSON	1.36
23	CA ALPINE	1.36
24	ID VALLEY	1.31
25	MT BEAVERHEAD	1.28
26	ID BLAINE	1.26
27	ID CAMAS	1.26
28	OR BAKER	1.23
29	CA BUTTE	1.14
30	MT MADISON	1.14
31	ID CLARK	1.11
32	WY PARK	1.08
33	NV CARSON CITY	1.04
34	NV DOUGLAS	1.04
35	NV LYON	1.04
36	NV STOREY	1.04
37	UT BOX ELDER2	1.04
38	UT DAVIS	1.04
39	UT MORGAN	1.04
40	UT RICH	1.04
41	UT TOOELE2	1.04
42	UT UTAH	1.04
43	UT WEBER	1.04
44	ID ADAMS	1.01
45	ID PAYETTE	1.00
46	NV CHURCHILL	0.99
47	NV ELKO	0.99
48	NV LANDER1	0.99
49	NV PERSHING	0.99
50	NV WASHOE	0.99
51	UT BOX ELDER1	0.99
52	UT CACHE	0.99
53	UT SUMMIT	0.99
54	UT TOOELE1	0.99
55	ID WASHINGTON	0.98
56	ID CANYON	0.98
57	ID CARIBOU	0.98
58	ID OWYHEE	0.98
59	ID ADA	0.92
60	ID FRANKLIN	0.92
61	CO JACKSON	0.91
62	CO ROUTT	0.91
63	NE SIOUX	0.91
64	MT VALLEY	0.89
65	ID MADISON	0.88
66	MT CASCADE	0.88
67	MT MISSOULA	0.87
68	CO BOULDER	0.82
69	CO GRAND	0.82
70	IA HARRISON	0.82
71	NE BOONE	0.82
72	NE BOX BUTTE	0.82
73	NE DODGE	0.82
74	NE MADISON	0.82

COUNTIES OF HIGHEST AVERAGE FALLOUT FROM SHOT:

BANEBERRY (UE-17) Dec 18, 1970 10 Kt

(microCuries per square meter)

Each Row Includes:

County Rank (most fallout = 1), State, County, Fallout Level in microCuries/square meter

Rank	County	Fallout
75	NE STANTON	0.82
76	NE WHEELER	0.82
77	MT CARBON	0.81
78	MT BIG HORN	0.80
79	IA SAC	0.75
80	ID CLEARWATER	0.69
81	MT FERGUS	0.69
82	MT LEWIS AND CL	0.69
83	NE BLAINE	0.68
84	MT MINERAL	0.68
85	ID IDAHO	0.67
86	OR UNION	0.67
87	ID LEWIS	0.66
88	MT MUSSELSHELL	0.66
89	MT PHILLIPS	0.66
90	MT PONDERA	0.66
91	MT ROSEBUD	0.66
92	MT SHERIDAN	0.66
93	MT TETON	0.66
94	NE THOMAS	0.64
95	WA SKAMANIA	0.64
96	MT HILL	0.61
97	MT STILLWATER	0.61
98	ND DUNN	0.60
99	MT LAKE	0.59
100	MT CHOUTEAU	0.58

U.S. COUNTIES RANKED BY TOTAL FALLOUT DEPOSITION
1951-1970
(microCuries per square meter)
Each Row Includes:
County Name, Fallout Level in microCuries/square meter and County Rank (1 = highest fallout in U.S.)

COUNTY	uCi/sqmeter	RANK
AL AUTAUGA	67.0	2726
AL BALDWIN	97.2	2066
AL BARBOUR	80.3	2387
AL BIBB	76.6	2480
AL BLOUNT	68.1	2693
AL BULLOCK	68.4	2685
AL BUTLER	66.4	2737
AL CALHOUN	69.3	2658
AL CHAMBERS	77.2	2458
AL CHEROKEE	66.8	2731
AL CHILTON	67.3	2715
AL CHOCTAW	66.5	2736
AL CLARKE	92.1	2143
AL CLAY	74.9	2519
AL CLEBURNE	69.7	2648
AL COFFEE	69.2	2660
AL COLBERT	85.7	2262
AL CONECUH	126.8	1721
AL COOSA	65.0	2761
AL COVINGTON	79.2	2410
AL CRENSHAW	62.9	2817
AL CULLMAN	76.7	2478
AL DALE	81.5	2359
AL DALLAS	74.2	2541
AL DE KALB	78.2	2437
AL ELMORE	73.8	2554
AL ESCAMBIA	88.5	2212
AL ETOWAH	79.3	2404
AL FAYETTE	64.5	2773
AL FRANKLIN	69.4	2656
AL GENEVA	98.6	2048
AL GREENE	98.7	2046
AL HALE	84.5	2291
AL HENRY	56.0	2894
AL HOUSTON	73.8	2553
AL JACKSON	75.1	2515
AL JEFFERSON	74.2	2540
AL LAMAR	62.9	2818
AL LAUDERDALE	78.7	2426
AL LAWRENCE	65.3	2756
AL LEE	76.8	2469
AL LIMESTONE	68.8	2669
AL LOWNDES	82.1	2342
AL MACON	64.1	2783
AL MADISON	81.1	2371
AL MARENGO	153.1	1425
AL MARION	64.8	2764
AL MARSHALL	75.4	2504
AL MOBILE	161.8	1331
AL MONROE	109.8	1915
AL MONTGOMERY	58.2	2872
AL MORGAN	75.3	2510
AL PERRY	80.5	2381
AL PICKENS	82.1	2343
AL PIKE	81.0	2374
AL RANDOLPH	75.2	2512
AL RUSSELL	69.2	2661
AL SHELBY	62.7	2821
AL ST CLAIR	68.0	2696
AL SUMTER	78.3	2433
AL TALLADEGA	76.0	2488
AL TALLAPOOSA	73.2	2566
AL TUSCALOOSA	88.5	2215
AL WALKER	66.8	2730
AL WASHINGTON	86.6	2246
AL WILCOX	111.3	1886
AL WINSTON	63.2	2810

U.S. COUNTIES RANKED BY TOTAL FALLOUT DEPOSITION
1951-1970
(microCuries per square meter)
Each Row Includes:
County Name, Fallout Level in microCuries/square meter and County Rank (1 = highest fallout in U.S.)

AR ARKANSAS	178.3	1176
AR ASHLEY	101.4	2010
AR BAXTER	200.6	978
AR BENTON	213.5	875
AR BOONE	169.6	1267
AR BRADLEY	107.0	1948
AR CALHOUN	84.1	2301
AR CARROLL	169.1	1270
AR CHICOT	96.3	2076
AR CLARK	150.3	1451
AR CLAY	169.0	1273
AR CLEBURNE	198.4	1000
AR CLEVELAND	121.6	1772
AR COLUMBIA	125.8	1734
AR CONWAY	172.4	1231
AR CRAIGHEAD	199.5	991
AR CRAWFORD	182.2	1144
AR CRITTENDEN	151.2	1446
AR CROSS	149.5	1457
AR DALLAS	127.3	1713
AR DESHA	143.2	1529
AR DREW	106.1	1963
AR FAULKNER	214.4	870
AR FRANKLIN	195.9	1024
AR FULTON	220.4	828
AR GARLAND	140.9	1560
AR GRANT	154.2	1412
AR GREENE	177.6	1188
AR HEMPSTEAD	134.0	1653
AR HOT SPRING	141.0	1558
AR HOWARD	153.5	1419
AR INDEPENDENCE	191.5	1056
AR IZARD	197.5	1013
AR JACKSON	206.8	926
AR JEFFERSON	176.7	1192
AR JOHNSON	195.5	1029
AR LAFAYETTE	142.2	1544
AR LAWRENCE	209.9	900
AR LEE	145.8	1498
AR LINCOLN	136.3	1616
AR LITTLE RIVER	157.8	1374
AR LOGAN	172.3	1232
AR LONOKE	172.5	1230
AR MADISON	206.9	925
AR MARION	191.6	1054
AR MILLER	140.5	1566
AR MISSISSIPPI	158.0	1370
AR MONROE	159.0	1359
AR MONTGOMERY	170.8	1251
AR NEVADA	138.8	1590
AR NEWTON	200.7	976
AR OUACHITA	121.3	1777
AR PERRY	204.2	946
AR PHILLIPS	139.2	1584
AR PIKE	147.9	1477
AR POINSETT	146.3	1494
AR POLK	182.2	1141
AR POPE	199.9	987
AR PRAIRIE	208.9	913
AR PULASKI	173.1	1228
AR RANDOLPH	178.5	1170
AR SALINE	182.7	1135
AR SCOTT	163.7	1315
AR SEARCY	179.1	1162
AR SEBASTIAN	170.2	1258
AR SEVIER	169.8	1262
AR SHARP	212.1	885
AR ST FRANCIS	165.3	1304

U.S. COUNTIES RANKED BY TOTAL FALLOUT DEPOSITION
1951-1970
(microCuries per square meter)
Each Row Includes:
County Name, Fallout Level in microCuries/square meter and County Rank (1 = highest fallout in U.S.)

AR STONE	166.9	1294
AR UNION	116.6	1836
AR VAN BUREN	197.7	1007
AR WASHINGTON	204.2	944
AR WHITE	190.5	1059
AR WOODRUFF	184.7	1116
AR YELL	176.2	1200
AZ APACHE	1424.3	33
AZ COCHISE	156.6	1382
AZ COCONINO1	2057.9	14
AZ COCONINO2	1591.2	26
AZ COCONINO3	401.3	315
AZ GILA	177.8	1184
AZ GRAHAM	173.5	1225
AZ GREENLEE	159.5	1356
AZ MARICOPA	173.9	1220
AZ MOHAVE1	1593.4	25
AZ MOHAVE2	2997.5	8
AZ MOHAVE3	598.7	174
AZ MOHAVE4	598.7	175
AZ NAVAJO	932.9	72
AZ PIMA	265.4	623
AZ PINAL	178.3	1174
AZ SANTA CRUZ	222.9	804
AZ YAVAPAI	220.6	826
AZ YUMA	151.3	1442
CA ALAMEDA	31.4	3027
CA ALPINE	111.2	1889
CA AMADOR	66.9	2728
CA BUTTE	52.3	2919
CA CALAVERAS	57.8	2875
CA COLUSA	39.7	2984
CA CONTRA COSTA	31.5	3024
CA DEL NORTE	32.6	3016
CA EL DORADO	52.9	2916
CA FRESNO	41.4	2976
CA GLENN	67.5	2709
CA HUMBOLDT	28.2	3045
CA IMPERIAL	25.8	3058
CA INYO1	275.9	578
CA INYO2	275.9	579
CA INYO3	1750.1	18
CA KERN	62.3	2829
CA KINGS	45.7	2955
CA LAKE	40.2	2982
CA LASSEN	66.8	2732
CA LOS ANGELES	114.6	1856
CA MADERA	35.9	2996
CA MARIN	41.8	2974
CA MARIPOSA	42.3	2972
CA MENDOCINO	25.7	3059
CA MERCED	30.5	3032
CA MODOC	46.8	2946
CA MONO	260.9	644
CA MONTEREY	49.7	2932
CA NAPA	33.8	3009
CA NEVADA	89.3	2197
CA ORANGE	43.2	2968
CA PLACER	63.4	2802
CA PLUMAS	62.5	2826
CA RIVERSIDE	25.1	3063
CA SACRAMENTO	36.3	2994
CA SAN BENITO	44.6	2961
CA SAN BERNADIN	313.6	443
CA SAN DIEGO	46.9	2944
CA SAN FRANCISC	26.4	3051
CA SAN JOAQUIN	30.9	3031
CA SAN LUIS OBI	46.8	2945

U.S. COUNTIES RANKED BY TOTAL FALLOUT DEPOSITION
1951-1970

(microCuries per square meter)

Each Row Includes:

County Name, Fallout Level in microCuries/square meter and County Rank (1 = highest fallout in U.S.)

County	Fallout	Rank
CA SAN MATEO	34.8	3002
CA SANTA BARBAR	52.9	2915
CA SANTA CLARA	34.4	3005
CA SANTA CRUZ	46.2	2952
CA SHASTA	25.6	3062
CA SIERRA	87.5	2230
CA SISKIYOU	29.5	3037
CA SOLANO	29.3	3038
CA SONOMA	52.3	2918
CA STANISLAUS	32.4	3020
CA SUTTER	33.4	3011
CA TEHAMA	55.0	2899
CA TRINITY	29.2	3039
CA TULARE	50.3	2929
CA TUOLUMNE	52.6	2917
CA VENTURA	63.0	2816
CA YOLO	32.5	3017
CA YUBA	57.8	2876
CO ADAMS	448.6	266
CO ALAMOSA	683.1	142
CO ARAPAHOE	434.5	278
CO ARCHULETA	1438.4	31
CO BACA	344.0	385
CO BENT	282.1	546
CO BOULDER	479.6	234
CO CHAFFEE	414.9	303
CO CHEYENNE	266.5	616
CO CLEAR CREEK	503.3	218
CO CONEJOS	1831.9	16
CO COSTILLA	698.7	126
CO CROWLEY	273.0	592
CO CUSTER	424.6	289
CO DELTA	1104.4	57
CO DENVER	290.0	516
CO DOLORES	889.9	78
CO DOUGLAS	473.0	243
CO EAGLE	717.5	115
CO EL PASO	276.5	570
CO ELBERT	271.3	601
CO FREMONT	376.0	342
CO GARFIELD	1200.0	43
CO GILPIN	476.8	236
CO GRAND	596.9	176
CO GUNNISON	1946.4	15
CO HINSDALE	1146.1	49
CO HUERFANO	377.3	339
CO JACKSON	319.9	428
CO JEFFERSON	485.8	227
CO KIOWA	337.6	400
CO KIT CARSON	290.1	514
CO LA PLATA	1111.6	55
CO LAKE	825.3	87
CO LARIMER	363.4	360
CO LAS ANIMAS	374.5	344
CO LINCOLN	290.2	513
CO LOGAN	250.1	682
CO MESA	1267.3	40
CO MINERAL	1156.1	47
CO MOFFAT	708.2	120
CO MONTEZUMA	801.4	93
CO MONTROSE	1050.3	62
CO MORGAN	463.5	253
CO OTERO	285.8	531
CO OURAY	802.9	90
CO PARK	313.6	442
CO PHILLIPS	363.2	361
CO PITKIN	1018.6	65
CO PROWERS	274.6	586

U.S. COUNTIES RANKED BY TOTAL FALLOUT DEPOSITION
1951-1970
(microCuries per square meter)
Each Row Includes:
County Name, Fallout Level in microCuries/square meter and County Rank (1 = highest fallout in U.S.)

CO PUEBLO	245.5	697
CO RIO BLANCO	973.9	69
CO RIO GRANDE	905.6	75
CO ROUTT	447.7	269
CO SAGUACHE	716.0	116
CO SAN JUAN	793.3	94
CO SAN MIGUEL	808.1	89
CO SEDGWICK	279.1	559
CO SUMMIT	641.4	158
CO TELLER	315.0	439
CO WASHINGTON	267.2	613
CO WELD	319.2	429
CO YUMA	280.7	553
CT FAIRFIELD	163.9	1314
CT HARTFORD	143.5	1524
CT LITCHFIELD	160.0	1350
CT MIDDLESEX	209.7	903
CT NEW HAVEN	178.7	1168
CT NEW LONDON	159.9	1353
CT TOLLAND	178.1	1180
CT WINDHAM	157.5	1378
DC WASHINGTON	81.4	2364
DE KENT	98.6	2051
DE NEW CASTLE	99.2	2035
DE SUSSEX	99.5	2031
FL ALACHUA	76.1	2486
FL BAKER	63.0	2815
FL BAY	65.2	2758
FL BRADFORD	58.4	2868
FL BREVARD	79.1	2412
FL BROWARD	50.7	2926
FL CALHOUN	75.9	2494
FL CHARLOTTE	44.9	2959
FL CITRUS	71.8	2608
FL CLAY	65.4	2755
FL COLLIER	57.2	2880
FL COLUMBIA	61.7	2834
FL DADE	50.3	2928
FL DE SOTO	43.7	2965
FL DIXIE	72.2	2595
FL DUVAL	44.6	2960
FL ESCAMBIA	85.1	2276
FL FLAGLER	78.6	2427
FL FRANKLIN	70.3	2635
FL GADSDEN	67.0	2727
FL GILCHRIST	74.5	2527
FL GLADES	49.0	2937
FL GULF	73.0	2572
FL HAMILTON	68.0	2697
FL HARDEE	40.2	2981
FL HENDRY	54.4	2902
FL HERNANDO	78.2	2441
FL HIGHLANDS	49.3	2934
FL HILLSBOROUGH	87.6	2226
FL HOLMES	96.3	2074
FL INDIAN RIVER	47.3	2943
FL JACKSON	63.5	2798
FL JEFFERSON	68.3	2689
FL LAFAYETTE	62.7	2822
FL LAKE	98.1	2055
FL LEE	49.8	2931
FL LEON	83.9	2305
FL LEVY	46.6	2949
FL LIBERTY	72.4	2590
FL MADISON	75.8	2498
FL MANATEE	51.7	2921
FL MARION	73.2	2567
FL MARTIN	49.0	2936

U.S. COUNTIES RANKED BY TOTAL FALLOUT DEPOSITION
1951-1970

(microCuries per square meter)
Each Row Includes:
County Name, Fallout Level in microCuries/square meter and County Rank (1 = highest fallout in U.S.)

County	Fallout	Rank
FL MONROE	53.5	2911
FL NASSAU	75.1	2517
FL OKALOOSA	136.7	1612
FL OKEECHOBEE	48.5	2940
FL ORANGE	74.0	2547
FL OSCEOLA	63.8	2793
FL PALM BEACH	55.8	2897
FL PASCO	58.5	2866
FL PINELLAS	72.1	2599
FL POLK	75.1	2514
FL PUTNAM	70.9	2622
FL SANTA ROSA	91.4	2155
FL SARASOTA	57.2	2879
FL SEMINOLE	73.6	2559
FL ST JOHNS	81.3	2366
FL ST LUCIE	53.5	2913
FL SUMTER	81.7	2353
FL SUWANNEE	70.8	2626
FL TAYLOR	62.7	2824
FL UNION	59.8	2857
FL VOLUSIA	85.0	2281
FL WAKULLA	71.8	2610
FL WALTON	62.8	2820
FL WASHINGTON	63.9	2788
GA APPLING	79.3	2406
GA ATKINSON	74.4	2532
GA BACON	77.9	2447
GA BAKER	81.3	2365
GA BALDWIN	74.5	2528
GA BANKS	72.4	2586
GA BARROW	63.8	2792
GA BARTOW	66.4	2739
GA BEN HILL	72.0	2605
GA BERRIEN	76.7	2473
GA BIBB	73.4	2564
GA BLECKLEY	77.8	2448
GA BRANTLEY	73.6	2556
GA BROOKS	66.7	2734
GA BRYAN	92.6	2133
GA BULLOCH	86.0	2253
GA BURKE	82.0	2344
GA BUTTS	68.3	2690
GA CALHOUN	84.5	2292
GA CAMDEN	76.0	2490
GA CANDLER	73.1	2570
GA CARROLL	81.7	2355
GA CATOOSA	68.4	2683
GA CHARLTON	68.5	2676
GA CHATHAM	83.4	2323
GA CHATTAHOOCHE	72.2	2598
GA CHATTOOGA	69.6	2651
GA CHEROKEE	68.4	2680
GA CLARKE	73.7	2555
GA CLAY	85.7	2263
GA CLAYTON	81.1	2370
GA CLINCH	74.3	2537
GA COBB	63.6	2796
GA COFFEE	65.0	2762
GA COLQUITT	83.5	2321
GA COLUMBIA	61.7	2833
GA COLUMBUS	76.0	2491
GA COOK	73.5	2560
GA COWETA	67.6	2706
GA CRAWFORD	81.9	2348
GA CRISP	75.4	2505
GA DADE	63.8	2791
GA DAWSON	64.2	2782
GA DE KALB	68.4	2684

U.S. COUNTIES RANKED BY TOTAL FALLOUT DEPOSITION
1951-1970
(microCuries per square meter)
Each Row Includes:
County Name, Fallout Level in microCuries/square meter and County Rank (1 = highest fallout in U.S.)

GA DECATUR	68.5	2677
GA DODGE	69.9	2643
GA DOOLY	71.8	2611
GA DOUGHERTY	80.7	2378
GA DOUGLAS	76.5	2481
GA EARLY	66.3	2743
GA ECHOLS	64.7	2767
GA EFFINGHAM	90.8	2165
GA ELBERT	62.8	2819
GA EMANUEL	75.3	2508
GA EVANS	86.3	2251
GA FANNIN	85.4	2266
GA FAYETTE	82.0	2345
GA FLOYD	75.7	2499
GA FORSYTH	64.6	2770
GA FRANKLIN	65.0	2759
GA FULTON	62.7	2823
GA GILMER	72.2	2597
GA GLASCOCK	74.4	2533
GA GLYNN	83.2	2325
GA GORDON	78.8	2424
GA GRADY	65.4	2754
GA GREENE	80.0	2391
GA GWINNETT	72.6	2583
GA HABERSHAM	72.3	2592
GA HALL	68.7	2672
GA HANCOCK	70.6	2631
GA HARALSON	69.9	2642
GA HARRIS	66.3	2742
GA HART	66.2	2746
GA HEARD	56.4	2890
GA HENRY	67.3	2716
GA HOUSTON	75.5	2503
GA IRWIN	71.5	2615
GA JACKSON	70.9	2623
GA JASPER	67.7	2703
GA JEFF DAVIS	79.9	2392
GA JEFFERSON	80.3	2384
GA JENKINS	85.1	2277
GA JOHNSON	78.8	2423
GA JONES	69.6	2652
GA LAMAR	67.1	2724
GA LANIER	77.9	2446
GA LAURENS	60.2	2852
GA LEE	78.3	2436
GA LIBERTY	105.8	1967
GA LINCOLN	63.4	2801
GA LONG	102.6	2000
GA LOWNDES	71.1	2621
GA LUMPKIN	66.2	2747
GA MACON	59.6	2858
GA MADISON	72.4	2587
GA MARION	60.0	2855
GA MCDUFFIE	74.4	2534
GA MCINTOSH	82.6	2334
GA MERIWETHER	69.4	2654
GA MILLER	56.5	2889
GA MITCHELL	78.9	2418
GA MONROE	76.9	2464
GA MONTGOMERY	90.1	2181
GA MORGAN	78.8	2421
GA MURRAY	71.4	2618
GA NEWTON	66.3	2741
GA OCONEE	69.6	2649
GA OGLETHORPE	68.1	2694
GA PAULDING	77.0	2463
GA PEACH	81.9	2347
GA PICKENS	64.4	2775

U.S. COUNTIES RANKED BY TOTAL FALLOUT DEPOSITION
1951-1970
(microCuries per square meter)
Each Row Includes:
County Name, Fallout Level in microCuries/square meter and County Rank (1 = highest fallout in U.S.)

GA PIERCE	68.7	2673
GA PIKE	70.8	2625
GA POLK	60.0	2854
GA PULASKI	85.8	2257
GA PUTNAM	68.3	2688
GA QUITMAN	85.9	2255
GA RABUN	107.4	1945
GA RANDOLPH	86.4	2249
GA RICHMOND	67.2	2720
GA ROCKDALE	62.5	2827
GA SCHLEY	84.0	2304
GA SCREVEN	90.6	2168
GA SEMINOLE	54.7	2901
GA SPALDING	71.6	2613
GA STEPHENS	70.3	2636
GA STEWART	76.0	2489
GA SUMTER	77.0	2461
GA TALBOT	63.8	2794
GA TALIAFERRO	74.9	2518
GA TATTNALL	83.9	2306
GA TAYLOR	67.2	2719
GA TELFAIR	77.0	2462
GA TERRELL	88.5	2211
GA THOMAS	76.8	2468
GA TIFT	73.9	2550
GA TOOMBS	78.1	2442
GA TOWNS	92.7	2131
GA TREUTLEN	59.9	2856
GA TROUP	74.3	2539
GA TURNER	72.7	2578
GA TWIGGS	78.8	2420
GA UNION	80.3	2386
GA UPSON	62.2	2830
GA WALKER	72.5	2585
GA WALTON	67.9	2698
GA WARE	64.4	2776
GA WARREN	74.5	2530
GA WASHINGTON	72.8	2574
GA WAYNE	96.0	2088
GA WEBSTER	84.4	2296
GA WHEELER	93.3	2124
GA WHITE	83.4	2322
GA WHITFIELD	67.3	2717
GA WILCOX	75.9	2495
GA WILKES	67.4	2713
GA WILKINSON	74.2	2542
GA WORTH	74.3	2536
IA ADAIR	595.3	178
IA ADAMS	595.0	179
IA ALLAMAKEE	321.6	425
IA APPANOOSE	662.6	154
IA AUDUBON	589.9	183
IA BENTON	765.1	99
IA BLACK HAWK	463.6	252
IA BOONE	766.5	98
IA BREMER	464.2	251
IA BUCHANAN	409.1	308
IA BUENA VISTA	697.5	127
IA BUTLER	391.4	325
IA CALHOUN	691.4	132
IA CARROLL	786.4	95
IA CASS	627.0	163
IA CEDAR	332.6	410
IA CERRO GORDO	368.0	350
IA CHEROKEE	684.5	141
IA CHICKASAW	196.6	1022
IA CLARKE	564.3	196
IA CLAY	687.4	138

U.S. COUNTIES RANKED BY TOTAL FALLOUT DEPOSITION
1951-1970

(microCuries per square meter)
Each Row Includes:
County Name, Fallout Level in microCuries/square meter and County Rank (1 = highest fallout in U.S.)

IA CLAYTON	343.2	389
IA CLINTON	334.3	408
IA CRAWFORD	672.6	150
IA DALLAS	376.3	340
IA DAVIS	638.7	160
IA DECATUR	569.2	195
IA DELAWARE	557.2	198
IA DES MOINES	294.8	496
IA DICKINSON	676.0	147
IA DUBUQUE	359.7	362
IA EMMET	657.6	156
IA FAYETTE	415.8	301
IA FLOYD	689.6	135
IA FRANKLIN	447.1	270
IA FREMONT	306.5	458
IA GREENE	715.1	117
IA GRUNDY	176.9	1191
IA GUTHRIE	526.3	208
IA HAMILTON	734.1	110
IA HANCOCK	684.8	140
IA HARDIN	404.5	312
IA HARRISON	316.8	434
IA HENRY	505.1	216
IA HOWARD	1133.0	51
IA HUMBOLDT	702.6	122
IA IDA	717.8	114
IA IOWA	691.0	133
IA JACKSON	352.4	371
IA JASPER	767.9	97
IA JEFFERSON	756.1	102
IA JOHNSON	477.2	235
IA JONES	295.0	495
IA KEOKUK	538.8	205
IA KOSSUTH	685.7	139
IA LEE	802.4	91
IA LINN	419.9	293
IA LOUISA	484.2	230
IA LUCAS	199.9	986
IA LYON	687.4	137
IA MADISON	601.1	172
IA MAHASKA	570.2	194
IA MARION	594.4	180
IA MARSHALL	764.0	100
IA MILLS	339.2	398
IA MITCHELL	1116.9	53
IA MONONA	700.1	124
IA MONROE	661.7	155
IA MONTGOMERY	415.1	302
IA MUSCATINE	276.4	573
IA O BRIEN	712.9	119
IA OSCEOLA	701.8	123
IA PAGE	352.4	370
IA PALO ALTO	677.8	146
IA PLYMOUTH	693.0	131
IA POCAHONTAS	694.9	129
IA POLK	335.3	405
IA POTTAWATTAMI	374.7	343
IA POWESHIEK	439.9	276
IA RINGGOLD	614.0	167
IA SAC	688.7	136
IA SCOTT	257.7	658
IA SHELBY	624.0	165
IA SIOUX	699.3	125
IA STORY	754.1	103
IA TAMA	498.2	221
IA TAYLOR	417.1	299
IA UNION	591.2	182
IA VAN BUREN	813.7	88

U.S. COUNTIES RANKED BY TOTAL FALLOUT DEPOSITION
1951-1970
(microCuries per square meter)

Each Row Includes:
County Name, Fallout Level in microCuries/square meter and County Rank (1 = highest fallout in U.S.)

County	Fallout	Rank
IA WAPELLO	608.5	170
IA WARREN	351.7	373
IA WASHINGTON	757.0	101
IA WAYNE	580.6	185
IA WEBSTER	349.0	377
IA WINNEBAGO	679.3	145
IA WINNESHIEK	693.3	130
IA WOODBURY	842.7	82
IA WORTH	1113.0	54
IA WRIGHT	343.7	387
ID ADA	419.4	296
ID ADAMS	236.9	737
ID BANNOCK	287.3	526
ID BEAR LAKE	864.6	81
ID BENEWAH	131.4	1679
ID BINGHAM	186.8	1089
ID BLAINE	908.3	74
ID BOISE	472.8	244
ID BONNER	106.1	1962
ID BONNEVILLE	249.6	683
ID BOUNDARY	85.0	2280
ID BUTTE	328.2	415
ID CAMAS	502.0	219
ID CANYON	233.5	751
ID CARIBOU	462.7	255
ID CASSIA	428.0	284
ID CLARK	262.1	637
ID CLEARWATER	305.0	461
ID CUSTER	1015.8	66
ID ELMORE	421.2	291
ID FRANKLIN	690.0	134
ID FREMONT	240.5	715
ID GEM	1030.8	63
ID GOODING	465.3	250
ID IDAHO	572.1	193
ID JEFFERSON	372.6	347
ID JEROME	347.3	380
ID KOOTENAI	96.1	2086
ID LATAH	171.3	1245
ID LEMHI	718.7	113
ID LEWIS	327.9	417
ID LINCOLN	391.1	327
ID MADISON	288.4	521
ID MINIDOKA	339.5	397
ID NEZ PERCE	184.2	1122
ID ONEIDA	524.1	209
ID OWYHEE	475.7	239
ID PAYETTE	323.7	422
ID POWER	280.3	555
ID SHOSHONE	228.0	774
ID TETON	299.9	478
ID TWIN FALLS	401.9	314
ID VALLEY	579.4	188
ID WASHINGTON	396.3	321
IL ADAMS	452.9	263
IL ALEXANDER	114.6	1855
IL BOND	195.6	1027
IL BOONE	265.2	625
IL BROWN	346.2	381
IL BUREAU	110.1	1905
IL CALHOUN	218.1	842
IL CARROLL	123.5	1755
IL CASS	358.0	363
IL CHAMPAIGN	238.2	729
IL CHRISTIAN	326.0	420
IL CLARK	206.3	933
IL CLAY	239.8	720
IL CLINTON	102.6	2001

U.S. COUNTIES RANKED BY TOTAL FALLOUT DEPOSITION
1951-1970

(microCuries per square meter)
Each Row Includes:
County Name, Fallout Level in microCuries/square meter and County Rank (1 = highest fallout in U.S.)

County	Fallout	Rank
IL COLES	343.4	388
IL COOK	72.3	2594
IL CRAWFORD	150.4	1450
IL CUMBERLAND	177.8	1183
IL DE KALB	131.3	1684
IL DE WITT	264.1	631
IL DOUGLAS	276.6	569
IL DU PAGE	81.2	2368
IL EDGAR	183.5	1130
IL EDWARDS	116.6	1837
IL EFFINGHAM	276.3	574
IL FAYETTE	188.5	1076
IL FORD	199.8	989
IL FRANKLIN	89.1	2205
IL FULTON	483.9	231
IL GALLATIN	116.4	1840
IL GREENE	334.9	407
IL GRUNDY	131.3	1683
IL HAMILTON	83.8	2308
IL HANCOCK	801.7	92
IL HARDIN	79.1	2414
IL HENDERSON	295.9	492
IL HENRY	276.9	567
IL IROQUOIS	141.7	1551
IL JACKSON	117.9	1810
IL JASPER	308.0	454
IL JEFFERSON	141.3	1554
IL JERSEY	234.7	745
IL JO DAVIESS	162.7	1321
IL JOHNSON	89.3	2198
IL KANE	120.0	1789
IL KANKAKEE	138.6	1594
IL KENDALL	81.5	2361
IL KNOX	287.8	524
IL LA SALLE	153.1	1430
IL LAKE	100.8	2017
IL LAWRENCE	159.8	1354
IL LEE	154.4	1408
IL LIVINGSTON	124.5	1744
IL LOGAN	251.5	673
IL MACON	294.1	499
IL MACOUPIN	324.5	421
IL MADISON	110.9	1893
IL MARION	94.3	2108
IL MARSHALL	307.3	456
IL MASON	279.4	558
IL MASSAC	87.5	2228
IL MCDONOUGH	496.3	223
IL MCHENRY	119.3	1797
IL MCLEAN	267.0	615
IL MENARD	204.8	940
IL MERCER	265.1	627
IL MONROE	116.3	1841
IL MONTGOMERY	331.6	412
IL MORGAN	238.3	728
IL MOULTRIE	402.6	313
IL OGLE	151.3	1445
IL PEORIA	299.0	480
IL PERRY	104.6	1981
IL PIATT	278.2	563
IL PIKE	408.8	309
IL POPE	104.0	1987
IL PULASKI	114.0	1862
IL PUTNAM	145.0	1507
IL RANDOLPH	136.0	1625
IL RICHLAND	200.9	970
IL ROCK ISLAND	278.4	562
IL SALINE	88.5	2214

U.S. COUNTIES RANKED BY TOTAL FALLOUT DEPOSITION
1951-1970
(microCuries per square meter)
Each Row Includes:
County Name, Fallout Level in microCuries/square meter and County Rank (1 = highest fallout in U.S.)

County	Fallout	Rank
IL SANGAMON	342.2	390
IL SCHUYLER	316.7	435
IL SCOTT	237.8	734
IL SHELBY	399.1	318
IL ST CLAIR	120.4	1784
IL STARK	284.6	537
IL STEPHENSON	154.0	1413
IL TAZEWELL	265.6	621
IL UNION	112.1	1878
IL VERMILION	215.7	859
IL WABASH	129.5	1697
IL WARREN	492.8	224
IL WASHINGTON	98.9	2042
IL WAYNE	123.4	1758
IL WHITE	90.7	2167
IL WHITESIDE	168.3	1280
IL WILL	134.6	1642
IL WILLIAMSON	91.5	2152
IL WINNEBAGO	291.5	511
IL WOODFORD	296.4	489
IN ADAMS	195.7	1026
IN ALLEN	390.0	328
IN BARTHOLOMEW	176.5	1194
IN BENTON	109.7	1918
IN BLACKFORD	67.9	2701
IN BOONE	117.5	1819
IN BROWN	109.2	1922
IN CARROLL	208.0	919
IN CASS	240.1	718
IN CLARK	182.3	1139
IN CLAY	117.5	1817
IN CLINTON	195.5	1030
IN CRAWFORD	239.4	722
IN DAVIESS	140.4	1570
IN DE KALB	63.2	2811
IN DEARBORN	104.1	1985
IN DECATUR	130.8	1687
IN DELAWARE	250.7	678
IN DUBOIS	193.8	1040
IN ELKHART	89.2	2203
IN FAYETTE	72.8	2575
IN FLOYD	181.0	1154
IN FOUNTAIN	116.6	1835
IN FRANKLIN	209.6	904
IN FULTON	195.0	1035
IN GIBSON	103.0	1996
IN GRANT	296.6	488
IN GREENE	240.8	712
IN HAMILTON	297.8	487
IN HANCOCK	89.1	2204
IN HARRISON	169.2	1269
IN HENDRICKS	249.0	688
IN HENRY	232.0	757
IN HOWARD	292.8	506
IN HUNTINGTON	154.6	1405
IN JACKSON	264.3	629
IN JASPER	134.2	1649
IN JAY	89.6	2193
IN JEFFERSON	229.0	770
IN JENNINGS	74.7	2522
IN JOHNSON	101.7	2009
IN KNOX	188.7	1073
IN KOSCIUSKO	100.5	2019
IN LA PORTE	96.1	2085
IN LAGRANGE	83.6	2319
IN LAKE	123.1	1761
IN LAWRENCE	338.6	399
IN MADISON	197.1	1016

U.S. COUNTIES RANKED BY TOTAL FALLOUT DEPOSITION
1951-1970
(microCuries per square meter)
Each Row Includes:
County Name, Fallout Level in microCuries/square meter and County Rank (1 = highest fallout in U.S.)

IN MARION	268.8	608
IN MARSHALL	92.4	2139
IN MARTIN	221.0	823
IN MIAMI	149.0	1468
IN MONROE	215.7	858
IN MONTGOMERY	126.3	1729
IN MORGAN	113.1	1867
IN NEWTON	122.4	1766
IN NOBLE	75.9	2492
IN OHIO	77.7	2450
IN ORANGE	209.1	910
IN OWEN	229.9	767
IN PARKE	112.4	1873
IN PERRY	146.6	1488
IN PIKE	100.9	2016
IN PORTER	123.2	1759
IN POSEY	132.8	1665
IN PULASKI	139.3	1583
IN PUTNAM	111.8	1884
IN RANDOLPH	169.5	1268
IN RIPLEY	109.1	1926
IN RUSH	110.3	1901
IN SCOTT	166.6	1298
IN SHELBY	185.8	1101
IN SPENCER	146.4	1493
IN ST JOSEPH	186.1	1096
IN STARKE	92.8	2128
IN STEUBEN	84.6	2287
IN SULLIVAN	144.1	1516
IN SWITZERLAND	241.6	708
IN TIPPECANOE	154.6	1404
IN TIPTON	289.9	517
IN UNION	196.6	1021
IN VANDERBURGH	149.2	1463
IN VERMILLION	188.0	1084
IN VIGO	140.0	1575
IN WABASH	158.9	1362
IN WARREN	153.3	1422
IN WARRICK	156.0	1386
IN WASHINGTON	168.2	1282
IN WAYNE	383.9	332
IN WELLS	289.7	518
IN WHITE	130.8	1688
IN WHITLEY	72.1	2600
KS ALLEN	166.1	1301
KS ANDERSON	147.9	1479
KS ATCHISON	167.7	1287
KS BARBER	161.7	1336
KS BARTON	235.3	743
KS BOURBON	144.3	1513
KS BROWN	158.1	1369
KS BUTLER	203.8	949
KS CHASE	202.7	956
KS CHAUTAUQUA	204.6	941
KS CHEROKEE	184.4	1120
KS CHEYENNE	284.1	538
KS CLARK	165.7	1303
KS CLAY	176.3	1198
KS CLOUD	165.2	1306
KS COFFEY	168.1	1283
KS COMANCHE	134.8	1640
KS COWLEY	211.8	889
KS CRAWFORD	206.5	931
KS DECATUR	266.0	618
KS DICKINSON	227.4	777
KS DONIPHAN	134.8	1641
KS DOUGLAS	161.1	1340
KS EDWARDS	183.6	1128

U.S. COUNTIES RANKED BY TOTAL FALLOUT DEPOSITION
1951-1970
(microCuries per square meter)
Each Row Includes:
County Name, Fallout Level in microCuries/square meter and County Rank (1 = highest fallout in U.S.)

KS ELK	190.0	1063
KS ELLIS	216.2	854
KS ELLSWORTH	220.0	830
KS FINNEY	269.8	604
KS FORD	295.8	493
KS FRANKLIN	161.7	1335
KS GEARY	200.8	974
KS GOVE	269.5	606
KS GRAHAM	282.1	547
KS GRANT	279.6	556
KS GRAY	252.9	668
KS GREELEY	249.2	686
KS GREENWOOD	200.3	980
KS HAMILTON	264.2	630
KS HARPER	189.8	1065
KS HARVEY	209.4	906
KS HASKELL	249.3	685
KS HODGEMAN	276.0	577
KS JACKSON	165.9	1302
KS JEFFERSON	162.2	1327
KS JEWELL	168.7	1275
KS JOHNSON	170.5	1255
KS KEARNY	277.5	566
KS KINGMAN	206.6	930
KS KIOWA	166.5	1299
KS LABETTE	188.5	1077
KS LANE	272.5	593
KS LEAVENWORTH	198.2	1003
KS LINCOLN	234.4	747
KS LINN	149.7	1454
KS LOGAN	265.9	620
KS LYON	203.1	953
KS MARION	228.9	771
KS MARSHALL	168.5	1277
KS MCPHERSON	222.7	806
KS MEADE	223.6	797
KS MIAMI	161.8	1332
KS MITCHELL	221.0	825
KS MONTGOMERY	174.4	1218
KS MORRIS	190.3	1060
KS MORTON	237.7	735
KS NEMAHA	155.1	1398
KS NEOSHO	187.3	1086
KS NESS	304.8	462
KS NORTON	287.8	525
KS OSAGE	159.2	1358
KS OSBORNE	229.6	769
KS OTTAWA	220.2	829
KS PAWNEE	185.5	1104
KS PHILLIPS	195.5	1031
KS POTTAWATOMIE	184.1	1125
KS PRATT	185.2	1111
KS RAWLINS	263.9	632
KS RENO	210.8	895
KS REPUBLIC	207.6	923
KS RICE	207.1	924
KS RILEY	194.5	1039
KS ROOKS	222.3	810
KS RUSH	209.1	909
KS RUSSELL	274.5	587
KS SALINE	258.7	652
KS SCOTT	268.8	609
KS SEDGWICK	171.0	1249
KS SEWARD	212.6	883
KS SHAWNEE	206.6	929
KS SHERIDAN	249.0	687
KS SHERMAN	182.2	1142
KS SMITH	195.2	1032

U.S. COUNTIES RANKED BY TOTAL FALLOUT DEPOSITION
1951-1970
(microCuries per square meter)
Each Row Includes:
County Name, Fallout Level in microCuries/square meter and County Rank (1 = highest fallout in U.S.)

County	Fallout	Rank
KS STAFFORD	189.0	1068
KS STANTON	285.1	534
KS STEVENS	391.2	326
KS SUMNER	203.5	951
KS THOMAS	248.9	689
KS TREGO	285.1	535
KS WABAUNSEE	221.1	822
KS WALLACE	273.8	590
KS WASHINGTON	178.9	1164
KS WICHITA	260.4	646
KS WILSON	184.6	1117
KS WOODSON	173.7	1222
KS WYANDOTTE	179.6	1159
KY ADAIR	70.6	2629
KY ALLEN	65.5	2753
KY ANDERSON	230.0	766
KY BALLARD	107.9	1939
KY BARREN	77.5	2455
KY BATH	298.5	486
KY BELL	123.9	1750
KY BOONE	181.3	1150
KY BOURBON	162.1	1328
KY BOYD	290.6	512
KY BOYLE	118.4	1804
KY BRACKEN	197.7	1008
KY BREATHITT	282.0	548
KY BRECKINRIDGE	80.8	2377
KY BULLITT	109.7	1919
KY BUTLER	96.3	2075
KY CALDWELL	85.4	2269
KY CALLOWAY	85.3	2272
KY CAMPBELL	385.0	330
KY CARLISLE	92.9	2127
KY CARROLL	365.5	356
KY CARTER	271.9	598
KY CASEY	91.1	2158
KY CHRISTIAN	96.1	2083
KY CLARK	310.9	446
KY CLAY	120.3	1785
KY CLINTON	69.8	2644
KY CRITTENDEN	74.6	2523
KY CUMBERLAND	128.4	1706
KY DAVIESS	131.6	1678
KY EDMONSON	83.7	2316
KY ELLIOTT	273.4	591
KY ESTILL	176.7	1193
KY FAYETTE	155.3	1395
KY FLEMING	275.5	581
KY FLOYD	260.7	645
KY FRANKLIN	307.1	457
KY FULTON	138.7	1593
KY GALLATIN	279.4	557
KY GARRARD	109.8	1914
KY GRANT	261.5	639
KY GRAVES	132.7	1671
KY GRAYSON	60.5	2848
KY GREEN	68.4	2678
KY GREENUP	446.9	271
KY HANCOCK	143.9	1519
KY HARDIN	68.4	2681
KY HARLAN	162.4	1324
KY HARRISON	336.1	403
KY HART	61.4	2839
KY HENDERSON	100.1	2025
KY HENRY	302.1	468
KY HICKMAN	143.1	1530
KY HOPKINS	73.4	2562
KY JACKSON	241.6	707

U.S. COUNTIES RANKED BY TOTAL FALLOUT DEPOSITION
1951-1970
(microCuries per square meter)
Each Row Includes:
County Name, Fallout Level in microCuries/square meter and County Rank (1 = highest fallout in U.S.)

County	Fallout	Rank
KY JEFFERSON	96.7	2071
KY JESSAMINE	181.9	1145
KY JOHNSON	265.3	624
KY KENTON	341.4	392
KY KNOTT	186.3	1095
KY KNOX	115.4	1850
KY LARUE	70.6	2630
KY LAUREL	85.5	2265
KY LAWRENCE	430.4	281
KY LEE	74.0	2546
KY LESLIE	214.6	867
KY LETCHER	176.1	1201
KY LEWIS	423.6	290
KY LINCOLN	99.5	2030
KY LIVINGSTON	99.1	2038
KY LOGAN	74.6	2526
KY LYON	76.9	2465
KY MADISON	169.0	1272
KY MAGOFFIN	276.1	576
KY MARION	63.3	2807
KY MARSHALL	116.9	1827
KY MARTIN	142.5	1539
KY MASON	272.4	594
KY MCCRACKEN	110.0	1907
KY MCCREARY	85.6	2264
KY MCLEAN	84.7	2286
KY MEADE	79.2	2407
KY MENIFEE	302.9	465
KY MERCER	115.5	1847
KY METCALFE	81.4	2362
KY MONROE	93.6	2118
KY MONTGOMERY	215.9	857
KY MORGAN	249.5	684
KY MUHLENBERG	79.2	2411
KY NELSON	56.6	2887
KY NICHOLAS	330.2	413
KY OHIO	109.9	1911
KY OLDHAM	218.5	839
KY OWEN	379.3	337
KY OWSLEY	241.5	710
KY PENDLETON	215.1	862
KY PERRY	192.1	1049
KY PIKE	244.7	701
KY POWELL	438.2	277
KY PULASKI	115.9	1843
KY ROBERTSON	286.1	530
KY ROCKCASTLE	168.3	1281
KY ROWAN	378.9	338
KY RUSSELL	115.5	1849
KY SCOTT	318.0	431
KY SHELBY	252.6	669
KY SIMPSON	72.7	2579
KY SPENCER	187.7	1085
KY TAYLOR	66.3	2744
KY TODD	116.9	1828
KY TRIGG	89.9	2185
KY TRIMBLE	429.0	283
KY UNION	119.7	1792
KY WARREN	95.3	2096
KY WASHINGTON	97.7	2057
KY WAYNE	90.4	2169
KY WEBSTER	84.6	2288
KY WHITLEY	88.0	2220
KY WOLFE	284.1	539
KY WOODFORD	315.2	438
LA ACADIA	253.9	666
LA ALLEN	222.3	807
LA ASCENSION	178.4	1173

U.S. COUNTIES RANKED BY TOTAL FALLOUT DEPOSITION
1951-1970
(microCuries per square meter)
Each Row Includes:
County Name, Fallout Level in microCuries/square meter and County Rank (1 = highest fallout in U.S.)

County	Fallout	Rank
LA ASSUMPTION	200.1	984
LA AVOYELLES	184.9	1113
LA BEAUREGARD	273.8	589
LA BIENVILLE	97.9	2056
LA BOSSIER	104.0	1988
LA CADDO	105.7	1969
LA CALCASIEU	253.3	667
LA CALDWELL	91.0	2163
LA CAMERON	250.9	677
LA CATAHOULA	163.2	1318
LA CLAIBORNE	83.6	2320
LA CONCORDIA	168.8	1274
LA DE SOTO	89.2	2202
LA EAST BATON R	208.8	914
LA EAST CARROLL	97.1	2067
LA EAST FELICIA	191.0	1057
LA EVANGELINE	257.8	657
LA FRANKLIN	94.0	2112
LA GRANT	298.6	485
LA IBERIA	219.7	832
LA IBERVILLE	209.8	901
LA JACKSON	98.3	2054
LA JEFFERSON	119.4	1793
LA JEFFERSON DA	217.9	844
LA LA SALLE	222.2	811
LA LAFAYETTE	270.7	602
LA LAFOURCHE	181.4	1149
LA LINCOLN	95.7	2091
LA LIVINGSTON	185.8	1103
LA MADISON	116.8	1832
LA MOREHOUSE	98.6	2050
LA NATCHITOCHES	93.5	2122
LA ORLEANS	80.4	2383
LA OUACHITA	110.4	1900
LA PLAQUEMINES	161.8	1330
LA POINTE COUPE	213.4	876
LA RAPIDES	291.9	510
LA RED RIVER	84.0	2303
LA RICHLAND	113.2	1866
LA SABINE	173.1	1226
LA ST BERNARD	151.3	1443
LA ST CHARLES	181.7	1148
LA ST HELENA	236.3	739
LA ST JAMES	148.8	1470
LA ST JOHN THE	153.6	1418
LA ST LANDRY	161.7	1333
LA ST MARTIN	199.9	988
LA ST MARY	201.5	964
LA ST TAMMANY	161.7	1334
LA TANGIPAHOA	230.3	761
LA TENSAS	223.5	799
LA TERREBONNE	210.2	899
LA UNION	99.3	2033
LA VERMILION	217.3	846
LA VERNON	171.6	1241
LA WASHINGTON	162.6	1322
LA WEBSTER	108.9	1930
LA WEST BATON R	211.6	891
LA WEST CARROLL	92.9	2126
LA WEST FELICIA	211.9	887
LA WINN	100.0	2027
MA BARNSTABLE	171.7	1238
MA BERKSHIRE	170.7	1252
MA BRISTOL	243.1	704
MA DUKES	172.1	1233
MA ESSEX	195.0	1033
MA FRANKLIN	131.7	1677
MA HAMPDEN	161.1	1339

U.S. COUNTIES RANKED BY TOTAL FALLOUT DEPOSITION
1951-1970
(microCuries per square meter)
Each Row Includes:
County Name, Fallout Level in microCuries/square meter and County Rank (1 = highest fallout in U.S.)

County	Fallout	Rank
MA HAMPSHIRE	147.9	1478
MA MIDDLESEX	245.1	699
MA NANTUCKET	120.2	1787
MA NORFOLK	299.9	477
MA PLYMOUTH	292.9	505
MA SUFFOLK	223.9	794
MA WORCESTER	158.8	1363
MD ALLEGANY	127.7	1712
MD ANNE ARUNDEL	81.5	2358
MD BALTIMORE	145.3	1503
MD CALVERT	84.7	2285
MD CAROLINE	74.3	2535
MD CARROLL	64.9	2763
MD CECIL	83.8	2309
MD CHARLES	81.0	2376
MD DORCHESTER	78.9	2419
MD FREDERICK	78.2	2439
MD GARRETT	175.4	1206
MD HARFORD	147.8	1480
MD HOWARD	82.4	2335
MD KENT	81.2	2367
MD MONTGOMERY	88.7	2210
MD PRINCE GEORG	87.3	2236
MD QUEEN ANNES	73.6	2558
MD SOMERSET	98.9	2043
MD ST MARYS	87.1	2239
MD TALBOT	80.3	2385
MD WASHINGTON	78.8	2422
MD WICOMICO	84.6	2290
MD WORCESTER	93.9	2116
ME ANDROSCOGGIN	103.0	1995
ME AROOSTOOK	184.1	1124
ME CUMBERLAND	108.1	1936
ME FRANKLIN	98.8	2045
ME HANCOCK	122.0	1769
ME KENNEBEC	104.4	1983
ME KNOX	112.9	1869
ME LINCOLN	108.9	1929
ME OXFORD	109.7	1916
ME PENOBSCOT	146.6	1491
ME PISCATAQUIS	137.5	1605
ME SAGADAHOC	105.1	1978
ME SOMERSET	92.1	2144
ME WALDO	109.0	1927
ME WASHINGTON	136.1	1621
ME YORK	105.6	1972
MI ALCONA	65.9	2751
MI ALGER	81.8	2352
MI ALLEGAN	91.1	2160
MI ALPENA	43.9	2963
MI ANTRIM	64.3	2778
MI ARENAC	70.7	2627
MI BARAGA	91.3	2156
MI BARRY	116.9	1826
MI BAY	67.5	2711
MI BENZIE	57.3	2878
MI BERRIEN	94.8	2103
MI BRANCH	109.7	1917
MI CALHOUN	133.5	1656
MI CASS	80.2	2388
MI CHARLEVOIX	68.0	2695
MI CHEBOYGAN	63.5	2799
MI CHIPPEWA	67.8	2702
MI CLARE	61.5	2837
MI CLINTON	133.3	1659
MI CRAWFORD	62.4	2828
MI DELTA	76.8	2467
MI DICKINSON	79.5	2397

U.S. COUNTIES RANKED BY TOTAL FALLOUT DEPOSITION
1951-1970
(microCuries per square meter)
Each Row Includes:
County Name, Fallout Level in microCuries/square meter and County Rank (1 = highest fallout in U.S.)

County	Fallout	Rank	County	Fallout	Rank
MI EATON	221.8	816	MI MISSAUKEE	57.1	2882
MI EMMET	57.1	2883	MI MONROE	147.4	1484
MI GENESEE	159.2	1357	MI MONTCALM	89.8	2189
MI GLADWIN	61.8	2831	MI MONTMORENCY	47.6	2942
MI GOGEBIC	114.3	1859	MI MUSKEGON	87.2	2238
MI GRAND TRAVER	54.4	2904	MI NEWAYGO	83.9	2307
MI GRATIOT	103.5	1990	MI OAKLAND	167.0	1292
MI HILLSDALE	145.2	1504	MI OCEANA	84.4	2298
MI HOUGHTON	113.3	1864	MI OGEMAW	58.1	2874
MI HURON	111.7	1885	MI ONTONAGON	108.9	1928
MI INGHAM	216.9	849	MI OSCEOLA	56.9	2885
MI IONIA	64.1	2784	MI OSCODA	48.6	2939
MI IOSCO	71.5	2617	MI OTSEGO	54.1	2905
MI IRON	94.6	2105	MI OTTAWA	121.3	1779
MI ISABELLA	68.7	2670	MI PRESQUE ISLE	50.4	2927
MI JACKSON	96.3	2077	MI ROSCOMMON	63.5	2800
MI KALAMAZOO	134.4	1647	MI SAGINAW	110.2	1903
MI KALKASKA	49.3	2935	MI SANILAC	120.1	1788
MI KENT	142.0	1547	MI SCHOOLCRAFT	92.6	2134
MI KEWEENAW	110.5	1899	MI SHIAWASSEE	146.6	1487
MI LAKE	51.6	2922	MI ST CLAIR	97.6	2061
MI LAPEER	148.4	1472	MI ST JOSEPH	78.6	2428
MI LEELANAU	47.9	2941	MI TUSCOLA	96.2	2079
MI LENAWEE	155.8	1389	MI VAN BUREN	132.9	1663
MI LIVINGSTON	151.7	1437	MI WASHTENAW	78.3	2435
MI LUCE	79.2	2408	MI WAYNE	143.6	1520
MI MACKINAC	64.3	2779	MI WEXFORD	58.3	2869
MI MACOMB	130.5	1690	MN AITKIN	107.2	1947
MI MANISTEE	58.4	2867	MN ANOKA	115.0	1852
MI MARQUETTE	72.6	2582	MN BECKER	74.6	2524
MI MASON	58.6	2865	MN BELTRAMI	78.0	2443
MI MECOSTA	60.8	2846	MN BENTON	105.8	1965
MI MENOMINEE	93.5	2119	MN BIG STONE	74.5	2529
MI MIDLAND	90.7	2166	MN BLUE EARTH	444.1	274

U.S. COUNTIES RANKED BY TOTAL FALLOUT DEPOSITION

1951-1970

(microCuries per square meter)

Each Row Includes:

County Name, Fallout Level in microCuries/square meter and County Rank (1 = highest fallout in U.S.)

County	Fallout	Rank
MN BROWN	446.8	272
MN CARLTON	134.6	1643
MN CARVER	123.5	1756
MN CASS	91.6	2150
MN CHIPPEWA	93.9	2114
MN CHISAGO	109.3	1921
MN CLAY	53.5	2912
MN CLEARWATER	66.7	2733
MN COOK	92.5	2136
MN COTTONWOOD	258.6	653
MN CROW WING	99.5	2029
MN DAKOTA	140.8	1562
MN DODGE	348.9	379
MN DOUGLAS	69.5	2653
MN FARIBAULT	729.3	111
MN FILLMORE	492.0	225
MN FREEBORN	129.2	1703
MN GOODHUE	340.7	394
MN GRANT	63.4	2805
MN HENNEPIN	97.3	2064
MN HOUSTON	322.1	424
MN HUBBARD	71.8	2609
MN ISANTI	104.0	1986
MN ITASCA	87.9	2222
MN JACKSON	300.0	476
MN KANABEC	102.5	2002
MN KANDIYOHI	102.3	2005
MN KITTSON	65.7	2752
MN KOOCHICHING	90.0	2183
MN LAC QUI PARL	94.8	2102
MN LAKE	99.0	2039
MN LAKE OF THE	58.3	2870
MN LE SUEUR	470.7	245
MN LINCOLN	300.6	474
MN LYON	303.1	464
MN MAHNOMEN	61.4	2838
MN MARSHALL	55.7	2898
MN MARTIN	462.0	257
MN MCLEOD	115.5	1848
MN MEEKER	111.1	1890
MN MILLE LACS	98.7	2047
MN MORRISON	91.9	2149
MN MOWER	752.6	105
MN MURRAY	451.0	265
MN NICOLLET	454.6	262
MN NOBLES	354.2	369
MN NORMAN	54.4	2903
MN OLMSTED	138.8	1591
MN OTTER TAIL	74.6	2525
MN PENNINGTON	67.3	2714
MN PINE	121.0	1780
MN PIPESTONE	462.3	256
MN POLK	68.9	2666
MN POPE	76.7	2472
MN RAMSEY	128.1	1710
MN RED LAKE	61.2	2842
MN REDWOOD	184.7	1115
MN RENVILLE	111.0	1891
MN RICE	485.7	228
MN ROCK	666.0	153
MN ROSEAU	63.0	2814
MN SCOTT	127.0	1719
MN SHERBURNE	117.1	1824
MN SIBLEY	298.6	483
MN ST LOUIS	111.9	1882
MN STEARNS	95.0	2098
MN STEELE	349.0	378
MN STEVENS	72.0	2603

U.S. COUNTIES RANKED BY TOTAL FALLOUT DEPOSITION
1951-1970
(microCuries per square meter)
Each Row Includes:
County Name, Fallout Level in microCuries/square meter and County Rank (1 = highest fallout in U.S.)

County	Fallout	Rank
MN SWIFT	84.5	2294
MN TODD	96.1	2084
MN TRAVERSE	68.4	2682
MN WABASHA	484.9	229
MN WADENA	89.3	2199
MN WASECA	465.9	248
MN WASHINGTON	127.8	1711
MN WATONWAN	721.1	112
MN WILKIN	56.0	2895
MN WINONA	351.3	374
MN WRIGHT	109.1	1925
MN YELLOW MEDIC	92.8	2129
MO ADAIR	317.0	432
MO ANDREW	128.3	1708
MO ATCHISON	233.2	752
MO AUDRAIN	368.4	349
MO BARRY	206.5	932
MO BARTON	174.5	1217
MO BATES	164.3	1312
MO BENTON	222.1	814
MO BOLLINGER	104.2	1984
MO BOONE	173.7	1221
MO BUCHANAN	126.0	1732
MO BUTLER	149.3	1460
MO CALDWELL	265.6	622
MO CALLAWAY	227.3	779
MO CAMDEN	245.6	696
MO CAPE GIRARDE	114.9	1853
MO CARROLL	185.3	1108
MO CARTER	154.3	1409
MO CASS	190.1	1061
MO CEDAR	182.3	1140
MO CHARITON	275.8	580
MO CHRISTIAN	200.0	985
MO CLARK	580.3	186
MO CLAY	165.2	1307
MO CLINTON	177.5	1189
MO COLE	201.1	969
MO COOPER	190.0	1064
MO CRAWFORD	167.3	1291
MO DADE	175.2	1208
MO DALLAS	178.8	1166
MO DAVIESS	366.4	355
MO DE KALB	230.2	763
MO DENT	192.0	1050
MO DOUGLAS	202.4	957
MO DUNKLIN	157.8	1375
MO FRANKLIN	218.9	837
MO GASCONADE	178.9	1165
MO GENTRY	166.7	1297
MO GREENE	195.5	1028
MO GRUNDY	268.7	610
MO HARRISON	255.4	664
MO HENRY	219.9	831
MO HICKORY	149.3	1458
MO HOLT	123.6	1754
MO HOWARD	179.2	1161
MO HOWELL	185.8	1100
MO IRON	134.3	1648
MO JACKSON	143.0	1532
MO JASPER	198.4	1001
MO JEFFERSON	162.9	1319
MO JOHNSON	178.0	1181
MO KNOX	752.1	106
MO LACLEDE	208.4	917
MO LAFAYETTE	205.7	937
MO LAWRENCE	221.4	817
MO LEWIS	577.0	189

U.S. COUNTIES RANKED BY TOTAL FALLOUT DEPOSITION
1951-1970
(microCuries per square meter)
Each Row Includes:
County Name, Fallout Level in microCuries/square meter and County Rank (1 = highest fallout in U.S.)

County	Fallout	Rank
MO LINCOLN	189.8	1066
MO LINN	336.8	401
MO LIVINGSTON	227.1	782
MO MACON	344.1	384
MO MADISON	108.6	1932
MO MARIES	188.2	1081
MO MARION	367.0	352
MO MCDONALD	215.6	861
MO MERCER	187.0	1088
MO MILLER	213.3	877
MO MISSISSIPPI	142.7	1535
MO MONITEAU	171.7	1237
MO MONROE	267.2	614
MO MONTGOMERY	143.2	1528
MO MORGAN	240.0	719
MO NEW MADRID	162.8	1320
MO NEWTON	222.2	813
MO NODAWAY	182.5	1137
MO OREGON	201.2	967
MO OSAGE	230.9	759
MO OZARK	197.6	1011
MO PEMISCOT	149.8	1453
MO PERRY	117.7	1814
MO PETTIS	240.6	714
MO PHELPS	190.8	1058
MO PIKE	393.6	322
MO PLATTE	153.7	1417
MO POLK	204.1	948
MO PULASKI	208.0	920
MO PUTNAM	483.4	232
MO RALLS	519.9	212
MO RANDOLPH	251.6	672
MO RAY	216.2	852
MO REYNOLDS	149.1	1466
MO RIPLEY	167.8	1286
MO SALINE	191.7	1053
MO SCHUYLER	458.1	260
MO SCOTLAND	680.0	144
MO SCOTT	109.1	1924
MO SHANNON	153.4	1420
MO SHELBY	440.9	275
MO ST CHARLES	144.2	1514
MO ST CLAIR	138.2	1598
MO ST FRANCOIS	138.3	1597
MO ST LOUIS	119.3	1794
MO STE GENEVIEV	102.3	2004
MO STODDARD	136.3	1618
MO STONE	171.6	1242
MO SULLIVAN	343.9	386
MO TANEY	130.0	1692
MO TEXAS	199.3	992
MO VERNON	188.9	1069
MO WARREN	134.9	1638
MO WASHINGTON	133.4	1658
MO WAYNE	160.0	1349
MO WEBSTER	194.9	1036
MO WORTH	419.8	294
MO WRIGHT	212.8	881
MS ADAMS	217.8	845
MS ALCORN	108.5	1933
MS AMITE	122.3	1767
MS ATTALA	83.7	2315
MS BENTON	126.1	1731
MS BOLIVAR	124.0	1748
MS CALHOUN	105.6	1974
MS CARROLL	118.9	1801
MS CHICKASAW	135.2	1634
MS CHOCTAW	83.3	2324

U.S. COUNTIES RANKED BY TOTAL FALLOUT DEPOSITION 1951-1970

(microCuries per square meter)

Each Row Includes:
County Name, Fallout Level in microCuries/square meter and County Rank (1 = highest fallout in U.S.)

MS CLAIBORNE	156.2	1384
MS CLARKE	107.9	1941
MS CLAY	89.8	2188
MS COAHOMA	134.1	1650
MS COPIAH	117.2	1822
MS COVINGTON	88.2	2219
MS DE SOTO	142.5	1540
MS FORREST	144.6	1511
MS FRANKLIN	197.4	1015
MS GEORGE	155.7	1390
MS GREENE	129.4	1699
MS GRENADA	122.0	1770
MS HANCOCK	160.3	1347
MS HARRISON	135.5	1633
MS HINDS	170.2	1259
MS HOLMES	74.7	2521
MS HUMPHREYS	106.2	1960
MS ISSAQUENA	112.0	1881
MS ITAWAMBA	86.7	2245
MS JACKSON	135.2	1637
MS JASPER	95.8	2090
MS JEFFERSON	153.4	1421
MS JEFFERSON DA	69.4	2655
MS JONES	76.9	2466
MS KEMPER	74.1	2543
MS LAFAYETTE	112.1	1879
MS LAMAR	121.5	1774
MS LAUDERDALE	143.2	1527
MS LAWRENCE	75.5	2502
MS LEAKE	81.9	2349
MS LEE	116.0	1842
MS LEFLORE	108.0	1937
MS LINCOLN	100.2	2023
MS LOWNDES	67.1	2723
MS MADISON	100.4	2022
MS MARION	149.1	1467
MS MARSHALL	140.5	1568
MS MONROE	89.9	2186
MS MONTGOMERY	121.3	1778
MS NESHOBA	68.7	2674
MS NEWTON	106.3	1958
MS NOXUBEE	68.9	2667
MS OKTIBBEHA	100.4	2021
MS PANOLA	132.7	1670
MS PEARL RIVER	164.7	1309
MS PERRY	112.4	1872
MS PIKE	158.6	1367
MS PONTOTOC	118.2	1809
MS PRENTISS	110.7	1894
MS QUITMAN	141.0	1559
MS RANKIN	78.4	2432
MS SCOTT	92.0	2147
MS SHARKEY	110.0	1906
MS SIMPSON	72.5	2584
MS SMITH	85.7	2261
MS STONE	140.5	1565
MS SUNFLOWER	102.6	1999
MS TALLAHATCHIE	131.8	1675
MS TATE	136.0	1623
MS TIPPAH	134.0	1652
MS TISHOMINGO	112.4	1874
MS TUNICA	126.4	1727
MS UNION	114.6	1858
MS WALTHALL	130.1	1691
MS WARREN	223.9	793
MS WASHINGTON	105.4	1977
MS WAYNE	105.6	1973
MS WEBSTER	96.4	2073

U.S. COUNTIES RANKED BY TOTAL FALLOUT DEPOSITION 1951-1970

(microCuries per square meter)

Each Row Includes:

County Name, Fallout Level in microCuries/square meter and County Rank (1 = highest fallout in U.S.)

MS WILKINSON	205.9	935
MS WINSTON	75.2	2513
MS YALOBUSHA	129.7	1696
MS YAZOO	114.6	1854
MT BEAVERHEAD	491.7	226
MT BIG HORN	221.4	818
MT BLAINE	251.3	674
MT BROADWATER	398.6	319
MT CARBON	258.4	655
MT CARTER	231.9	758
MT CASCADE	316.9	433
MT CHOUTEAU	451.5	264
MT CUSTER	356.2	366
MT DANIELS	201.3	965
MT DAWSON	333.3	409
MT DEER LODGE	639.7	159
MT FALLON	272.1	597
MT FERGUS	302.5	466
MT FLATHEAD	212.8	879
MT GALLATIN	399.9	316
MT GARFIELD	206.6	928
MT GLACIER	140.4	1569
MT GOLDEN VALLE	201.7	963
MT GRANITE	244.7	702
MT HILL	238.3	727
MT JEFFERSON	417.9	298
MT JUDITH BASIN	504.8	217
MT LAKE	200.7	975
MT LEWIS AND CL	398.0	320
MT LIBERTY	193.7	1041
MT LINCOLN	126.8	1722
MT MADISON	426.5	285
MT MCCONE	269.6	605
MT MEAGHER	497.0	222
MT MINERAL	271.5	600
MT MISSOULA	351.7	372
MT MUSSELSHELL	228.5	773
MT PARK	309.0	451
MT PETROLEUM	199.5	990
MT PHILLIPS	182.5	1136
MT PONDERA	190.1	1062
MT POWDER RIVER	373.2	346
MT POWELL	419.9	292
MT PRAIRIE	252.4	670
MT RAVALLI	232.8	756
MT RICHLAND	301.5	473
MT ROOSEVELT	260.2	647
MT ROSEBUD	200.8	973
MT SANDERS	183.0	1134
MT SHERIDAN	237.9	732
MT SILVER BOW	444.5	273
MT STILLWATER	233.8	750
MT SWEET GRASS	223.0	802
MT TETON	293.8	500
MT TOOLE	205.8	936
MT TREASURE	176.1	1202
MT VALLEY	203.1	952
MT WHEATLAND	381.3	336
MT WIBAUX	287.9	523
MT YELLOWSTONE	140.2	1571
NC ALAMANCE	83.6	2318
NC ALEXANDER	87.6	2227
NC ALLEGHANY	90.1	2180
NC ANSON	64.2	2780
NC ASHE	117.8	1813
NC AVERY	90.1	2182
NC BEAUFORT	143.2	1526
NC BERTIE	87.0	2241

U.S. COUNTIES RANKED BY TOTAL FALLOUT DEPOSITION
1951-1970
(microCuries per square meter)
Each Row Includes:
County Name, Fallout Level in microCuries/square meter and County Rank (1 = highest fallout in U.S.)

County	Fallout	Rank
NC BLADEN	72.0	2604
NC BRUNSWICK	67.2	2722
NC BUNCOMBE	91.9	2148
NC BURKE	94.9	2101
NC CABARRUS	60.9	2845
NC CALDWELL	90.2	2172
NC CAMDEN	276.5	571
NC CARTERET	87.5	2232
NC CASWELL	63.1	2812
NC CATAWBA	83.6	2317
NC CHATHAM	123.4	1757
NC CHEROKEE	100.1	2024
NC CHOWAN	172.6	1229
NC CLAY	106.5	1956
NC CLEVELAND	81.7	2354
NC COLUMBUS	76.1	2487
NC CRAVEN	87.6	2225
NC CUMBERLAND	80.6	2380
NC CURRITUCK	277.6	565
NC DARE	89.7	2192
NC DAVIDSON	85.8	2256
NC DAVIE	66.4	2738
NC DUPLIN	77.1	2460
NC DURHAM	91.6	2151
NC EDGECOMBE	69.6	2650
NC FORSYTH	76.8	2470
NC FRANKLIN	58.9	2863
NC GASTON	69.8	2646
NC GATES	82.0	2346
NC GRAHAM	92.1	2146
NC GRANVILLE	124.5	1746
NC GREENE	69.1	2664
NC GUILFORD	87.3	2237
NC HALIFAX	72.9	2573
NC HARNETT	74.0	2548
NC HAYWOOD	104.8	1979
NC HENDERSON	100.9	2014
NC HERTFORD	84.5	2293
NC HOKE	68.4	2686
NC HYDE	105.8	1966
NC IREDELL	86.2	2252
NC JACKSON	109.9	1913
NC JOHNSTON	66.3	2745
NC JONES	112.4	1875
NC LEE	74.1	2544
NC LENOIR	80.7	2379
NC LINCOLN	75.6	2501
NC MACON	110.0	1910
NC MADISON	90.9	2164
NC MARTIN	121.4	1775
NC MCDOWELL	92.3	2140
NC MECKLENBURG	64.0	2787
NC MITCHELL	96.9	2069
NC MONTGOMERY	70.2	2637
NC MOORE	88.3	2216
NC NASH	70.7	2628
NC NEW HANOVER	89.0	2207
NC NORTHAMPTON	146.2	1496
NC ONSLOW	89.7	2191
NC ORANGE	102.8	1997
NC PAMLICO	92.1	2145
NC PASQUOTANK	276.7	568
NC PENDER	81.5	2360
NC PERQUIMANS	278.7	561
NC PERSON	135.8	1631
NC PITT	85.4	2268
NC POLK	86.7	2244
NC RANDOLPH	84.6	2289

U.S. COUNTIES RANKED BY TOTAL FALLOUT DEPOSITION
1951-1970
(microCuries per square meter)
Each Row Includes:
County Name, Fallout Level in microCuries/square meter and County Rank (1 = highest fallout in U.S.)

County	Fallout	Rank
NC RICHMOND	65.9	2750
NC ROBESON	73.9	2552
NC ROCKINGHAM	59.2	2860
NC ROWAN	73.6	2557
NC RUTHERFORD	79.5	2396
NC SAMPSON	73.2	2568
NC SCOTLAND	67.2	2721
NC STANLY	68.2	2691
NC STOKES	67.5	2710
NC SURRY	89.2	2201
NC SWAIN	105.7	1968
NC TRANSYLVANIA	105.7	1970
NC TYRRELL	63.4	2803
NC UNION	69.7	2647
NC VANCE	132.9	1664
NC WAKE	91.1	2161
NC WARREN	79.1	2413
NC WASHINGTON	176.5	1197
NC WATAUGA	96.2	2080
NC WAYNE	79.3	2403
NC WILKES	122.1	1768
NC WILSON	76.7	2477
NC YADKIN	92.5	2137
NC YANCEY	96.0	2087
ND ADAMS	225.3	789
ND BARNES	181.8	1147
ND BENSON	165.3	1305
ND BILLINGS	384.1	331
ND BOTTINEAU	271.9	599
ND BOWMAN	222.0	815
ND BURKE	277.8	564
ND BURLEIGH	226.7	784
ND CASS	200.1	982
ND CAVALIER	137.4	1607
ND DICKEY	209.1	912
ND DIVIDE	281.3	549
ND DUNN	310.8	447
ND EDDY	167.5	1290
ND EMMONS	223.7	795
ND FOSTER	199.0	997
ND GOLDEN VALLE	306.3	459
ND GRAND FORKS	174.8	1211
ND GRANT	227.4	778
ND GRIGGS	174.7	1213
ND HETTINGER	240.4	716
ND KIDDER	154.0	1414
ND LA MOURE	184.4	1119
ND LOGAN	177.8	1185
ND MCHENRY	272.3	595
ND MCINTOSH	183.6	1129
ND MCKENZIE	298.6	484
ND MCLEAN	255.8	663
ND MERCER	282.6	545
ND MORTON	240.7	713
ND MOUNTRAIL	257.5	659
ND NELSON	171.2	1246
ND OLIVER	280.9	551
ND PEMBINA	161.4	1337
ND PIERCE	117.6	1816
ND RAMSEY	155.2	1396
ND RANSOM	221.3	820
ND RENVILLE	250.4	679
ND RICHLAND	261.5	640
ND ROLETTE	121.0	1781
ND SARGENT	212.8	880
ND SHERIDAN	212.9	878
ND SIOUX	174.5	1216
ND SLOPE	265.0	628

U.S. COUNTIES RANKED BY TOTAL FALLOUT DEPOSITION
1951-1970

(microCuries per square meter)
Each Row Includes:
County Name, Fallout Level in microCuries/square meter and County Rank (1 = highest fallout in U.S.)

County	Fallout	Rank
ND STARK	332.4	411
ND STEELE	163.7	1316
ND STUTSMAN	177.7	1187
ND TOWNER	125.6	1739
ND TRAILL	181.3	1151
ND WALSH	180.1	1158
ND WARD	286.7	528
ND WELLS	136.5	1614
ND WILLIAMS	180.5	1156
NE ADAMS	258.5	654
NE ANTELOPE	426.4	286
NE ARTHUR	230.0	765
NE BANNER	207.9	922
NE BLAINE	206.0	934
NE BOONE	203.5	950
NE BOX BUTTE	171.6	1239
NE BOYD	187.2	1087
NE BROWN	200.8	972
NE BUFFALO	295.9	491
NE BURT	169.7	1266
NE BUTLER	322.5	423
NE CASS	285.6	533
NE CEDAR	373.7	345
NE CHASE	294.6	497
NE CHERRY	310.1	449
NE CHEYENNE	251.0	675
NE CLAY	282.7	544
NE COLFAX	341.5	391
NE CUMING	412.1	304
NE CUSTER	328.2	416
NE DAKOTA	681.2	143
NE DAWES	202.9	954
NE DAWSON	283.2	541
NE DEUEL	225.7	787
NE DIXON	746.8	107
NE DODGE	169.8	1263
NE DOUGLAS	296.0	490
NE DUNDY	259.4	649
NE FILLMORE	238.8	724
NE FRANKLIN	180.8	1155
NE FRONTIER	288.8	519
NE FURNAS	281.2	550
NE GAGE	250.3	680
NE GARDEN	209.2	908
NE GARFIELD	424.6	288
NE GOSPER	276.4	572
NE GRANT	201.2	968
NE GREELEY	354.6	368
NE HALL	208.4	916
NE HAMILTON	250.3	681
NE HARLAN	196.8	1019
NE HAYES	261.0	643
NE HITCHCOCK	252.0	671
NE HOLT	591.4	181
NE HOOKER	241.5	709
NE HOWARD	552.4	200
NE JEFFERSON	192.7	1046
NE JOHNSON	241.6	706
NE KEARNEY	204.2	947
NE KEITH	292.5	508
NE KEYA PAHA	171.6	1240
NE KIMBALL	193.7	1042
NE KNOX	174.7	1214
NE LANCASTER	230.2	764
NE LINCOLN	314.9	440
NE LOGAN	293.3	502
NE LOUP	600.8	173
NE MADISON	458.6	259

U.S. COUNTIES RANKED BY TOTAL FALLOUT DEPOSITION
1951-1970
(microCuries per square meter)
Each Row Includes:
County Name, Fallout Level in microCuries/square meter and County Rank (1 = highest fallout in U.S.)

County	Fallout	Rank
NE MCPHERSON	236.2	740
NE MERRICK	351.0	375
NE MORRILL	227.5	776
NE NANCE	349.2	376
NE NEMAHA	221.3	819
NE NUCKOLLS	237.1	736
NE OTOE	214.2	871
NE PAWNEE	157.7	1376
NE PERKINS	290.0	515
NE PHELPS	285.7	532
NE PIERCE	178.2	1178
NE PLATTE	335.2	406
NE POLK	326.5	419
NE RED WILLOW	269.4	607
NE RICHARDSON	160.6	1345
NE ROCK	144.4	1512
NE SALINE	276.3	575
NE SARPY	149.7	1455
NE SAUNDERS	301.8	471
NE SCOTTS BLUFF	137.5	1606
NE SEWARD	257.2	660
NE SHERIDAN	200.1	983
NE SHERMAN	540.4	204
NE SIOUX	300.1	475
NE STANTON	704.9	121
NE THAYER	212.7	882
NE THOMAS	259.3	650
NE THURSTON	696.5	128
NE VALLEY	336.0	404
NE WASHINGTON	476.4	237
NE WAYNE	383.3	333
NE WEBSTER	211.3	892
NE WHEELER	191.5	1055
NE YORK	248.6	690
NH BELKNAP	118.8	1802
NH CARROLL	141.9	1549
NH CHESHIRE	133.7	1654
NH COOS	110.7	1895
NH GRAFTON	123.1	1760
NH HILLSBOROUGH	128.3	1707
NH MERRIMACK	118.9	1799
NH ROCKINGHAM	155.0	1400
NH STRAFFORD	127.0	1720
NH SULLIVAN	121.6	1773
NJ ATLANTIC	176.5	1196
NJ BERGEN	141.6	1552
NJ BURLINGTON	215.1	863
NJ CAMDEN	114.1	1861
NJ CAPE MAY	132.7	1668
NJ CUMBERLAND	109.9	1912
NJ ESSEX	149.2	1464
NJ GLOUCESTER	112.2	1876
NJ HUDSON	152.7	1431
NJ HUNTERDON	101.0	2013
NJ MERCER	131.8	1676
NJ MIDDLESEX	138.4	1595
NJ MONMOUTH	174.6	1215
NJ MORRIS	135.9	1627
NJ OCEAN	196.3	1023
NJ PASSAIC	136.7	1611
NJ SALEM	102.7	1998
NJ SOMERSET	120.9	1782
NJ SUSSEX	118.3	1806
NJ UNION	143.5	1522
NJ WARREN	122.8	1763
NM BERNALILLO	834.3	84
NM CATRON	341.3	393
NM CHAVES	579.6	187

U.S. COUNTIES RANKED BY TOTAL FALLOUT DEPOSITION
1951-1970
(microCuries per square meter)
Each Row Includes:
County Name, Fallout Level in microCuries/square meter and County Rank (1 = highest fallout in U.S.)

County	Fallout	Rank
NM COLFAX	549.5	202
NM CURRY	483.1	233
NM DE BACA	511.0	215
NM DONA ANA	128.5	1705
NM EDDY	234.5	746
NM GRANT	150.4	1449
NM GUADALUPE	613.3	168
NM HARDING	523.2	210
NM HIDALGO	133.1	1661
NM LEA	223.6	796
NM LINCOLN	576.4	190
NM LOS ALAMOS	672.4	151
NM LUNA	139.6	1580
NM MCKINLEY	1166.9	46
NM MORA	552.5	199
NM OTERO	148.2	1473
NM QUAY	537.9	206
NM RIO ARRIBA	573.7	191
NM ROOSEVELT	463.2	254
NM SAN JUAN	715.0	118
NM SAN MIGUEL	582.8	184
NM SANDOVAL	1013.7	67
NM SANTA FE	835.3	83
NM SIERRA	214.9	865
NM SOCORRO	418.2	297
NM TAOS	517.3	214
NM TORRANCE	827.5	86
NM UNION	527.6	207
NM VALENCIA	903.9	76
NV CARSON CITY	469.3	247
NV CHURCHILL	612.7	169
NV CLARK1	3315.8	6
NV CLARK2	448.5	267
NV CLARK3	171.8	1236
NV DOUGLAS	457.1	261
NV ELKO	752.7	104
NV ESMERALDA1	366.6	354
NV ESMERALDA2	674.3	149
NV EUREKA	1196.3	44
NV HUMBOLDT	321.0	427
NV LANDER1	475.4	240
NV LANDER2	735.9	109
NV LINCOLN1	3420.7	5
NV LINCOLN2	2191.8	12
NV LYON	475.8	238
NV MINERAL	672.1	152
NV NYE1	624.6	164
NV NYE2	3822.2	3
NV NYE3	118.3	1805
NV PERSHING	366.8	353
NV STOREY	461.3	258
NV WASHOE	473.3	242
NV WHITE PINE1	1599.7	24
NV WHITE PINE2	2451.0	10
NV WHITE PINE3	2170.9	13
NY ALBANY	275.0	584
NY ALLEGANY	153.1	1427
NY BRONX	133.2	1660
NY BROOME	73.4	2563
NY CATTARAUGUS	146.6	1489
NY CAYUGA	99.6	2028
NY CHAUTAUQUA	178.0	1182
NY CHEMUNG	100.6	2018
NY CHENANGO	98.6	2052
NY CLINTON	154.6	1406
NY COLUMBIA	303.4	463
NY CORTLAND	104.7	1980
NY DELAWARE	142.7	1536

U.S. COUNTIES RANKED BY TOTAL FALLOUT DEPOSITION 1951-1970

(microCuries per square meter)
Each Row Includes:
County Name, Fallout Level in microCuries/square meter and County Rank (1 = highest fallout in U.S.)

County	Fallout	Rank
NY DUTCHESS	134.1	1651
NY ERIE	186.7	1091
NY ESSEX	201.8	961
NY FRANKLIN	140.5	1567
NY FULTON	406.1	310
NY GENESEE	178.3	1175
NY GREENE	201.7	962
NY HAMILTON	183.9	1126
NY HERKIMER	137.3	1608
NY JEFFERSON	139.7	1577
NY KINGS	155.2	1397
NY LEWIS	107.5	1944
NY LIVINGSTON	164.2	1313
NY MADISON	102.5	2003
NY MONROE	104.5	1982
NY MONTGOMERY	188.8	1072
NY NASSAU	170.2	1257
NY NEW YORK	141.5	1553
NY NIAGARA	166.8	1296
NY ONEIDA	106.6	1954
NY ONONDAGA	87.0	2240
NY ONTARIO	152.0	1434
NY ORANGE	130.9	1686
NY ORLEANS	143.5	1523
NY OSWEGO	136.3	1617
NY OTSEGO	147.7	1482
NY PUTNAM	146.3	1495
NY QUEENS	157.9	1371
NY RENSSELAER	606.8	171
NY RICHMOND	138.2	1600
NY ROCKLAND	118.7	1803
NY SARATOGA	411.9	305
NY SCHENECTADY	315.7	437
NY SCHOHARIE	178.3	1177
NY SCHUYLER	84.9	2282
NY SENECA	127.1	1716
NY ST LAWRENCE	161.0	1341
NY STEUBEN	147.0	1485
NY SUFFOLK	160.0	1351
NY SULLIVAN	148.0	1475
NY TIOGA	82.1	2341
NY TOMPKINS	122.7	1764
NY ULSTER	170.9	1250
NY WARREN	614.5	166
NY WASHINGTON	637.0	161
NY WAYNE	153.1	1426
NY WESTCHESTER	155.5	1394
NY WYOMING	175.8	1204
NY YATES	106.7	1952
OH ADAMS	280.8	552
OH ALLEN	302.1	467
OH ASHLAND	209.3	907
OH ASHTABULA	107.9	1938
OH ATHENS	243.7	703
OH AUGLAIZE	174.9	1209
OH BELMONT	197.6	1010
OH BROWN	222.8	805
OH BUTLER	405.2	311
OH CARROLL	88.9	2209
OH CHAMPAIGN	313.5	444
OH CLARK	345.2	383
OH CLERMONT	224.7	791
OH CLINTON	336.4	402
OH COLUMBIANA	195.0	1034
OH COSHOCTON	227.2	781
OH CRAWFORD	173.1	1227
OH CUYAHOGA	175.4	1207
OH DARKE	209.8	902

U.S. COUNTIES RANKED BY TOTAL FALLOUT DEPOSITION
1951-1970
(microCuries per square meter)
Each Row Includes:
County Name, Fallout Level in microCuries/square meter and County Rank (1 = highest fallout in U.S.)

County	Fallout	Rank
OH DEFIANCE	70.5	2633
OH DELAWARE	184.1	1123
OH ERIE	178.7	1167
OH FAIRFIELD	166.9	1295
OH FAYETTE	170.3	1256
OH FRANKLIN	107.7	1943
OH FULTON	282.8	543
OH GALLIA	230.8	760
OH GEAUGA	156.7	1381
OH GREENE	162.2	1326
OH GUERNSEY	217.2	848
OH HAMILTON	195.9	1025
OH HANCOCK	292.8	507
OH HARDIN	111.2	1888
OH HARRISON	184.4	1118
OH HENRY	93.5	2120
OH HIGHLAND	308.8	453
OH HOCKING	181.0	1153
OH HOLMES	102.3	2006
OH HURON	148.1	1474
OH JACKSON	295.2	494
OH JEFFERSON	85.0	2279
OH KNOX	302.0	469
OH LAKE	198.3	1002
OH LAWRENCE	268.7	611
OH LICKING	311.1	445
OH LOGAN	224.5	792
OH LORAIN	151.4	1439
OH LUCAS	183.5	1131
OH MADISON	238.9	723
OH MAHONING	188.6	1075
OH MARION	183.8	1127
OH MEDINA	381.5	335
OH MEIGS	186.5	1094
OH MERCER	106.8	1950
OH MIAMI	310.6	448
OH MONROE	185.2	1110
OH MONTGOMERY	376.1	341
OH MORGAN	293.2	503
OH MORROW	179.4	1160
OH MUSKINGUM	216.8	851
OH NOBLE	157.5	1379
OH OTTAWA	114.1	1860
OH PAULDING	215.0	864
OH PERRY	301.9	470
OH PICKAWAY	77.6	2453
OH PIKE	292.4	509
OH PORTAGE	237.8	733
OH PREBLE	283.0	542
OH PUTNAM	139.5	1581
OH RICHLAND	307.3	455
OH ROSS	177.4	1190
OH SANDUSKY	192.2	1048
OH SCIOTO	432.5	280
OH SENECA	367.4	351
OH SHELBY	75.3	2509
OH STARK	189.4	1067
OH SUMMIT	126.4	1728
OH TRUMBULL	204.2	945
OH TUSCARAWAS	199.2	993
OH UNION	219.5	835
OH VAN WERT	159.9	1352
OH VINTON	162.3	1325
OH WARREN	339.5	396
OH WASHINGTON	186.8	1090
OH WAYNE	106.2	1961
OH WILLIAMS	62.6	2825
OH WOOD	197.0	1017

U.S. COUNTIES RANKED BY TOTAL FALLOUT DEPOSITION
1951-1970

(microCuries per square meter)
Each Row Includes:
County Name, Fallout Level in microCuries/square meter and County Rank (1 = highest fallout in U.S.)

County	Fallout	Rank
OH WYANDOT	279.0	560
OK ADAIR	185.1	1112
OK ALFALFA	186.1	1097
OK ATOKA	221.3	821
OK BEAVER	197.5	1014
OK BECKHAM	183.2	1132
OK BLAINE	134.8	1639
OK BRYAN	185.3	1109
OK CADDO	132.7	1667
OK CANADIAN	143.6	1521
OK CARTER	144.1	1515
OK CHEROKEE	240.2	717
OK CHOCTAW	198.8	998
OK CIMARRON	305.6	460
OK CLEVELAND	126.6	1724
OK COAL	145.1	1506
OK COMANCHE	141.1	1557
OK COTTON	118.9	1800
OK CRAIG	256.5	661
OK CREEK	248.2	692
OK CUSTER	117.5	1818
OK DELAWARE	227.7	775
OK DEWEY	137.7	1603
OK ELLIS	214.4	869
OK GARFIELD	174.8	1210
OK GARVIN	124.3	1747
OK GRADY	124.5	1745
OK GRANT	186.0	1099
OK GREER	179.0	1163
OK HARMON	164.3	1311
OK HARPER	139.1	1585
OK HASKELL	183.2	1133
OK HUGHES	204.9	939
OK JACKSON	117.9	1811
OK JEFFERSON	126.5	1726
OK JOHNSTON	153.9	1416
OK KAY	210.7	896
OK KINGFISHER	158.7	1365
OK KIOWA	120.7	1783
OK LATIMER	185.8	1102
OK LE FLORE	167.5	1288
OK LINCOLN	169.0	1271
OK LOGAN	151.5	1438
OK LOVE	155.9	1388
OK MAJOR	136.6	1613
OK MARSHALL	140.7	1563
OK MAYES	221.0	824
OK MCCLAIN	129.1	1704
OK MCCURTAIN	191.9	1052
OK MCINTOSH	202.4	959
OK MURRAY	125.3	1740
OK MUSKOGEE	216.1	855
OK NOBLE	188.2	1082
OK NOWATA	214.1	872
OK OKFUSKEE	171.5	1243
OK OKLAHOMA	124.6	1743
OK OKMULGEE	185.5	1106
OK OSAGE	209.1	911
OK OTTAWA	236.3	738
OK PAWNEE	188.4	1078
OK PAYNE	156.1	1385
OK PITTSBURG	220.6	827
OK PONTOTOC	178.5	1171
OK POTTAWATOMIE	153.3	1423
OK PUSHMATAHA	186.1	1098
OK ROGER MILLS	223.5	800
OK ROGERS	214.8	866
OK SEMINOLE	191.9	1051

U.S. COUNTIES RANKED BY TOTAL FALLOUT DEPOSITION
1951-1970
(microCuries per square meter)
Each Row Includes:
County Name, Fallout Level in microCuries/square meter and County Rank (1 = highest fallout in U.S.)

County	Fallout	Rank
OK SEQUOYAH	210.9	893
OK STEPHENS	127.0	1717
OK TEXAS	355.9	367
OK TILLMAN	103.7	1989
OK TULSA	274.2	588
OK WAGONER	216.0	856
OK WASHINGTON	211.9	886
OK WASHITA	177.7	1186
OK WOODS	150.4	1448
OK WOODWARD	124.7	1742
OR BAKER	309.4	450
OR BENTON	19.9	3087
OR CLACKAMAS	31.3	3029
OR CLATSOP	27.7	3047
OR COLUMBIA	24.0	3068
OR COOS	25.6	3061
OR CROOK	36.3	2995
OR CURRY	28.9	3042
OR DESCHUTES	22.8	3075
OR DOUGLAS	27.8	3046
OR GILLIAM	25.7	3060
OR GRANT	193.6	1044
OR HARNEY	110.0	1908
OR HOOD RIVER	23.3	3070
OR JACKSON	19.9	3088
OR JEFFERSON	25.9	3056
OR JOSEPHINE	26.2	3053
OR KLAMATH	29.6	3035
OR LAKE	40.3	2979
OR LANE	29.1	3040
OR LINCOLN	29.7	3034
OR LINN	26.2	3052
OR MALHEUR	135.9	1629
OR MARION	28.9	3041
OR MORROW	35.3	3001
OR MULTNOMAH	21.8	3081
OR POLK	20.3	3085
OR SHERMAN	23.6	3069
OR TILLAMOOK	27.7	3048
OR UMATILLA	105.5	1975
OR UNION	172.0	1235
OR WALLOWA	330.1	414
OR WASCO	22.1	3079
OR WASHINGTON	20.7	3084
OR WHEELER	29.9	3033
OR YAMHILL	19.2	3090
PA ADAMS	136.0	1624
PA ALLEGHENY	159.0	1360
PA ARMSTRONG	198.5	999
PA BEAVER	202.2	960
PA BEDFORD	135.9	1628
PA BERKS	93.3	2125
PA BLAIR	103.2	1994
PA BRADFORD	84.5	2295
PA BUCKS	95.7	2093
PA BUTLER	172.0	1234
PA CAMBRIA	194.7	1037
PA CAMERON	143.0	1531
PA CARBON	95.0	2099
PA CENTRE	148.8	1469
PA CHESTER	173.6	1224
PA CLARION	197.5	1012
PA CLEARFIELD	208.5	915
PA CLINTON	154.6	1403
PA COLUMBIA	76.4	2484
PA CRAWFORD	202.4	958
PA CUMBERLAND	63.7	2795
PA DAUPHIN	61.5	2835

U.S. COUNTIES RANKED BY TOTAL FALLOUT DEPOSITION
1951-1970
(microCuries per square meter)
Each Row Includes:
County Name, Fallout Level in microCuries/square meter and County Rank (1 = highest fallout in U.S.)

County	Fallout	Rank
PA DELAWARE	157.9	1373
PA ELK	258.8	651
PA ERIE	157.6	1377
PA FAYETTE	215.7	860
PA FOREST	222.2	812
PA FRANKLIN	75.9	2497
PA FULTON	223.6	798
PA GREENE	176.2	1199
PA HUNTINGDON	148.7	1471
PA INDIANA	199.1	995
PA JEFFERSON	185.5	1105
PA JUNIATA	66.2	2748
PA LACKAWANNA	140.9	1561
PA LANCASTER	135.8	1630
PA LAWRENCE	182.4	1138
PA LEBANON	61.3	2841
PA LEHIGH	96.9	2068
PA LUZERNE	117.9	1812
PA LYCOMING	72.0	2606
PA MCKEAN	125.9	1733
PA MERCER	204.2	943
PA MIFFLIN	70.0	2641
PA MONROE	112.9	1868
PA MONTGOMERY	97.2	2065
PA MONTOUR	82.7	2330
PA NORTHAMPTON	99.3	2034
PA NORTHUMBERLA	142.3	1542
PA PERRY	60.4	2849
PA PHILADELPHIA	81.6	2356
PA PIKE	123.9	1751
PA POTTER	131.4	1680
PA SCHUYLKILL	84.3	2299
PA SNYDER	234.9	744
PA SOMERSET	167.9	1285
PA SULLIVAN	86.3	2250
PA SUSQUEHANNA	94.5	2106
PA TIOGA	98.8	2044
PA UNION	82.2	2337
PA VENANGO	327.8	418
PA WARREN	158.7	1366
PA WASHINGTON	188.2	1080
PA WAYNE	138.2	1601
PA WESTMORELAND	193.7	1043
PA WYOMING	115.9	1844
PA YORK	145.8	1499
RI BRISTOL	181.9	1146
RI KENT	176.5	1195
RI NEWPORT	138.3	1596
RI PROVIDENCE	149.2	1462
RI WASHINGTON	192.3	1047
SC ABBEVILLE	67.6	2708
SC AIKEN	59.1	2861
SC ALLENDALE	67.6	2707
SC ANDERSON	75.9	2496
SC BAMBERG	75.1	2516
SC BARNWELL	68.7	2671
SC BEAUFORT	81.6	2357
SC BERKELEY	84.1	2300
SC CALHOUN	89.0	2206
SC CHARLESTON	82.1	2339
SC CHEROKEE	72.6	2580
SC CHESTER	68.4	2679
SC CHESTERFIELD	79.4	2399
SC CLARENDON	92.2	2141
SC COLLETON	82.9	2326
SC DARLINGTON	82.6	2332
SC DILLON	77.4	2457
SC DORCHESTER	83.8	2311

U.S. COUNTIES RANKED BY TOTAL FALLOUT DEPOSITION
1951-1970
(microCuries per square meter)
Each Row Includes:
County Name, Fallout Level in microCuries/square meter and County Rank (1 = highest fallout in U.S.)

County	Fallout	Rank
SC EDGEFIELD	68.9	2665
SC FAIRFIELD	66.3	2740
SC FLORENCE	112.7	1871
SC GEORGETOWN	72.4	2588
SC GREENVILLE	88.3	2218
SC GREENWOOD	70.9	2624
SC HAMPTON	71.1	2620
SC HORRY	87.4	2233
SC JASPER	93.5	2123
SC KERSHAW	77.4	2456
SC LANCASTER	67.3	2718
SC LAURENS	71.2	2619
SC LEE	94.1	2110
SC LEXINGTON	63.8	2790
SC MARION	89.2	2200
SC MARLBORO	81.2	2369
SC MCCORMICK	70.4	2634
SC NEWBERRY	69.1	2663
SC OCONEE	106.6	1953
SC ORANGEBURG	86.4	2248
SC PICKENS	94.4	2107
SC RICHLAND	59.4	2859
SC SALUDA	67.9	2699
SC SPARTANBURG	81.0	2375
SC SUMTER	90.1	2178
SC UNION	68.2	2692
SC WILLIAMSBURG	78.2	2438
SC YORK	72.7	2577
SD AURORA	254.2	665
SD BEADLE	211.8	890
SD BENNETT	233.1	754
SD BON HOMME	158.7	1364
SD BROOKINGS	399.8	317
SD BROWN	169.8	1264
SD BRULE	261.4	641
SD BUFFALO	214.1	873
SD BUTTE	219.6	833
SD CAMPBELL	196.8	1018
SD CHARLES MIX	209.5	905
SD CLARK	199.1	994
SD CLAY	356.7	365
SD CODINGTON	206.7	927
SD CORSON	233.2	753
SD CUSTER	783.7	96
SD DAVISON	411.8	306
SD DAY	200.2	981
SD DEUEL	90.2	2177
SD DEWEY	272.2	596
SD DOUGLAS	208.3	918
SD EDMUNDS	184.8	1114
SD FALL RIVER	552.0	201
SD FAULK	245.4	698
SD GRANT	72.2	2596
SD GREGORY	197.7	1009
SD HAAKON	316.0	436
SD HAMLIN	652.9	157
SD HAND	200.4	979
SD HANSON	415.8	300
SD HARDING	256.1	662
SD HUGHES	166.9	1293
SD HUTCHINSON	572.1	192
SD HYDE	227.3	780
SD JACKSON	283.5	540
SD JERAULD	356.7	364
SD JONES	216.2	853
SD KINGSBURY	596.4	177
SD LAKE	547.2	203
SD LAWRENCE	219.5	834

U.S. COUNTIES RANKED BY TOTAL FALLOUT DEPOSITION
1951-1970
(microCuries per square meter)
Each Row Includes:
County Name, Fallout Level in microCuries/square meter and County Rank (1 = highest fallout in U.S.)

County	Fallout	Rank
SD LINCOLN	392.7	323
SD LYMAN	275.2	583
SD MARSHALL	188.1	1083
SD MCCOOK	558.0	197
SD MCPHERSON	167.5	1289
SD MEADE	261.8	638
SD MELLETTE	274.7	585
SD MINER	517.8	213
SD MINNEHAHA	364.6	358
SD MOODY	473.9	241
SD PENNINGTON	218.2	841
SD PERKINS	223.0	803
SD POTTER	236.1	741
SD ROBERTS	223.4	801
SD SANBORN	214.6	868
SD SHANNON	523.0	211
SD SPINK	225.9	786
SD STANLEY	631.6	162
SD SULLY	238.7	725
SD TODD	234.0	749
SD TRIPP	205.2	938
SD TURNER	739.5	108
SD UNION	385.8	329
SD WALWORTH	219.4	836
SD WASHABAUGH	1071.2	58
SD YANKTON	411.8	307
SD ZIEBACH	265.9	619
TN ANDERSON	105.7	1971
TN BEDFORD	81.8	2350
TN BENTON	72.0	2602
TN BLEDSOE	95.8	2089
TN BLOUNT	97.5	2062
TN BRADLEY	89.9	2187
TN CAMPBELL	125.1	1741
TN CANNON	67.0	2725
TN CARROLL	96.2	2082
TN CARTER	121.8	1771
TN CHEATHAM	90.4	2170
TN CHESTER	103.4	1992
TN CLAIBORNE	126.2	1730
TN CLAY	112.8	1870
TN COCKE	99.2	2036
TN COFFEE	96.3	2078
TN CROCKETT	135.2	1635
TN CUMBERLAND	92.7	2130
TN DAVIDSON	86.6	2247
TN DE KALB	77.6	2452
TN DECATUR	90.2	2176
TN DICKSON	95.6	2094
TN DYER	147.5	1483
TN FAYETTE	133.5	1657
TN FENTRESS	91.1	2162
TN FRANKLIN	68.8	2668
TN GIBSON	163.3	1317
TN GILES	72.3	2593
TN GRAINGER	73.1	2569
TN GREENE	90.4	2171
TN GRUNDY	95.7	2092
TN HAMBLEN	80.5	2382
TN HAMILTON	83.7	2314
TN HANCOCK	123.9	1749
TN HARDEMAN	128.2	1709
TN HARDIN	85.7	2258
TN HAWKINS	92.7	2132
TN HAYWOOD	145.8	1500
TN HENDERSON	87.7	2224
TN HENRY	95.1	2097
TN HICKMAN	84.8	2283

U.S. COUNTIES RANKED BY TOTAL FALLOUT DEPOSITION
1951-1970
(microCuries per square meter)
Each Row Includes:
County Name, Fallout Level in microCuries/square meter and County Rank (1 = highest fallout in U.S.)

TN HOUSTON	75.9	2493
TN HUMPHREYS	90.2	2175
TN JACKSON	71.5	2616
TN JEFFERSON	84.7	2284
TN JOHNSON	85.3	2273
TN KNOX	81.1	2372
TN LAKE	154.9	1402
TN LAUDERDALE	151.7	1436
TN LAWRENCE	73.3	2565
TN LEWIS	63.6	2797
TN LINCOLN	71.8	2612
TN LOUDON	86.7	2243
TN MACON	58.8	2864
TN MADISON	149.1	1465
TN MARION	70.2	2638
TN MARSHALL	82.6	2333
TN MAURY	81.8	2351
TN MCMINN	93.9	2115
TN MCNAIRY	110.9	1892
TN MEIGS	78.9	2416
TN MONROE	94.0	2111
TN MONTGOMERY	98.5	2053
TN MOORE	67.4	2712
TN MORGAN	89.0	2208
TN OBION	151.3	1444
TN OVERTON	78.6	2430
TN PERRY	81.0	2373
TN PICKETT	108.3	1935
TN POLK	89.3	2196
TN PUTNAM	85.3	2270
TN RHEA	90.2	2173
TN ROANE	99.0	2040
TN ROBERTSON	79.8	2394
TN RUTHERFORD	61.4	2840
TN SCOTT	109.5	1920
TN SEQUATCHIE	90.2	2174
TN SEVIER	112.1	1880
TN SHELBY	116.8	1830
TN SMITH	71.9	2607
TN STEWART	80.0	2390
TN SULLIVAN	115.8	1846
TN SUMNER	78.6	2429
TN TIPTON	155.7	1391
TN TROUSDALE	58.1	2873
TN UNICOI	85.7	2259
TN UNION	75.2	2511
TN VAN BUREN	84.4	2297
TN WARREN	85.2	2274
TN WASHINGTON	158.3	1368
TN WAYNE	74.1	2545
TN WEAKLEY	156.7	1380
TN WHITE	79.0	2415
TN WILLIAMSON	77.5	2454
TN WILSON	72.4	2589
TX ANDERSON	82.7	2331
TX ANDREWS	169.8	1265
TX ANGELINA	155.9	1387
TX ARANSAS	36.6	2992
TX ARCHER	92.2	2142
TX ARMSTRONG	210.5	897
TX ATASCOSA	35.8	2997
TX AUSTIN	132.7	1672
TX BAILEY	288.7	520
TX BANDERA	58.9	2862
TX BASTROP	41.8	2973
TX BAYLOR	83.8	2310
TX BEE	38.9	2987
TX BELL	69.8	2645

U.S. COUNTIES RANKED BY TOTAL FALLOUT DEPOSITION
1951-1970
(microCuries per square meter)
Each Row Includes:
County Name, Fallout Level in microCuries/square meter and County Rank (1 = highest fallout in U.S.)

County	Fallout	Rank
TX BEXAR	46.3	2951
TX BLANCO	61.0	2844
TX BORDEN	93.5	2121
TX BOSQUE	82.1	2340
TX BOWIE	144.7	1508
TX BRAZORIA	141.1	1555
TX BRAZOS	66.6	2735
TX BREWSTER	46.8	2947
TX BRISCOE	243.1	705
TX BROOKS	35.5	2999
TX BROWN	74.3	2538
TX BURLESON	61.8	2832
TX BURNET	54.8	2900
TX CALDWELL	42.5	2970
TX CALHOUN	40.0	2983
TX CALLAHAN	88.3	2217
TX CAMERON	38.7	2988
TX CAMP	139.7	1579
TX CARSON	211.8	888
TX CASS	137.6	1604
TX CASTRO	270.6	603
TX CHAMBERS	147.8	1481
TX CHEROKEE	100.5	2020
TX CHILDRESS	178.2	1179
TX CLAY	116.7	1834
TX COCHRAN	246.8	695
TX COKE	56.7	2886
TX COLEMAN	65.0	2760
TX COLLIN	178.5	1169
TX COLLINGSWORT	188.7	1074
TX COLORADO	50.2	2930
TX COMAL	49.6	2933
TX COMANCHE	67.7	2704
TX CONCHO	70.5	2632
TX COOKE	174.7	1212
TX CORYELL	63.3	2808
TX COTTLE	92.6	2135
TX CRANE	53.7	2908
TX CROCKETT	45.1	2958
TX CROSBY	171.1	1247
TX CULBERSON	160.6	1344
TX DALLAM	425.8	287
TX DALLAS	160.4	1346
TX DAWSON	106.6	1955
TX DE WITT	41.3	2977
TX DEAF SMITH	298.8	482
TX DELTA	132.6	1673
TX DENTON	171.4	1244
TX DICKENS	106.3	1957
TX DIMMIT	42.4	2971
TX DONLEY	213.9	874
TX DUVAL	31.8	3022
TX EASTLAND	79.6	2395
TX ECTOR	116.4	1839
TX EDWARDS	33.9	3008
TX EL PASO	216.9	850
TX ELLIS	127.0	1718
TX ERATH	76.6	2479
TX FALLS	68.3	2687
TX FANNIN	210.9	894
TX FAYETTE	53.7	2909
TX FISHER	96.8	2070
TX FLOYD	199.0	996
TX FOARD	97.4	2063
TX FORT BEND	132.9	1662
TX FRANKLIN	117.2	1823
TX FREESTONE	74.9	2520
TX FRIO	53.7	2910

U.S. COUNTIES RANKED BY TOTAL FALLOUT DEPOSITION
1951-1970

(microCuries per square meter)
Each Row Includes:
County Name, Fallout Level in microCuries/square meter and County Rank (1 = highest fallout in U.S.)

County	Fallout	Rank
TX GAINES	200.9	971
TX GALVESTON	129.2	1702
TX GARZA	103.3	1993
TX GILLESPIE	56.5	2888
TX GLASSCOCK	54.0	2907
TX GOLIAD	32.4	3019
TX GONZALES	43.4	2966
TX GRAY	184.2	1121
TX GRAYSON	201.3	966
TX GREGG	99.4	2032
TX GRIMES	144.0	1518
TX GUADALUPE	31.5	3025
TX HALE	241.1	711
TX HALL	198.0	1005
TX HAMILTON	75.4	2507
TX HANSFORD	288.1	522
TX HARDEMAN	116.5	1838
TX HARDIN	138.9	1587
TX HARRIS	142.6	1537
TX HARRISON	98.9	2041
TX HARTLEY	340.3	395
TX HASKELL	85.2	2275
TX HAYS	40.3	2980
TX HEMPHILL	176.1	1203
TX HENDERSON	99.1	2037
TX HIDALGO	33.1	3013
TX HILL	88.0	2221
TX HOCKLEY	248.4	691
TX HOOD	90.0	2184
TX HOPKINS	149.3	1461
TX HOUSTON	77.7	2451
TX HOWARD	74.4	2531
TX HUDSPETH	186.5	1093
TX HUNT	142.8	1534
TX HUTCHINSON	287.2	527
TX IRION	56.0	2892
TX JACK	97.7	2058
TX JACKSON	43.3	2967
TX JASPER	161.1	1338
TX JEFF DAVIS	146.7	1486
TX JEFFERSON	85.1	2278
TX JIM HOGG	39.5	2985
TX JIM WELLS	43.0	2969
TX JOHNSON	106.3	1959
TX JONES	69.2	2662
TX KARNES	35.7	2998
TX KAUFMAN	110.7	1896
TX KENDALL	48.9	2938
TX KENEDY	33.0	3014
TX KENT	92.4	2138
TX KERR	51.9	2920
TX KIMBLE	45.9	2953
TX KING	78.6	2431
TX KINNEY	31.6	3023
TX KLEBERG	36.5	2993
TX KNOX	94.9	2100
TX LA SALLE	32.6	3015
TX LAMAR	194.6	1038
TX LAMB	265.2	626
TX LAMPASAS	64.6	2769
TX LAVACA	41.7	2975
TX LEE	60.1	2853
TX LEON	73.0	2571
TX LIBERTY	142.6	1538
TX LIMESTONE	87.9	2223
TX LIPSCOMB	263.3	634
TX LIVE OAK	34.5	3003
TX LLANO	61.2	2843

U.S. COUNTIES RANKED BY TOTAL FALLOUT DEPOSITION
1951-1970
(microCuries per square meter)
Each Row Includes:
County Name, Fallout Level in microCuries/square meter and County Rank (1 = highest fallout in U.S.)

County	Fallout	Rank
TX LOVING	168.5	1278
TX LUBBOCK	218.3	840
TX LYNN	197.8	1006
TX MADISON	107.9	1940
TX MARION	103.4	1991
TX MARTIN	89.5	2194
TX MASON	63.2	2809
TX MATAGORDA	46.7	2948
TX MAVERICK	34.2	3007
TX MCCULLOCH	72.6	2581
TX MCLENNAN	84.0	2302
TX MCMULLEN	45.3	2956
TX MEDINA	56.9	2884
TX MENARD	53.4	2914
TX MIDLAND	64.7	2765
TX MILAM	64.2	2781
TX MILLS	56.1	2891
TX MITCHELL	64.5	2772
TX MONTAGUE	129.8	1694
TX MONTGOMERY	139.4	1582
TX MOORE	247.5	693
TX MORRIS	107.8	1942
TX MOTLEY	186.6	1092
TX NACOGDOCHES	79.2	2409
TX NAVARRO	91.2	2157
TX NEWTON	160.9	1343
TX NOLAN	72.3	2591
TX NUECES	31.4	3026
TX OCHILTREE	280.4	554
TX OLDHAM	429.9	282
TX ORANGE	250.9	676
TX PALO PINTO	91.1	2159
TX PANOLA	89.7	2190
TX PARKER	110.0	1909
TX PARMER	301.6	472
TX PECOS	51.2	2924
TX POLK	150.9	1447
TX POTTER	364.5	359
TX PRESIDIO	58.2	2871
TX RAINS	147.9	1476
TX RANDALL	200.6	977
TX REAGAN	56.0	2893
TX REAL	35.3	3000
TX RED RIVER	142.0	1548
TX REEVES	138.7	1592
TX REFUGIO	1106.7	56
TX ROBERTS	188.9	1070
TX ROBERTSON	67.6	2705
TX ROCKWALL	134.4	1646
TX RUNNELS	64.7	2766
TX RUSK	109.2	1923
TX SABINE	155.1	1399
TX SAN AUGUSTIN	155.0	1401
TX SAN JACINTO	133.6	1655
TX SAN PATRICIO	34.4	3006
TX SAN SABA	63.1	2813
TX SCHLEICHER	45.7	2954
TX SCURRY	82.8	2327
TX SHACKELFORD	73.9	2549
TX SHELBY	70.2	2639
TX SHERMAN	226.7	785
TX SMITH	123.0	1762
TX SOMERVELL	79.3	2402
TX STARR	37.2	2991
TX STEPHENS	76.7	2474
TX STERLING	64.3	2777
TX STONEWALL	94.0	2113
TX SUTTON	37.5	2990

U.S. COUNTIES RANKED BY TOTAL FALLOUT DEPOSITION
1951-1970
(microCuries per square meter)
Each Row Includes:
County Name, Fallout Level in microCuries/square meter and County Rank (1 = highest fallout in U.S.)

County	Fallout	Rank
TX SWISHER	235.6	742
TX TARRANT	146.1	1497
TX TAYLOR	45.3	2957
TX TERRELL	34.4	3004
TX TERRY	229.8	768
TX THROCKMORTON	76.4	2483
TX TITUS	125.7	1738
TX TOM GREEN	56.0	2896
TX TRAVIS	46.4	2950
TX TRINITY	143.4	1525
TX TYLER	142.0	1546
TX UPSHUR	110.6	1898
TX UPTON	44.4	2962
TX UVALDE	39.2	2986
TX VAL VERDE	24.9	3065
TX VAN ZANDT	142.9	1533
TX VICTORIA	33.7	3010
TX WALKER	151.3	1441
TX WALLER	136.4	1615
TX WARD	123.6	1753
TX WASHINGTON	129.3	1700
TX WEBB	31.1	3030
TX WHARTON	51.5	2923
TX WHEELER	222.3	809
TX WICHITA	98.6	2049
TX WILBARGER	87.5	2231
TX WILLACY	37.7	2989
TX WILLIAMSON	69.3	2657
TX WILSON	32.5	3018
TX WINKLER	145.6	1502
TX WISE	131.4	1681
TX WOOD	140.1	1572
TX YOAKUM	218.7	838
TX YOUNG	77.7	2449
TX ZAPATA	33.2	3012
TX ZAVALA	40.6	2978
UT BEAVER	1376.5	35
UT BOX ELDER1	675.4	148
UT BOX ELDER2	1280.6	38
UT CACHE	1241.6	41
UT CARBON	886.6	79
UT DAGGETT	914.2	73
UT DAVIS	1708.9	20
UT DUCHESNE	893.7	77
UT EMERY	1167.9	45
UT GARFIELD	2304.2	11
UT GRAND	1133.0	50
UT IRON1	1468.9	28
UT IRON2	1220.9	42
UT IRON3	1297.8	37
UT JUAB	1267.5	39
UT KANE1	2832.7	9
UT KANE2	3114.7	7
UT MILLARD	1438.1	32
UT MORGAN	1718.9	19
UT PIUTE	1148.3	48
UT RICH	1377.3	34
UT SALT LAKE	1601.6	23
UT SAN JUAN	957.1	71
UT SANPETE	1070.3	59
UT SEVIER	1058.6	60
UT SUMMIT	1452.9	30
UT TOOELE1	1519.1	27
UT TOOELE2	1678.8	21
UT UINTAH	986.3	68
UT UTAH	1760.9	17
UT WASATCH	1452.9	29
UT WASHINGTON1	3547.0	4

U.S. COUNTIES RANKED BY TOTAL FALLOUT DEPOSITION
1951-1970
(microCuries per square meter)
Each Row Includes:
County Name, Fallout Level in microCuries/square meter and County Rank (1 = highest fallout in U.S.)

County	Fallout	Rank
UT WASHINGTON2	6788.4	1
UT WASHINGTON3	4952.8	2
UT WAYNE	874.7	80
UT WEBER	1645.0	22
VA ACCOMACK	181.3	1152
VA ALBEMARLE	54.1	2906
VA ALLEGHANY	132.8	1666
VA AMELIA	225.1	790
VA AMHERST	122.5	1765
VA APPOMATTOX	119.3	1796
VA ARLINGTON	87.3	2235
VA AUGUSTA	120.3	1786
VA BATH	261.2	642
VA BEDFORD	132.3	1674
VA BLAND	57.1	2881
VA BOTETOURT	258.1	656
VA BRUNSWICK	78.3	2434
VA BUCHANAN	228.6	772
VA BUCKINGHAM	64.5	2774
VA CAMPBELL	118.2	1808
VA CAROLINE	127.2	1714
VA CARROLL	166.4	1300
VA CHARLES CITY	144.0	1517
VA CHARLOTTE	60.4	2850
VA CHESTERFIELD	218.0	843
VA CLARKE	64.5	2771
VA CRAIG	136.9	1610
VA CULPEPER	86.0	2254
VA CUMBERLAND	73.9	2551
VA DICKENSON	174.2	1219
VA DINWIDDIE	82.7	2328
VA ESSEX	72.0	2601
VA FAIRFAX	142.1	1545
VA FAUQUIER	136.3	1619
VA FLOYD	135.7	1632
VA FLUVANNA	66.9	2729
VA FRANKLIN	208.0	921
VA FREDERICK	115.9	1845
VA GILES	111.2	1887
VA GLOUCESTER	76.8	2471
VA GOOCHLAND	225.6	788
VA GRAYSON	108.5	1934
VA GREENE	110.6	1897
VA GREENSVILLE	88.5	2213
VA HALIFAX	67.9	2700
VA HAMPTON	78.9	2417
VA HANOVER	226.9	783
VA HENRICO	94.7	2104
VA HENRY	75.6	2500
VA HIGHLAND	136.0	1622
VA ISLE OF WIGH	83.7	2313
VA JAMES CITY	76.7	2475
VA KING AND QUE	152.1	1433
VA KING GEORGE	79.9	2393
VA KING WILLIAM	159.0	1361
VA LANCASTER	86.8	2242
VA LEE	119.7	1791
VA LOUDOUN	140.0	1574
VA LOUISA	70.2	2640
VA LUNENBURG	125.8	1735
VA MADISON	125.7	1736
VA MATHEWS	82.3	2336
VA MECKLENBURG	131.1	1685
VA MIDDLESEX	91.5	2154
VA MONTGOMERY	116.9	1829
VA NELSON	117.6	1815
VA NEW KENT	145.7	1501
VA NEWPORT NEWS	85.4	2267

U.S. COUNTIES RANKED BY TOTAL FALLOUT DEPOSITION
1951-1970
(microCuries per square meter)
Each Row Includes:
County Name, Fallout Level in microCuries/square meter and County Rank (1 = highest fallout in U.S.)

County	Fallout	Rank
VA NORFOLK/CHES	82.2	2338
VA NORTHAMPTON	78.0	2444
VA NORTHUMBERLA	87.3	2234
VA NOTTOWAY	78.2	2440
VA ORANGE	135.2	1636
VA PAGE	60.6	2847
VA PATRICK	117.4	1820
VA PITTSYLVANIA	131.4	1682
VA POWHATAN	230.3	762
VA PRINCE EDWAR	126.5	1725
VA PRINCE GEORG	95.4	2095
VA PRINCE WILLI	78.8	2425
VA PULASKI	119.3	1795
VA RAPPAHANNOCK	126.7	1723
VA RICHMOND	72.8	2576
VA ROANOKE	130.0	1693
VA ROCKBRIDGE	113.2	1865
VA ROCKINGHAM	134.4	1645
VA RUSSELL	118.2	1807
VA SCOTT	116.8	1833
VA SHENANDOAH	108.6	1931
VA SMYTH	146.6	1490
VA SOUTHAMPTON	79.5	2398
VA SPOTSYLVANIA	129.7	1695
VA STAFFORD	85.3	2271
VA SUFFOLK/NANS	125.7	1737
VA SURRY	79.4	2400
VA SUSSEX	79.3	2405
VA TAZEWELL	153.2	1424
VA VIRGINIA BEA	107.0	1949
VA WARREN	141.1	1556
VA WASHINGTON	105.4	1976
VA WESTMORELAND	77.9	2445
VA WISE	100.9	2015
VA WYTHE	64.1	2785
VA YORK	76.4	2482
VT ADDISON	299.8	479
VT BENNINGTON	391.7	324
VT CALEDONIA	113.6	1863
VT CHITTENDEN	217.2	847
VT ESSEX	119.9	1790
VT FRANKLIN	153.1	1428
VT GRAND ISLE	168.6	1276
VT LAMOILLE	136.2	1620
VT ORANGE	129.4	1698
VT ORLEANS	123.7	1752
VT RUTLAND	193.4	1045
VT WASHINGTON	138.2	1599
VT WINDHAM	198.1	1004
VT WINDSOR	142.3	1543
WA ADAMS	27.6	3049
WA ASOTIN	160.2	1348
WA BENTON	26.0	3054
WA CHELAN	26.8	3050
WA CLALLAM	20.0	3086
WA CLARK	22.9	3074
WA COLUMBIA	111.9	1883
WA COWLITZ	26.0	3055
WA DOUGLAS	31.9	3021
WA FERRY	57.7	2877
WA FRANKLIN	22.9	3073
WA GARFIELD	202.8	955
WA GRANT	24.2	3067
WA GRAYS HARBOR	23.2	3071
WA ISLAND	12.4	3094
WA JEFFERSON	21.7	3082
WA KING	24.5	3066
WA KITSAP	18.5	3091

U.S. COUNTIES RANKED BY TOTAL FALLOUT DEPOSITION
1951-1970
(microCuries per square meter)
Each Row Includes:
County Name, Fallout Level in microCuries/square meter and County Rank (1 = highest fallout in U.S.)

County	Fallout	Rank
WA KITTITAS	25.9	3057
WA KLICKITAT	21.2	3083
WA LEWIS	22.6	3076
WA LINCOLN	29.5	3036
WA MASON	19.3	3089
WA OKANOGAN	43.7	2964
WA PACIFIC	22.3	3078
WA PEND OREILLE	112.2	1877
WA PIERCE	28.7	3044
WA SAN JUAN	13.5	3093
WA SKAGIT	21.8	3080
WA SKAMANIA	31.4	3028
WA SNOHOMISH	28.9	3043
WA SPOKANE	76.7	2476
WA STEVENS	50.8	2925
WA THURSTON	17.5	3092
WA WAHKIAKUM	24.9	3064
WA WALLA WALLA	90.1	2179
WA WHATCOM	23.1	3072
WA WHITMAN	89.4	2195
WA YAKIMA	22.4	3077
WI ADAMS	151.4	1440
WI ASHLAND	139.0	1586
WI BARRON	138.9	1588
WI BAYFIELD	130.7	1689
WI BROWN	69.3	2659
WI BUFFALO	470.5	246
WI BURNETT	135.9	1626
WI CALUMET	101.1	2012
WI CHIPPEWA	138.1	1602
WI CLARK	134.5	1644
WI COLUMBIA	142.4	1541
WI CRAWFORD	244.9	700
WI DANE	210.3	898
WI DODGE	97.6	2060
WI DOOR	79.4	2401
WI DOUGLAS	140.0	1573
WI DUNN	154.5	1407
WI EAU CLAIRE	141.7	1550
WI FLORENCE	77.2	2459
WI FOND DU LAC	106.8	1951
WI FOREST	87.5	2229
WI GRANT	364.7	357
WI GREEN	262.6	636
WI GREEN LAKE	110.3	1902
WI IOWA	498.9	220
WI IRON	132.7	1669
WI JACKSON	154.2	1411
WI JEFFERSON	180.3	1157
WI JUNEAU	170.1	1261
WI KENOSHA	83.7	2312
WI KEWAUNEE	81.4	2363
WI LA CROSSE	160.9	1342
WI LAFAYETTE	286.4	529
WI LANGLADE	91.5	2153
WI LINCOLN	115.3	1851
WI MANITOWOC	82.7	2329
WI MARATHON	110.1	1904
WI MARINETTE	102.0	2007
WI MARQUETTE	137.2	1609
WI MENOMINEE	96.5	2072
WI MILWAUKEE	63.4	2804
WI MONROE	162.0	1329
WI OCONTO	101.2	2011
WI ONEIDA	117.0	1825
WI OUTAGAMIE	76.3	2485
WI OZAUKEE	85.7	2260
WI PEPIN	175.6	1205

U.S. COUNTIES RANKED BY TOTAL FALLOUT DEPOSITION
1951-1970

(microCuries per square meter)
Each Row Includes:
County Name, Fallout Level in microCuries/square meter and County Rank (1 = highest fallout in U.S.)

WI PIERCE	139.9	1576
WI POLK	138.9	1589
WI PORTAGE	119.1	1798
WI PRICE	127.1	1715
WI RACINE	75.4	2506
WI RICHLAND	238.0	731
WI ROCK	259.6	648
WI RUSK	144.7	1510
WI SAUK	234.2	748
WI SAWYER	157.9	1372
WI SHAWANO	97.7	2059
WI SHEBOYGAN	100.0	2026
WI ST CROIX	140.6	1564
WI TAYLOR	129.3	1701
WI TREMPEALEAU	168.5	1279
WI VERNON	321.1	426
WI VILAS	114.6	1857
WI WALWORTH	293.2	504
WI WASHBURN	153.1	1429
WI WASHINGTON	93.8	2117
WI WAUKESHA	106.0	1964
WI WAUPACA	121.4	1776
WI WAUSHARA	139.7	1578
WI WINNEBAGO	101.9	2008
WI WOOD	145.2	1505
WV BARBOUR	65.3	2757
WV BERKELEY	63.3	2806
WV BOONE	419.4	295
WV BRAXTON	159.6	1355
WV BROOKE	156.5	1383
WV CABELL	94.3	2109
WV CALHOUN	171.1	1248
WV CLAY	149.3	1459
WV DODDRIDGE	64.6	2768
WV FAYETTE	155.6	1393
WV GILMER	170.6	1254
WV GRANT	150.2	1452
WV GREENBRIER	170.6	1253
WV HAMPSHIRE	61.5	2836
WV HANCOCK	96.2	2081
WV HARDY	116.8	1831
WV HARRISON	173.6	1223
WV JACKSON	80.1	2389
WV JEFFERSON	68.6	2675
WV KANAWHA	275.3	582
WV LEWIS	164.4	1310
WV LINCOLN	298.8	481
WV LOGAN	447.7	268
WV MARION	152.5	1432
WV MARSHALL	188.9	1071
WV MASON	308.8	452
WV MCDOWELL	263.9	633
WV MERCER	233.0	755
WV MINERAL	60.3	2851
WV MINGO	239.8	721
WV MONONGALIA	149.6	1456
WV MONROE	268.6	612
WV MORGAN	71.6	2614
WV NICHOLAS	152.0	1435
WV OHIO	117.2	1821
WV PENDLETON	263.0	635
WV PLEASANTS	107.3	1946
WV POCAHONTAS	146.4	1492
WV PRESTON	73.5	2561
WV PUTNAM	154.3	1410
WV RALEIGH	266.3	617
WV RANDOLPH	155.7	1392
WV RITCHIE	66.1	2749

U.S. COUNTIES RANKED BY TOTAL FALLOUT DEPOSITION
1951-1970
(microCuries per square meter)
Each Row Includes:
County Name, Fallout Level in microCuries/square meter and County Rank (1 = highest fallout in U.S.)

County	Fallout	Rank
WV ROANE	63.9	2789
WV SUMMERS	284.8	536
WV TAYLOR	64.1	2786
WV TUCKER	162.5	1323
WV TYLER	185.3	1107
WV UPSHUR	182.2	1143
WV WAYNE	293.6	501
WV WEBSTER	144.7	1509
WV WETZEL	164.8	1308
WV WIRT	170.1	1260
WV WOOD	178.4	1172
WV WYOMING	222.3	808
WY ALBANY	345.9	382
WY BIG HORN	154.0	1415
WY CAMPBELL	370.4	348
WY CARBON	829.6	85
WY CONVERSE	465.7	249
WY CROOK	238.1	730
WY FREMONT	1374.2	36
WY GOSHEN	212.6	884
WY HOT SPRINGS	168.0	1284
WY JOHNSON	381.6	334
WY LARAMIE	204.5	942
WY LINCOLN	1053.4	61
WY NATRONA	294.3	498
WY NIOBRARA	434.4	279
WY PARK	238.7	726
WY PLATTE	319.0	430
WY SHERIDAN	188.3	1079
WY SUBLETTE	1132.1	52
WY SWEETWATER	963.7	70
WY TETON	247.2	694
WY UINTA	1030.3	64
WY WASHAKIE	314.2	441
WY WESTON	196.7	1020

U.S. COUNTIES RANKED BY TOTAL FALLOUT DEPOSITION II
1951-1970
(microCuries per square meter)
Each Row Includes:
County Name, Fallout Level in microCuries/square meter and County Rank (1 = highest fallout in U.S.)

RANK	COUNTY	uCi/sqmeter
1	UT WASHINGTON2	6788.45
2	UT WASHINGTON3	4952.76
3	NV NYE2	3822.15
4	UT WASHINGTON1	3547.02
5	NV LINCOLN1	3420.75
6	NV CLARK1	3315.80
7	UT KANE2	3114.65
8	AZ MOHAVE2	2997.49
9	UT KANE1	2832.65
10	NV WHITE PINE2	2451.01
11	UT GARFIELD	2304.25
12	NV LINCOLN2	2191.79
13	NV WHITE PINE3	2170.94
14	AZ COCONINO1	2057.89
15	CO GUNNISON	1946.42
16	CO CONEJOS	1831.89
17	UT UTAH	1760.93
18	CA INYO3	1750.08
19	UT MORGAN	1718.89
20	UT DAVIS	1708.92
21	UT TOOELE2	1678.75
22	UT WEBER	1644.96
23	UT SALT LAKE	1601.58
24	NV WHITE PINE1	1599.72
25	AZ MOHAVE1	1593.40
26	AZ COCONINO2	1591.21
27	UT TOOELE1	1519.07
28	UT IRON1	1468.87
29	UT WASATCH	1452.94
30	UT SUMMIT	1452.86
31	CO ARCHULETA	1438.43
32	UT MILLARD	1438.13
33	AZ APACHE	1424.32
34	UT RICH	1377.29
35	UT BEAVER	1376.48
36	WY FREMONT	1374.23
37	UT IRON3	1297.81
38	UT BOX ELDER2	1280.59
39	UT JUAB	1267.54
40	CO MESA	1267.33
41	UT CACHE	1241.60
42	UT IRON2	1220.93
43	CO GARFIELD	1199.96
44	NV EUREKA	1196.26
45	UT EMERY	1167.87
46	NM MCKINLEY	1166.85
47	CO MINERAL	1156.07
48	UT PIUTE	1148.28
49	CO HINSDALE	1146.12
50	UT GRAND	1133.04
51	IA HOWARD	1133.00
52	WY SUBLETTE	1132.07
53	IA MITCHELL	1116.94
54	IA WORTH	1113.05
55	CO LA PLATA	1111.60
56	TX REFUGIO	1106.74
57	CO DELTA	1104.38
58	SD WASHABAUGH	1071.25
59	UT SANPETE	1070.27
60	UT SEVIER	1058.64
61	WY LINCOLN	1053.35
62	CO MONTROSE	1050.29
63	ID GEM	1030.75
64	WY UINTA	1030.29
65	CO PITKIN	1018.63
66	ID CUSTER	1015.84
67	NM SANDOVAL	1013.70

U.S. COUNTIES RANKED BY TOTAL FALLOUT DEPOSITION II
1951-1970
(microCuries per square meter)
Each Row Includes:
County Name, Fallout Level in microCuries/square meter and County Rank (1 = highest fallout in U.S.)

68	UT UINTAH	986.33
69	CO RIO BLANCO	973.92
70	WY SWEETWATER	963.72
71	UT SAN JUAN	957.07
72	AZ NAVAJO	932.91
73	UT DAGGETT	914.18
74	ID BLAINE	908.33
75	CO RIO GRANDE	905.59
76	NM VALENCIA	903.87
77	UT DUCHESNE	893.73
78	CO DOLORES	889.93
79	UT CARBON	886.62
80	UT WAYNE	874.74
81	ID BEAR LAKE	864.58
82	IA WOODBURY	842.74
83	NM SANTA FE	835.30
84	NM BERNALILLO	834.31
85	WY CARBON	829.62
86	NM TORRANCE	827.51
87	CO LAKE	825.34
88	IA VAN BUREN	813.71
89	CO SAN MIGUEL	808.07
90	CO OURAY	802.94
91	IA LEE	802.44
92	IL HANCOCK	801.68
93	CO MONTEZUMA	801.36
94	CO SAN JUAN	793.25
95	IA CARROLL	786.42
96	SD CUSTER	783.73
97	IA JASPER	767.92
98	IA BOONE	766.49
99	IA BENTON	765.06
100	IA MARSHALL	763.97
101	IA WASHINGTON	756.97
102	IA JEFFERSON	756.13
103	IA STORY	754.09
104	NV ELKO	752.66
105	MN MOWER	752.61
106	MO KNOX	752.10
107	NE DIXON	746.75
108	SD TURNER	739.52
109	NV LANDER2	735.91
110	IA HAMILTON	734.10
111	MN FARIBAULT	729.28
112	MN WATONWAN	721.07
113	ID LEMHI	718.73
114	IA IDA	717.75
115	CO EAGLE	717.55
116	CO SAGUACHE	715.97
117	IA GREENE	715.09
118	NM SAN JUAN	715.04
119	IA O BRIEN	712.89
120	CO MOFFAT	708.22
121	NE STANTON	704.90
122	IA HUMBOLDT	702.65
123	IA OSCEOLA	701.76
124	IA MONONA	700.07
125	IA SIOUX	699.28
126	CO COSTILLA	698.66
127	IA BUENA VISTA	697.46
128	NE THURSTON	696.54
129	IA POCAHONTAS	694.93
130	IA WINNESHIEK	693.28
131	IA PLYMOUTH	692.97
132	IA CALHOUN	691.41
133	IA IOWA	691.02
134	ID FRANKLIN	690.01
135	IA FLOYD	689.63

U.S. COUNTIES RANKED BY TOTAL FALLOUT DEPOSITION II
1951-1970

(microCuries per square meter)
Each Row Includes:
County Name, Fallout Level in microCuries/square meter and County Rank (1 = highest fallout in U.S.)

136	IA SAC	688.68
137	IA LYON	687.43
138	IA CLAY	687.38
139	IA KOSSUTH	685.66
140	IA HANCOCK	684.75
141	IA CHEROKEE	684.50
142	CO ALAMOSA	683.08
143	NE DAKOTA	681.23
144	MO SCOTLAND	680.01
145	IA WINNEBAGO	679.33
146	IA PALO ALTO	677.76
147	IA DICKINSON	676.03
148	UT BOX ELDER1	675.37
149	NV ESMERALDA2	674.28
150	IA CRAWFORD	672.60
151	NM LOS ALAMOS	672.40
152	NV MINERAL	672.09
153	MN ROCK	666.03
154	IA APPANOOSE	662.58
155	IA MONROE	661.74
156	IA EMMET	657.61
157	SD HAMLIN	652.91
158	CO SUMMIT	641.44
159	MT DEER LODGE	639.75
160	IA DAVIS	638.73
161	NY WASHINGTON	636.95
162	SD STANLEY	631.65
163	IA CASS	626.99
164	NV NYE1	624.58
165	IA SHELBY	623.95
166	NY WARREN	614.55
167	IA RINGGOLD	613.99
168	NM GUADALUPE	613.32
169	NV CHURCHILL	612.70
170	IA WAPELLO	608.54
171	NY RENSSELAER	606.82
172	IA MADISON	601.13
173	NE LOUP	600.80
174	AZ MOHAVE3	598.72
175	AZ MOHAVE4	598.72
176	CO GRAND	596.92
177	SD KINGSBURY	596.37
178	IA ADAIR	595.29
179	IA ADAMS	595.04
180	IA MARION	594.37
181	NE HOLT	591.41
182	IA UNION	591.19
183	IA AUDUBON	589.86
184	NM SAN MIGUEL	582.80
185	IA WAYNE	580.62
186	MO CLARK	580.35
187	NM CHAVES	579.58
188	ID VALLEY	579.41
189	MO LEWIS	577.00
190	NM LINCOLN	576.40
191	NM RIO ARRIBA	573.74
192	SD HUTCHINSON	572.09
193	ID IDAHO	572.08
194	IA MAHASKA	570.21
195	IA DECATUR	569.18
196	IA CLARKE	564.31
197	SD MCCOOK	557.97
198	IA DELAWARE	557.21
199	NM MORA	552.54
200	NE HOWARD	552.42
201	SD FALL RIVER	552.01
202	NM COLFAX	549.53
203	SD LAKE	547.18

U.S. COUNTIES RANKED BY TOTAL FALLOUT DEPOSITION II
1951-1970

(microCuries per square meter)
Each Row Includes:
County Name, Fallout Level in microCuries/square meter and County Rank (1 = highest fallout in U.S.)

204	NE SHERMAN	540.41
205	IA KEOKUK	538.84
206	NM QUAY	537.89
207	NM UNION	527.59
208	IA GUTHRIE	526.32
209	ID ONEIDA	524.09
210	NM HARDING	523.15
211	SD SHANNON	523.04
212	MO RALLS	519.93
213	SD MINER	517.85
214	NM TAOS	517.26
215	NM DE BACA	510.97
216	IA HENRY	505.05
217	MT JUDITH BASIN	504.76
218	CO CLEAR CREEK	503.31
219	ID CAMAS	501.99
220	WI IOWA	498.94
221	IA TAMA	498.16
222	MT MEAGHER	496.96
223	IL MCDONOUGH	496.30
224	IL WARREN	492.77
225	MN FILLMORE	492.01
226	MT BEAVERHEAD	491.65
227	CO JEFFERSON	485.81
228	MN RICE	485.73
229	MN WABASHA	484.91
230	IA LOUISA	484.20
231	IL FULTON	483.86
232	MO PUTNAM	483.39
233	NM CURRY	483.12
234	CO BOULDER	479.57
235	IA JOHNSON	477.17
236	CO GILPIN	476.81
237	NE WASHINGTON	476.43
238	NV LYON	475.84
239	ID OWYHEE	475.67
240	NV LANDER1	475.40
241	SD MOODY	473.91
242	NV WASHOE	473.32
243	CO DOUGLAS	472.96
244	ID BOISE	472.78
245	MN LE SUEUR	470.75
246	WI BUFFALO	470.52
247	NV CARSON CITY	469.25
248	MN WASECA	465.92
249	WY CONVERSE	465.71
250	ID GOODING	465.33
251	IA BREMER	464.18
252	IA BLACK HAWK	463.64
253	CO MORGAN	463.49
254	NM ROOSEVELT	463.20
255	ID CARIBOU	462.68
256	MN PIPESTONE	462.27
257	MN MARTIN	461.98
258	NV STOREY	461.29
259	NE MADISON	458.60
260	MO SCHUYLER	458.11
261	NV DOUGLAS	457.11
262	MN NICOLLET	454.61
263	IL ADAMS	452.94
264	MT CHOUTEAU	451.47
265	MN MURRAY	451.02
266	CO ADAMS	448.57
267	NV CLARK2	448.45
268	WV LOGAN	447.71
269	CO ROUTT	447.71
270	IA FRANKLIN	447.05
271	KY GREENUP	446.90

U.S. COUNTIES RANKED BY TOTAL FALLOUT DEPOSITION II
1951-1970
(microCuries per square meter)
Each Row Includes:
County Name, Fallout Level in microCuries/square meter and County Rank (1 = highest fallout in U.S.)

272	MN BROWN	446.85
273	MT SILVER BOW	444.52
274	MN BLUE EARTH	444.10
275	MO SHELBY	440.85
276	IA POWESHIEK	439.89
277	KY POWELL	438.24
278	CO ARAPAHOE	434.52
279	WY NIOBRARA	434.45
280	OH SCIOTO	432.52
281	KY LAWRENCE	430.44
282	TX OLDHAM	429.87
283	KY TRIMBLE	428.98
284	ID CASSIA	428.02
285	MT MADISON	426.45
286	NE ANTELOPE	426.39
287	TX DALLAM	425.79
288	NE GARFIELD	424.61
289	CO CUSTER	424.59
290	KY LEWIS	423.58
291	ID ELMORE	421.16
292	MT POWELL	419.89
293	IA LINN	419.87
294	MO WORTH	419.80
295	WV BOONE	419.44
296	ID ADA	419.43
297	NM SOCORRO	418.16
298	MT JEFFERSON	417.89
299	IA TAYLOR	417.13
300	SD HANSON	415.77
301	IA FAYETTE	415.77
302	IA MONTGOMERY	415.13
303	CO CHAFFEE	414.92
304	NE CUMING	412.08
305	NY SARATOGA	411.91
306	SD DAVISON	411.81
307	SD YANKTON	411.77
308	IA BUCHANAN	409.10
309	IL PIKE	408.83
310	NY FULTON	406.06
311	OH BUTLER	405.21
312	IA HARDIN	404.51
313	IL MOULTRIE	402.59
314	ID TWIN FALLS	401.92
315	AZ COCONINO3	401.28
316	MT GALLATIN	399.89
317	SD BROOKINGS	399.83
318	IL SHELBY	399.15
319	MT BROADWATER	398.63
320	MT LEWIS AND CL	398.03
321	ID WASHINGTON	396.33
322	MO PIKE	393.60
323	SD LINCOLN	392.72
324	VT BENNINGTON	391.69
325	IA BUTLER	391.42
326	KS STEVENS	391.16
327	ID LINCOLN	391.10
328	IN ALLEN	389.97
329	SD UNION	385.78
330	KY CAMPBELL	384.98
331	ND BILLINGS	384.09
332	IN WAYNE	383.88
333	NE WAYNE	383.32
334	WY JOHNSON	381.57
335	OH MEDINA	381.46
336	MT WHEATLAND	381.33
337	KY OWEN	379.35
338	KY ROWAN	378.91
339	CO HUERFANO	377.27

U.S. COUNTIES RANKED BY TOTAL FALLOUT DEPOSITION II
1951-1970
(microCuries per square meter)

Each Row Includes:
County Name, Fallout Level in microCuries/square meter and County Rank (1 = highest fallout in U.S.)

340	IA DALLAS	376.32
341	OH MONTGOMERY	376.08
342	CO FREMONT	376.02
343	IA POTTAWATTAMI	374.69
344	CO LAS ANIMAS	374.45
345	NE CEDAR	373.69
346	MT POWDER RIVER	373.25
347	ID JEFFERSON	372.64
348	WY CAMPBELL	370.39
349	MO AUDRAIN	368.39
350	IA CERRO GORDO	368.00
351	OH SENECA	367.41
352	MO MARION	367.02
353	NV PERSHING	366.83
354	NV ESMERALDA1	366.58
355	MO DAVIESS	366.44
356	KY CARROLL	365.47
357	WI GRANT	364.66
358	SD MINNEHAHA	364.60
359	TX POTTER	364.51
360	CO LARIMER	363.36
361	CO PHILLIPS	363.22
362	IA DUBUQUE	359.66
363	IL CASS	358.02
364	SD JERAULD	356.73
365	SD CLAY	356.72
366	MT CUSTER	356.19
367	OK TEXAS	355.90
368	NE GREELEY	354.57
369	MN NOBLES	354.17
370	IA PAGE	352.38
371	IA JACKSON	352.36
372	MT MISSOULA	351.68
373	IA WARREN	351.68
374	MN WINONA	351.25
375	NE MERRICK	351.04
376	NE NANCE	349.17
377	IA WEBSTER	349.03
378	MN STEELE	348.96
379	MN DODGE	348.90
380	ID JEROME	347.26
381	IL BROWN	346.19
382	WY ALBANY	345.87
383	OH CLARK	345.23
384	MO MACON	344.05
385	CO BACA	344.02
386	MO SULLIVAN	343.93
387	IA WRIGHT	343.74
388	IL COLES	343.44
389	IA CLAYTON	343.16
390	IL SANGAMON	342.19
391	NE COLFAX	341.55
392	KY KENTON	341.43
393	NM CATRON	341.26
394	MN GOODHUE	340.66
395	TX HARTLEY	340.26
396	OH WARREN	339.51
397	ID MINIDOKA	339.50
398	IA MILLS	339.17
399	IN LAWRENCE	338.63
400	CO KIOWA	337.59
401	MO LINN	336.83
402	OH CLINTON	336.37
403	KY HARRISON	336.08
404	NE VALLEY	335.99
405	IA POLK	335.33
406	NE PLATTE	335.17
407	IL GREENE	334.93

U.S. COUNTIES RANKED BY TOTAL FALLOUT DEPOSITION II
1951-1970
(microCuries per square meter)
Each Row Includes:
County Name, Fallout Level in microCuries/square meter and County Rank (1 = highest fallout in U.S.)

Rank	County	Fallout
408	IA CLINTON	334.31
409	MT DAWSON	333.33
410	IA CEDAR	332.60
411	ND STARK	332.40
412	IL MONTGOMERY	331.57
413	KY NICHOLAS	330.17
414	OR WALLOWA	330.15
415	ID BUTTE	328.23
416	NE CUSTER	328.23
417	ID LEWIS	327.89
418	PA VENANGO	327.78
419	NE POLK	326.46
420	IL CHRISTIAN	326.02
421	IL MACOUPIN	324.48
422	ID PAYETTE	323.74
423	NE BUTLER	322.46
424	MN HOUSTON	322.07
425	IA ALLAMAKEE	321.59
426	WI VERNON	321.13
427	NV HUMBOLDT	321.05
428	CO JACKSON	319.89
429	CO WELD	319.24
430	WY PLATTE	319.02
431	KY SCOTT	318.02
432	MO ADAIR	317.01
433	MT CASCADE	316.85
434	IA HARRISON	316.77
435	IL SCHUYLER	316.66
436	SD HAAKON	315.99
437	NY SCHENECTADY	315.70
438	KY WOODFORD	315.17
439	CO TELLER	315.04
440	NE LINCOLN	314.86
441	WY WASHAKIE	314.19
442	CO PARK	313.63
443	CA SAN BERNADIN	313.57
444	OH CHAMPAIGN	313.47
445	OH LICKING	311.08
446	KY CLARK	310.87
447	ND DUNN	310.82
448	OH MIAMI	310.64
449	NE CHERRY	310.14
450	OR BAKER	309.42
451	MT PARK	309.04
452	WV MASON	308.84
453	OH HIGHLAND	308.83
454	IL JASPER	307.99
455	OH RICHLAND	307.34
456	IL MARSHALL	307.29
457	KY FRANKLIN	307.06
458	IA FREMONT	306.47
459	ND GOLDEN VALLE	306.28
460	OK CIMARRON	305.57
461	ID CLEARWATER	304.95
462	KS NESS	304.84
463	NY COLUMBIA	303.39
464	MN LYON	303.09
465	KY MENIFEE	302.92
466	MT FERGUS	302.46
467	OH ALLEN	302.13
468	KY HENRY	302.12
469	OH KNOX	302.02
470	OH PERRY	301.86
471	NE SAUNDERS	301.76
472	TX PARMER	301.61
473	MT RICHLAND	301.52
474	MN LINCOLN	300.62
475	NE SIOUX	300.12

U.S. COUNTIES RANKED BY TOTAL FALLOUT DEPOSITION II
1951-1970
(microCuries per square meter)
Each Row Includes:
County Name, Fallout Level in microCuries/square meter and County Rank (1 = highest fallout in U.S.)

476	MN JACKSON	299.96
477	MA NORFOLK	299.85
478	ID TETON	299.85
479	VT ADDISON	299.82
480	IL PEORIA	298.99
481	WV LINCOLN	298.83
482	TX DEAF SMITH	298.81
483	MN SIBLEY	298.64
484	ND MCKENZIE	298.57
485	LA GRANT	298.56
486	KY BATH	298.51
487	IN HAMILTON	297.84
488	IN GRANT	296.61
489	IL WOODFORD	296.35
490	NE DOUGLAS	295.98
491	NE BUFFALO	295.92
492	IL HENDERSON	295.91
493	KS FORD	295.76
494	OH JACKSON	295.19
495	IA JONES	295.04
496	IA DES MOINES	294.80
497	NE CHASE	294.60
498	WY NATRONA	294.32
499	IL MACON	294.08
500	MT TETON	293.78
501	WV WAYNE	293.62
502	NE LOGAN	293.35
503	OH MORGAN	293.18
504	WI WALWORTH	293.18
505	MA PLYMOUTH	292.95
506	IN HOWARD	292.78
507	OH HANCOCK	292.78
508	NE KEITH	292.52
509	OH PIKE	292.40
510	LA RAPIDES	291.94
511	IL WINNEBAGO	291.52
512	KY BOYD	290.65
513	CO LINCOLN	290.25
514	CO KIT CARSON	290.13
515	NE PERKINS	290.04
516	CO DENVER	289.95
517	IN TIPTON	289.95
518	IN WELLS	289.66
519	NE FRONTIER	288.81
520	TX BAILEY	288.73
521	ID MADISON	288.44
522	TX HANSFORD	288.07
523	MT WIBAUX	287.92
524	IL KNOX	287.84
525	KS NORTON	287.81
526	ID BANNOCK	287.31
527	TX HUTCHINSON	287.17
528	ND WARD	286.69
529	WI LAFAYETTE	286.41
530	KY ROBERTSON	286.09
531	CO OTERO	285.78
532	NE PHELPS	285.70
533	NE CASS	285.64
534	KS STANTON	285.09
535	KS TREGO	285.09
536	WV SUMMERS	284.76
537	IL STARK	284.64
538	KS CHEYENNE	284.15
539	KY WOLFE	284.06
540	SD JACKSON	283.53
541	NE DAWSON	283.22
542	OH PREBLE	283.02
543	OH FULTON	282.76

U.S. COUNTIES RANKED BY TOTAL FALLOUT DEPOSITION II
1951-1970
(microCuries per square meter)
Each Row Includes:
County Name, Fallout Level in microCuries/square meter and County Rank (1 = highest fallout in U.S.)

544	NE CLAY	282.72
545	ND MERCER	282.55
546	CO BENT	282.15
547	KS GRAHAM	282.11
548	KY BREATHITT	281.97
549	ND DIVIDE	281.26
550	NE FURNAS	281.18
551	ND OLIVER	280.87
552	OH ADAMS	280.85
553	CO YUMA	280.74
554	TX OCHILTREE	280.36
555	ID POWER	280.32
556	KS GRANT	279.58
557	KY GALLATIN	279.44
558	IL MASON	279.40
559	CO SEDGWICK	279.14
560	OH WYANDOT	278.97
561	NC PERQUIMANS	278.75
562	IL ROCK ISLAND	278.35
563	IL PIATT	278.23
564	ND BURKE	277.80
565	NC CURRITUCK	277.62
566	KS KEARNY	277.54
567	IL HENRY	276.87
568	NC PASQUOTANK	276.73
569	IL DOUGLAS	276.62
570	CO EL PASO	276.54
571	NC CAMDEN	276.48
572	NE GOSPER	276.45
573	IA MUSCATINE	276.43
574	IL EFFINGHAM	276.34
575	NE SALINE	276.31
576	KY MAGOFFIN	276.07
577	KS HODGEMAN	275.98
578	CA INYO1	275.94
579	CA INYO2	275.94
580	MO CHARITON	275.82
581	KY FLEMING	275.51
582	WV KANAWHA	275.26
583	SD LYMAN	275.23
584	NY ALBANY	275.00
585	SD MELLETTE	274.71
586	CO PROWERS	274.58
587	KS RUSSELL	274.53
588	OK TULSA	274.19
589	LA BEAUREGARD	273.81
590	KS WALLACE	273.76
591	KY ELLIOTT	273.37
592	CO CROWLEY	273.02
593	KS LANE	272.51
594	KY MASON	272.40
595	ND MCHENRY	272.28
596	SD DEWEY	272.15
597	MT FALLON	272.15
598	KY CARTER	271.91
599	ND BOTTINEAU	271.89
600	MT MINERAL	271.46
601	CO ELBERT	271.32
602	LA LAFAYETTE	270.68
603	TX CASTRO	270.56
604	KS FINNEY	269.76
605	MT MCCONE	269.59
606	KS GOVE	269.47
607	NE RED WILLOW	269.36
608	IN MARION	268.81
609	KS SCOTT	268.81
610	MO GRUNDY	268.72
611	OH LAWRENCE	268.69

U.S. COUNTIES RANKED BY TOTAL FALLOUT DEPOSITION II
1951-1970
(microCuries per square meter)
Each Row Includes:
County Name, Fallout Level in microCuries/square meter and County Rank (1 = highest fallout in U.S.)

Rank	County	Fallout	Rank	County	Fallout
612	WV MONROE	268.56	646	KS WICHITA	260.37
613	CO WASHINGTON	267.24	647	MT ROOSEVELT	260.23
614	MO MONROE	267.15	648	WI ROCK	259.55
615	IL MCLEAN	267.04	649	NE DUNDY	259.36
616	CO CHEYENNE	266.48	650	NE THOMAS	259.31
617	WV RALEIGH	266.26	651	PA ELK	258.76
618	KS DECATUR	265.98	652	KS SALINE	258.72
619	SD ZIEBACH	265.94	653	MN COTTONWOOD	258.55
620	KS LOGAN	265.89	654	NE ADAMS	258.47
621	IL TAZEWELL	265.64	655	MT CARBON	258.42
622	MO CALDWELL	265.60	656	VA BOTETOURT	258.06
623	AZ PIMA	265.35	657	LA EVANGELINE	257.84
624	KY JOHNSON	265.28	658	IA SCOTT	257.67
625	IL BOONE	265.21	659	ND MOUNTRAIL	257.51
626	TX LAMB	265.16	660	NE SEWARD	257.20
627	IL MERCER	265.10	661	OK CRAIG	256.47
628	ND SLOPE	265.02	662	SD HARDING	256.05
629	IN JACKSON	264.34	663	ND MCLEAN	255.79
630	KS HAMILTON	264.21	664	MO HARRISON	255.44
631	IL DE WITT	264.13	665	SD AURORA	254.20
632	KS RAWLINS	263.92	666	LA ACADIA	253.86
633	WV MCDOWELL	263.91	667	LA CALCASIEU	253.35
634	TX LIPSCOMB	263.33	668	KS GRAY	252.87
635	WV PENDLETON	262.98	669	KY SHELBY	252.62
636	WI GREEN	262.55	670	MT PRAIRIE	252.44
637	ID CLARK	262.09	671	NE HITCHCOCK	252.01
638	SD MEADE	261.79	672	MO RANDOLPH	251.59
639	KY GRANT	261.54	673	IL LOGAN	251.47
640	ND RICHLAND	261.48	674	MT BLAINE	251.26
641	SD BRULE	261.41	675	NE CHEYENNE	251.00
642	VA BATH	261.17	676	TX ORANGE	250.94
643	NE HAYES	260.96	677	LA CAMERON	250.94
644	CA MONO	260.93	678	IN DELAWARE	250.72
645	KY FLOYD	260.65	679	ND RENVILLE	250.37

U.S. COUNTIES RANKED BY TOTAL FALLOUT DEPOSITION II
1951-1970
(microCuries per square meter)
Each Row Includes:
County Name, Fallout Level in microCuries/square meter and County Rank (1 = highest fallout in U.S.)

680	NE GAGE	250.33
681	NE HAMILTON	250.28
682	CO LOGAN	250.13
683	ID BONNEVILLE	249.64
684	KY MORGAN	249.46
685	KS HASKELL	249.31
686	KS GREELEY	249.20
687	KS SHERIDAN	249.00
688	IN HENDRICKS	248.96
689	KS THOMAS	248.87
690	NE YORK	248.60
691	TX HOCKLEY	248.43
692	OK CREEK	248.25
693	TX MOORE	247.50
694	WY TETON	247.18
695	TX COCHRAN	246.80
696	MO CAMDEN	245.57
697	CO PUEBLO	245.45
698	SD FAULK	245.38
699	MA MIDDLESEX	245.10
700	WI CRAWFORD	244.85
701	KY PIKE	244.73
702	MT GRANITE	244.72
703	OH ATHENS	243.67
704	MA BRISTOL	243.10
705	TX BRISCOE	243.08
706	NE JOHNSON	241.64
707	KY JACKSON	241.60
708	IN SWITZERLAND	241.57
709	NE HOOKER	241.52
710	KY OWSLEY	241.50
711	TX HALE	241.08
712	IN GREENE	240.78
713	ND MORTON	240.70
714	MO PETTIS	240.58
715	ID FREMONT	240.54
716	ND HETTINGER	240.36
717	OK CHEROKEE	240.18
718	IN CASS	240.10
719	MO MORGAN	240.03
720	IL CLAY	239.80
721	WV MINGO	239.78
722	IN CRAWFORD	239.42
723	OH MADISON	238.86
724	NE FILLMORE	238.83
725	SD SULLY	238.73
726	WY PARK	238.71
727	MT HILL	238.29
728	IL MORGAN	238.26
729	IL CHAMPAIGN	238.17
730	WY CROOK	238.10
731	WI RICHLAND	238.04
732	MT SHERIDAN	237.86
733	OH PORTAGE	237.84
734	IL SCOTT	237.82
735	KS MORTON	237.66
736	NE NUCKOLLS	237.14
737	ID ADAMS	236.86
738	OK OTTAWA	236.33
739	LA ST HELENA	236.29
740	NE MCPHERSON	236.20
741	SD POTTER	236.13
742	TX SWISHER	235.61
743	KS BARTON	235.30
744	PA SNYDER	234.87
745	IL JERSEY	234.66
746	NM EDDY	234.51
747	KS LINCOLN	234.41

U.S. COUNTIES RANKED BY TOTAL FALLOUT DEPOSITION II

1951-1970

(microCuries per square meter)

Each Row Includes:

County Name, Fallout Level in microCuries/square meter and County Rank (1 = highest fallout in U.S.)

748	WI SAUK	234.20
749	SD TODD	234.05
750	MT STILLWATER	233.82
751	ID CANYON	233.53
752	MO ATCHISON	233.25
753	SD CORSON	233.18
754	SD BENNETT	233.14
755	WV MERCER	233.01
756	MT RAVALLI	232.76
757	IN HENRY	231.98
758	MT CARTER	231.88
759	MO OSAGE	230.87
760	OH GALLIA	230.78
761	LA TANGIPAHOA	230.30
762	VA POWHATAN	230.28
763	MO DE KALB	230.24
764	NE LANCASTER	230.16
765	NE ARTHUR	230.04
766	KY ANDERSON	230.02
767	IN OWEN	229.91
768	TX TERRY	229.84
769	KS OSBORNE	229.57
770	IN JEFFERSON	228.95
771	KS MARION	228.94
772	VA BUCHANAN	228.60
773	MT MUSSELSHELL	228.53
774	ID SHOSHONE	228.01
775	OK DELAWARE	227.67
776	NE MORRILL	227.49
777	KS DICKINSON	227.38
778	ND GRANT	227.38
779	MO CALLAWAY	227.26
780	SD HYDE	227.26
781	OH COSHOCTON	227.16
782	MO LIVINGSTON	227.05
783	VA HANOVER	226.93
784	ND BURLEIGH	226.72
785	TX SHERMAN	226.66
786	SD SPINK	225.92
787	NE DEUEL	225.70
788	VA GOOCHLAND	225.62
789	ND ADAMS	225.25
790	VA AMELIA	225.09
791	OH CLERMONT	224.75
792	OH LOGAN	224.51
793	MS WARREN	223.93
794	MA SUFFOLK	223.89
795	ND EMMONS	223.70
796	NM LEA	223.61
797	KS MEADE	223.61
798	PA FULTON	223.57
799	LA TENSAS	223.49
800	OK ROGER MILLS	223.49
801	SD ROBERTS	223.44
802	MT SWEET GRASS	223.04
803	SD PERKINS	223.01
804	AZ SANTA CRUZ	222.89
805	OH BROWN	222.75
806	KS MCPHERSON	222.68
807	LA ALLEN	222.35
808	WV WYOMING	222.33
809	TX WHEELER	222.30
810	KS ROOKS	222.28
811	LA LA SALLE	222.24
812	PA FOREST	222.23
813	MO NEWTON	222.19
814	MO BENTON	222.08
815	ND BOWMAN	222.04

U.S. COUNTIES RANKED BY TOTAL FALLOUT DEPOSITION II
1951-1970
(microCuries per square meter)
Each Row Includes:
County Name, Fallout Level in microCuries/square meter and County Rank (1 = highest fallout in U.S.)

Rank	County	Fallout
816	MI EATON	221.79
817	MO LAWRENCE	221.40
818	MT BIG HORN	221.36
819	NE NEMAHA	221.34
820	ND RANSOM	221.34
821	OK ATOKA	221.30
822	KS WABAUNSEE	221.06
823	IN MARTIN	221.04
824	OK MAYES	220.97
825	KS MITCHELL	220.97
826	AZ YAVAPAI	220.60
827	OK PITTSBURG	220.59
828	AR FULTON	220.35
829	KS OTTAWA	220.20
830	KS ELLSWORTH	220.03
831	MO HENRY	219.86
832	LA IBERIA	219.69
833	SD BUTTE	219.59
834	SD LAWRENCE	219.50
835	OH UNION	219.50
836	SD WALWORTH	219.41
837	MO FRANKLIN	218.88
838	TX YOAKUM	218.67
839	KY OLDHAM	218.52
840	TX LUBBOCK	218.33
841	SD PENNINGTON	218.16
842	IL CALHOUN	218.14
843	VA CHESTERFIELD	217.97
844	LA JEFFERSON DA	217.90
845	MS ADAMS	217.81
846	LA VERMILION	217.26
847	VT CHITTENDEN	217.23
848	OH GUERNSEY	217.20
849	MI INGHAM	216.88
850	TX EL PASO	216.86
851	OH MUSKINGUM	216.81
852	MO RAY	216.24
853	SD JONES	216.17
854	KS ELLIS	216.16
855	OK MUSKOGEE	216.06
856	OK WAGONER	216.01
857	KY MONTGOMERY	215.87
858	IN MONROE	215.73
859	IL VERMILION	215.72
860	PA FAYETTE	215.72
861	MO MCDONALD	215.60
862	KY PENDLETON	215.09
863	NJ BURLINGTON	215.08
864	OH PAULDING	214.96
865	NM SIERRA	214.95
866	OK ROGERS	214.77
867	KY LESLIE	214.64
868	SD SANBORN	214.57
869	OK ELLIS	214.43
870	AR FAULKNER	214.41
871	NE OTOE	214.21
872	OK NOWATA	214.13
873	SD BUFFALO	214.06
874	TX DONLEY	213.89
875	AR BENTON	213.49
876	LA POINTE COUPE	213.38
877	MO MILLER	213.30
878	ND SHERIDAN	212.87
879	MT FLATHEAD	212.79
880	ND SARGENT	212.77
881	MO WRIGHT	212.77
882	NE THAYER	212.68
883	KS SEWARD	212.65

U.S. COUNTIES RANKED BY TOTAL FALLOUT DEPOSITION II
1951-1970
(microCuries per square meter)
Each Row Includes:
County Name, Fallout Level in microCuries/square meter and County Rank (1 = highest fallout in U.S.)

Rank	County	Fallout
884	WY GOSHEN	212.60
885	AR SHARP	212.11
886	OK WASHINGTON	211.87
887	LA WEST FELICIA	211.86
888	TX CARSON	211.85
889	KS COWLEY	211.79
890	SD BEADLE	211.77
891	LA WEST BATON R	211.55
892	NE WEBSTER	211.34
893	OK SEQUOYAH	210.89
894	TX FANNIN	210.85
895	KS RENO	210.80
896	OK KAY	210.73
897	TX ARMSTRONG	210.48
898	WI DANE	210.26
899	LA TERREBONNE	210.21
900	AR LAWRENCE	209.90
901	LA IBERVILLE	209.84
902	OH DARKE	209.77
903	CT MIDDLESEX	209.66
904	IN FRANKLIN	209.64
905	SD CHARLES MIX	209.53
906	KS HARVEY	209.43
907	OH ASHLAND	209.30
908	NE GARDEN	209.21
909	KS RUSH	209.14
910	IN ORANGE	209.13
911	OK OSAGE	209.12
912	ND DICKEY	209.06
913	AR PRAIRIE	208.86
914	LA EAST BATON R	208.78
915	PA CLEARFIELD	208.49
916	NE HALL	208.39
917	MO LACLEDE	208.36
918	SD DOUGLAS	208.25
919	IN CARROLL	208.04
920	MO PULASKI	208.03
921	VA FRANKLIN	208.00
922	NE BANNER	207.95
923	KS REPUBLIC	207.57
924	KS RICE	207.14
925	AR MADISON	206.89
926	AR JACKSON	206.84
927	SD CODINGTON	206.67
928	MT GARFIELD	206.65
929	KS SHAWNEE	206.62
930	KS KINGMAN	206.59
931	KS CRAWFORD	206.50
932	MO BARRY	206.50
933	IL CLARK	206.29
934	NE BLAINE	205.99
935	MS WILKINSON	205.95
936	MT TOOLE	205.78
937	MO LAFAYETTE	205.70
938	SD TRIPP	205.21
939	OK HUGHES	204.92
940	IL MENARD	204.83
941	KS CHAUTAUQUA	204.61
942	WY LARAMIE	204.46
943	PA MERCER	204.25
944	AR WASHINGTON	204.23
945	OH TRUMBULL	204.18
946	AR PERRY	204.15
947	NE KEARNEY	204.15
948	MO POLK	204.13
949	KS BUTLER	203.82
950	NE BOONE	203.52
951	KS SUMNER	203.46

U.S. COUNTIES RANKED BY TOTAL FALLOUT DEPOSITION II
1951-1970
(microCuries per square meter)
Each Row Includes:
County Name, Fallout Level in microCuries/square meter and County Rank (1 = highest fallout in U.S.)

Rank	County	Fallout
952	MT VALLEY	203.08
953	KS LYON	203.06
954	NE DAWES	202.92
955	WA GARFIELD	202.80
956	KS CHASE	202.69
957	MO DOUGLAS	202.38
958	PA CRAWFORD	202.37
959	OK MCINTOSH	202.36
960	PA BEAVER	202.16
961	NY ESSEX	201.81
962	NY GREENE	201.74
963	MT GOLDEN VALLE	201.73
964	LA ST MARY	201.52
965	MT DANIELS	201.30
966	TX GRAYSON	201.27
967	MO OREGON	201.24
968	NE GRANT	201.22
969	MO COLE	201.12
970	IL RICHLAND	200.94
971	TX GAINES	200.86
972	NE BROWN	200.85
973	MT ROSEBUD	200.85
974	KS GEARY	200.84
975	MT LAKE	200.74
976	AR NEWTON	200.71
977	TX RANDALL	200.63
978	AR BAXTER	200.56
979	SD HAND	200.44
980	KS GREENWOOD	200.31
981	SD DAY	200.18
982	ND CASS	200.14
983	NE SHERIDAN	200.08
984	LA ASSUMPTION	200.05
985	MO CHRISTIAN	200.00
986	IA LUCAS	199.92
987	AR POPE	199.88
988	LA ST MARTIN	199.87
989	IL FORD	199.75
990	MT PETROLEUM	199.49
991	AR CRAIGHEAD	199.46
992	MO TEXAS	199.34
993	OH TUSCARAWAS	199.16
994	SD CLARK	199.10
995	PA INDIANA	199.07
996	TX FLOYD	199.03
997	ND FOSTER	199.01
998	OK CHOCTAW	198.82
999	PA ARMSTRONG	198.49
1000	AR CLEBURNE	198.43
1001	MO JASPER	198.36
1002	OH LAKE	198.27
1003	KS LEAVENWORTH	198.20
1004	VT WINDHAM	198.12
1005	TX HALL	197.99
1006	TX LYNN	197.85
1007	AR VAN BUREN	197.73
1008	KY BRACKEN	197.71
1009	SD GREGORY	197.67
1010	OH BELMONT	197.63
1011	MO OZARK	197.59
1012	PA CLARION	197.53
1013	AR IZARD	197.52
1014	OK BEAVER	197.52
1015	MS FRANKLIN	197.42
1016	IN MADISON	197.12
1017	OH WOOD	196.96
1018	SD CAMPBELL	196.85
1019	NE HARLAN	196.76

U.S. COUNTIES RANKED BY TOTAL FALLOUT DEPOSITION II 1951-1970

(microCuries per square meter)
Each Row Includes:
County Name, Fallout Level in microCuries/square meter and County Rank (1 = highest fallout in U.S.)

1020	WY WESTON	196.73
1021	IN UNION	196.61
1022	IA CHICKASAW	196.56
1023	NJ OCEAN	196.25
1024	AR FRANKLIN	195.88
1025	OH HAMILTON	195.85
1026	IN ADAMS	195.71
1027	IL BOND	195.56
1028	MO GREENE	195.52
1029	AR JOHNSON	195.50
1030	IN CLINTON	195.46
1031	KS PHILLIPS	195.46
1032	KS SMITH	195.17
1033	MA ESSEX	195.04
1034	OH COLUMBIANA	195.02
1035	IN FULTON	194.99
1036	MO WEBSTER	194.85
1037	PA CAMBRIA	194.73
1038	TX LAMAR	194.60
1039	KS RILEY	194.53
1040	IN DUBOIS	193.79
1041	MT LIBERTY	193.74
1042	NE KIMBALL	193.74
1043	PA WESTMORELAND	193.73
1044	OR GRANT	193.57
1045	VT RUTLAND	193.38
1046	NE JEFFERSON	192.68
1047	RI WASHINGTON	192.32
1048	OH SANDUSKY	192.16
1049	KY PERRY	192.11
1050	MO DENT	191.99
1051	OK SEMINOLE	191.93
1052	OK MCCURTAIN	191.92
1053	MO SALINE	191.74
1054	AR MARION	191.56
1055	NE WHEELER	191.49
1056	AR INDEPENDENCE	191.48
1057	LA EAST FELICIA	190.99
1058	MO PHELPS	190.82
1059	AR WHITE	190.48
1060	KS MORRIS	190.34
1061	MO CASS	190.10
1062	MT PONDERA	190.06
1063	KS ELK	190.03
1064	MO COOPER	189.98
1065	KS HARPER	189.77
1066	MO LINCOLN	189.76
1067	OH STARK	189.37
1068	KS STAFFORD	188.98
1069	MO VERNON	188.93
1070	TX ROBERTS	188.87
1071	WV MARSHALL	188.85
1072	NY MONTGOMERY	188.81
1073	IN KNOX	188.70
1074	TX COLLINGSWORT	188.68
1075	OH MAHONING	188.57
1076	IL FAYETTE	188.47
1077	KS LABETTE	188.47
1078	OK PAWNEE	188.44
1079	WY SHERIDAN	188.26
1080	PA WASHINGTON	188.17
1081	MO MARIES	188.17
1082	OK NOBLE	188.16
1083	SD MARSHALL	188.05
1084	IN VERMILLION	188.01
1085	KY SPENCER	187.71
1086	KS NEOSHO	187.31
1087	NE BOYD	187.17

U.S. COUNTIES RANKED BY TOTAL FALLOUT DEPOSITION II
1951-1970
(microCuries per square meter)
Each Row Includes:
County Name, Fallout Level in microCuries/square meter and County Rank (1 = highest fallout in U.S.)

1088	MO MERCER	187.00
1089	ID BINGHAM	186.83
1090	OH WASHINGTON	186.80
1091	NY ERIE	186.65
1092	TX MOTLEY	186.60
1093	TX HUDSPETH	186.48
1094	OH MEIGS	186.45
1095	KY KNOTT	186.28
1096	IN ST JOSEPH	186.12
1097	OK ALFALFA	186.09
1098	OK PUSHMATAHA	186.08
1099	OK GRANT	186.02
1100	MO HOWELL	185.83
1101	IN SHELBY	185.80
1102	OK LATIMER	185.79
1103	LA LIVINGSTON	185.76
1104	KS PAWNEE	185.54
1105	PA JEFFERSON	185.47
1106	OK OKMULGEE	185.46
1107	WV TYLER	185.35
1108	MO CARROLL	185.33
1109	OK BRYAN	185.31
1110	OH MONROE	185.22
1111	KS PRATT	185.16
1112	OK ADAIR	185.07
1113	LA AVOYELLES	184.92
1114	SD EDMUNDS	184.79
1115	MN REDWOOD	184.71
1116	AR WOODRUFF	184.69
1117	KS WILSON	184.60
1118	OH HARRISON	184.39
1119	ND LA MOURE	184.37
1120	KS CHEROKEE	184.35
1121	TX GRAY	184.20
1122	ID NEZ PERCE	184.19
1123	OH DELAWARE	184.12
1124	ME AROOSTOOK	184.10
1125	KS POTTAWATOMIE	184.08
1126	NY HAMILTON	183.93
1127	OH MARION	183.82
1128	KS EDWARDS	183.58
1129	ND MCINTOSH	183.56
1130	IL EDGAR	183.52
1131	OH LUCAS	183.46
1132	OK BECKHAM	183.23
1133	OK HASKELL	183.18
1134	MT SANDERS	183.01
1135	AR SALINE	182.66
1136	MT PHILLIPS	182.51
1137	MO NODAWAY	182.47
1138	PA LAWRENCE	182.37
1139	IN CLARK	182.28
1140	MO CEDAR	182.26
1141	AR POLK	182.22
1142	KS SHERMAN	182.19
1143	WV UPSHUR	182.17
1144	AR CRAWFORD	182.17
1145	KY JESSAMINE	181.89
1146	RI BRISTOL	181.88
1147	ND BARNES	181.81
1148	LA ST CHARLES	181.70
1149	LA LAFOURCHE	181.37
1150	KY BOONE	181.34
1151	ND TRAILL	181.29
1152	VA ACCOMACK	181.29
1153	OH HOCKING	181.01
1154	IN FLOYD	180.96
1155	NE FRANKLIN	180.79

U.S. COUNTIES RANKED BY TOTAL FALLOUT DEPOSITION II
1951-1970

(microCuries per square meter)
Each Row Includes:
County Name, Fallout Level in microCuries/square meter and County Rank (1 = highest fallout in U.S.)

Rank	County	Fallout
1156	ND WILLIAMS	180.53
1157	WI JEFFERSON	180.27
1158	ND WALSH	180.09
1159	KS WYANDOTTE	179.62
1160	OH MORROW	179.37
1161	MO HOWARD	179.18
1162	AR SEARCY	179.10
1163	OK GREER	178.99
1164	KS WASHINGTON	178.90
1165	MO GASCONADE	178.86
1166	MO DALLAS	178.80
1167	OH ERIE	178.71
1168	CT NEW HAVEN	178.70
1169	TX COLLIN	178.52
1170	AR RANDOLPH	178.51
1171	OK PONTOTOC	178.45
1172	WV WOOD	178.44
1173	LA ASCENSION	178.38
1174	AZ PINAL	178.28
1175	NY GENESEE	178.27
1176	AR ARKANSAS	178.26
1177	NY SCHOHARIE	178.26
1178	NE PIERCE	178.21
1179	TX CHILDRESS	178.17
1180	CT TOLLAND	178.07
1181	MO JOHNSON	178.00
1182	NY CHAUTAUQUA	177.97
1183	IL CUMBERLAND	177.82
1184	AZ GILA	177.81
1185	ND LOGAN	177.79
1186	OK WASHITA	177.74
1187	ND STUTSMAN	177.69
1188	AR GREENE	177.56
1189	MO CLINTON	177.46
1190	OH ROSS	177.37
1191	IA GRUNDY	176.93
1192	AR JEFFERSON	176.73
1193	KY ESTILL	176.69
1194	IN BARTHOLOMEW	176.52
1195	RI KENT	176.49
1196	NJ ATLANTIC	176.48
1197	NC WASHINGTON	176.46
1198	KS CLAY	176.29
1199	PA GREENE	176.23
1200	AR YELL	176.21
1201	KY LETCHER	176.11
1202	MT TREASURE	176.09
1203	TX HEMPHILL	176.06
1204	NY WYOMING	175.81
1205	WI PEPIN	175.56
1206	MD GARRETT	175.44
1207	OH CUYAHOGA	175.41
1208	MO DADE	175.22
1209	OH AUGLAIZE	174.86
1210	OK GARFIELD	174.83
1211	ND GRAND FORKS	174.75
1212	TX COOKE	174.71
1213	ND GRIGGS	174.70
1214	NE KNOX	174.69
1215	NJ MONMOUTH	174.65
1216	ND SIOUX	174.55
1217	MO BARTON	174.46
1218	KS MONTGOMERY	174.40
1219	VA DICKENSON	174.22
1220	AZ MARICOPA	173.90
1221	MO BOONE	173.71
1222	KS WOODSON	173.67
1223	WV HARRISON	173.63

U.S. COUNTIES RANKED BY TOTAL FALLOUT DEPOSITION II
1951-1970
(microCuries per square meter)
Each Row Includes:
County Name, Fallout Level in microCuries/square meter and County Rank (1 = highest fallout in U.S.)

Rank	County	Fallout
1224	PA CHESTER	173.60
1225	AZ GRAHAM	173.51
1226	LA SABINE	173.15
1227	OH CRAWFORD	173.14
1228	AR PULASKI	173.11
1229	NC CHOWAN	172.58
1230	AR LONOKE	172.49
1231	AR CONWAY	172.37
1232	AR LOGAN	172.28
1233	MA DUKES	172.14
1234	PA BUTLER	172.02
1235	OR UNION	171.97
1236	NV CLARK3	171.84
1237	MO MONITEAU	171.75
1238	MA BARNSTABLE	171.73
1239	NE BOX BUTTE	171.60
1240	NE KEYA PAHA	171.59
1241	LA VERNON	171.58
1242	MO STONE	171.57
1243	OK OKFUSKEE	171.54
1244	TX DENTON	171.44
1245	ID LATAH	171.31
1246	ND NELSON	171.20
1247	TX CROSBY	171.10
1248	WV CALHOUN	171.07
1249	KS SEDGWICK	170.95
1250	NY ULSTER	170.88
1251	AR MONTGOMERY	170.78
1252	MA BERKSHIRE	170.66
1253	WV GREENBRIER	170.62
1254	WV GILMER	170.58
1255	KS JOHNSON	170.51
1256	OH FAYETTE	170.27
1257	NY NASSAU	170.19
1258	AR SEBASTIAN	170.18
1259	MS HINDS	170.17
1260	WV WIRT	170.12
1261	WI JUNEAU	170.08
1262	AR SEVIER	169.83
1263	NE DODGE	169.83
1264	SD BROWN	169.78
1265	TX ANDREWS	169.76
1266	NE BURT	169.73
1267	AR BOONE	169.60
1268	IN RANDOLPH	169.54
1269	IN HARRISON	169.16
1270	AR CARROLL	169.11
1271	OK LINCOLN	168.99
1272	KY MADISON	168.99
1273	AR CLAY	168.96
1274	LA CONCORDIA	168.79
1275	KS JEWELL	168.75
1276	VT GRAND ISLE	168.63
1277	KS MARSHALL	168.48
1278	TX LOVING	168.47
1279	WI TREMPEALEAU	168.46
1280	IL WHITESIDE	168.31
1281	KY ROCKCASTLE	168.28
1282	IN WASHINGTON	168.18
1283	KS COFFEY	168.14
1284	WY HOT SPRINGS	168.03
1285	PA SOMERSET	167.86
1286	MO RIPLEY	167.83
1287	KS ATCHISON	167.67
1288	OK LE FLORE	167.53
1289	SD MCPHERSON	167.52
1290	ND EDDY	167.47
1291	MO CRAWFORD	167.25

U.S. COUNTIES RANKED BY TOTAL FALLOUT DEPOSITION II
1951-1970
(microCuries per square meter)

Each Row Includes:
County Name, Fallout Level in microCuries/square meter and County Rank (1 = highest fallout in U.S.)

Rank	County	Fallout
1292	MI OAKLAND	166.96
1293	SD HUGHES	166.94
1294	AR STONE	166.92
1295	OH FAIRFIELD	166.86
1296	NY NIAGARA	166.76
1297	MO GENTRY	166.68
1298	IN SCOTT	166.59
1299	KS KIOWA	166.55
1300	VA CARROLL	166.41
1301	KS ALLEN	166.05
1302	KS JACKSON	165.86
1303	KS CLARK	165.68
1304	AR ST FRANCIS	165.32
1305	ND BENSON	165.29
1306	KS CLOUD	165.20
1307	MO CLAY	165.17
1308	WV WETZEL	164.79
1309	MS PEARL RIVER	164.73
1310	WV LEWIS	164.36
1311	OK HARMON	164.30
1312	MO BATES	164.25
1313	NY LIVINGSTON	164.20
1314	CT FAIRFIELD	163.91
1315	AR SCOTT	163.75
1316	ND STEELE	163.69
1317	TN GIBSON	163.28
1318	LA CATAHOULA	163.15
1319	MO JEFFERSON	162.88
1320	MO NEW MADRID	162.78
1321	IL JO DAVIESS	162.73
1322	LA WASHINGTON	162.57
1323	WV TUCKER	162.50
1324	KY HARLAN	162.37
1325	OH VINTON	162.26
1326	OH GREENE	162.17
1327	KS JEFFERSON	162.16
1328	KY BOURBON	162.06
1329	WI MONROE	162.01
1330	LA PLAQUEMINES	161.84
1331	AL MOBILE	161.77
1332	KS MIAMI	161.75
1333	LA ST LANDRY	161.72
1334	LA ST TAMMANY	161.68
1335	KS FRANKLIN	161.67
1336	KS BARBER	161.67
1337	ND PEMBINA	161.36
1338	TX JASPER	161.14
1339	MA HAMPDEN	161.11
1340	KS DOUGLAS	161.07
1341	NY ST LAWRENCE	160.96
1342	WI LA CROSSE	160.88
1343	TX NEWTON	160.87
1344	TX CULBERSON	160.64
1345	NE RICHARDSON	160.59
1346	TX DALLAS	160.40
1347	MS HANCOCK	160.33
1348	WA ASOTIN	160.23
1349	MO WAYNE	160.05
1350	CT LITCHFIELD	160.04
1351	NY SUFFOLK	159.96
1352	OH VAN WERT	159.88
1353	CT NEW LONDON	159.86
1354	IL LAWRENCE	159.85
1355	WV BRAXTON	159.64
1356	AZ GREENLEE	159.50
1357	MI GENESEE	159.25
1358	KS OSAGE	159.23
1359	AR MONROE	159.00

U.S. COUNTIES RANKED BY TOTAL FALLOUT DEPOSITION II
1951-1970
(microCuries per square meter)
Each Row Includes:
County Name, Fallout Level in microCuries/square meter and County Rank (1 = highest fallout in U.S.)

1360	PA ALLEGHENY	158.99
1361	VA KING WILLIAM	158.97
1362	IN WABASH	158.87
1363	MA WORCESTER	158.84
1364	SD BON HOMME	158.72
1365	OK KINGFISHER	158.66
1366	PA WARREN	158.65
1367	MS PIKE	158.59
1368	TN WASHINGTON	158.29
1369	KS BROWN	158.14
1370	AR MISSISSIPPI	158.00
1371	NY QUEENS	157.91
1372	WI SAWYER	157.91
1373	PA DELAWARE	157.87
1374	AR LITTLE RIVER	157.85
1375	MO DUNKLIN	157.78
1376	NE PAWNEE	157.69
1377	PA ERIE	157.65
1378	CT WINDHAM	157.52
1379	OH NOBLE	157.46
1380	TN WEAKLEY	156.74
1381	OH GEAUGA	156.65
1382	AZ COCHISE	156.55
1383	WV BROOKE	156.49
1384	MS CLAIBORNE	156.23
1385	OK PAYNE	156.08
1386	IN WARRICK	155.96
1387	TX ANGELINA	155.89
1388	OK LOVE	155.86
1389	MI LENAWEE	155.80
1390	MS GEORGE	155.74
1391	TN TIPTON	155.71
1392	WV RANDOLPH	155.70
1393	WV FAYETTE	155.64
1394	NY WESTCHESTER	155.54
1395	KY FAYETTE	155.29
1396	ND RAMSEY	155.22
1397	NY KINGS	155.17
1398	KS NEMAHA	155.14
1399	TX SABINE	155.12
1400	NH ROCKINGHAM	155.02
1401	TX SAN AUGUSTIN	155.00
1402	TN LAKE	154.93
1403	PA CLINTON	154.62
1404	IN TIPPECANOE	154.61
1405	IN HUNTINGTON	154.57
1406	NY CLINTON	154.56
1407	WI DUNN	154.45
1408	IL LEE	154.39
1409	MO CARTER	154.27
1410	WV PUTNAM	154.26
1411	WI JACKSON	154.21
1412	AR GRANT	154.21
1413	IL STEPHENSON	154.05
1414	ND KIDDER	154.02
1415	WY BIG HORN	153.97
1416	OK JOHNSTON	153.92
1417	MO PLATTE	153.70
1418	LA ST JOHN THE	153.59
1419	AR HOWARD	153.48
1420	MO SHANNON	153.44
1421	MS JEFFERSON	153.40
1422	IN WARREN	153.32
1423	OK POTTAWATOMIE	153.31
1424	VA TAZEWELL	153.18
1425	AL MARENGO	153.13
1426	NY WAYNE	153.12
1427	NY ALLEGANY	153.11

U.S. COUNTIES RANKED BY TOTAL FALLOUT DEPOSITION II
1951-1970
(microCuries per square meter)
Each Row Includes:
County Name, Fallout Level in microCuries/square meter and County Rank (1 = highest fallout in U.S.)

Rank	County	Fallout
1428	VT FRANKLIN	153.10
1429	WI WASHBURN	153.10
1430	IL LA SALLE	153.06
1431	NJ HUDSON	152.72
1432	WV MARION	152.48
1433	VA KING AND QUE	152.08
1434	NY ONTARIO	152.05
1435	WV NICHOLAS	152.02
1436	TN LAUDERDALE	151.74
1437	MI LIVINGSTON	151.69
1438	OK LOGAN	151.46
1439	OH LORAIN	151.45
1440	WI ADAMS	151.37
1441	TX WALKER	151.35
1442	AZ YUMA	151.35
1443	LA ST BERNARD	151.32
1444	TN OBION	151.32
1445	IL OGLE	151.25
1446	AR CRITTENDEN	151.20
1447	TX POLK	150.93
1448	OK WOODS	150.44
1449	NM GRANT	150.44
1450	IL CRAWFORD	150.35
1451	AR CLARK	150.33
1452	WV GRANT	150.18
1453	MO PEMISCOT	149.83
1454	KS LINN	149.75
1455	NE SARPY	149.74
1456	WV MONONGALIA	149.60
1457	AR CROSS	149.50
1458	MO HICKORY	149.32
1459	WV CLAY	149.31
1460	MO BUTLER	149.27
1461	TX HOPKINS	149.27
1462	RI PROVIDENCE	149.24
1463	IN VANDERBURGH	149.21
1464	NJ ESSEX	149.16
1465	TN MADISON	149.11
1466	MO REYNOLDS	149.11
1467	MS MARION	149.09
1468	IN MIAMI	148.98
1469	PA CENTRE	148.83
1470	LA ST JAMES	148.83
1471	PA HUNTINGDON	148.74
1472	MI LAPEER	148.38
1473	NM OTERO	148.19
1474	OH HURON	148.13
1475	NY SULLIVAN	148.02
1476	TX RAINS	147.92
1477	AR PIKE	147.91
1478	MA HAMPSHIRE	147.89
1479	KS ANDERSON	147.89
1480	MD HARFORD	147.77
1481	TX CHAMBERS	147.77
1482	NY OTSEGO	147.71
1483	TN DYER	147.51
1484	MI MONROE	147.43
1485	NY STEUBEN	147.03
1486	TX JEFF DAVIS	146.70
1487	MI SHIAWASSEE	146.61
1488	IN PERRY	146.60
1489	NY CATTARAUGUS	146.59
1490	VA SMYTH	146.57
1491	ME PENOBSCOT	146.55
1492	WV POCAHONTAS	146.41
1493	IN SPENCER	146.39
1494	AR POINSETT	146.32
1495	NY PUTNAM	146.26

U.S. COUNTIES RANKED BY TOTAL FALLOUT DEPOSITION II
1951-1970

(microCuries per square meter)
Each Row Includes:
County Name, Fallout Level in microCuries/square meter and County Rank (1 = highest fallout in U.S.)

1496	NC NORTHAMPTON	146.21
1497	TX TARRANT	146.10
1498	AR LEE	145.82
1499	PA YORK	145.76
1500	TN HAYWOOD	145.76
1501	VA NEW KENT	145.75
1502	TX WINKLER	145.63
1503	MD BALTIMORE	145.29
1504	MI HILLSDALE	145.16
1505	WI WOOD	145.15
1506	OK COAL	145.06
1507	IL PUTNAM	145.04
1508	TX BOWIE	144.72
1509	WV WEBSTER	144.69
1510	WI RUSK	144.68
1511	MS FORREST	144.58
1512	NE ROCK	144.42
1513	KS BOURBON	144.28
1514	MO ST CHARLES	144.23
1515	OK CARTER	144.13
1516	IN SULLIVAN	144.10
1517	VA CHARLES CITY	144.04
1518	TX GRIMES	144.02
1519	KY HANCOCK	143.94
1520	MI WAYNE	143.62
1521	OK CANADIAN	143.59
1522	NJ UNION	143.52
1523	NY ORLEANS	143.48
1524	CT HARTFORD	143.46
1525	TX TRINITY	143.39
1526	NC BEAUFORT	143.23
1527	MS LAUDERDALE	143.20
1528	MO MONTGOMERY	143.19
1529	AR DESHA	143.17
1530	KY HICKMAN	143.07
1531	PA CAMERON	143.05
1532	MO JACKSON	142.98
1533	TX VAN ZANDT	142.89
1534	TX HUNT	142.77
1535	MO MISSISSIPPI	142.73
1536	NY DELAWARE	142.68
1537	TX HARRIS	142.59
1538	TX LIBERTY	142.59
1539	KY MARTIN	142.50
1540	MS DE SOTO	142.45
1541	WI COLUMBIA	142.36
1542	PA NORTHUMBERLA	142.30
1543	VT WINDSOR	142.26
1544	AR LAFAYETTE	142.21
1545	VA FAIRFAX	142.11
1546	TX TYLER	141.98
1547	MI KENT	141.98
1548	TX RED RIVER	141.95
1549	NH CARROLL	141.90
1550	WI EAU CLAIRE	141.72
1551	IL IROQUOIS	141.65
1552	NJ BERGEN	141.56
1553	NY NEW YORK	141.49
1554	IL JEFFERSON	141.30
1555	TX BRAZORIA	141.14
1556	VA WARREN	141.13
1557	OK COMANCHE	141.10
1558	AR HOT SPRING	141.04
1559	MS QUITMAN	141.00
1560	AR GARLAND	140.88
1561	PA LACKAWANNA	140.86
1562	MN DAKOTA	140.82
1563	OK MARSHALL	140.70

U.S. COUNTIES RANKED BY TOTAL FALLOUT DEPOSITION II

1951-1970

(microCuries per square meter)

Each Row Includes:

County Name, Fallout Level in microCuries/square meter and County Rank (1 = highest fallout in U.S.)

Rank	County	Fallout
1564	WI ST CROIX	140.58
1565	MS STONE	140.50
1566	AR MILLER	140.49
1567	NY FRANKLIN	140.46
1568	MS MARSHALL	140.46
1569	MT GLACIER	140.42
1570	IN DAVIESS	140.39
1571	MT YELLOWSTONE	140.22
1572	TX WOOD	140.05
1573	WI DOUGLAS	140.01
1574	VA LOUDOUN	140.00
1575	IN VIGO	139.98
1576	WI PIERCE	139.92
1577	NY JEFFERSON	139.75
1578	WI WAUSHARA	139.70
1579	TX CAMP	139.70
1580	NM LUNA	139.55
1581	OH PUTNAM	139.50
1582	TX MONTGOMERY	139.45
1583	IN PULASKI	139.30
1584	AR PHILLIPS	139.18
1585	OK HARPER	139.12
1586	WI ASHLAND	138.99
1587	TX HARDIN	138.94
1588	WI BARRON	138.94
1589	WI POLK	138.85
1590	AR NEVADA	138.77
1591	MN OLMSTED	138.76
1592	TX REEVES	138.74
1593	KY FULTON	138.66
1594	IL KANKAKEE	138.61
1595	NJ MIDDLESEX	138.37
1596	RI NEWPORT	138.35
1597	MO ST FRANCOIS	138.33
1598	MO ST CLAIR	138.19
1599	VT WASHINGTON	138.19
1600	NY RICHMOND	138.18
1601	PA WAYNE	138.17
1602	WI CHIPPEWA	138.06
1603	OK DEWEY	137.71
1604	TX CASS	137.63
1605	ME PISCATAQUIS	137.55
1606	NE SCOTTS BLUFF	137.52
1607	ND CAVALIER	137.38
1608	NY HERKIMER	137.29
1609	WI MARQUETTE	137.23
1610	VA CRAIG	136.90
1611	NJ PASSAIC	136.71
1612	FL OKALOOSA	136.68
1613	OK MAJOR	136.56
1614	ND WELLS	136.52
1615	TX WALLER	136.37
1616	AR LINCOLN	136.33
1617	NY OSWEGO	136.31
1618	MO STODDARD	136.31
1619	VA FAUQUIER	136.26
1620	VT LAMOILLE	136.20
1621	ME WASHINGTON	136.13
1622	VA HIGHLAND	136.03
1623	MS TATE	136.02
1624	PA ADAMS	136.00
1625	IL RANDOLPH	135.99
1626	WI BURNETT	135.93
1627	NJ MORRIS	135.90
1628	PA BEDFORD	135.90
1629	OR MALHEUR	135.88
1630	PA LANCASTER	135.81
1631	NC PERSON	135.79

U.S. COUNTIES RANKED BY TOTAL FALLOUT DEPOSITION II
1951-1970

(microCuries per square meter)
Each Row Includes:
County Name, Fallout Level in microCuries/square meter and County Rank (1 = highest fallout in U.S.)

Rank	County	Fallout
1632	VA FLOYD	135.67
1633	MS HARRISON	135.50
1634	MS CHICKASAW	135.23
1635	TN CROCKETT	135.22
1636	VA ORANGE	135.21
1637	MS JACKSON	135.17
1638	MO WARREN	134.94
1639	OK BLAINE	134.82
1640	KS COMANCHE	134.80
1641	KS DONIPHAN	134.77
1642	IL WILL	134.58
1643	MN CARLTON	134.57
1644	WI CLARK	134.48
1645	VA ROCKINGHAM	134.43
1646	TX ROCKWALL	134.40
1647	MI KALAMAZOO	134.39
1648	MO IRON	134.33
1649	IN JASPER	134.16
1650	MS COAHOMA	134.11
1651	NY DUTCHESS	134.11
1652	MS TIPPAH	134.04
1653	AR HEMPSTEAD	134.01
1654	NH CHESHIRE	133.70
1655	TX SAN JACINTO	133.59
1656	MI CALHOUN	133.55
1657	TN FAYETTE	133.49
1658	MO WASHINGTON	133.40
1659	MI CLINTON	133.31
1660	NY BRONX	133.23
1661	NM HIDALGO	133.13
1662	TX FORT BEND	132.89
1663	MI VAN BUREN	132.87
1664	NC VANCE	132.86
1665	IN POSEY	132.79
1666	VA ALLEGHANY	132.75
1667	OK CADDO	132.75
1668	NJ CAPE MAY	132.71
1669	WI IRON	132.70
1670	MS PANOLA	132.70
1671	KY GRAVES	132.70
1672	TX AUSTIN	132.69
1673	TX DELTA	132.59
1674	VA BEDFORD	132.34
1675	MS TALLAHATCHIE	131.78
1676	NJ MERCER	131.77
1677	MA FRANKLIN	131.67
1678	KY DAVIESS	131.64
1679	ID BENEWAH	131.43
1680	PA POTTER	131.42
1681	TX WISE	131.40
1682	VA PITTSYLVANIA	131.38
1683	IL GRUNDY	131.27
1684	IL DE KALB	131.26
1685	VA MECKLENBURG	131.14
1686	NY ORANGE	130.85
1687	IN DECATUR	130.84
1688	IN WHITE	130.76
1689	WI BAYFIELD	130.68
1690	MI MACOMB	130.54
1691	MS WALTHALL	130.07
1692	MO TANEY	130.02
1693	VA ROANOKE	130.01
1694	TX MONTAGUE	129.79
1695	VA SPOTSYLVANIA	129.74
1696	MS YALOBUSHA	129.67
1697	IL WABASH	129.47
1698	VT ORANGE	129.44
1699	MS GREENE	129.37

U.S. COUNTIES RANKED BY TOTAL FALLOUT DEPOSITION II
1951-1970
(microCuries per square meter)
Each Row Includes:
County Name, Fallout Level in microCuries/square meter and County Rank (1 = highest fallout in U.S.)

1700	TX WASHINGTON	129.31
1701	WI TAYLOR	129.31
1702	TX GALVESTON	129.22
1703	MN FREEBORN	129.19
1704	OK MCCLAIN	129.09
1705	NM DONA ANA	128.49
1706	KY CUMBERLAND	128.36
1707	NH HILLSBOROUGH	128.34
1708	MO ANDREW	128.26
1709	TN HARDEMAN	128.21
1710	MN RAMSEY	128.08
1711	MN WASHINGTON	127.84
1712	MD ALLEGANY	127.72
1713	AR DALLAS	127.34
1714	VA CAROLINE	127.15
1715	WI PRICE	127.14
1716	NY SENECA	127.13
1717	OK STEPHENS	127.03
1718	TX ELLIS	127.02
1719	MN SCOTT	127.01
1720	NH STRAFFORD	126.96
1721	AL CONECUH	126.83
1722	MT LINCOLN	126.79
1723	VA RAPPAHANNOCK	126.67
1724	OK CLEVELAND	126.62
1725	VA PRINCE EDWAR	126.54
1726	OK JEFFERSON	126.52
1727	MS TUNICA	126.44
1728	OH SUMMIT	126.40
1729	IN MONTGOMERY	126.31
1730	TN CLAIBORNE	126.25
1731	MS BENTON	126.13
1732	MO BUCHANAN	125.98
1733	PA MCKEAN	125.94
1734	AR COLUMBIA	125.79
1735	VA LUNENBURG	125.78
1736	VA MADISON	125.75
1737	VA SUFFOLK/NANS	125.74
1738	TX TITUS	125.68
1739	ND TOWNER	125.65
1740	OK MURRAY	125.33
1741	TN CAMPBELL	125.07
1742	OK WOODWARD	124.73
1743	OK OKLAHOMA	124.65
1744	IL LIVINGSTON	124.55
1745	OK GRADY	124.53
1746	NC GRANVILLE	124.51
1747	OK GARVIN	124.31
1748	MS BOLIVAR	124.04
1749	TN HANCOCK	123.94
1750	KY BELL	123.89
1751	PA PIKE	123.85
1752	VT ORLEANS	123.70
1753	TX WARD	123.63
1754	MO HOLT	123.63
1755	IL CARROLL	123.50
1756	MN CARVER	123.47
1757	NC CHATHAM	123.39
1758	IL WAYNE	123.36
1759	IN PORTER	123.18
1760	NH GRAFTON	123.15
1761	IN LAKE	123.13
1762	TX SMITH	122.97
1763	NJ WARREN	122.76
1764	NY TOMPKINS	122.71
1765	VA AMHERST	122.52
1766	IN NEWTON	122.43
1767	MS AMITE	122.32

U.S. COUNTIES RANKED BY TOTAL FALLOUT DEPOSITION II 1951-1970

(microCuries per square meter)

Each Row Includes:

County Name, Fallout Level in microCuries/square meter and County Rank (1 = highest fallout in U.S.)

Rank	County	Fallout	Rank	County	Fallout
1768	NC WILKES	122.07	1802	NH BELKNAP	118.83
1769	ME HANCOCK	122.02	1803	NY ROCKLAND	118.68
1770	MS GRENADA	122.02	1804	KY BOYLE	118.36
1771	TN CARTER	121.78	1805	NV NYE3	118.32
1772	AR CLEVELAND	121.64	1806	NJ SUSSEX	118.30
1773	NH SULLIVAN	121.63	1807	VA RUSSELL	118.22
1774	MS LAMAR	121.50	1808	VA CAMPBELL	118.20
1775	NC MARTIN	121.41	1809	MS PONTOTOC	118.17
1776	WI WAUPACA	121.41	1810	IL JACKSON	117.94
1777	AR OUACHITA	121.31	1811	OK JACKSON	117.93
1778	MS MONTGOMERY	121.29	1812	PA LUZERNE	117.92
1779	MI OTTAWA	121.28	1813	NC ASHE	117.82
1780	MN PINE	121.04	1814	MO PERRY	117.74
1781	ND ROLETTE	120.98	1815	VA NELSON	117.57
1782	NJ SOMERSET	120.89	1816	ND PIERCE	117.55
1783	OK KIOWA	120.73	1817	IN CLAY	117.54
1784	IL ST CLAIR	120.40	1818	OK CUSTER	117.53
1785	KY CLAY	120.33	1819	IN BOONE	117.49
1786	VA AUGUSTA	120.30	1820	VA PATRICK	117.38
1787	MA NANTUCKET	120.20	1821	WV OHIO	117.24
1788	MI SANILAC	120.11	1822	MS COPIAH	117.20
1789	IL KANE	120.02	1823	TX FRANKLIN	117.20
1790	VT ESSEX	119.87	1824	MN SHERBURNE	117.09
1791	VA LEE	119.74	1825	WI ONEIDA	117.00
1792	KY UNION	119.67	1826	MI BARRY	116.93
1793	LA JEFFERSON	119.38	1827	KY MARSHALL	116.90
1794	MO ST LOUIS	119.34	1828	KY TODD	116.86
1795	VA PULASKI	119.32	1829	VA MONTGOMERY	116.85
1796	VA APPOMATTOX	119.32	1830	TN SHELBY	116.83
1797	IL MCHENRY	119.31	1831	WV HARDY	116.83
1798	WI PORTAGE	119.09	1832	LA MADISON	116.82
1799	NH MERRIMACK	118.92	1833	VA SCOTT	116.81
1800	OK COTTON	118.91	1834	TX CLAY	116.67
1801	MS CARROLL	118.87	1835	IN FOUNTAIN	116.61

U.S. COUNTIES RANKED BY TOTAL FALLOUT DEPOSITION II
1951-1970
(microCuries per square meter)
Each Row Includes:
County Name, Fallout Level in microCuries/square meter and County Rank (1 = highest fallout in U.S.)

Rank	County	Fallout
1836	AR UNION	116.60
1837	IL EDWARDS	116.56
1838	TX HARDEMAN	116.46
1839	TX ECTOR	116.44
1840	IL GALLATIN	116.41
1841	IL MONROE	116.34
1842	MS LEE	116.04
1843	KY PULASKI	115.93
1844	PA WYOMING	115.89
1845	VA FREDERICK	115.87
1846	TN SULLIVAN	115.78
1847	KY MERCER	115.55
1848	MN MCLEOD	115.54
1849	KY RUSSELL	115.46
1850	KY KNOX	115.44
1851	WI LINCOLN	115.25
1852	MN ANOKA	114.97
1853	MO CAPE GIRARDE	114.92
1854	MS YAZOO	114.63
1855	IL ALEXANDER	114.59
1856	CA LOS ANGELES	114.58
1857	WI VILAS	114.58
1858	MS UNION	114.56
1859	MI GOGEBIC	114.28
1860	OH OTTAWA	114.09
1861	NJ CAMDEN	114.07
1862	IL PULASKI	114.02
1863	VT CALEDONIA	113.65
1864	MI HOUGHTON	113.33
1865	VA ROCKBRIDGE	113.21
1866	LA RICHLAND	113.16
1867	IN MORGAN	113.07
1868	PA MONROE	112.89
1869	ME KNOX	112.88
1870	TN CLAY	112.80
1871	SC FLORENCE	112.70
1872	MS PERRY	112.43
1873	IN PARKE	112.42
1874	MS TISHOMINGO	112.37
1875	NC JONES	112.35
1876	NJ GLOUCESTER	112.18
1877	WA PEND OREILLE	112.18
1878	IL UNION	112.14
1879	MS LAFAYETTE	112.14
1880	TN SEVIER	112.12
1881	MS ISSAQUENA	112.00
1882	MN ST LOUIS	111.94
1883	WA COLUMBIA	111.87
1884	IN PUTNAM	111.78
1885	MI HURON	111.71
1886	AL WILCOX	111.28
1887	VA GILES	111.22
1888	OH HARDIN	111.20
1889	CA ALPINE	111.18
1890	MN MEEKER	111.10
1891	MN RENVILLE	111.03
1892	TN MCNAIRY	110.95
1893	IL MADISON	110.91
1894	MS PRENTISS	110.75
1895	NH COOS	110.75
1896	TX KAUFMAN	110.70
1897	VA GREENE	110.57
1898	TX UPSHUR	110.57
1899	MI KEWEENAW	110.52
1900	LA OUACHITA	110.37
1901	IN RUSH	110.33
1902	WI GREEN LAKE	110.31
1903	MI SAGINAW	110.23

U.S. COUNTIES RANKED BY TOTAL FALLOUT DEPOSITION II
1951-1970
(microCuries per square meter)
Each Row Includes:
County Name, Fallout Level in microCuries/square meter and County Rank (1 = highest fallout in U.S.)

1904	WI MARATHON	110.14
1905	IL BUREAU	110.07
1906	MS SHARKEY	110.02
1907	KY MCCRACKEN	109.99
1908	OR HARNEY	109.99
1909	TX PARKER	109.96
1910	NC MACON	109.95
1911	KY OHIO	109.91
1912	NJ CUMBERLAND	109.87
1913	NC JACKSON	109.86
1914	KY GARRARD	109.85
1915	AL MONROE	109.78
1916	ME OXFORD	109.74
1917	MI BRANCH	109.73
1918	IN BENTON	109.69
1919	KY BULLITT	109.66
1920	TN SCOTT	109.49
1921	MN CHISAGO	109.33
1922	IN BROWN	109.20
1923	TX RUSK	109.16
1924	MO SCOTT	109.15
1925	MN WRIGHT	109.12
1926	IN RIPLEY	109.10
1927	ME WALDO	108.99
1928	MI ONTONAGON	108.94
1929	ME LINCOLN	108.87
1930	LA WEBSTER	108.86
1931	VA SHENANDOAH	108.58
1932	MO MADISON	108.56
1933	MS ALCORN	108.51
1934	VA GRAYSON	108.51
1935	TN PICKETT	108.30
1936	ME CUMBERLAND	108.08
1937	MS LEFLORE	108.02
1938	OH ASHTABULA	107.95
1939	KY BALLARD	107.93
1940	TX MADISON	107.90
1941	MS CLARKE	107.88
1942	TX MORRIS	107.82
1943	OH FRANKLIN	107.66
1944	NY LEWIS	107.53
1945	GA RABUN	107.40
1946	WV PLEASANTS	107.27
1947	MN AITKIN	107.18
1948	AR BRADLEY	107.04
1949	VA VIRGINIA BEA	107.00
1950	OH MERCER	106.84
1951	WI FOND DU LAC	106.77
1952	NY YATES	106.69
1953	SC OCONEE	106.65
1954	NY ONEIDA	106.62
1955	TX DAWSON	106.60
1956	NC CLAY	106.48
1957	TX DICKENS	106.28
1958	MS NEWTON	106.27
1959	TX JOHNSON	106.27
1960	MS HUMPHREYS	106.24
1961	OH WAYNE	106.21
1962	ID BONNER	106.14
1963	AR DREW	106.13
1964	WI WAUKESHA	106.01
1965	MN BENTON	105.81
1966	NC HYDE	105.78
1967	GA LIBERTY	105.78
1968	NC SWAIN	105.71
1969	LA CADDO	105.71
1970	NC TRANSYLVANIA	105.70
1971	TN ANDERSON	105.68

U.S. COUNTIES RANKED BY TOTAL FALLOUT DEPOSITION II
1951-1970
(microCuries per square meter)
Each Row Includes:
County Name, Fallout Level in microCuries/square meter and County Rank (1 = highest fallout in U.S.)

1972	ME YORK	105.65
1973	MS WAYNE	105.61
1974	MS CALHOUN	105.61
1975	OR UMATILLA	105.49
1976	VA WASHINGTON	105.44
1977	MS WASHINGTON	105.39
1978	ME SAGADAHOC	105.09
1979	NC HAYWOOD	104.82
1980	NY CORTLAND	104.71
1981	IL PERRY	104.65
1982	NY MONROE	104.46
1983	ME KENNEBEC	104.37
1984	MO BOLLINGER	104.16
1985	IN DEARBORN	104.11
1986	MN ISANTI	104.03
1987	IL POPE	104.00
1988	LA BOSSIER	103.98
1989	OK TILLMAN	103.73
1990	MI GRATIOT	103.53
1991	TX MARION	103.38
1992	TN CHESTER	103.36
1993	TX GARZA	103.28
1994	PA BLAIR	103.25
1995	ME ANDROSCOGGIN	103.00
1996	IN GIBSON	102.98
1997	NC ORANGE	102.76
1998	NJ SALEM	102.74
1999	MS SUNFLOWER	102.65
2000	GA LONG	102.65
2001	IL CLINTON	102.64
2002	MN KANABEC	102.52
2003	NY MADISON	102.47
2004	MO STE GENEVIEV	102.28
2005	MN KANDIYOHI	102.26
2006	OH HOLMES	102.25
2007	WI MARINETTE	101.98
2008	WI WINNEBAGO	101.93
2009	IN JOHNSON	101.73
2010	AR ASHLEY	101.39
2011	WI OCONTO	101.22
2012	WI CALUMET	101.09
2013	NJ HUNTERDON	101.00
2014	NC HENDERSON	100.95
2015	VA WISE	100.85
2016	IN PIKE	100.85
2017	IL LAKE	100.76
2018	NY CHEMUNG	100.60
2019	IN KOSCIUSKO	100.52
2020	TX CHEROKEE	100.47
2021	MS OKTIBBEHA	100.44
2022	MS MADISON	100.44
2023	MS LINCOLN	100.19
2024	NC CHEROKEE	100.10
2025	KY HENDERSON	100.07
2026	WI SHEBOYGAN	99.99
2027	LA WINN	99.98
2028	NY CAYUGA	99.60
2029	MN CROW WING	99.54
2030	KY LINCOLN	99.52
2031	DE SUSSEX	99.50
2032	TX GREGG	99.43
2033	LA UNION	99.34
2034	PA NORTHAMPTON	99.27
2035	DE NEW CASTLE	99.25
2036	TN COCKE	99.23
2037	TX HENDERSON	99.13
2038	KY LIVINGSTON	99.05
2039	MN LAKE	99.03

U.S. COUNTIES RANKED BY TOTAL FALLOUT DEPOSITION II
1951-1970
(microCuries per square meter)
Each Row Includes:
County Name, Fallout Level in microCuries/square meter and County Rank (1 = highest fallout in U.S.)

2040	TN ROANE	98.99
2041	TX HARRISON	98.95
2042	IL WASHINGTON	98.91
2043	MD SOMERSET	98.87
2044	PA TIOGA	98.85
2045	ME FRANKLIN	98.81
2046	AL GREENE	98.75
2047	MN MILLE LACS	98.66
2048	AL GENEVA	98.64
2049	TX WICHITA	98.63
2050	LA MOREHOUSE	98.63
2051	DE KENT	98.60
2052	NY CHENANGO	98.57
2053	TN MONTGOMERY	98.53
2054	LA JACKSON	98.26
2055	FL LAKE	98.13
2056	LA BIENVILLE	97.91
2057	KY WASHINGTON	97.73
2058	TX JACK	97.67
2059	WI SHAWANO	97.66
2060	WI DODGE	97.64
2061	MI ST CLAIR	97.60
2062	TN BLOUNT	97.52
2063	TX FOARD	97.38
2064	MN HENNEPIN	97.32
2065	PA MONTGOMERY	97.23
2066	AL BALDWIN	97.20
2067	LA EAST CARROLL	97.08
2068	PA LEHIGH	96.91
2069	NC MITCHELL	96.88
2070	TX FISHER	96.84
2071	KY JEFFERSON	96.73
2072	WI MENOMINEE	96.53
2073	MS WEBSTER	96.42
2074	FL HOLMES	96.34
2075	KY BUTLER	96.34
2076	AR CHICOT	96.31
2077	MI JACKSON	96.26
2078	TN COFFEE	96.25
2079	MI TUSCOLA	96.22
2080	NC WATAUGA	96.21
2081	WV HANCOCK	96.17
2082	TN CARROLL	96.16
2083	KY CHRISTIAN	96.15
2084	MN TODD	96.12
2085	IN LA PORTE	96.10
2086	ID KOOTENAI	96.10
2087	NC YANCEY	96.04
2088	GA WAYNE	96.01
2089	TN BLEDSOE	95.78
2090	MS JASPER	95.76
2091	LA LINCOLN	95.75
2092	TN GRUNDY	95.70
2093	PA BUCKS	95.66
2094	TN DICKSON	95.59
2095	VA PRINCE GEORG	95.44
2096	KY WARREN	95.34
2097	TN HENRY	95.07
2098	MN STEARNS	95.03
2099	PA CARBON	94.97
2100	TX KNOX	94.93
2101	NC BURKE	94.91
2102	MN LAC QUI PARL	94.85
2103	MI BERRIEN	94.75
2104	VA HENRICO	94.70
2105	MI IRON	94.57
2106	PA SUSQUEHANNA	94.53
2107	SC PICKENS	94.35

U.S. COUNTIES RANKED BY TOTAL FALLOUT DEPOSITION II
1951-1970
(microCuries per square meter)
Each Row Includes:
County Name, Fallout Level in microCuries/square meter and County Rank (1 = highest fallout in U.S.)

Rank	County	Fallout
2108	IL MARION	94.28
2109	WV CABELL	94.27
2110	SC LEE	94.10
2111	TN MONROE	94.01
2112	LA FRANKLIN	94.00
2113	TX STONEWALL	93.98
2114	MN CHIPPEWA	93.93
2115	TN MCMINN	93.87
2116	MD WORCESTER	93.86
2117	WI WASHINGTON	93.82
2118	KY MONROE	93.60
2119	MI MENOMINEE	93.51
2120	OH HENRY	93.49
2121	TX BORDEN	93.49
2122	LA NATCHITOCHES	93.47
2123	SC JASPER	93.45
2124	GA WHEELER	93.35
2125	PA BERKS	93.26
2126	LA WEST CARROLL	92.93
2127	KY CARLISLE	92.88
2128	IN STARKE	92.81
2129	MN YELLOW MEDIC	92.79
2130	TN CUMBERLAND	92.75
2131	GA TOWNS	92.74
2132	TN HAWKINS	92.70
2133	GA BRYAN	92.65
2134	MI SCHOOLCRAFT	92.59
2135	TX COTTLE	92.57
2136	MN COOK	92.53
2137	NC YADKIN	92.53
2138	TX KENT	92.36
2139	IN MARSHALL	92.35
2140	NC MCDOWELL	92.30
2141	SC CLARENDON	92.25
2142	TX ARCHER	92.20
2143	AL CLARKE	92.15
2144	ME SOMERSET	92.10
2145	NC PAMLICO	92.07
2146	NC GRAHAM	92.06
2147	MS SCOTT	92.01
2148	NC BUNCOMBE	91.89
2149	MN MORRISON	91.88
2150	MN CASS	91.62
2151	NC DURHAM	91.59
2152	IL WILLIAMSON	91.54
2153	WI LANGLADE	91.53
2154	VA MIDDLESEX	91.47
2155	FL SANTA ROSA	91.40
2156	MI BARAGA	91.29
2157	TX NAVARRO	91.20
2158	KY CASEY	91.15
2159	TX PALO PINTO	91.11
2160	MI ALLEGAN	91.08
2161	NC WAKE	91.07
2162	TN FENTRESS	91.06
2163	LA CALDWELL	91.03
2164	NC MADISON	90.90
2165	GA EFFINGHAM	90.82
2166	MI MIDLAND	90.71
2167	IL WHITE	90.69
2168	GA SCREVEN	90.60
2169	KY WAYNE	90.38
2170	TN CHEATHAM	90.37
2171	TN GREENE	90.36
2172	NC CALDWELL	90.25
2173	TN RHEA	90.24
2174	TN SEQUATCHIE	90.23
2175	TN HUMPHREYS	90.23

U.S. COUNTIES RANKED BY TOTAL FALLOUT DEPOSITION II
1951-1970
(microCuries per square meter)
Each Row Includes:
County Name, Fallout Level in microCuries/square meter and County Rank (1 = highest fallout in U.S.)

2176	TN DECATUR	90.21
2177	SD DEUEL	90.19
2178	SC SUMTER	90.14
2179	WA WALLA WALLA	90.09
2180	NC ALLEGHANY	90.09
2181	GA MONTGOMERY	90.07
2182	NC AVERY	90.06
2183	MN KOOCHICHING	89.98
2184	TX HOOD	89.97
2185	KY TRIGG	89.91
2186	MS MONROE	89.91
2187	TN BRADLEY	89.87
2188	MS CLAY	89.84
2189	MI MONTCALM	89.80
2190	TX PANOLA	89.66
2191	NC ONSLOW	89.65
2192	NC DARE	89.65
2193	IN JAY	89.62
2194	TX MARTIN	89.46
2195	WA WHITMAN	89.44
2196	TN POLK	89.34
2197	CA NEVADA	89.33
2198	IL JOHNSON	89.30
2199	MN WADENA	89.29
2200	SC MARION	89.24
2201	NC SURRY	89.23
2202	LA DE SOTO	89.19
2203	IN ELKHART	89.17
2204	IN HANCOCK	89.13
2205	IL FRANKLIN	89.10
2206	SC CALHOUN	89.03
2207	NC NEW HANOVER	88.99
2208	TN MORGAN	88.96
2209	OH CARROLL	88.87
2210	MD MONTGOMERY	88.72
2211	GA TERRELL	88.53
2212	AL ESCAMBIA	88.51
2213	VA GREENSVILLE	88.48
2214	IL SALINE	88.46
2215	AL TUSCALOOSA	88.45
2216	NC MOORE	88.35
2217	TX CALLAHAN	88.31
2218	SC GREENVILLE	88.31
2219	MS COVINGTON	88.19
2220	KY WHITLEY	88.04
2221	TX HILL	88.01
2222	MN ITASCA	87.88
2223	TX LIMESTONE	87.87
2224	TN HENDERSON	87.71
2225	NC CRAVEN	87.62
2226	FL HILLSBOROUGH	87.59
2227	NC ALEXANDER	87.55
2228	IL MASSAC	87.54
2229	WI FOREST	87.53
2230	CA SIERRA	87.53
2231	TX WILBARGER	87.48
2232	NC CARTERET	87.47
2233	SC HORRY	87.36
2234	VA NORTHUMBERLA	87.28
2235	VA ARLINGTON	87.28
2236	MD PRINCE GEORG	87.26
2237	NC GUILFORD	87.26
2238	MI MUSKEGON	87.16
2239	MD ST MARYS	87.12
2240	NY ONONDAGA	86.99
2241	NC BERTIE	86.99
2242	VA LANCASTER	86.76
2243	TN LOUDON	86.74

U.S. COUNTIES RANKED BY TOTAL FALLOUT DEPOSITION II

1951-1970

(microCuries per square meter)

Each Row Includes:

County Name, Fallout Level in microCuries/square meter and County Rank (1 = highest fallout in U.S.)

2244	NC POLK	86.73
2245	MS ITAWAMBA	86.73
2246	AL WASHINGTON	86.64
2247	TN DAVIDSON	86.63
2248	SC ORANGEBURG	86.44
2249	GA RANDOLPH	86.42
2250	PA SULLIVAN	86.35
2251	GA EVANS	86.28
2252	NC IREDELL	86.18
2253	GA BULLOCH	86.03
2254	VA CULPEPER	85.97
2255	GA QUITMAN	85.90
2256	NC DAVIDSON	85.81
2257	GA PULASKI	85.78
2258	TN HARDIN	85.74
2259	TN UNICOI	85.72
2260	WI OZAUKEE	85.70
2261	MS SMITH	85.70
2262	AL COLBERT	85.67
2263	GA CLAY	85.66
2264	KY MCCREARY	85.56
2265	KY LAUREL	85.48
2266	GA FANNIN	85.45
2267	VA NEWPORT NEWS	85.43
2268	NC PITT	85.40
2269	KY CALDWELL	85.39
2270	TN PUTNAM	85.33
2271	VA STAFFORD	85.32
2272	KY CALLOWAY	85.31
2273	TN JOHNSON	85.28
2274	TN WARREN	85.23
2275	TX HASKELL	85.20
2276	FL ESCAMBIA	85.11
2277	GA JENKINS	85.09
2278	TX JEFFERSON	85.08
2279	OH JEFFERSON	85.01
2280	ID BOUNDARY	85.01
2281	FL VOLUSIA	84.99
2282	NY SCHUYLER	84.85
2283	TN HICKMAN	84.76
2284	TN JEFFERSON	84.74
2285	MD CALVERT	84.68
2286	KY MCLEAN	84.66
2287	IN STEUBEN	84.63
2288	KY WEBSTER	84.63
2289	NC RANDOLPH	84.56
2290	MD WICOMICO	84.55
2291	AL HALE	84.50
2292	GA CALHOUN	84.50
2293	NC HERTFORD	84.49
2294	MN SWIFT	84.48
2295	PA BRADFORD	84.48
2296	GA WEBSTER	84.44
2297	TN VAN BUREN	84.37
2298	MI OCEANA	84.36
2299	PA SCHUYLKILL	84.26
2300	SC BERKELEY	84.14
2301	AR CALHOUN	84.09
2302	TX MCLENNAN	84.03
2303	LA RED RIVER	84.02
2304	GA SCHLEY	83.98
2305	FL LEON	83.94
2306	GA TATTNALL	83.93
2307	MI NEWAYGO	83.85
2308	IL HAMILTON	83.83
2309	MD CECIL	83.83
2310	TX BAYLOR	83.81
2311	SC DORCHESTER	83.76

U.S. COUNTIES RANKED BY TOTAL FALLOUT DEPOSITION II
1951-1970

(microCuries per square meter)

Each Row Includes:

County Name, Fallout Level in microCuries/square meter and County Rank (1 = highest fallout in U.S.)

2312	WI KENOSHA	83.72
2313	VA ISLE OF WIGH	83.70
2314	TN HAMILTON	83.68
2315	MS ATTALA	83.68
2316	KY EDMONSON	83.67
2317	NC CATAWBA	83.63
2318	NC ALAMANCE	83.62
2319	IN LAGRANGE	83.60
2320	LA CLAIBORNE	83.55
2321	GA COLQUITT	83.47
2322	GA WHITE	83.42
2323	GA CHATHAM	83.37
2324	MS CHOCTAW	83.25
2325	GA GLYNN	83.17
2326	SC COLLETON	82.92
2327	TX SCURRY	82.78
2328	VA DINWIDDIE	82.72
2329	WI MANITOWOC	82.71
2330	PA MONTOUR	82.70
2331	TX ANDERSON	82.65
2332	SC DARLINGTON	82.60
2333	TN MARSHALL	82.60
2334	GA MCINTOSH	82.59
2335	MD HOWARD	82.42
2336	VA MATHEWS	82.28
2337	PA UNION	82.19
2338	VA NORFOLK/CHES	82.15
2339	SC CHARLESTON	82.12
2340	TX BOSQUE	82.11
2341	NY TIOGA	82.11
2342	AL LOWNDES	82.10
2343	AL PICKENS	82.05
2344	GA BURKE	82.04
2345	GA FAYETTE	82.01
2346	NC GATES	81.97
2347	GA PEACH	81.95
2348	GA CRAWFORD	81.94
2349	MS LEAKE	81.92
2350	TN BEDFORD	81.84
2351	TN MAURY	81.81
2352	MI ALGER	81.75
2353	FL SUMTER	81.74
2354	NC CLEVELAND	81.73
2355	GA CARROLL	81.72
2356	PA PHILADELPHIA	81.64
2357	SC BEAUFORT	81.64
2358	MD ANNE ARUNDEL	81.55
2359	AL DALE	81.53
2360	NC PENDER	81.53
2361	IL KENDALL	81.52
2362	KY METCALFE	81.38
2363	WI KEWAUNEE	81.38
2364	DC WASHINGTON	81.36
2365	GA BAKER	81.34
2366	FL ST JOHNS	81.30
2367	MD KENT	81.21
2368	IL DU PAGE	81.19
2369	SC MARLBORO	81.17
2370	GA CLAYTON	81.11
2371	AL MADISON	81.11
2372	TN KNOX	81.10
2373	TN PERRY	81.04
2374	AL PIKE	81.04
2375	SC SPARTANBURG	81.01
2376	MD CHARLES	80.96
2377	KY BRECKINRIDGE	80.79
2378	GA DOUGHERTY	80.67
2379	NC LENOIR	80.66

U.S. COUNTIES RANKED BY TOTAL FALLOUT DEPOSITION II
1951-1970

(microCuries per square meter)
Each Row Includes:
County Name, Fallout Level in microCuries/square meter and County Rank (1 = highest fallout in U.S.)

2380	NC CUMBERLAND	80.65
2381	AL PERRY	80.53
2382	TN HAMBLEN	80.51
2383	LA ORLEANS	80.38
2384	GA JEFFERSON	80.30
2385	MD TALBOT	80.27
2386	GA UNION	80.27
2387	AL BARBOUR	80.25
2388	MI CASS	80.17
2389	WV JACKSON	80.10
2390	TN STEWART	80.02
2391	GA GREENE	80.01
2392	GA JEFF DAVIS	79.92
2393	VA KING GEORGE	79.86
2394	TN ROBERTSON	79.78
2395	TX EASTLAND	79.56
2396	NC RUTHERFORD	79.55
2397	MI DICKINSON	79.50
2398	VA SOUTHAMPTON	79.48
2399	SC CHESTERFIELD	79.39
2400	VA SURRY	79.39
2401	WI DOOR	79.38
2402	TX SOMERVELL	79.33
2403	NC WAYNE	79.31
2404	AL ETOWAH	79.30
2405	VA SUSSEX	79.27
2406	GA APPLING	79.26
2407	KY MEADE	79.25
2408	MI LUCE	79.24
2409	TX NACOGDOCHES	79.24
2410	AL COVINGTON	79.21
2411	KY MUHLENBERG	79.19
2412	FL BREVARD	79.15
2413	NC WARREN	79.09
2414	IL HARDIN	79.05
2415	TN WHITE	78.96
2416	TN MEIGS	78.94
2417	VA HAMPTON	78.94
2418	GA MITCHELL	78.92
2419	MD DORCHESTER	78.87
2420	GA TWIGGS	78.85
2421	GA MORGAN	78.84
2422	MD WASHINGTON	78.83
2423	GA JOHNSON	78.80
2424	GA GORDON	78.79
2425	VA PRINCE WILLI	78.79
2426	AL LAUDERDALE	78.67
2427	FL FLAGLER	78.60
2428	MI ST JOSEPH	78.58
2429	TN SUMNER	78.56
2430	TN OVERTON	78.55
2431	TX KING	78.55
2432	MS RANKIN	78.41
2433	AL SUMTER	78.34
2434	VA BRUNSWICK	78.31
2435	MI WASHTENAW	78.31
2436	GA LEE	78.28
2437	AL DE KALB	78.25
2438	SC WILLIAMSBURG	78.23
2439	MD FREDERICK	78.20
2440	VA NOTTOWAY	78.19
2441	FL HERNANDO	78.16
2442	GA TOOMBS	78.13
2443	MN BELTRAMI	78.02
2444	VA NORTHAMPTON	78.00
2445	VA WESTMORELAND	77.91
2446	GA LANIER	77.90
2447	GA BACON	77.87

U.S. COUNTIES RANKED BY TOTAL FALLOUT DEPOSITION II
1951-1970
(microCuries per square meter)
Each Row Includes:
County Name, Fallout Level in microCuries/square meter and County Rank (1 = highest fallout in U.S.)

Rank	County	Fallout	Rank	County	Fallout
2448	GA BLECKLEY	77.84	2482	VA YORK	76.43
2449	TX YOUNG	77.70	2483	TX THROCKMORTON	76.41
2450	IN OHIO	77.66	2484	PA COLUMBIA	76.37
2451	TX HOUSTON	77.66	2485	WI OUTAGAMIE	76.31
2452	TN DE KALB	77.64	2486	FL ALACHUA	76.14
2453	OH PICKAWAY	77.56	2487	NC COLUMBUS	76.09
2454	TN WILLIAMSON	77.51	2488	AL TALLADEGA	76.02
2455	KY BARREN	77.49	2489	GA STEWART	75.98
2456	SC KERSHAW	77.43	2490	GA CAMDEN	75.96
2457	SC DILLON	77.40	2491	GA COLUMBUS	75.96
2458	AL CHAMBERS	77.16	2492	IN NOBLE	75.95
2459	WI FLORENCE	77.16	2493	TN HOUSTON	75.94
2460	NC DUPLIN	77.15	2494	FL CALHOUN	75.93
2461	GA SUMTER	77.01	2495	GA WILCOX	75.90
2462	GA TELFAIR	76.95	2496	SC ANDERSON	75.89
2463	GA PAULDING	76.95	2497	PA FRANKLIN	75.88
2464	GA MONROE	76.91	2498	FL MADISON	75.80
2465	KY LYON	76.89	2499	GA FLOYD	75.70
2466	MS JONES	76.87	2500	VA HENRY	75.63
2467	MI DELTA	76.80	2501	NC LINCOLN	75.55
2468	GA THOMAS	76.76	2502	MS LAWRENCE	75.53
2469	AL LEE	76.76	2503	GA HOUSTON	75.46
2470	NC FORSYTH	76.76	2504	AL MARSHALL	75.45
2471	VA GLOUCESTER	76.76	2505	GA CRISP	75.37
2472	MN POPE	76.74	2506	WI RACINE	75.36
2473	GA BERRIEN	76.72	2507	TX HAMILTON	75.36
2474	TX STEPHENS	76.71	2508	GA EMANUEL	75.35
2475	VA JAMES CITY	76.69	2509	OH SHELBY	75.34
2476	WA SPOKANE	76.68	2510	AL MORGAN	75.29
2477	NC WILSON	76.67	2511	TN UNION	75.18
2478	AL CULLMAN	76.66	2512	AL RANDOLPH	75.17
2479	TX ERATH	76.64	2513	MS WINSTON	75.17
2480	AL BIBB	76.56	2514	FL POLK	75.12
2481	GA DOUGLAS	76.55	2515	AL JACKSON	75.11

U.S. COUNTIES RANKED BY TOTAL FALLOUT DEPOSITION II
1951-1970
(microCuries per square meter)
Each Row Includes:
County Name, Fallout Level in microCuries/square meter and County Rank (1 = highest fallout in U.S.)

Rank	County	Fallout
2516	SC BAMBERG	75.10
2517	FL NASSAU	75.08
2518	GA TALIAFERRO	74.94
2519	AL CLAY	74.89
2520	TX FREESTONE	74.86
2521	MS HOLMES	74.66
2522	IN JENNINGS	74.65
2523	KY CRITTENDEN	74.63
2524	MN BECKER	74.62
2525	MN OTTER TAIL	74.60
2526	KY LOGAN	74.60
2527	FL GILCHRIST	74.53
2528	GA BALDWIN	74.52
2529	MN BIG STONE	74.52
2530	GA WARREN	74.49
2531	TX HOWARD	74.41
2532	GA ATKINSON	74.39
2533	GA GLASCOCK	74.38
2534	GA MCDUFFIE	74.36
2535	MD CAROLINE	74.35
2536	GA WORTH	74.32
2537	GA CLINCH	74.31
2538	TX BROWN	74.29
2539	GA TROUP	74.27
2540	AL JEFFERSON	74.23
2541	AL DALLAS	74.23
2542	GA WILKINSON	74.20
2543	MS KEMPER	74.09
2544	NC LEE	74.09
2545	TN WAYNE	74.08
2546	KY LEE	74.02
2547	FL ORANGE	73.98
2548	NC HARNETT	73.95
2549	TX SHACKELFORD	73.91
2550	GA TIFT	73.91
2551	VA CUMBERLAND	73.88
2552	NC ROBESON	73.87
2553	AL HOUSTON	73.84
2554	AL ELMORE	73.82
2555	GA CLARKE	73.67
2556	GA BRANTLEY	73.63
2557	NC ROWAN	73.59
2558	MD QUEEN ANNES	73.59
2559	FL SEMINOLE	73.56
2560	GA COOK	73.54
2561	WV PRESTON	73.48
2562	KY HOPKINS	73.43
2563	NY BROOME	73.42
2564	GA BIBB	73.42
2565	TN LAWRENCE	73.32
2566	AL TALLAPOOSA	73.24
2567	FL MARION	73.20
2568	NC SAMPSON	73.16
2569	TN GRAINGER	73.13
2570	GA CANDLER	73.10
2571	TX LEON	72.99
2572	FL GULF	72.95
2573	NC HALIFAX	72.90
2574	GA WASHINGTON	72.82
2575	IN FAYETTE	72.80
2576	VA RICHMOND	72.75
2577	SC YORK	72.75
2578	GA TURNER	72.70
2579	KY SIMPSON	72.67
2580	SC CHEROKEE	72.65
2581	TX MCCULLOCH	72.62
2582	MI MARQUETTE	72.58
2583	GA GWINNETT	72.58

U.S. COUNTIES RANKED BY TOTAL FALLOUT DEPOSITION II
1951-1970

(microCuries per square meter)
Each Row Includes:
County Name, Fallout Level in microCuries/square meter and County Rank (1 = highest fallout in U.S.)

Rank	County	Fallout
2584	MS SIMPSON	72.51
2585	GA WALKER	72.46
2586	GA BANKS	72.40
2587	GA MADISON	72.38
2588	SC GEORGETOWN	72.38
2589	TN WILSON	72.37
2590	FL LIBERTY	72.35
2591	TX NOLAN	72.31
2592	GA HABERSHAM	72.27
2593	TN GILES	72.27
2594	IL COOK	72.27
2595	FL DIXIE	72.24
2596	SD GRANT	72.23
2597	GA GILMER	72.20
2598	GA CHATTAHOOCHE	72.20
2599	FL PINELLAS	72.12
2600	IN WHITLEY	72.07
2601	VA ESSEX	72.05
2602	TN BENTON	72.01
2603	MN STEVENS	71.99
2604	NC BLADEN	71.98
2605	GA BEN HILL	71.98
2606	PA LYCOMING	71.97
2607	TN SMITH	71.87
2608	FL CITRUS	71.84
2609	MN HUBBARD	71.83
2610	FL WAKULLA	71.82
2611	GA DOOLY	71.79
2612	TN LINCOLN	71.78
2613	GA SPALDING	71.59
2614	WV MORGAN	71.57
2615	GA IRWIN	71.49
2616	TN JACKSON	71.48
2617	MI IOSCO	71.45
2618	GA MURRAY	71.43
2619	SC LAURENS	71.23
2620	SC HAMPTON	71.08
2621	GA LOWNDES	71.06
2622	FL PUTNAM	70.93
2623	GA JACKSON	70.89
2624	SC GREENWOOD	70.89
2625	GA PIKE	70.80
2626	FL SUWANNEE	70.77
2627	MI ARENAC	70.71
2628	NC NASH	70.67
2629	KY ADAIR	70.59
2630	KY LARUE	70.59
2631	GA HANCOCK	70.56
2632	TX CONCHO	70.53
2633	OH DEFIANCE	70.46
2634	SC MCCORMICK	70.39
2635	FL FRANKLIN	70.31
2636	GA STEPHENS	70.25
2637	NC MONTGOMERY	70.25
2638	TN MARION	70.23
2639	TX SHELBY	70.17
2640	VA LOUISA	70.16
2641	PA MIFFLIN	70.05
2642	GA HARALSON	69.90
2643	GA DODGE	69.86
2644	KY CLINTON	69.84
2645	TX BELL	69.81
2646	NC GASTON	69.75
2647	NC UNION	69.71
2648	AL CLEBURNE	69.68
2649	GA OCONEE	69.64
2650	NC EDGECOMBE	69.61
2651	GA CHATTOOGA	69.58

U.S. COUNTIES RANKED BY TOTAL FALLOUT DEPOSITION II
1951-1970
(microCuries per square meter)
Each Row Includes:
County Name, Fallout Level in microCuries/square meter and County Rank (1 = highest fallout in U.S.)

2652	GA JONES	69.57
2653	MN DOUGLAS	69.49
2654	GA MERIWETHER	69.44
2655	MS JEFFERSON DA	69.36
2656	AL FRANKLIN	69.36
2657	TX WILLIAMSON	69.32
2658	AL CALHOUN	69.31
2659	WI BROWN	69.25
2660	AL COFFEE	69.24
2661	AL RUSSELL	69.22
2662	TX JONES	69.16
2663	SC NEWBERRY	69.10
2664	NC GREENE	69.09
2665	SC EDGEFIELD	68.94
2666	MN POLK	68.91
2667	MS NOXUBEE	68.91
2668	TN FRANKLIN	68.84
2669	AL LIMESTONE	68.82
2670	MI ISABELLA	68.75
2671	SC BARNWELL	68.75
2672	GA HALL	68.74
2673	GA PIERCE	68.73
2674	MS NESHOBA	68.67
2675	WV JEFFERSON	68.65
2676	GA CHARLTON	68.52
2677	GA DECATUR	68.46
2678	KY GREEN	68.44
2679	SC CHESTER	68.42
2680	GA CHEROKEE	68.42
2681	KY HARDIN	68.40
2682	MN TRAVERSE	68.37
2683	GA CATOOSA	68.37
2684	GA DE KALB	68.37
2685	AL BULLOCK	68.35
2686	NC HOKE	68.35
2687	TX FALLS	68.34
2688	GA PUTNAM	68.31
2689	FL JEFFERSON	68.30
2690	GA BUTTS	68.26
2691	NC STANLY	68.23
2692	SC UNION	68.16
2693	AL BLOUNT	68.06
2694	GA OGLETHORPE	68.06
2695	MI CHARLEVOIX	68.04
2696	AL ST CLAIR	68.02
2697	FL HAMILTON	67.98
2698	GA WALTON	67.93
2699	SC SALUDA	67.92
2700	VA HALIFAX	67.87
2701	IN BLACKFORD	67.86
2702	MI CHIPPEWA	67.75
2703	GA JASPER	67.70
2704	TX COMANCHE	67.65
2705	TX ROBERTSON	67.63
2706	GA COWETA	67.59
2707	SC ALLENDALE	67.58
2708	SC ABBEVILLE	67.58
2709	CA GLENN	67.53
2710	NC STOKES	67.51
2711	MI BAY	67.47
2712	TN MOORE	67.44
2713	GA WILKES	67.39
2714	MN PENNINGTON	67.32
2715	AL CHILTON	67.28
2716	GA HENRY	67.27
2717	GA WHITFIELD	67.27
2718	SC LANCASTER	67.26
2719	GA TAYLOR	67.24

U.S. COUNTIES RANKED BY TOTAL FALLOUT DEPOSITION II
1951-1970
(microCuries per square meter)
Each Row Includes:
County Name, Fallout Level in microCuries/square meter and County Rank (1 = highest fallout in U.S.)

Rank	County	Fallout
2720	GA RICHMOND	67.20
2721	NC SCOTLAND	67.19
2722	NC BRUNSWICK	67.15
2723	MS LOWNDES	67.12
2724	GA LAMAR	67.10
2725	TN CANNON	67.03
2726	AL AUTAUGA	67.00
2727	FL GADSDEN	66.98
2728	CA AMADOR	66.94
2729	VA FLUVANNA	66.88
2730	AL WALKER	66.82
2731	AL CHEROKEE	66.79
2732	CA LASSEN	66.76
2733	MN CLEARWATER	66.75
2734	GA BROOKS	66.72
2735	TX BRAZOS	66.64
2736	AL CHOCTAW	66.51
2737	AL BUTLER	66.43
2738	NC DAVIE	66.39
2739	GA BARTOW	66.37
2740	SC FAIRFIELD	66.33
2741	GA NEWTON	66.33
2742	GA HARRIS	66.33
2743	GA EARLY	66.31
2744	KY TAYLOR	66.30
2745	NC JOHNSTON	66.25
2746	GA HART	66.24
2747	GA LUMPKIN	66.17
2748	PA JUNIATA	66.16
2749	WV RITCHIE	66.14
2750	NC RICHMOND	65.94
2751	MI ALCONA	65.90
2752	MN KITTSON	65.68
2753	KY ALLEN	65.47
2754	GA GRADY	65.44
2755	FL CLAY	65.43
2756	AL LAWRENCE	65.33
2757	WV BARBOUR	65.28
2758	FL BAY	65.20
2759	GA FRANKLIN	65.05
2760	TX COLEMAN	65.01
2761	AL COOSA	65.01
2762	GA COFFEE	64.99
2763	MD CARROLL	64.94
2764	AL MARION	64.78
2765	TX MIDLAND	64.72
2766	TX RUNNELS	64.72
2767	GA ECHOLS	64.67
2768	WV DODDRIDGE	64.61
2769	TX LAMPASAS	64.58
2770	GA FORSYTH	64.57
2771	VA CLARKE	64.55
2772	TX MITCHELL	64.53
2773	AL FAYETTE	64.47
2774	VA BUCKINGHAM	64.47
2775	GA PICKENS	64.44
2776	GA WARE	64.38
2777	TX STERLING	64.34
2778	MI ANTRIM	64.32
2779	MI MACKINAC	64.29
2780	NC ANSON	64.24
2781	TX MILAM	64.24
2782	GA DAWSON	64.19
2783	AL MACON	64.12
2784	MI IONIA	64.09
2785	VA WYTHE	64.08
2786	WV TAYLOR	64.05
2787	NC MECKLENBURG	64.03

U.S. COUNTIES RANKED BY TOTAL FALLOUT DEPOSITION II
1951-1970
(microCuries per square meter)
Each Row Includes:
County Name, Fallout Level in microCuries/square meter and County Rank (1 = highest fallout in U.S.)

Rank	County	Fallout
2788	FL WASHINGTON	63.95
2789	WV ROANE	63.92
2790	SC LEXINGTON	63.85
2791	GA DADE	63.85
2792	GA BARROW	63.84
2793	FL OSCEOLA	63.83
2794	GA TALBOT	63.78
2795	PA CUMBERLAND	63.69
2796	GA COBB	63.62
2797	TN LEWIS	63.62
2798	FL JACKSON	63.53
2799	MI CHEBOYGAN	63.52
2800	MI ROSCOMMON	63.46
2801	GA LINCOLN	63.42
2802	CA PLACER	63.38
2803	NC TYRRELL	63.38
2804	WI MILWAUKEE	63.37
2805	MN GRANT	63.35
2806	WV BERKELEY	63.34
2807	KY MARION	63.30
2808	TX CORYELL	63.26
2809	TX MASON	63.24
2810	AL WINSTON	63.23
2811	IN DE KALB	63.21
2812	NC CASWELL	63.13
2813	TX SAN SABA	63.06
2814	MN ROSEAU	63.04
2815	FL BAKER	62.97
2816	CA VENTURA	62.96
2817	AL CRENSHAW	62.93
2818	AL LAMAR	62.89
2819	GA ELBERT	62.79
2820	FL WALTON	62.78
2821	AL SHELBY	62.72
2822	FL LAFAYETTE	62.72
2823	GA FULTON	62.70
2824	FL TAYLOR	62.70
2825	OH WILLIAMS	62.58
2826	CA PLUMAS	62.51
2827	GA ROCKDALE	62.46
2828	MI CRAWFORD	62.38
2829	CA KERN	62.29
2830	GA UPSON	62.20
2831	MI GLADWIN	61.81
2832	TX BURLESON	61.79
2833	GA COLUMBIA	61.73
2834	FL COLUMBIA	61.72
2835	PA DAUPHIN	61.52
2836	WV HAMPSHIRE	61.51
2837	MI CLARE	61.46
2838	MN MAHNOMEN	61.42
2839	KY HART	61.41
2840	TN RUTHERFORD	61.35
2841	PA LEBANON	61.27
2842	MN RED LAKE	61.23
2843	TX LLANO	61.22
2844	TX BLANCO	61.02
2845	NC CABARRUS	60.93
2846	MI MECOSTA	60.78
2847	VA PAGE	60.56
2848	KY GRAYSON	60.49
2849	PA PERRY	60.44
2850	VA CHARLOTTE	60.35
2851	WV MINERAL	60.31
2852	GA LAURENS	60.24
2853	TX LEE	60.06
2854	GA POLK	60.05
2855	GA MARION	60.01

U.S. COUNTIES RANKED BY TOTAL FALLOUT DEPOSITION II
1951-1970
(microCuries per square meter)
Each Row Includes:
County Name, Fallout Level in microCuries/square meter and County Rank (1 = highest fallout in U.S.)

Rank	County	Fallout
2856	GA TREUTLEN	59.88
2857	FL UNION	59.75
2858	GA MACON	59.56
2859	SC RICHLAND	59.43
2860	NC ROCKINGHAM	59.23
2861	SC AIKEN	59.08
2862	TX BANDERA	58.87
2863	NC FRANKLIN	58.85
2864	TN MACON	58.83
2865	MI MASON	58.63
2866	FL PASCO	58.55
2867	MI MANISTEE	58.39
2868	FL BRADFORD	58.35
2869	MI WEXFORD	58.33
2870	MN LAKE OF THE	58.31
2871	TX PRESIDIO	58.24
2872	AL MONTGOMERY	58.20
2873	TN TROUSDALE	58.13
2874	MI OGEMAW	58.09
2875	CA CALAVERAS	57.81
2876	CA YUBA	57.77
2877	WA FERRY	57.70
2878	MI BENZIE	57.31
2879	FL SARASOTA	57.18
2880	FL COLLIER	57.16
2881	VA BLAND	57.12
2882	MI MISSAUKEE	57.09
2883	MI EMMET	57.08
2884	TX MEDINA	56.94
2885	MI OSCEOLA	56.85
2886	TX COKE	56.71
2887	KY NELSON	56.56
2888	TX GILLESPIE	56.51
2889	GA MILLER	56.46
2890	GA HEARD	56.36
2891	TX MILLS	56.14
2892	TX IRION	56.03
2893	TX REAGAN	55.99
2894	AL HENRY	55.99
2895	MN WILKIN	55.96
2896	TX TOM GREEN	55.96
2897	FL PALM BEACH	55.81
2898	MN MARSHALL	55.72
2899	CA TEHAMA	54.96
2900	TX BURNET	54.77
2901	GA SEMINOLE	54.74
2902	FL HENDRY	54.44
2903	MN NORMAN	54.42
2904	MI GRAND TRAVER	54.38
2905	MI OTSEGO	54.13
2906	VA ALBEMARLE	54.10
2907	TX GLASSCOCK	53.99
2908	TX CRANE	53.73
2909	TX FAYETTE	53.71
2910	TX FRIO	53.69
2911	FL MONROE	53.50
2912	MN CLAY	53.49
2913	FL ST LUCIE	53.46
2914	TX MENARD	53.45
2915	CA SANTA BARBAR	52.88
2916	CA EL DORADO	52.88
2917	CA TUOLUMNE	52.64
2918	CA SONOMA	52.32
2919	CA BUTTE	52.30
2920	TX KERR	51.93
2921	FL MANATEE	51.71
2922	MI LAKE	51.63
2923	TX WHARTON	51.52

U.S. COUNTIES RANKED BY TOTAL FALLOUT DEPOSITION II
1951-1970

(microCuries per square meter)
Each Row Includes:
County Name, Fallout Level in microCuries/square meter and County Rank (1 = highest fallout in U.S.)

Rank	County	Fallout
2924	TX PECOS	51.20
2925	WA STEVENS	50.84
2926	FL BROWARD	50.65
2927	MI PRESQUE ISLE	50.41
2928	FL DADE	50.34
2929	CA TULARE	50.29
2930	TX COLORADO	50.19
2931	FL LEE	49.85
2932	CA MONTEREY	49.72
2933	TX COMAL	49.56
2934	FL HIGHLANDS	49.30
2935	MI KALKASKA	49.25
2936	FL MARTIN	49.01
2937	FL GLADES	48.99
2938	TX KENDALL	48.88
2939	MI OSCODA	48.56
2940	FL OKEECHOBEE	48.54
2941	MI LEELANAU	47.93
2942	MI MONTMORENCY	47.55
2943	FL INDIAN RIVER	47.34
2944	CA SAN DIEGO	46.88
2945	CA SAN LUIS OBI	46.84
2946	CA MODOC	46.82
2947	TX BREWSTER	46.76
2948	TX MATAGORDA	46.75
2949	FL LEVY	46.60
2950	TX TRAVIS	46.39
2951	TX BEXAR	46.34
2952	CA SANTA CRUZ	46.22
2953	TX KIMBLE	45.90
2954	TX SCHLEICHER	45.71
2955	CA KINGS	45.69
2956	TX MCMULLEN	45.35
2957	TX TAYLOR	45.26
2958	TX CROCKETT	45.13
2959	FL CHARLOTTE	44.87
2960	FL DUVAL	44.65
2961	CA SAN BENITO	44.58
2962	TX UPTON	44.36
2963	MI ALPENA	43.91
2964	WA OKANOGAN	43.74
2965	FL DE SOTO	43.68
2966	TX GONZALES	43.42
2967	TX JACKSON	43.33
2968	CA ORANGE	43.20
2969	TX JIM WELLS	42.97
2970	TX CALDWELL	42.49
2971	TX DIMMIT	42.40
2972	CA MARIPOSA	42.34
2973	TX BASTROP	41.79
2974	CA MARIN	41.75
2975	TX LAVACA	41.67
2976	CA FRESNO	41.38
2977	TX DE WITT	41.32
2978	TX ZAVALA	40.58
2979	OR LAKE	40.32
2980	TX HAYS	40.30
2981	FL HARDEE	40.25
2982	CA LAKE	40.23
2983	TX CALHOUN	39.98
2984	CA COLUSA	39.65
2985	TX JIM HOGG	39.53
2986	TX UVALDE	39.21
2987	TX BEE	38.93
2988	TX CAMERON	38.73
2989	TX WILLACY	37.74
2990	TX SUTTON	37.53
2991	TX STARR	37.24

U.S. COUNTIES RANKED BY TOTAL FALLOUT DEPOSITION II
1951-1970
(microCuries per square meter)
Each Row Includes:
County Name, Fallout Level in microCuries/square meter and County Rank (1 = highest fallout in U.S.)

Rank	County	Fallout
2992	TX ARANSAS	36.63
2993	TX KLEBERG	36.54
2994	CA SACRAMENTO	36.29
2995	OR CROOK	36.25
2996	CA MADERA	35.95
2997	TX ATASCOSA	35.79
2998	TX KARNES	35.66
2999	TX BROOKS	35.47
3000	TX REAL	35.29
3001	OR MORROW	35.28
3002	CA SAN MATEO	34.79
3003	TX LIVE OAK	34.46
3004	TX TERRELL	34.42
3005	CA SANTA CLARA	34.42
3006	TX SAN PATRICIO	34.37
3007	TX MAVERICK	34.23
3008	TX EDWARDS	33.93
3009	CA NAPA	33.78
3010	TX VICTORIA	33.75
3011	CA SUTTER	33.36
3012	TX ZAPATA	33.19
3013	TX HIDALGO	33.07
3014	TX KENEDY	33.01
3015	TX LA SALLE	32.65
3016	CA DEL NORTE	32.61
3017	CA YOLO	32.52
3018	TX WILSON	32.52
3019	TX GOLIAD	32.39
3020	CA STANISLAUS	32.37
3021	WA DOUGLAS	31.91
3022	TX DUVAL	31.78
3023	TX KINNEY	31.60
3024	CA CONTRA COSTA	31.53
3025	TX GUADALUPE	31.51
3026	TX NUECES	31.44
3027	CA ALAMEDA	31.39
3028	WA SKAMANIA	31.37
3029	OR CLACKAMAS	31.33
3030	TX WEBB	31.08
3031	CA SAN JOAQUIN	30.91
3032	CA MERCED	30.51
3033	OR WHEELER	29.89
3034	OR LINCOLN	29.66
3035	OR KLAMATH	29.58
3036	WA LINCOLN	29.53
3037	CA SISKIYOU	29.46
3038	CA SOLANO	29.28
3039	CA TRINITY	29.20
3040	OR LANE	29.11
3041	OR MARION	28.94
3042	OR CURRY	28.94
3043	WA SNOHOMISH	28.88
3044	WA PIERCE	28.74
3045	CA HUMBOLDT	28.15
3046	OR DOUGLAS	27.82
3047	OR CLATSOP	27.72
3048	OR TILLAMOOK	27.70
3049	WA ADAMS	27.63
3050	WA CHELAN	26.78
3051	CA SAN FRANCISC	26.35
3052	OR LINN	26.24
3053	OR JOSEPHINE	26.16
3054	WA BENTON	26.02
3055	WA COWLITZ	25.97
3056	OR JEFFERSON	25.87
3057	WA KITTITAS	25.87
3058	CA IMPERIAL	25.78
3059	CA MENDOCINO	25.74

U.S. COUNTIES RANKED BY TOTAL FALLOUT DEPOSITION II
1951-1970

(microCuries per square meter)
Each Row Includes:
County Name, Fallout Level in microCuries/square meter and County Rank (1 = highest fallout in U.S.)

3060	OR GILLIAM	25.67
3061	OR COOS	25.58
3062	CA SHASTA	25.58
3063	CA RIVERSIDE	25.13
3064	WA WAHKIAKUM	24.94
3065	TX VAL VERDE	24.90
3066	WA KING	24.55
3067	WA GRANT	24.19
3068	OR COLUMBIA	24.01
3069	OR SHERMAN	23.60
3070	OR HOOD RIVER	23.31
3071	WA GRAYS HARBOR	23.15
3072	WA WHATCOM	23.06
3073	WA FRANKLIN	22.94
3074	WA CLARK	22.90
3075	OR DESCHUTES	22.77
3076	WA LEWIS	22.55
3077	WA YAKIMA	22.41
3078	WA PACIFIC	22.27
3079	OR WASCO	22.13
3080	WA SKAGIT	21.85
3081	OR MULTNOMAH	21.83
3082	WA JEFFERSON	21.71
3083	WA KLICKITAT	21.18
3084	OR WASHINGTON	20.75
3085	OR POLK	20.32
3086	WA CLALLAM	20.01
3087	OR BENTON	19.91
3088	OR JACKSON	19.87
3089	WA MASON	19.27
3090	OR YAMHILL	19.15
3091	WA KITSAP	18.54
3092	WA THURSTON	17.53
3093	WA SAN JUAN	13.49
3094	WA ISLAND	12.37

TECHNICAL

SECTION 10

RADIONUCLIDES

1951-1962

RADIONUCLIDES

Each nuclear test produces a number of radioactive elements called radionuclides. Chemically, they are similar to their non-radioactive counterparts, but they vary in the amount and type of radiation they emit. They are also known as radioactive isotopes or radioisotopes.

ISOTOPES

An atom is comprised of neutrons, protons and electrons. The protons supply a positive charge, the electrons supply a negative charge, and the neutrons supply no charge at all, but contribute to the mass—how heavy the element is. In classical atomic theory, there are an equal number of electrons for each proton, and the number of protons (and corresponding electrons) determines the element's chemical characteristics. If there are six protons, then the element will be known forever as carbon and will always behave chemically as carbon. Add a proton and a corresponding electron and the element becomes nitrogen—and will behave chemically as nitrogen.

Neutrons contribute essentially nothing to the chemical characteristics of the element, but they contribute significantly to the element's mass. In fact, how "heavy" an element is depends on how many neutrons and protons it has. For example, two atoms of iodine might be chemically similar, but because one has more neutrons, it would have a higher mass number. Thus, while the two atoms would both be the same element and equivalent behavior chemically, they have different masses, and are thus *different isotopes of the same element.*

Iodine 132 is slightly heavier than iodine-131 and they are both heavier than iodine-127. But the three atoms are all chemically the same. All three isotopes are chemically indistinguishable from one another. Carbon-14 is slightly heavier than carbon-12, but they both behave *chemically* just like any carbon anywhere.

While all isotopes of an element act the same chemically, they are unique from the perspective of nuclear physics.

Radioactive Iodine-131, for example, has a half-life of 8.04 days—which means, half of the original mass decays in 8.04 days, then half of that decays in another 8.04 days, and so on. In the process of decay, this radionuclide (another name for radioactive isotope) emits some powerful beta particles and gamma rays. A close, somewhat lighter cousin to iodine-131, iodine-*130* is somewhat hotter radiologically. It has a half-life of only 12½ hours and emits a gamma ray as well as a beta particle that, measured in electron volts, is more powerful than the one emitted by iodine-131. Iodine-128, by contrast emits a gamma ray and a beta particle that is just *one-tenth* as hot as the one emitted by iodine-131. But iodine-128 has a half-life of 17 million years.

The radioisotopes discussed in this book are generally referred by their chemical symbols. For example, iodine-131 is written as I-131; the radioactive isotope manganese-54 is Mn-54, and Beryllium-7 is written as Be-7.

ESTIMATING THE RADIONUCLIDES

Nuclear fallout is composed of more than one element. A nuclear blast is usually hot enough to vaporize just about anything, reducing it to the size of small pellets no more than the size of fog particles. Some elements, however, resist vaporization and pretty much stay in solid form throughout the experience. Scientists term these materials *refractory*. Others, like the 700-ton steel towers supporting the nuclear canister, will vaporize easily, and then solidify into fog-sized spherical particles that can be carried east with the wind. Since the atomic fireball usually touches the ground shortly after detonation, material from the desert often joins the atomic debris cloud.

Elements caught up in the debris cloud are also made radioactive through a process called neutron activation. As a result, the debris cloud is a mixture of upwards of 145 different isotopes of elements, from beryllium to americium. And, of course, since a nuclear explosion is not particularly efficient, there might even be some plutonium trailing along amid the debris.

The problem of guessing what is in the cloud is complicated by several factors. For one thing, a particular radioactive element will change from one element to another over time. For example, the iodine-131 that the NCI evaluated began its life in the nuclear explosion as an entirely different element. *indium*-131. In a complicated series of transformations, the indium-131 atoms change to tin-131, then to antimony-131—all while the nuclear fireball is still churning above ground zero. About an hour after detonation, while the debris cloud is still visible, the atoms that were once indium-131 are now tellurium-131. After another hour, most of these atoms will change from tellurium-131 to iodine-131. Since I-131 has a half-life of eight days, this means that the fallout that reaches the ground will contain a substantial amount of this element.

All the elements found in fallout have similarly complicated histories. This means that the nuclear fallout coming to earth on the first day after the detonation will have a different mixture of isotopes than the fallout coming down on day two. Each day post-shot has it's own unique mixture of radioactive elements. In addition, the mixture of radioactive elements varies drastically with the particular detonation. The debris cloud from the underground shot SEDAN, for example, contained elements from the desert soil, while the debris cloud from HARRY probably contained a significant amount of the steel tower supporting the device. Some airborne detonations produced relatively little fallout. One shot, Teapot HA (for high-altitude) resulted in such a small amount of fallout that it was not included in the NCI's study.

All values found in this book are calculated estimates based upon ratios between Hicks Tables values for I-131 and the particular radionuclide. Since the original I-131 values were based upon the lognormal distribution, the range of values was often quite wide. It should be understood that this is also the case with the derived values appearing in this book for both total fallout and the various radionuclides

For this Volume the following radionuclides were evaluated for each of the major aboveground tests from 1951-1962:

Beryllium-7, Sodium-24, Manganese-54, Iron-59, Cobalt-60, Tungsten-185, Gold-198, Gold-199, Lead-203, Uranium-237, Uranium-240, Neptunium-237, Americium-241, Curium-242, Bromine-82, Bromine-83, Strontium-89, Strontium-90, Yttrium-90, Yttrium-91, Zirconium-95, Niobium-95m, Niobium-95, Molybdenum-99, Technetium-99m, Ruthenium-106, Rhodium-106, Silver-109m, Silver-111, Palladium-112, Cadmium-115, Indium-115m, Indium-117, Tellurium-127m, Iodine-130, Iodine-131, Iodine-132, Iodine-133, Iodine-135, Cesium-137, Barium-137m, Lanthanum-141, Cerium-144, Praesodymium-144, Europium-155, Terbium-161.

Generally, deposition calculations are made for the day of deposition. However, for those radionuclides marked with an asterisk (*) the deposition value is estimated for H+ 365 days, or one year post-shot. These radionuclides include: **Be7, Mn54, Fe59, Co60, Sr89, Sr90, Y90, Y91, Zr95, Nb95m, Nb95, Rh106, Ru106, Te127m, Cs137, Ba137m, Eu155, W185, Am241, and Cm242**.

In the book the following radionuclides are given as their chemical symbol:

SYMBOL	RADIONUCLIDE
Ag109m	Silver-109m (meta)
Ag111	Silver-111
Am241*	Americium-241
Au198, 199	Gold 198, 199
Ba137m*	Barium-137m
Be7*	Beryllium-7
Br82, Br83	Bromine-82, 83

Cd115	Cadmium-115
Ce144	Cerium-144
Cm242*	Curium-242
Co60*	Cobalt-60
Cs137*	Cesium-137
Cu64	Copper-64
Eu155*	Europium-155
Fe59*	Iron-59
I-130, I-131, I-132, I-133, I135	Iodine-130, 131, 132, 133, 135
In115m	Indium-115m
In117	Indium-117
Kr85, Kr87, Kr88	Krypton-85, 87, 88
La141	Lanthanum-141
Mn54*	Manganese 54
Mo99	Molybdenum-99
Na24	Sodium-24
Nb95*	Niobium-95
Nb95m*	Niobium-95m (meta)
Np237	Neptunium-237
Pb203	Lead-203
Pd111, Pd112	Palladium, 111, 112

Pr144	Praesodymium-144
Rh106*	Rhodium-106
Ru106*	Ruthenium-106
Sm156	Samarium-156
Sr89*, Sr90*	Strontium 89, 90
Tb161	Terbium-161
Tc99m	Technetium-99m
Te127m*	Tellurium-127m
U237	Uranium-237
W181, W185*, W188	Tungsten 181,185, 188
Y90*, Y91*	Yttrium-90, 91
Zr95*	Zirconium-95

As with all radionuclides one can calculate the activity at any time post-detonation given the half-life.

The formula: $$A_t = A_o e^{-\left(\frac{0.693t}{T}\right)}$$

Where: A_t = Activity at time t

A_o = Activity at original time (time of detonation or fallout deposition)

T = Published Half-Life of the radionuclide

t = time of interest. Note that T and t must be in the same units of measure.

MAPS AND TABLES
OF
ESTIMATED RADIONUCLIDE DEPOSITION

The following maps and tables are estimates of deposition values based upon Hicks Tables ratios applied to the National Cancer Institute's deposition values. Unless otherwise indicated, the fallout values are expressed in nanoCuries per square meter. One Curie/sq meter is equivalent to 3.7×10^{10} disintegrations or radioactive events per second. Each of these events can produce a particle or ray. Unless otherwise indicated fallout radioactivity values in this book and in this section are expressed in nanoCuries/square meter, a level equal to one billionth of a Curie.

Maps displaying radionuclide values as columns in the z-axis only show relative values for that particular radionuclide. Estimates of for different radionuclides may vary by many orders of magnitude, thus it is inappropriate to compare the relative fallout values based upon these illustrations.

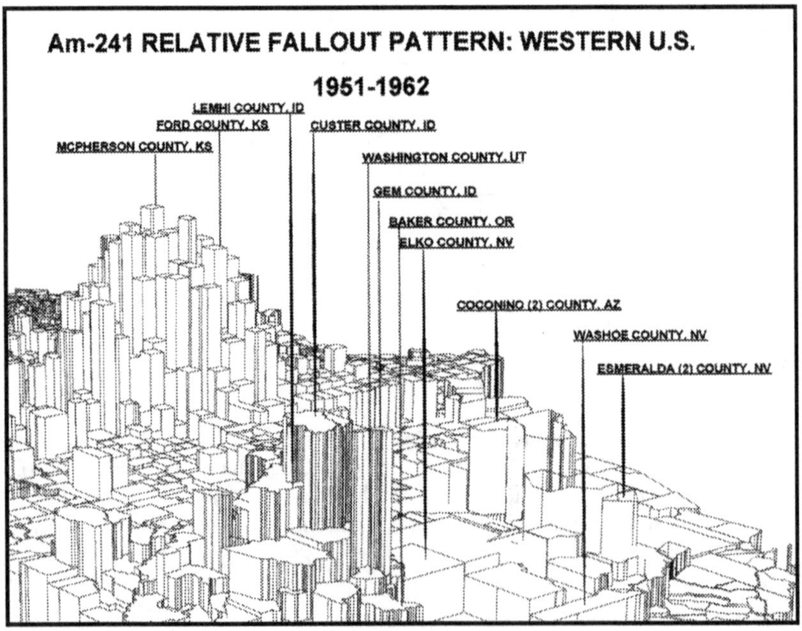

Americium-241 (Am-241)

Americium-241 is considered by the International Atomic Energy Commission to be a Class I very highly toxic radionuclide[1 2]. It is produced primarily by the disintegration of plutonium-241. Radiation includes alpha particles, gamma rays and X-rays[3]. It is a long-lived radionuclide with a half-life of 432.2 years.[4] For that reason, americium-241 will have retained 92% of its activity 50 years after deposition. Americium-241 was produced primarily by shots Tumbler-Snapper HOW (TS8), Plumbbob WHITNEY (PB16) and shot SEDAN. Counties affected included the group located northwest of the Nevada Test Site as well as an area in the Midwest surrounding McPherson County, Kansas.

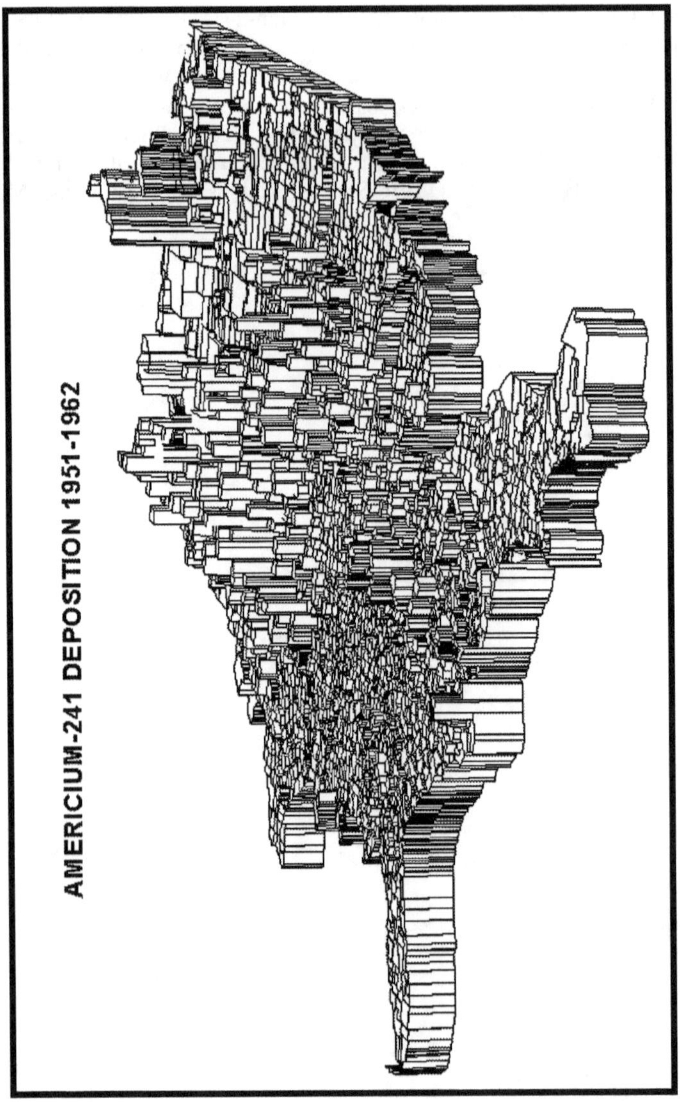
AMERICIUM-241 DEPOSITION 1951-1962

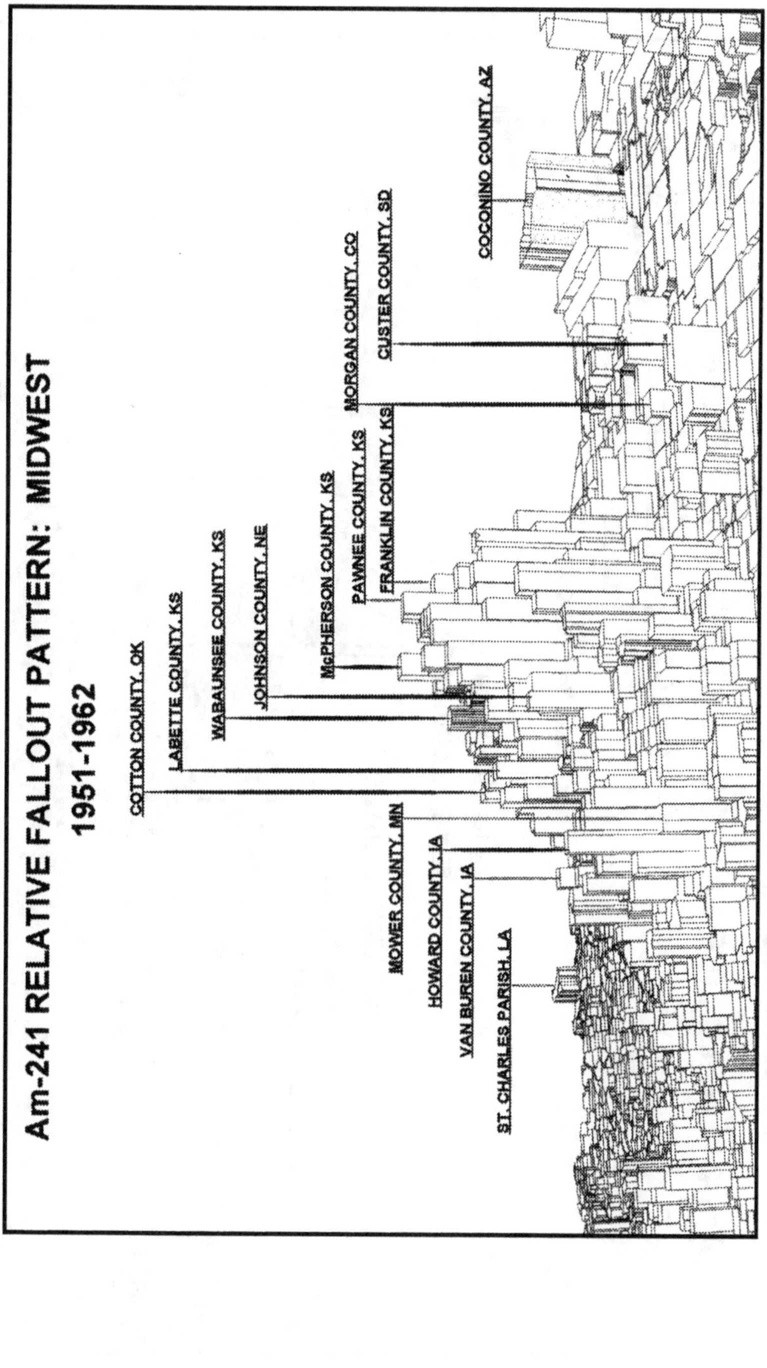
Am-241 RELATIVE FALLOUT PATTERN: MIDWEST
1951-1962
COTTON COUNTY, OK
LABETTE COUNTY, KS
WABAUNSEE COUNTY, KS
JOHNSON COUNTY, NE
McPHERSON COUNTY, KS
PAWNEE COUNTY, KS
FRANKLIN COUNTY, KS
MORGAN COUNTY, CO
CUSTER COUNTY, SD
COCONINO COUNTY, AZ
MOWER COUNTY, MN
HOWARD COUNTY, IA
VAN BUREN COUNTY, IA
ST. CHARLES PARISH, LA

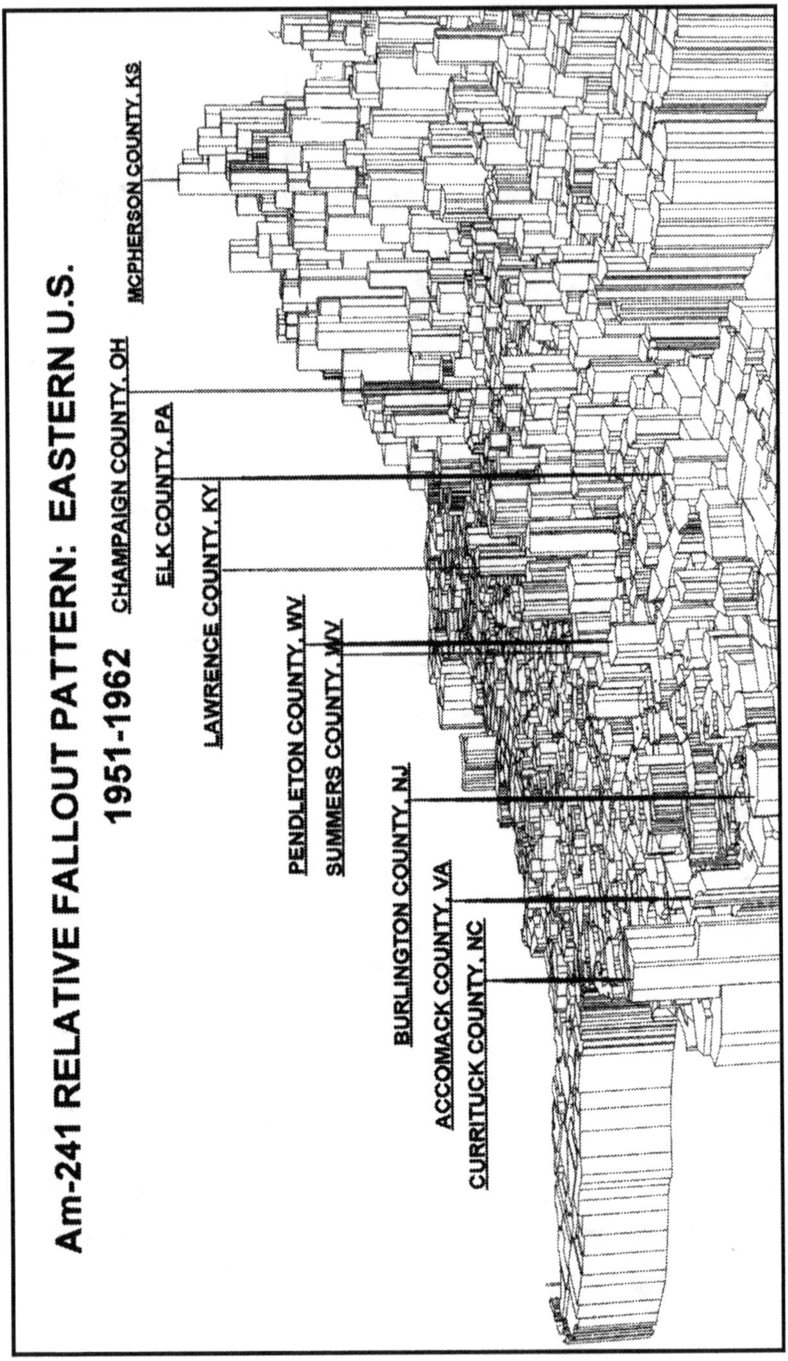
Am-241 RELATIVE FALLOUT PATTERN: EASTERN U.S.
1951-1962
CHAMPAIGN COUNTY, OH
MCPHERSON COUNTY, KS
ELK COUNTY, PA
LAWRENCE COUNTY, KY
PENDLETON COUNTY, WV
SUMMERS COUNTY, WV
BURLINGTON COUNTY, NJ
ACCOMACK COUNTY, VA
CURRITUCK COUNTY, NC

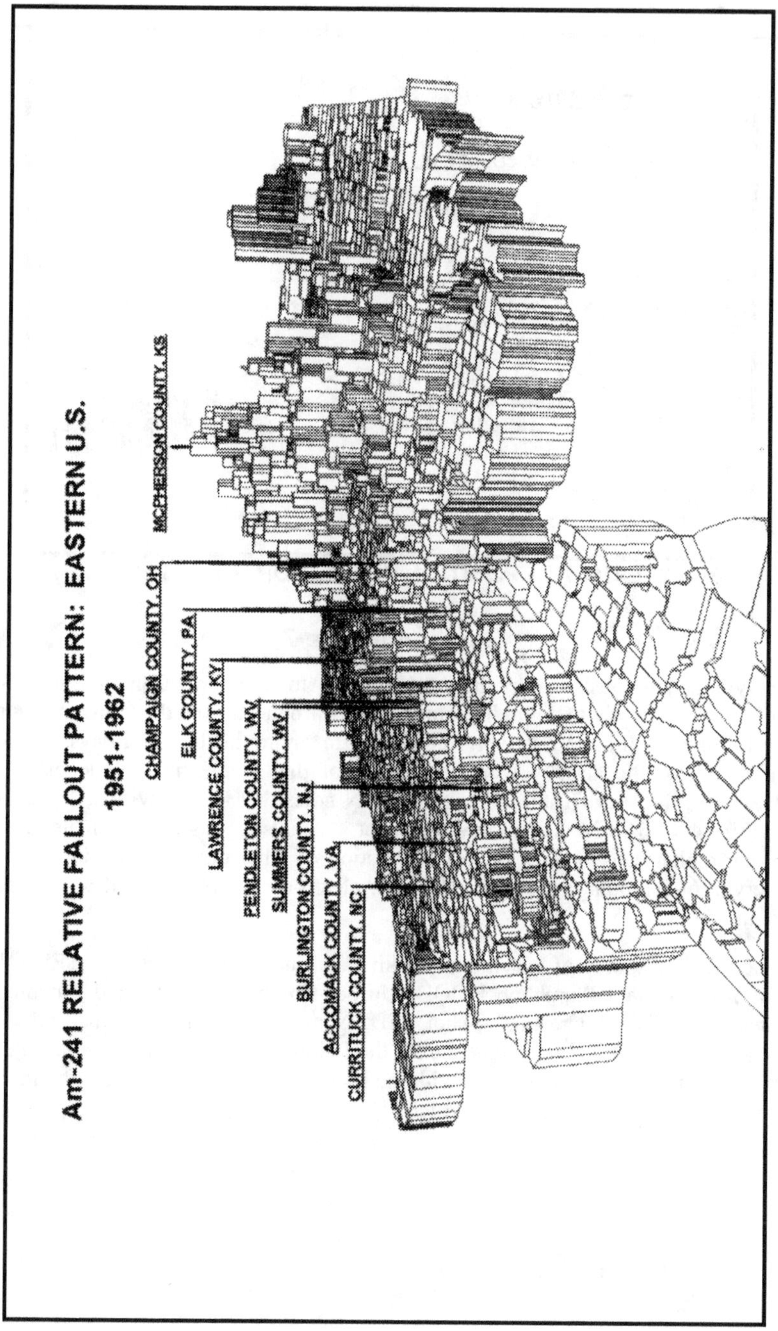
Am-241 RELATIVE FALLOUT PATTERN: EASTERN U.S.
1951-1962
MCPHERSON COUNTY, KS
CHAMPAIGN COUNTY, OH
ELK COUNTY, PA
LAWRENCE COUNTY, KY
PENDLETON COUNTY, WV
SUMMERS COUNTY, WV
BURLINGTON COUNTY, NJ
ACCOMACK COUNTY, VA
CURRITUCK COUNTY, NC

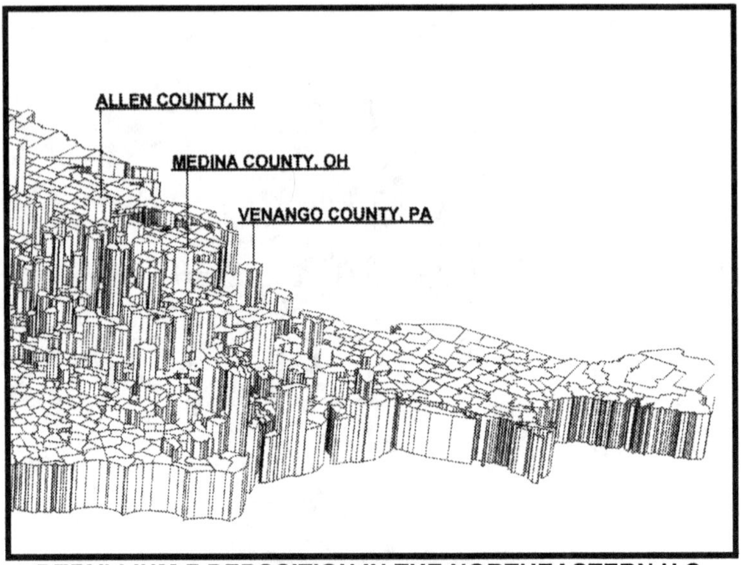

BERYLLIUM-7 DEPOSITION IN THE NORTHEASTERN U.S.

Beryllium-7 (Be-7)

Beryllium-7 is considered by the International Atomic Energy Commission to be a Class IV (slightly toxic) radioisotope[5].[6] It is produced from the transformation of Lithium-6. It has a relatively short half-life, approximately 53.1 days[7] Thus, after 1 year the activity would be 0.008 of that at the time of deposition. However, the Hicks Table for shot SEDAN notes that Be7 activity at 1 year is 5.58E-04 vs 6.38E-2 at one day post shot.[8] This suggests that the material is being replenished through the transformation of other radionuclides in fallout. Beryllium-7 produces electrons (through electron capture), as well as gamma radiation.

The primary source of beryllium-7 from U.S. nuclear tests was shot SEDAN, followed by shot Plumbbob HOOD, Upshot-Knothole SIMON and Tumbler-Snapper HOW. Interestingly, shot SEDAN produced more Be-7 than all the other shots combined. For this reason, the fallout pattern for Be-7 is associated with that for SEDAN. Counties with the highest Be-7 fallout included the Iowa counties of Howard, Worth and Mitchell.

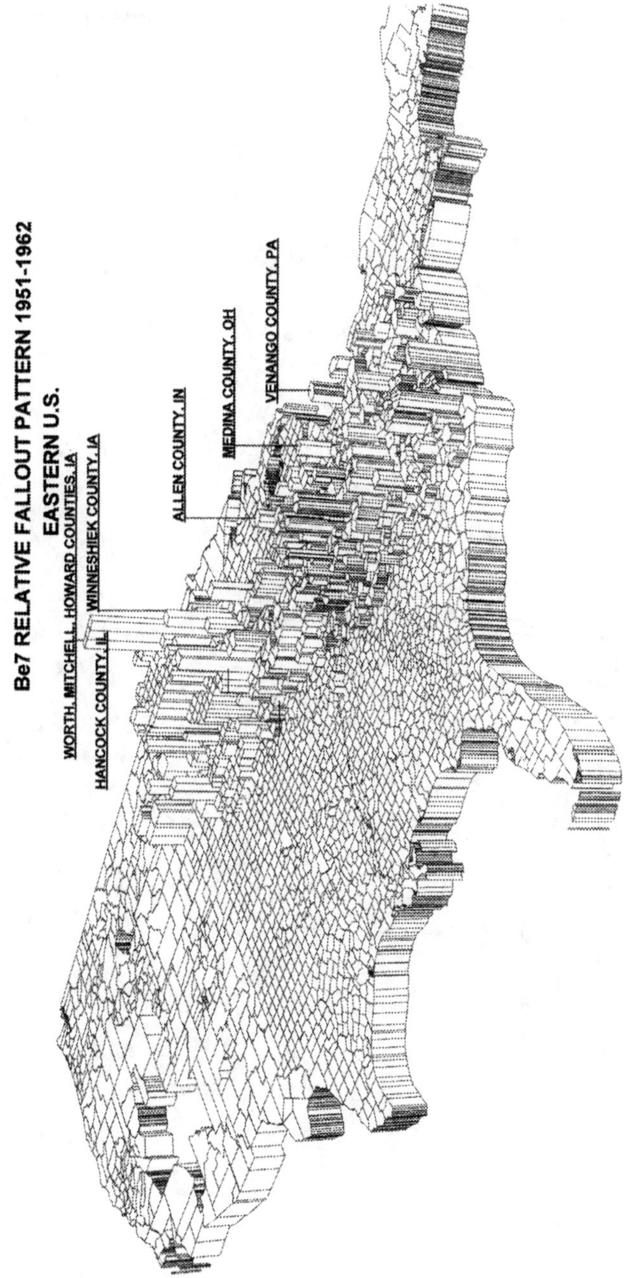
Be7 RELATIVE FALLOUT PATTERN 1951-1962
EASTERN U.S.
WORTH, MITCHELL, HOWARD COUNTIES, IA
HANCOCK COUNTY, IL
WINNESHIEK COUNTY, IA
ALLEN COUNTY, IN
MEDINA COUNTY, OH
VENANGO COUNTY, PA

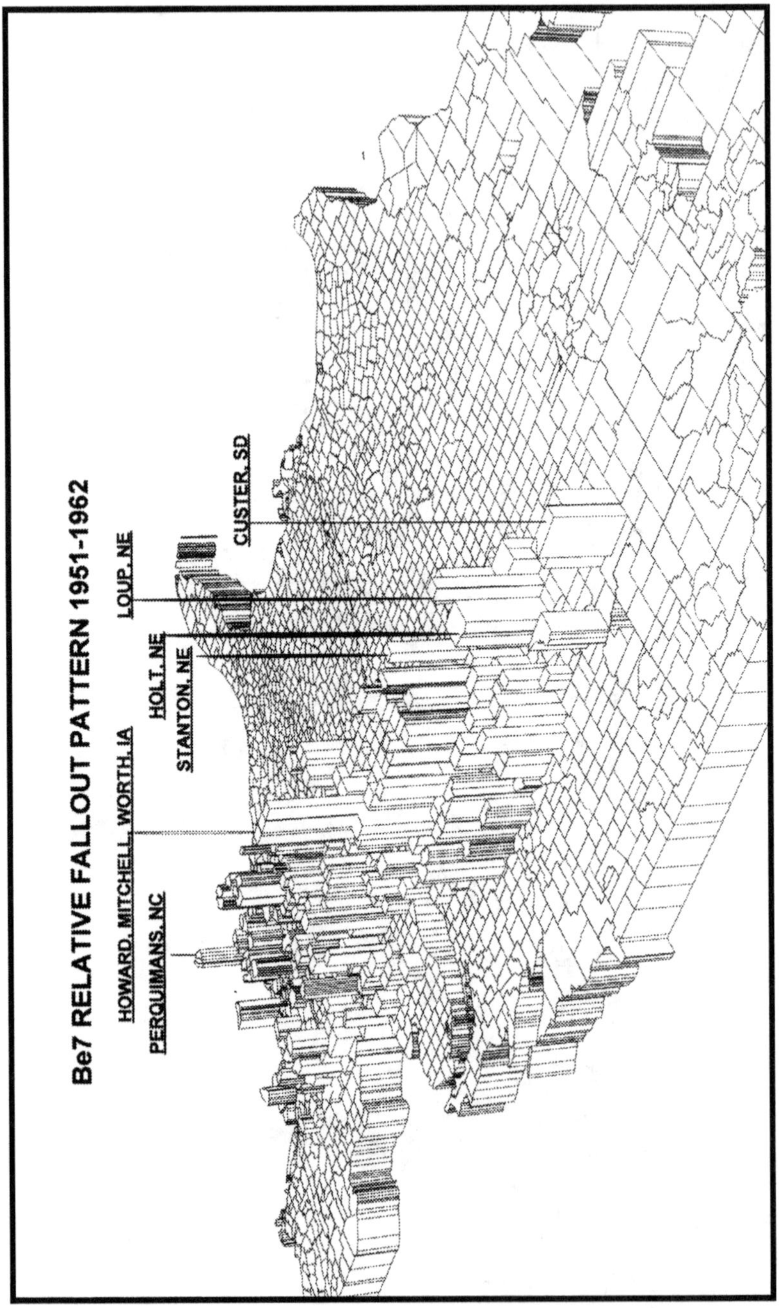
Be7 RELATIVE FALLOUT PATTERN 1951-1962
HOWARD, MITCHELL, WORTH, IA
PERQUIMANS, NC
LOUP, NE
HOLT, NE
STANTON, NE
CUSTER, SD

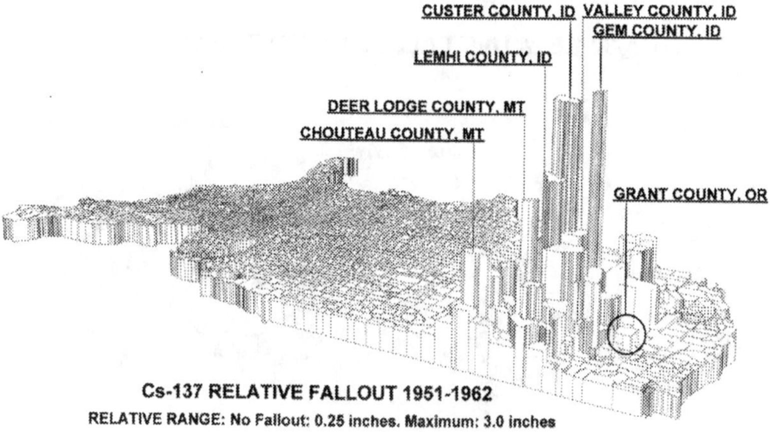

Cesium-137 (Cs-137)

Cesium-137 is considered to be a Class III or moderately toxic radionuclide[9]. [10] It is a fission product with a half-life of approximately 30 years. Twenty years after deposition, cesium-137 retains 63% of its original activity, while after 50 years the activity reduces to 31.5% of the deposition level. Cesium-137 produces beta particles and X-rays.[11]

Most of the 3D maps in this volume were calculated using the zero rate equal to 0.25 virtual inches in height (z-axis) against a maximum of 1.5 virtual height. This is equal to a range ratio of 1:6. Even by expanding the range from 0.25-1.5 to 0.25 to 5.0—an expansion from 1:6 to 1:20, it is clear that most of the Cs-137 remained in the Northwest U.S. The same kind of pattern is seen with strontium-89, strontium-90, yttrium 90, barium 137m and many more. Nuclear tests high in production of Cs-137 include Tumbler-Snapper HOW, Plumbbob BOLTZMANN and Tumbler-Snapper EASY.

Co-60 RELATIVE FALLOUT PATTERN 1951-1962

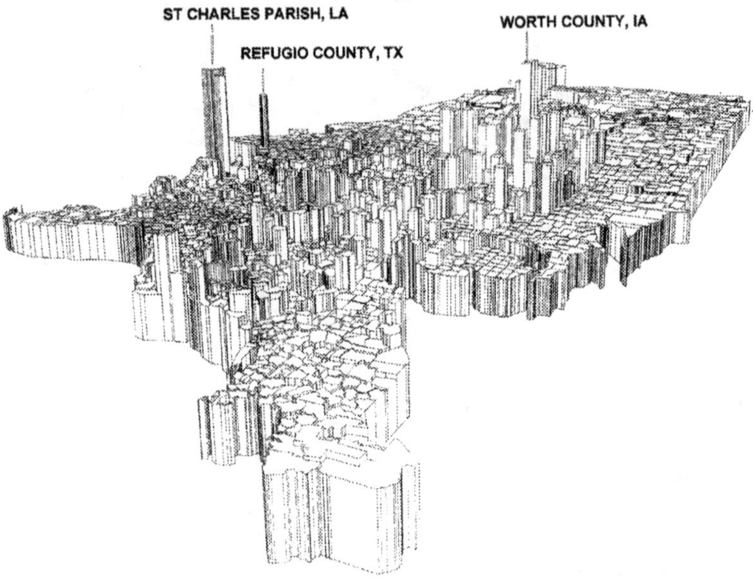

Cobalt-60 (Co-60)

Cobalt-60 is considered to be a Class III (moderately toxic) radionuclide[12,13]. It has a relatively long half-life of 5.27 years. During the decay process Co-60 produces Beta and Gamma radiation Ten years after deposition the activity of Co60 has decreased to 0.072 of the original. The nuclear detonation highest in Co-60 was shot Galileo, PB-12, followed by shot SEDAN. For that reason, the primary deposition areas for this radionuclide are found in the Southwest and Midwestern United States.

EASTERN U.S.
Co-60 RELATIVE FALLOUT PATTERN 1951-1962

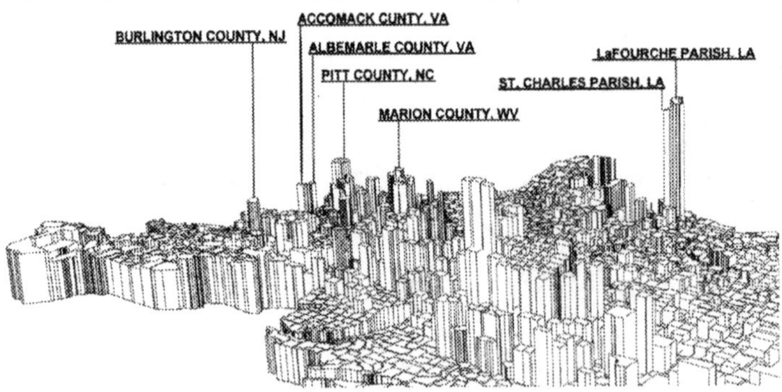

Co-60 RELATIVE FALLOUT PATTERN 1951-1962
EASTERN U.S.

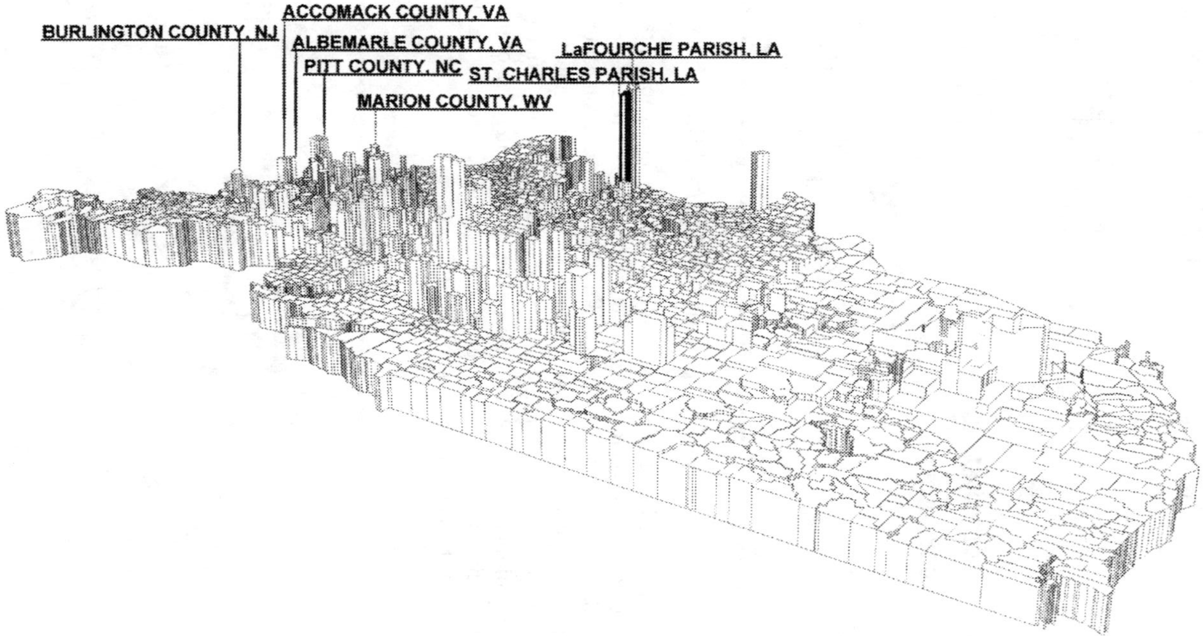

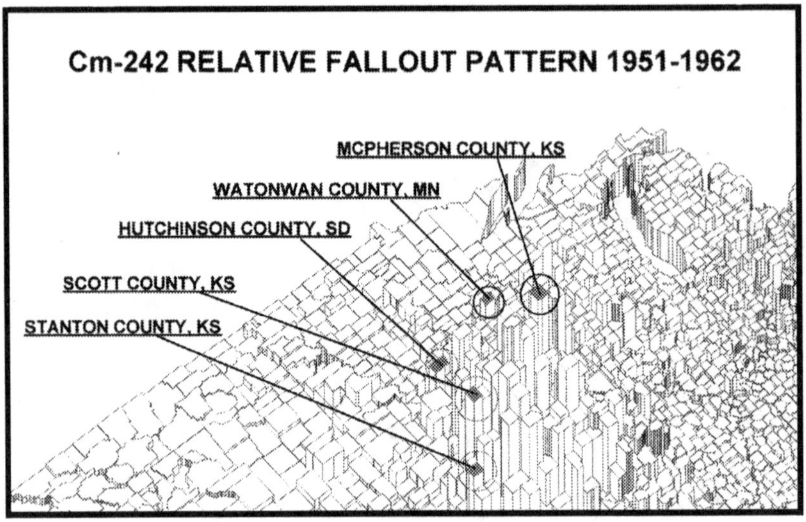

Curium-242 (Cm-242)

Curium-242 is considered to be a Class I (very highly toxic) radionuclide by the International Atomic Energy Commission[14]. [15] It can be produced either through transformation of americium-241 or by neutron capture from uranium-238 and plutonium-239. It produces alpha particles, gamma rays, X-rays and electrons. It has a half-life of approximately 163 days. After 1 year the activity level of this curium-242 has dropped to approximately 1/5 of the original. Shots producing relatively high levels of curium-242 include SEDAN, Plumbbob CHARLESTON and Plumbbob WHITNEY.

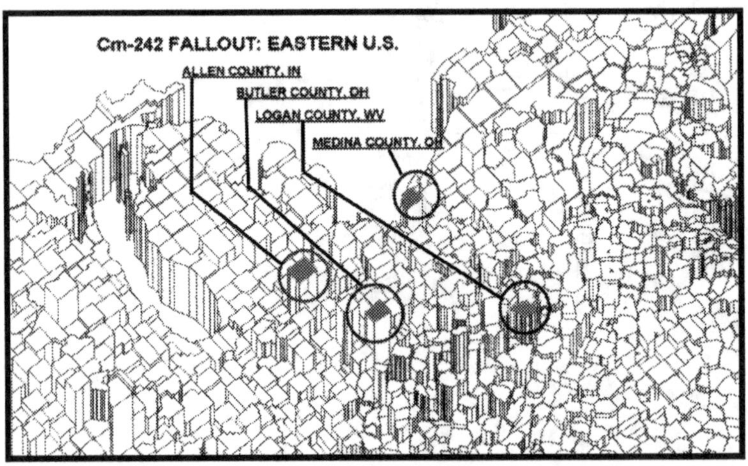

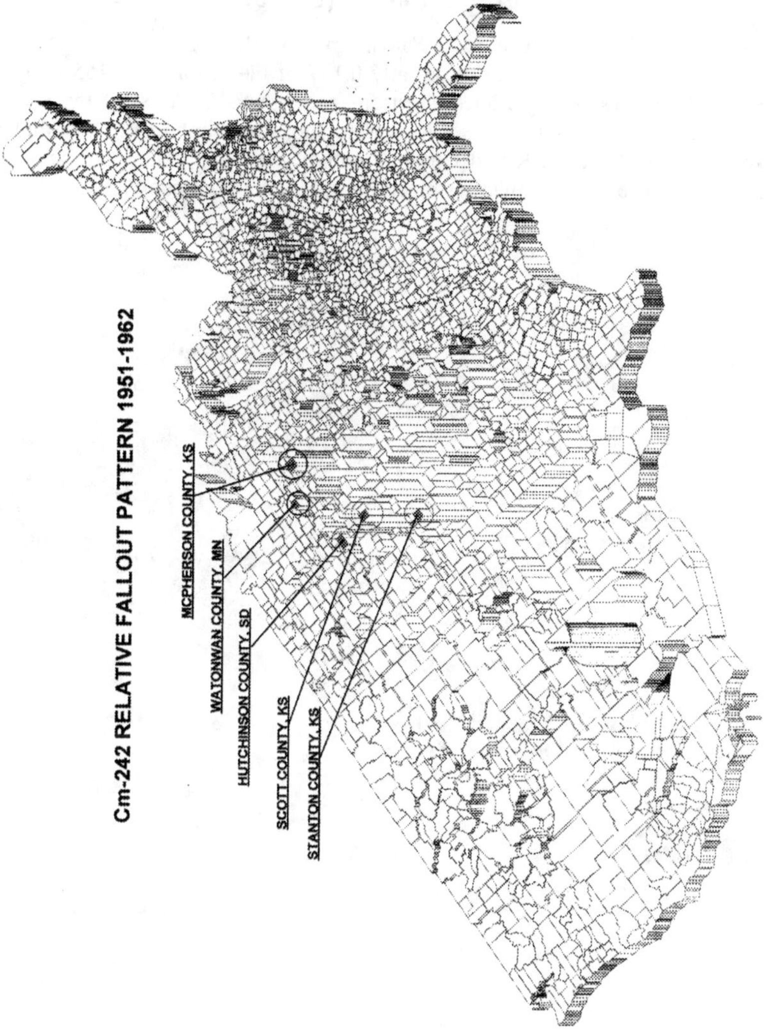

Cm-242 RELATIVE FALLOUT PATTERN 1951-1962

Europium-155 (Eu-155)

Europium-155 is produced from Samarium-154. It has a half-life of 1.811 years. Five years after deposition only 0.147 of the original Eu-155 remains. Eu-155 produces beta particles, electrons and X-rays[16]. Most of the Eu-155 deposited on counties in the U.S. was from shot Plumbbob OWENS, followed by Plumbbob DOPPLER and Upshot-Knothole SIMON. Plumbbob OWENS produced approximately twice as much Eu-155 as did Plumbbob KEPLER.

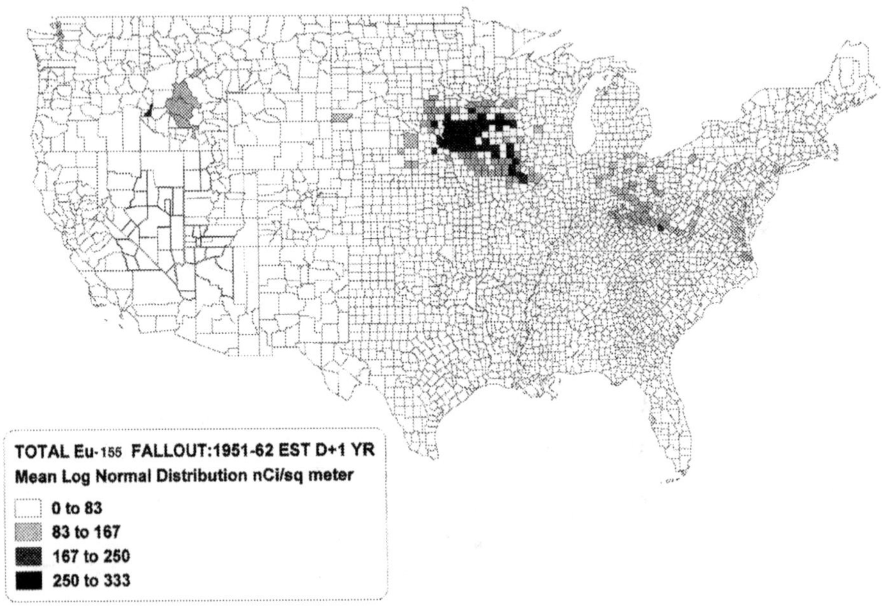

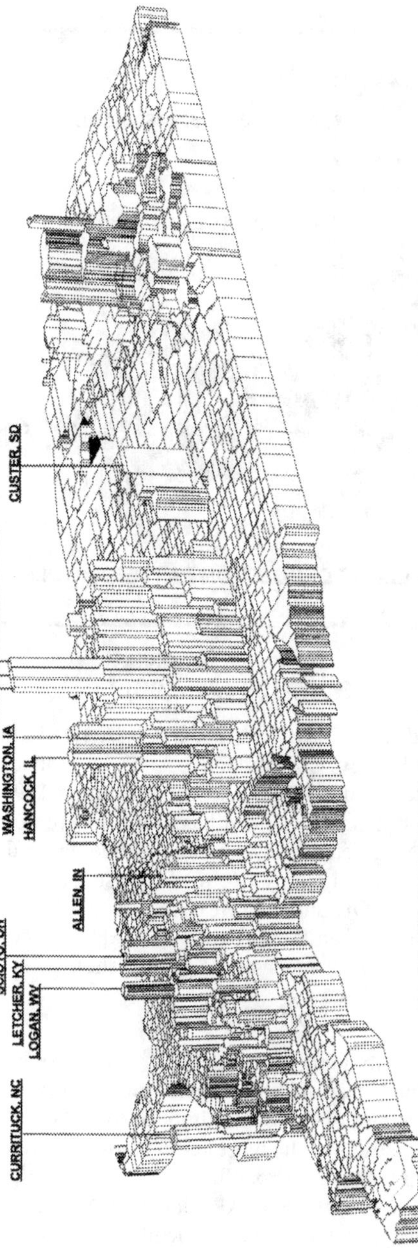

Eu-155 RELATIVE FALLOUT 1951-1962

RELATIVE RANGE: No Fallout: 0.25 inches. Maximum: 2.00 inches (1:8)

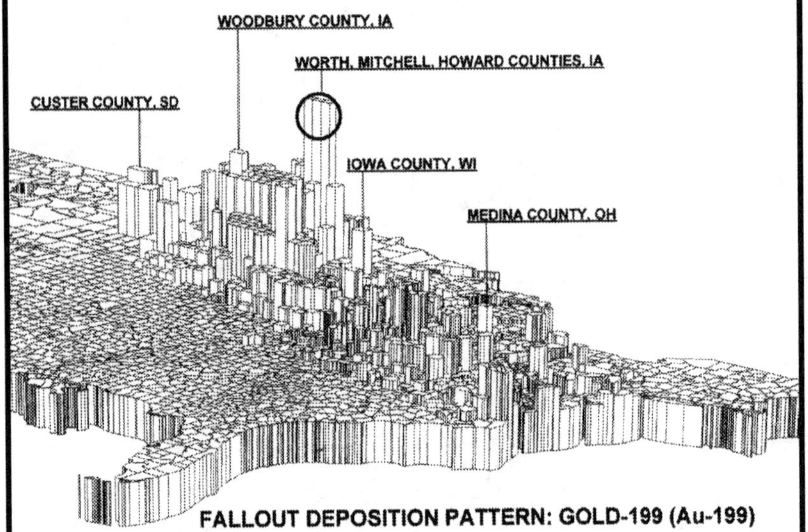

FALLOUT DEPOSITION PATTERN: GOLD-199 (Au-199)

Gold-199 (Au-199)

Gold 199 is classified as a Class III (moderately toxic) radionuclide by the International Atomic Energy Commission[17].[18] It is produced in a nuclear detonation as a daughter product of gold-197 and platinum-199. It produces beta particles, electrons and gamma rays, and has a half life of approximately 2.68 days. Thus, ten days after detonation, the radionuclide will reduced to 0.075 of the original level. Interestingly, only two nuclear tests produced either Au-198 or Au-199: Plumbbob HOOD and shot SEDAN. And while HOOD deposited fallout over more counties than SEDAN, the SEDAN shot deposited far more radioactive gold—11.2 times as much as the HOOD shot.

Iron-59 (Fe-59)

Iron-59 is a Class II (highly toxic) radionuclide.[19] [20] It is produced from iron-58 in a nuclear reaction and has a half-life of approximately 45 days. Thus, 90 days after deposition, the amount of Fe-59 has dropped to 25% of the original level. Plumbbob GALILEO distributed the most Fe-59 across the country, followed by Upshot-Knothole SIMON and Storax SEDAN. Some nuclear tests, such as Tumbler-Snapper HOW, Jangle SUGAR, Jangle UNCLE, Tumbler-Snapper EASY and Tumbler-Snapper FOX produced negligible Fe-59.

Fe-59 RELATIVE FALLOUT 1951-1962

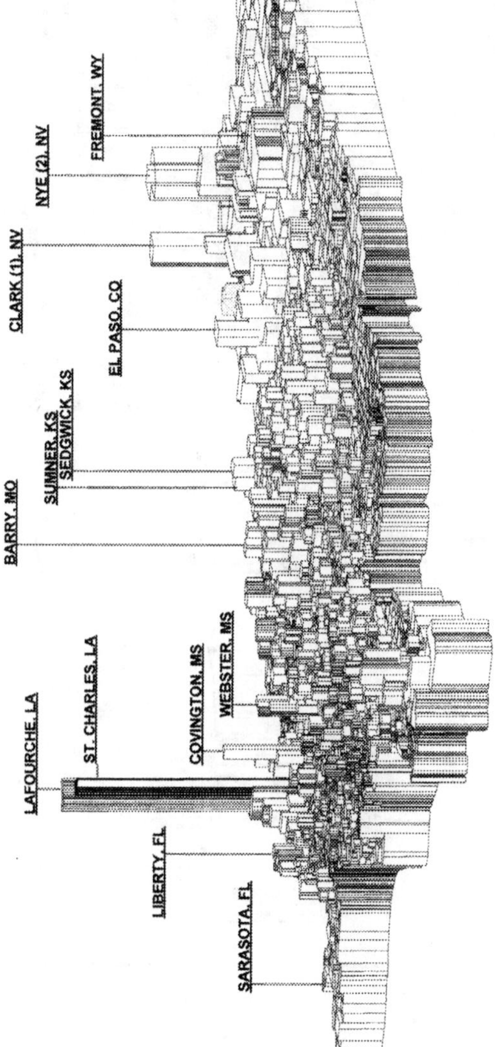

RELATIVE RANGE: No Fallout: 0.25 inches. Maximum: 2.00 inches (1:8)

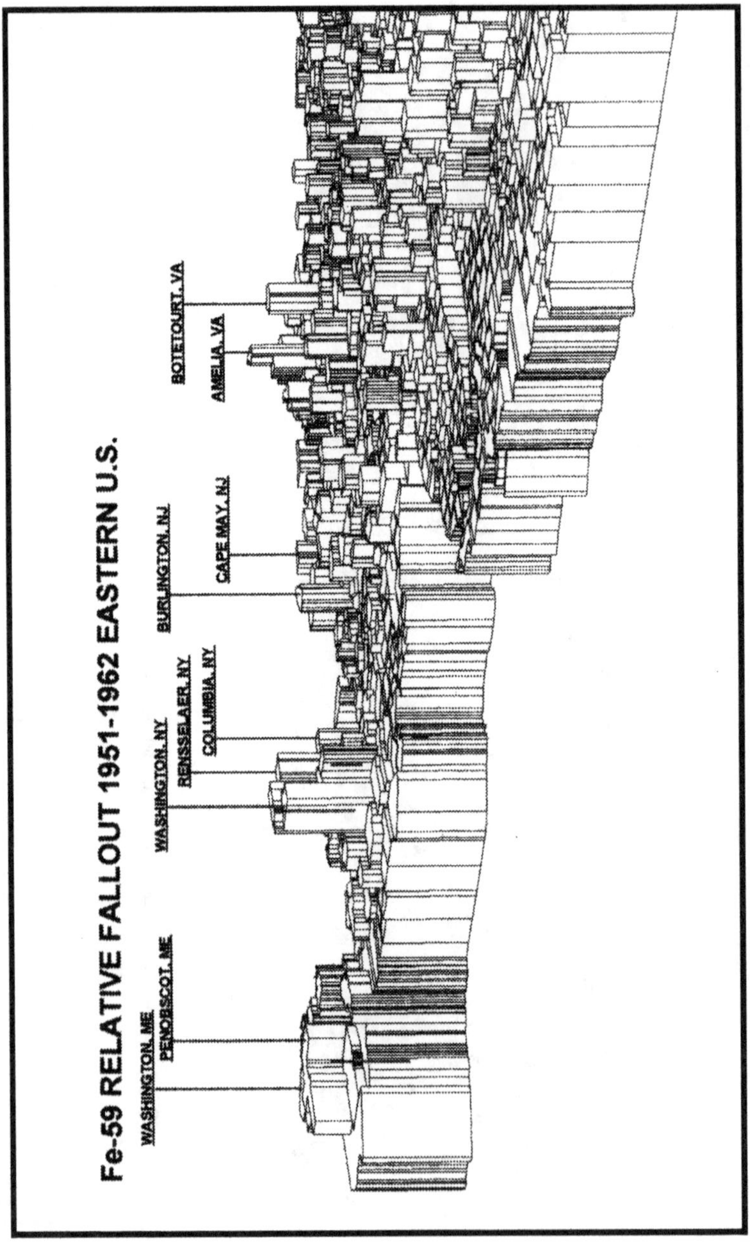
Fe-59 RELATIVE FALLOUT 1951-1962 EASTERN U.S.
WASHINGTON, ME
PENOBSCOT, ME
WASHINGTON, NY
RENSSELAER, NY
COLUMBIA, NY
BURLINGTON, NJ
CAPE MAY, NJ
BOTETOURT, VA
AMELIA, VA

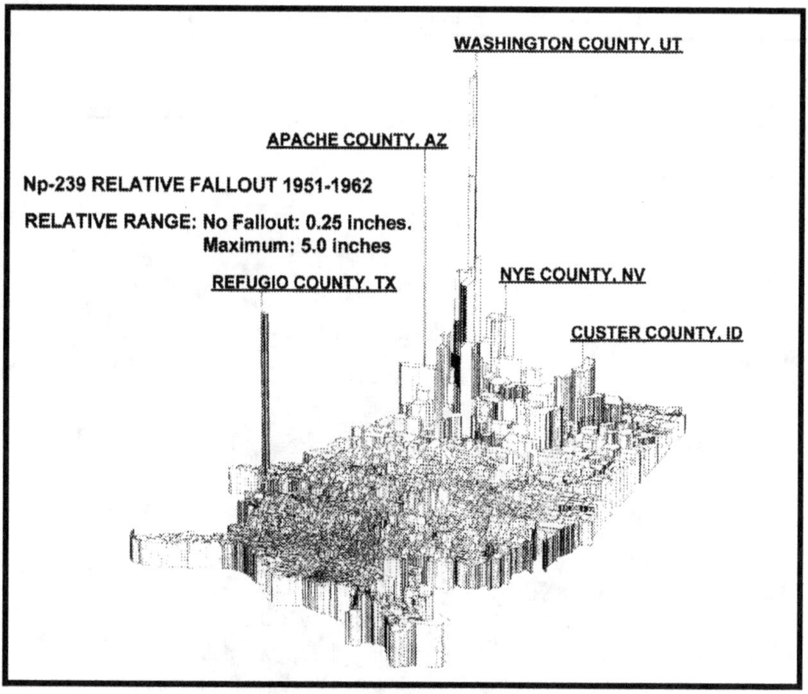

Neptunium-239 (Np-239)

Neptunium is a daughter product of uranium-238 and uranium-239. It produces beta particles, electrons and gamma radiation (as X-rays)[21]. It has a half-life of approximately 2.34 days. The nuclear test with the greatest ratio of Np-239 in fallout was Teapot ESS, followed by Tumbler-Snapper ABLE, Tumbler-Snapper BAKER and Ranger BAKER and BAKER-2. Plumbbob shots OWEN and COULOMB B produced *no* Np-239.

Interestingly, even though Upshot-Knothole HARRY ranked 20th among the nuclear tests for percent of Np-239 in fallout, because it produced so much fallout, HARRY ranked number one in deposition of Np-239. Other shots depositing high levels of this radionuclide included Tumbler-Snapper GEORGE (TS7) and Upshot-Knothole SIMON (UK7). As the expanded range (1:20) 3D map above shows, the vast majority of Np-239 was deposited on the counties surrounding the Nevada Test Site. Still, thanks primarily to the 1953 nuclear test Upshot-Knothole SIMON, significant amounts of this short-lived radionuclide was deposited in the Eastern United States.

Pd-111m RELATIVE FALLOUT PATTERN 1951-1962

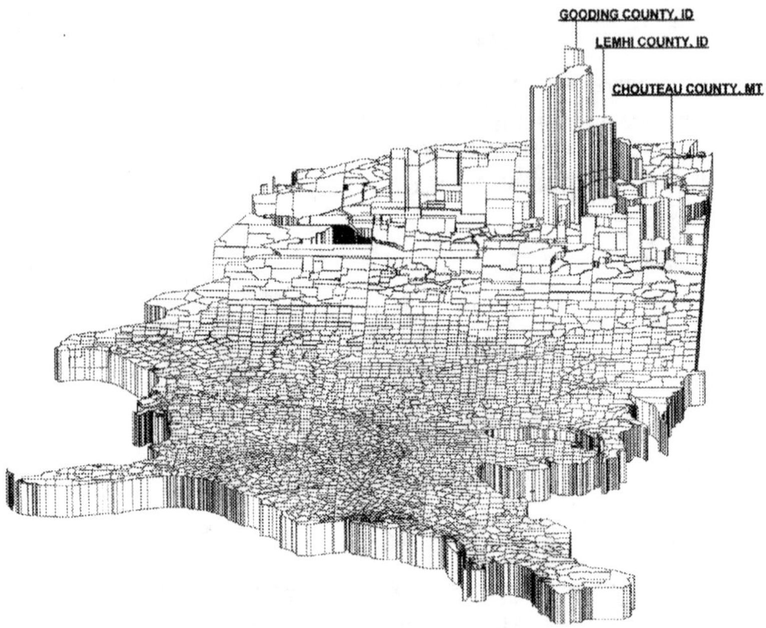

Palladium-111m (Pd-111m)

Palladium-111m is a fission product of palladium-110. It produces beta particles, electrons and gamma rays and has a relatively short half-life of only 5.5 hours. Thus, within 24 hours of detonation, less than 0.05 of the original amount of Pd-111m remains in the fallout. For that reason, exposure to Pd-111m was generally limited to counties close to the Nevada Test Site. Shots Plumbbob KEPLER, Teapot POST and Plumbbob OWENS had the greatest ratio of Pd-111m in fallout, while the Upshot-Knothole shots HARRY, NANCY and SIMON were responsible for the greatest amounts of Pd-111m deposited.

Strontium-90

Strontium-90 is considered a Class I (very highly toxic) radionuclide by the International Atomic Energy Commission[22]. [23] It is a fission product with a half-life of approximately 29 years.[24] Thus, 40 years after deposition, over a third of the Sr-90 deposited is still active. Strontium-90 produces no gamma radiation, only beta particles. Most of the Sr-90 deposited across the country came from Upshot-Knothole SIMON, followed by Tumbler-Snapper GEORGE and Upshot-Knothole HARRY. Still, as the expanded range (1:20) 3D map on the next page suggests, the vast majority of Sr-90 was deposited in the counties north of the Nevada Test Site. Sr-90 deposition levels are similar to those for Sr-89, Y-90, and Y-91.

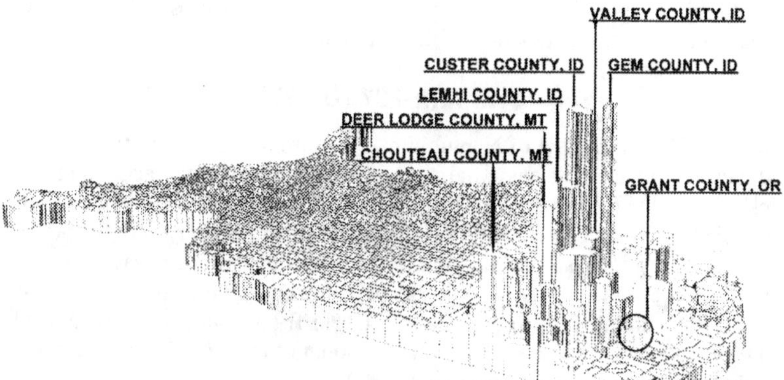

Sr-90, Y-90 RELATIVE FALLOUT 1951-1962

RELATIVE RANGE: No Fallout: 0.25 inches. Maximum: 3.0 inches

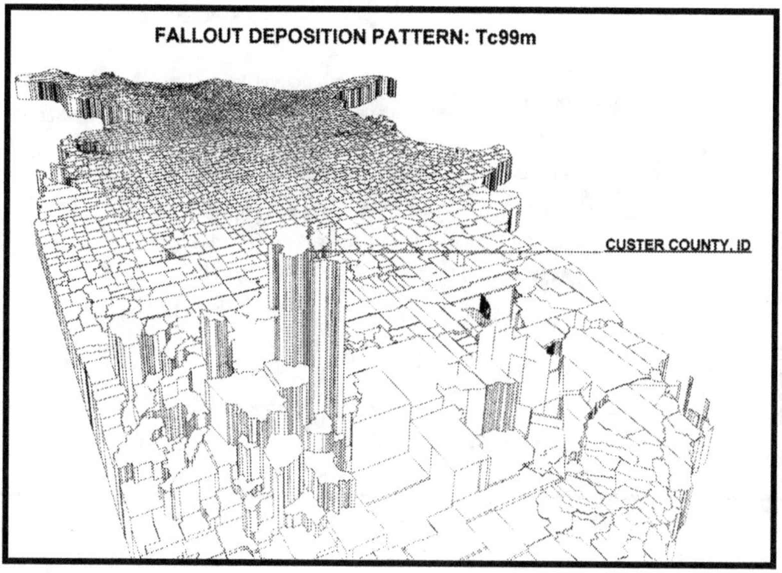

Technetium-99m (Tc99m)

Technectium-99m has an extremely short half-life of only 6 hours, but because it is constantly being produced from the longer-lived Mo-99 (half-life: 66 hours)[25] it remains in the environment until the parent radionuclides are depleted. Technectium-99m produces X-rays and electrons. Shots Buster BAKER and Plumbbob OWENS produced the greatest ratio of Tc99m in fallout, while shot Tumbler-Snapper HOW and Upshot-Knothole HARRY produced the greatest deposition of this radionuclide.

Uranium-237 (U-237)

Uranium 237 is a product of Neptunium-237. It produces beta particles, electrons and gamma radiation. It has a half-life of only 6.75 days, yet according to the Hicks Tables the overall activity level for U-237 from many detonations such as shot TESLA remained relatively constant for up to fifty days post shot[26]. For that reason, despite it's short half-life, U-237 was deposited in areas far from the Nevada Test Site. Upshot-Knothole BADGER produced the greatest percentage of U-237 in fallout, followed by Jangle UNCLE and Plumbbob HOOD. Shot SEDAN, however, deposited the greatest amount of U-237, followed by Upshot-Knothole SIMON and Tumbler-Snapper HOW.

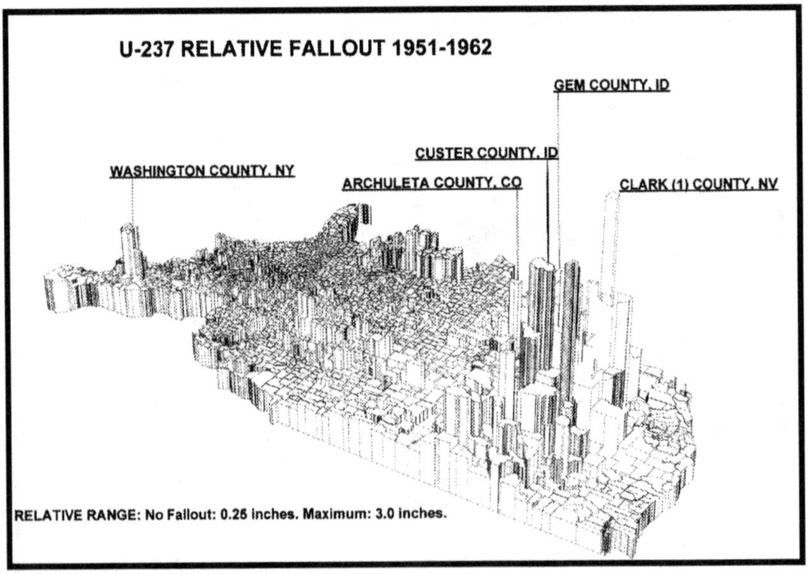

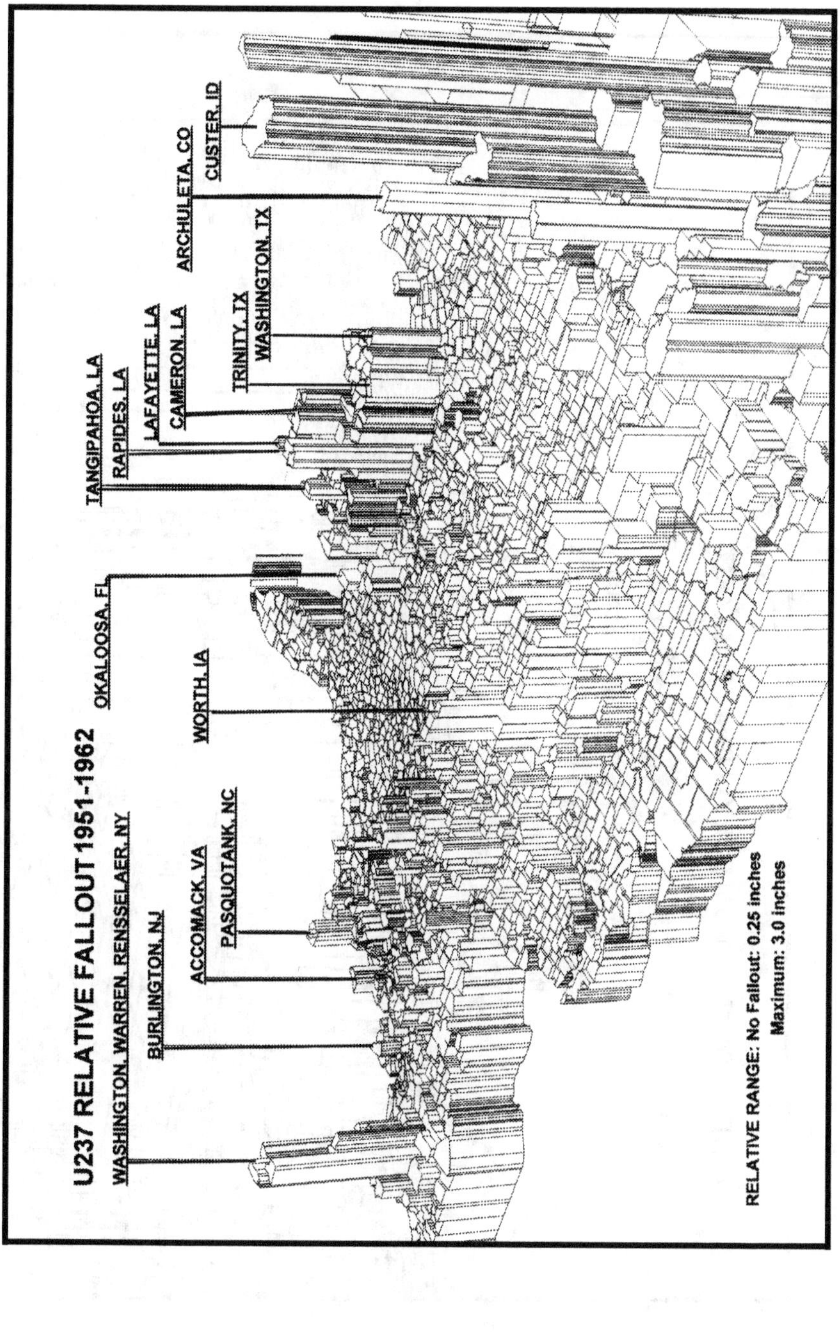
U237 RELATIVE FALLOUT 1951-1962
WASHINGTON, WARREN, RENSSELAER, NY
BURLINGTON, NJ
ACCOMACK, VA
PASQUOTANK, NC
WORTH, IA
OKALOOSA, FL
TANGIPAHOA, LA
RAPIDES, LA
LAFAYETTE, LA
CAMERON, LA
TRINITY, TX
WASHINGTON, TX
ARCHULETA, CO
CUSTER, ID
RELATIVE RANGE: No Fallout: 0.25 inches
Maximum: 3.0 inches

AREA OF MAXIMUM BERYLLIUM-7 FALLOUT
1951-1962
HOWARD COUNTY, IOWA
MITCHELL COUNTY, IOWA
WORTH COUNTY, IOWA

COUNTIES WITH HIGHEST RADIONUCLIDE DEPOSITION AMOUNTS

U.S. RANK WITH ASSOCIATED DEPOSITION ACTIVITY: nCi/Sq meter

RANK	Be7*		Na24		Mn54*	
1	HOWARD IA	17.6	NYE2 NV	1849.1	MITCHELL IA	17.4
2	WORTH IA	17.6	GARFIELD UT	1197.5	HOWARD IA	17.4
3	MITCHELL IA	17.6	INYO3 CA	853.7	WORTH IA	17.3
4	LEE IA	11.5	NYE1 NV	849.1	VAN BUREN IA	11.8
5	HANCOCK IL	11.5	WHITE PINE2 NV	796.1	HANCOCK IL	11.6
6	WASHINGTON IA	11.5	MINERAL NV	765.9	JEFFERSON IA	11.6
7	JEFFERSON IA	11.5	WASHABAUGH SD	739.8	LEE IA	11.6
8	VAN BUREN IA	11.5	WASHINGTON1 UT	673.5	WASHINGTON IA	11.5
9	WOODBURY IA	11.5	WASHINGTON2 UT	672.5	WOODBURY IA	11.5
10	WATONWAN MN	10.8	CHURCHILL NV	636.9	WATONWAN MN	10.8
11	FARIBAULT MN	10.8	WASHINGTON3 UT	588.8	FARIBAULT MN	10.8
12	MOWER MN	10.8	LINCOLN1 NV	584.0	MOWER MN	10.8
13	DIXON NE	10.5	SWEETWATER WY	526.7	PLYMOUTH IA	10.5
14	WASHABAUGH SD	10.4	LANDER2 NV	519.8	DIXON NE	10.5
15	PLYMOUTH IA	10.4	LYON NV	516.9	WASHABAUGH SD	10.4

RANK	Fe59*		Co60*		Cu64	
1	LAFOURCHE LA	0.16	MITCHELL IA	7.4	WASHINGTON2 UT	52784
2	ST CHARLES LA	0.15	HOWARD IA	7.3	WASHINGTON3 UT	42969
3	CLARK1 NV	0.07	WORTH IA	7.3	WASHINGTON1 UT	32574
4	NYE2 NV	0.07	LAFOURCHE LA	6.1	KANE1 UT	31455
5	COVINGTON MS	0.06	ST CHARLES LA	5.8	KANE2 UT	29399
6	SEDGWICK KS	0.05	VAN BUREN IA	5.3	MOHAVE2 AZ	27960
7	SUMNER KS	0.05	WOODBURY IA	5.1	CLARK1 NV	27907
8	BARRY MO	0.05	HANCOCK IL	5.1	LINCOLN1 NV	26943
9	EL PASO CO	0.05	LEE IA	5.0	COCONINO1 AZ	23722
10	BALDWIN AL	0.05	JEFFERSON IA	4.9	MOHAVE1 AZ	18376
11	WEBSTER MS	0.05	WASHINGTON IA	4.9	GUNNISON CO	17836
12	VERNON MO	0.05	PLYMOUTH IA	4.7	CONEJOS CO	16612
13	SHAWNEE KS	0.05	CRAWFORD IA	4.7	GARFIELD UT	15251
14	WILSON KS	0.05	IDA IA	4.7	COCONINO2 AZ	11594

NOTES TO SECTION 10

[1] International Atomic Energy Commission, Vienna, *Safe Handling of Radioactive Nuclides*, Safety Series No. 1.1973.
[2]Wilson, William Feathergail. NORM: A Guide to Naturally Occurring Radioactive Material. Pennwell Books. Pennwell Pub. Tulsa, OK. p. 147.
[3] Radiological Health Handbook. US Department of Health, Education and Welfare, Public Health Service. Jan. 1970. p.374.
[4] Emsley, John, The Elements, 3rd Edition. Clarendon Press, Oxford. 1998. p.21.
[5] International Atomic Energy Commission. Op cit.
[6] Wilson. Op cit. p. 147.
[7] Radiological Health Handbook. Ibid. p. 232.
[8] Hicks, Harry G. *Results of Calculations of External Gamma Radiation Exposure Rates from Fallout and the Related Radionuclide Compositions. Operation Nougat Through Bowline, 1962-1968.* Lawrence Livermore National Lab, CA Jul 1981. DE82007515. p. B-8, 10
[9] International Atomic Energy Commission Op cit.
[10] Wilson. Op cit. p. 147.
[11] Radiological Health Handbook. Op cit. p. 303.
[12] International Atomic Energy Commission. Op cit.
[13] Wilson, Op cit. p. 147.
[14] International Atomic Energy Commission. Op cit.
[15] Wilson. Op cit. p. 147.
[16] Radiological Health Handbook. Op cit. p. 316.
[17] International Atomic Energy Commission. Op cit.
[18] Wilson. Op cit. p. 147.
[19] International Atomic Energy Commission. Op cit.
[20] Wilson. Op cit. p. 147.
[21] Radiological Health Handbook. Op cit. p. 371.
[22] International Atomic Energy Commission. Op cit.
[23] Wilson. Op cit. p. 147.
[24] Radiological Health Handbook. Op cit. p. 266.
[25] Radiological Health Handbook. Op cit. pp 273-275.
[26] Hicks, Op cit.DE82005601. p.C-8

TABLE: TOXICITY OF SELECTED RADIONUCLIDES

Radionuclide	Half-Life	RADIATION (SEE NOTES)	TARGET ORGAN[1]	TARGET ORGANS AND MORTALITY AND MORBITITY RISK		
				AIR	WATER	FOOD
Am241	432.2 years	α(5.637 MeV); e^-; γ		No single cancer type accounts for more than 40% of the total cancer mortality. Lung	No single cancer type accounts for more than 40% of the total cancer mortality.	No single cancer type accounts for more than 40% of the total cancer mortality.
Au198	2.694 days	β^-(0.812 MeV); e^-; γ	Soluble: GI (Lower Large Intestine) Insoluble: GI (Lower Large Intestine)	-------	-------	-------
Au199	3.14 days	β^-(0.453 MeV); e^-; γ	Soluble: GI (Lower Large Intestine) Insoluble: GI (Lower Large Intestine)	-------	-------	
Be7	53.82 days	EC (Electron Capture: 0.862 MeV); γ	Soluble: GI,: Lower Large Intestine; Total Body; Insoluble: Lung	-------	-------	
Br82	1.471 days	β^-(3.093 MeV); γ				
Cd115m	44.6 days	β^-(1.629 MeV); γ	Soluble: GI (Lower Large Intestine, Liver, Kidney; Insoluble: Lung, GI (Lower Large			

TABLE: TOXICITY OF SELECTED RADIONUCLIDES

Radionuclide	Half-Life	RADIATION (SEE NOTES)	TARGET ORGAN[1]	TARGET ORGANS AND MORTALITY AND MORBITITY RISK		
				AIR	WATER	FOOD
			Intestine)			
Ce144	284.6 days	β^-(0.310 MeV max), e^-, γ	Soluble: GI (Lower Large Intestine), Bone, Liver; Insoluble: Lung, GI (Lower Large Intestine)	No single cancer type accounts for more than 40% of the total cancer mortality.; Lung	Colon	Colon
Cm242	162.8 days	α(6.126 MeV 74%; 6.07 MeV 25%)); e^-; γ Pu X-rays. (NOTE: Parent: Pu238; daughters: Am242 and Cf246)[i]	Soluble: GI (Lower Large Intestine), Liver Insoluble: Lung, GI (Lower Large Intestine)	Liver, Lung	Colon	Colon
Co60	5.27 years	β^-(1.322 MeV); γ	Soluble: GI (Lower Large Intestine); Total Body; Insoluble: Lung, GI, (Lower Large Intestine).	Lung	No single cancer type accounts for more than 40% of the total cancer mortality.	No single cancer type accounts for more than 40% of the total cancer mortality.
Cs137	30.3 years	β^-(1.175 MeV); γ	Soluble: Total Body, Liver, Spleen, Muscle; Insoluble: Lung, GI (Lower Large Intestine)	Lung, No single cancer type accounts for more than 40% of the total cancer mortality.	-------	No single cancer type accounts for more than 40% of the total cancer mortality.

TABLE: TOXICITY OF SELECTED RADIONUCLIDES

Radionuclide	Half-Life	RADIATION (SEE NOTES)	TARGET ORGAN[1]	TARGET ORGANS AND MORTALITY AND MORBITITY RISK		
				AIR	WATER	FOOD
Eu155	4.71 years	β^-(1.969 MeV) e^-; γ	Soluble: GI (Lower Large Intestine), Kidney, bone; Insoluble: Lung: GI (Lower Large Intestine)			
Fe59	44.51 days	β^-(1.565 MeV); γ	Soluble: GI (Lower Large Intestine), Spleen; Insoluble: Lung, GI (Lower Large Intestine)	Lung	Colon	
I-131	8.04 days	β^-(0.606 Mev average); e^-; γ	Soluble: Thyroid; Insoluble: GI (Lower Large Intestine), Lung	Lung, Thyroid	Thyroid	No single cancer type accounts for more than 40% of the total cancer mortality.
I-133	20.8 hours	β^-(1.27 MeV max); γ	Soluble: Thyroid; Insoluble: GI (Lower Large Intestine), Lung	Thyroid, Lung	Thyroid	Thyroid
Mn54	312.2 days	EC (Electron Capture) (1.377 MeV); e^-; γ	Soluble: GI (Lower Large Intestine), Liver; Insol. Lung, GI	-------	-------	-------
Mo99	2.748 days	β^-(1.357 MeV); γ	Soluble: Kidney, GI (Lower Large Intestine);	-------	No single cancer type accounts for more than 40% of	No single cancer type accounts for more than 40% of

TABLE: TOXICITY OF SELECTED RADIONUCLIDES

Radionuclide	Half-Life	RADIATION (SEE NOTES)	TARGET ORGAN[1]	TARGET ORGANS AND MORTALITY AND MORBITITY RISK		
				AIR	WATER	FOOD
			Insoluble: GI (Lower Large Intestine)		the total cancer mortality.	the total cancer mortality.
Na24	14.96 years	β^-(5.514 MeV); γ	Soluble: GI (Small Intestine); Insoluble: GI: (Lower Large Intestine)			
Nb95	34.97 days	β^-(0.926 MeV); γ	Soluble: GI (Lower Large Intestine), Total Body; Insoluble: Lung, GI (Lower Large Intestine)	-------	Colon	Colon
Nb95m	90.0 hours	IT; e^-; γ	-------	-------	Colon	Colon
Np239	2,140,000 years	α (4.957 MeV); e^-; γ	Soluble: GI (Lower Large Intestine) Insoluble: GI (Lower Large Intestine)	Colon, Lung	Colon	Colon
Pb203	2.162 days	EC (Electron Capture: 0.97 MeV); e^-; γ	Soluble: GI (Lower Large Intestine) Insoluble: GI (Lower Large Intestine)	-------	-------	-------
Ru106	1.02 years	β^-0.039 MeV); **no γ**	Soluble: GI (Lower Large Intestine);	-------	Colon	Colon

TABLE: TOXICITY OF SELECTED RADIONUCLIDES						
Radionuclide	**Half-Life**	**RADIATION (SEE NOTES)**	**TARGET ORGAN[1]**	**TARGET ORGANS AND MORTALITY AND MORBITITY RISK**		
				AIR	**WATER**	**FOOD**
			Insoluble: Lung, GI, (Lower Large Intestine)			
Sb126	12.4 days	β^-(3.67 MeV); γ	-------	-------	Colon	
Sr89	50.52 days	β^-((1.492 MeV); γ	Soluble: Bone Insoluble: Lung, GI (Lower Large Intestine)	-------	Colon	Colon
Sr90	29.1 years	β^-((0.546 Mev) **no γ**	Soluble: Bone Insoluble: GI (Lower Large Intestine)	-------	Leukemia	Leukemia
Tb161	6.91 days	β^-(0.593 MeV); e^-; γ	Soluble: GI (Lower Large Intestine), Bone, Kidney, Total Body; Insoluble: Lung, GI (Lower Large Intestine)	-------	-------	-------

TABLE: TOXICITY OF SELECTED RADIONUCLIDES

Radionuclide	Half-Life	RADIATION (SEE NOTES)	TARGET ORGAN[1]	TARGET ORGANS AND MORTALITY AND MORBITITY RISK		
				AIR	WATER	FOOD
Te127m	109 days	IT(0.088 Mev 98%)β^-(0.770 MeV 2%); e^-; γ	Soluble: Kidney, Testis, GI (Lower Large Intestine); Insoluble: Lung, GI, (Lower Large Intestine)	Colon, Lungs	Colon	Colon
U237	6.75 days	β^-(0.519 MeV); e^-; γ	-------	-------	-------	-------
U240	14.1 hours	β^-(0.360 MeV max) e^-;γ	-------	-------	-------	-------
W185	74.8 days	β^-(1.312 MeV); **no** γ[ii]	Soluble: GI (Lower Large Intestine); Insoluble: GI (Lower Large Intestine)	-------	-------	-------
Y90	2.67 days	β^-(2.283 MeV) **no** γ	Soluble: GI (Lower Large Intestine) Insoluble: GI (Lower Large Intestine)	-------	Colon	-------

TABLE: TOXICITY OF SELECTED RADIONUCLIDES

Radionuclide	Half-Life	RADIATION (SEE NOTES)	TARGET ORGAN[1]	TARGET ORGANS AND MORTALITY AND MORBITITY RISK		
				AIR	WATER	FOOD
Y91	58.5 days	β^-(1.545 MeV); γ	Soluble: GI (Lower Large Intestine), Bone Insoluble: Lung, GI (Lower Large Intestine)		-------	-------
Zr95	64.02 days	β^-(1.125 MeV); γ	-------	-------	Colon	Colon

REFERENCES:

1. Derived Concentration Limits of Radionuclides in Air and Water for Occupational Exposure (40 h/week). Table A1:Iia. P. 59. Safe Handling of Radionuclides 1973 Edition. International Atomic Energy Agency, Vienna, 1973.
2. Safe Handling of Radionuclides 1973 Edition. Code of Practice : Sponsored by the International Atomic Energy Agency and the World Health Organization. International Atomic Energy Agency, Vienna, 1973. Table AI-IIa.
3. Federal Guidance Report No. 13, Part I Interim Version. Health Risks from Low-Level Environmental Exposure to Radionuclides. Radionuclide-Specific Lifetime Ragiogenic Cancer Risk Coefficients for the U.S. Population, Based on Age-Dependent Intake, Dosimetry, and Risk Models. Table2.1, 2.2, 2.3. Oak Ridge National Laboratory, Office of Radiation and Indoor Air, United States Environmental Protection Agency, Washington, DC. 20460. EPA 402-R-97-014. Jan, 1998.[1] Radiological Health Handbook. Ibid. p.336.
4. Radiological Health Handbook: US Department of Health, Education and Welfare. January, 1970.P.275

NOTES AND REFERENCES

RADIATION SYMBOLS

EC:Electron Capture
IT: Isomeric Transition (decay from an excited metastable state to a lower state)
α: Alpha Particle emission
β^- Negative Beta Particle emission
γ: Gamma Ray emission
e^- electron production

NOTES TO APPENDIX B

1. Table A1:Iia. Derived Concentration Limits of Radionuclides in Air and Water for Occupational Exposure (40 h/week). P. 59. Safe Handling of Radionuclides 1973 Edition. International Atomic Energy Agency, Vienna, 1973.
2. Safe Handling of Radionuclides 1973 Edition. Code of Practice : Sponsored by the International Atomic Energy Agency and the World Health Organization. International Atomic Energy Agency, Vienna, 1973. Table AI-IIa.
3. Table2.1, 2.2, 2.3. Federal Guidance Report No. 13, Part I Interim Version. Health Risks from Low-Level Environmental Exposure to Radionuclides. Radionuclide-Specific Lifetime Ragiogenic Cancer Risk Coefficients for the U.S. Population, Based on Age-Dependent Intake, Dosimetry, and Risk Models. Oak Ridge National Laboratory, Office of Radiation and Indoor Air, United States Environmental Protection Agency, Washington, DC. 20460. EPA 402-R-97-014. Jan, 1998.[1] Radiological Health Handbook. Ibid. p.336.
4. Radiological Health Handbook: US Department of Health, Education and Welfare. January, 1970.P.275

BIBLIOGRAPHY

1. Allen, Philip W. and Lester Machta, *Operation Buster: Transport of Radioactive Debris from Operations Buster and Jangle, Project 7.1.* Armed Forces Special Weapons Project, Washington, D.C. March 15, 1952, WT 308.
2. *Announced United States Nuclear Tests July 1945 Through December 1982.* NVO-209 (Rev 3) Office of Public Affairs, U.S. Department of Energy, Nevada Operations Office. in Cooperation with Los Alamos National Laboratory, Lawrence Livermore National Laboratory, Sandia National Laboratories. January, 1983.
3. Bogen, Kenneth T. and Abraham S. Goldin. *Population Exposure to External Natural Radiation Background in the United States*, ORP/SEPD-80-12. Surveillance and Emergency Preparedness Division, Office of Radiation Programs, U.S. Environmental Protection Agency, Washington, DC. 20460. April, 1981.
4. Breiman, Leo, Jerome Friedman, Richard Olshen, and Charles Stone. *Classification and Regression Trees.* Pacific Grove: Wadsworth, 1984.
5. Cotton, William R. and Richard A. Anthes, *Storm and Cloud Dynamics*, Vol 44 in International Geophysics Series. Academic Press, Inc. Harcourt Brace Jovanovich, Publishers. San Diego, CA. 1989.
6. Devesa, Susan S. , Dan J. Grauman, William J. Blot, Gene A. Pennello, Robert N. Hoover, Joseph F. Fraumeni Jr; *.Atlas of Cancer Mortality in the United States 1950-94.* National Institutes of Health, National Cancer Institute NIH Publication No. 99-4564, September, 1999.
7. Beck, Harold L., Irenke K. Helfer, Andre Bouville, and Mona Dreicer. "Estimates of Fallout in the Continental U.S. from Nevada Weapons Testing Based on Gummed-Film Monitoring Data" *Health Physics* Vol. 59. No. 5. (November) pp 565-576, 1990.
8. Bouville, Andre et al. *Estimated Exposures and Thyroid Doses Received by the American People from Iodine-I-131 in Fallout Following Nevada Atmospheric Nuclear Bomb Tests. A Report from the National Cancer Institute.* U.S. Department of Health and Human Services, National Institutes of Health, National Cancer Institute. October, 1997.
9. *Cancer Epidemiology and Prevention.*, Editors: Schottenfeld, David and Joseph F. Fraumeni, Jr... Second Edition.Oxford University Press. 1996.
10. Centers for Disease Control: http://wonder.cdc.gov/
11. Coleman, Hugh W. and W. Glenn Steele, *Experimentation and Uncertainty Analysis for Engineers.* 2nd Ed. John Wiley and Sons, Inc. New York. 1999.
12. Corcoran, Brent Coull and Aneesh Patel. *EGRET for Windows. User Manual.* Cytel Statistical Software, Cytel Software Corporation 765 Massachusetts Avenue, Cambridge, MA 02139. August 1999.
13. Defense Nuclear Agency 6001F-6008F.
14. *Defense Nuclear Agency Fact Sheet*: Subject: Teapot Series DNA 6010F-6013F.
15. *Operation Tumbler-Snapper.* Defense Nuclear Agency 6020F. "
16. Doviak, Richard J. and Dušan S. Zrnić. Doppler Radar and Weather Observations. 2nd Ed. Academic Press, Inc. Harcourt Brace Jovanovich, Publishers, San Diego. , 1993.
17. Eckerman, Keith F. et al. *Federal Guidance Report No. 13: Health Risks from Low-Level Environmental Exposure to Radionuclides. Part I. Interim Version.* EPA 402-R-97-014. Oak Ridge Natonal Laboratory, Office of Radiation and Indoor Air, United States Environmental Protection Agency, Washington, DC. 20460. Jan 1998.
18. Eckerman, Keith F. and Jeffrey C. Ryman. *Federal Guidance Report No. 12: "External Exposure to Radionuclides in Air, Water and Soil" Exposure-to-Dose Coefficients for General Application*, Based on the 1987 Federal Radiation Protection Guidance. Oak Ridge National Laboratry, Office of Radiation and Indoor Air. U.S. Environmenal Protection Agency, Washington DC. 20460. 1993.

19. *Effects on Populations of Exposure to Low Levels of Ionizing Radiation:* 1980. National Research Council, Washington DC. November 1980

20. Elwood, Mark, *Critical Appraisal of Epidemiological Studies and Clinical Trials* 2nd Ed. Oxford University Press, Oxford, UK, 1998.

21. Emsley, John, *The Elements*. 3rd Ed. Clarendon Press. Oxford UK, 1998.

22. *Exposure of the American People To Iodine-131 From Nevada Nuclear-Bomb Tests. Review of the National Cancer Institute Report and Public Health Implications.* Institute of Medicine., National Research Council. National Academy Press. Washington DC 1999.

23. Friedman J.H. *Multivariate Adaptive Regression Splines,* Annals of Statistics, 19, 1-141 (March). Salford Systems. San Diego, CA. 1999.

24. "Fallout From Nuclear Weapons Tests": *Hearings Before the Special Subcommittee on Radiation of the Joint Committee on Atomic Energy, Congress of the United States, 86th Congress, First Session, On the Fallout From Nuclear Weapons Tests.* Washington, DC. 1959.

25. Ferber, Gilbert J. et al. *Long-Range Transport and Diffusion Experiments*. AD-764 893. Advanced Research Project Agency. May, 1973.

26. Francis, Magnus B. *Toxic Substances in the Environment.* Wiley-Interscience. New York, 1994.

27. *General and Applied Toxicology*. Ed: Bryan Ballantine, Timothy Marrs, Paul Turner, Stockton Press, New York, NY., 1995.

28. Glasstone, Samuel and Philip J. Dolan, *The Effects of Nuclear Weapons*., U.S. Department of Energy and Department of Defense, 1977.

29. HASL-142, *Quarterly Summary Report*, Jan 1, 1964, United States Atomic Energy Commission, Health and Safety Laboratory, New York, NY. 1964.

30. *Health Impacts of Large Releases of Radionuclides.* Ciba Foundation Symposium 203. John Wiley & Sons, New York, NY, 1997.

31. Hicks, Harry G. *Results of Calculations of External Gamma-Radiation Exposure Rates from Fallout and the Related Radionuclide Compositions. Operation Ranger 1951..* Lawrence Livermore National Lab. CA. Jul 1981. UCRL-53152 Pt. 1.

32. Hicks, Harry G. *Results of Calculations of External Gamma-Radiation Exposure Rates from Fallout and the Related Radionuclide Compositions. Operation Buster-Jangle 1951.* Lawrence Livermore National Lab. CA. Jul 1981. UCRL-53152 Pt. 2

33. Hicks, Harry G. *Results of Calculations of External Gamma-Radiation Exposure Rates from Fallout and the Related Radionuclide Compositions. Operation Tumbler-Snapper 1952.* Lawrence Livermore National Lab. CA. Jul 1981. UCRL-53152 Pt. 3

34. Hicks, Harry G. *Results of Calculations of External Gamma-Radiation Exposure Rates from Fallout and the Related Radionuclide Compositions. Operation Upshot-Knothole 1953* Lawrence Livermore National Lab. CA. Jul 1981. UCRL-53152 Pt. 4

35. Hicks, Harry G. *Results of Calculations of External Gamma-Radiation Exposure Rates from Fallout and the Related Radionuclide Compositions. Operation Teapot 1955.* Lawrence Livermore National Lab. CA. Jul 1981. UCRL-53152 Pt. 5.

36. Hicks, Harry G. *Results of Calculations of External Gamma-Radiation Exposure Rates from Fallout and the Related Radionuclide Compositions. Operation Plumbbob 1957.* Lawrence Livermore National Lab. CA. Jul 1981. UCRL-53152 Pt. 6.

37. Hicks, Harry G. *Results of Calculations of External Gamma-Radiation Exposure Rates from Fallout and the Related Radionuclide Compositions. Operations Nougat through Bowline, 1962-1968* Lawrence Livermore National Lab. CA. Jul 1981. UCRL-53152 Pt. 8.

38. Hicks, Harry G. "Calculation of the Concentration of Any Radionuclide Deposited on the Ground by Offsite Fallout from a Nuclear Detonation," *Health Physics. Vol 42. No 5. (May), pp 585-600. 1982.*

39. Hicks, Harry G. "Additional Calculations of Radionuclide Production Following Nuclear Explosions and Pu Isotopic Ratios for Nevada Test Site Events. " *Health Physics.* Vol 59. No. 5 (November) pp 515-523. 1990.

40. Hirsch, Robert P. and Richard K. Riegelman: *Statistical First Aid: Interpretation of Health Research Data.* Blackwell Science. Cambridge, MA, 1992.

41. Houze, Robert A. Jr. *Cloud Dynamics.* Volume 53 in the International Geophysics Series. Academic Press, Inc. Harcourt Brace and Company. San Diego, CA, 1993.

42. *ICD 9 CM Code Book.* Volumes 1,2,3. 1998. St. Anthony Publishing, Inc. Reston VA. September 1997.

43. Kachigan, Sam Kash *Multivariate Statistical Analysis: A Conceptual Introduction.* 2nd Ed. Radius Press New York. 1991.

44. Keller, Edward A. *Environmental Geology*, 2nd Ed. Charles E. Merrill Publishing Co. Columbus, OH, 1979.

45. Knapp, Rebecca G. and M. Clinton Miller III. *Clinical Epidemiology and Biostatistics*, Williams and Wilkins. Baltimore MD, Philadelphia PA, 1992.

46. Krey, Philip W. , Merrill Heit and Kevin M. Miller. "Radioactive Fallout Reconstruction from Contemporary Measurements of Reservoir Sediments." *Health Physics* Vol. 59. No. 5. (November) pp 541-554, 1990.

47. Larson, Kermit H., Howard A. Hawthorne and Keith R. Price. *Preliminary Report on Rain Out from Shots Diablo and Stokes in the Belle Fource-Rapid City, South Dakota Area. Program 37.* August 12-August 17, 1957. Date of Report: Aug 27, 1957.

48. Lippman, Morton and Richard B. Schlesinger, *Chemical Contamination in the Human Environment.* Oxford University Press. New York, 1979.

49. List, Robert J. *Radioactive Debris from Operations Tumbler and Snapper," NYO-4512.* U.S. Weather Bureau, February 25, 1953.

50. List, Robert J. *The Transport of Atomic Debris from Operation Upshot-Knothole" NYO-4602.* United States Atomic Energy Commission, Technical Information Services, Oak Ridge, TN. Weather Bureau, Washington, DC. Jun 25, 1954.

51. List, Robert J. *Radioactive Fallout in North America from Operation Teapot, NYO-4696*, U.S. Atomic Energy Commission, Washington, DC. February, 1956.

52. Miller, Richard L. *Under The Cloud: The Decades of Nuclear Testing* Macmillan. New York, 1986; Two-Sixty Press 1998.

53. Mason, Thomas J., Frank W. McKay, Robert Hoover, William J. Blot, Joseph F. Fraumeni Jr. *Atlas of Cancer Mortality for U.S. Counties: 1950-1969.* DHEW Publication No. (NIH) 75-780. . Epidemiology Branch, National Cancer Institute. 1975.

54. Mason, Thomas J. , Wilson B. Riggan, and John F. Acquavella *U.S. Cancer Mortality Rates and Trends 1950-1979* Vol I, II, III. NCI/EPA Interagency Agreement on Environmental Carcinogenesis. EPA-600/ 1-83-015b, September, 1983.

55. *Microsoft Excel 97 Worksheet Function Reference.* Microsoft Press. 1997.

56. Morganthau, Manfred; Harvey Meieran, Richard Showers, Jeffrey Morse, Norman Dombeck, Arnoldo Garcia., *Local Fallout from Nuclear Test Detonations*, and *Compilation of Fallout Patterns and Related Test Data.* DASA 1251 Vols 1 and Vol 2.. U.S. Army Nuclear Defense Laboratory., Defense Atomic Support Agency. Undated.

57. NOAA Technical Memorandum ERL ARL-36: March 1972.

58. Norman, Geoffrey G. and David L. Streiner, *Biostatistics: The Bare Essentials.* B.C. Decker, Inc. Hamilton, Ontario, Canada. 1998.

59. *Occupational, Industrial, and Environmental Toxicology.* Hamilton, Richard J MD, Ed. Scott D Phillips MD ,Ed. Michael I Greenberg MD, MPH Ed. in Chief: Mosby Publishing Co. St. Louis. 1997.

60. Page, Randy M. ; Galen E. Cole and Thomas C. Timmreck. *Basic Epidemiological Methods and Biostatistics.,* Jones and Bartlett Publishers, Sudbury, MA. 1998.

61. Petrie, Aviva and Caroline Sabin, *Medical Statistics at a Glance*. Blackwell Science, Oxford, UK. 2000.

62. Proctor, Robert N. *Cancer Wars: How Politics Shapes What We Know and Don't Know About Cancer*. Perseus Books LLC. New York. 1995.

63. *Radioactive Debris from Operations Upshot and Knothole NYO-4552* . Health and Safety Laboratory, New York Operations Office. Jun 25, 1954.

64. *Radiological Effluents Released From U.S. Continental Tests 1961 Through 1992*. United States Department of Energy Nevada Operations Office.DOE/NV-3117 (Rev. 1) UC-702. August 1996.

65. *Radiation Protection Theory and Practice* Proceedings of the 25th Anniversary Symposium of the Society for Radiological Protection, Malvern 4-9. Edited by E.P. Goldfinch. IOP Publishing Ltd. Bristol, UK.1989.

66. Roman, Steven, *Writing Excel Macros*, O'Reilly Publications, Cambridge, MA. May, 1999.

67. *Safe Handling of Radionuclides*, 1973 Edition. International Atomic Energy Agency, Vienna, 1973.

68. Snedecor, George W. and William G. Cochran. *Statistical Methods*. Iowa State University Press, Ames IA 1989.

69. *Spatial Epidemiology Methods and Applications*. , Ed: P. Elliott, J.C. Wakefield, N.G. Best and D.J. Briggs. Oxford University Press. Oxford, UK. 2000.

70. Steinberg, Dan and Phillip Colla. *CART: A Tree-Structured Non-Parametric Data Analysis*. San Diego, CA. Salford Systems, 1995.

71. Szklo, Moyses and E. Javier Nieto. *Epidemiology: Beyond the Basics*. Aspen, Gaithersburg, MD 2000.

72. *Thunderstorm Morphology and Dynamics*: Vol II of *Thunderstorms: A Social, Scientific and Technological Documentary*. 2nd Ed. Edited by Edwin Kessler, University of Oklahoma Press, Norman, OK 1983.

73. *Toxicological Profile for Ionizing Radiation. Draft for Public Comment*. U.S. Department of health and Human Services, Public Health Service, Agency for Toxic Substances and Disease Registry,. September, 1997.

74. *Transport, Deposition and Meteorological Studies of Nuclear Debris in the Atmosphere*. R.J. List (Ed). T.W. Ashenfelter, G.J. Ferber, J.L. Heffter, L. Machta, A.D. Taylor and K. Telegadas. AEC, Division of Biology and Medicine Fallout Studies Branch, Washington, DC.

75. *United States Nuclear Tests July 1945 through September 1992*. Department of Energy, DOE/NV-209 (Rev 14) December 1994.

76. U.S. *Cancer Mortality Rates and Trends 1950-1969* U.S. Environmental Protection Agency and National Cancer Institute William B. Riggan and Thomas J. Mason Eds. EPA-600/1-83-015a. Sep 1983.

77. *U.S. Cancer Atlas*. National Cancer Institute. 2000. http://www2.nci.nih.gov

78. Yu C. et al. . *Manual for Implementing Residual Radioactive Material Guidelines Using RESRAD, Version 5.0*. Working Draft for Comment. Environmental Assessment Division, Argonne National Laboratory. Operated by The University of Chicago under Contract W-31-109-Eng-38 for the United States Department of Energy. September, 1993.

79. Webb, Jeff: *Using Excel Visual Basic* 2nd Edition. Que. Corporation. Indianapolis, IN. 1996.

80. *What Risk? Science, Politics and Public Health*. Ed. Roger Bate. Butterworth Heinemann. Oxford, UK. 1997.

81. Wilhelmson, Robert. et al. *Study of a Numerically Modeled Severe Storm*. PATHFINDER Sampler, COMMAS Images and Multimedia Teaching Modules. National Center for Supercomputing Applications, Champaign IL. 1988.

About the Author

Richard L. Miller is an industrial health specialist with field experience in onsite coordination of more than 500 safety and health investigations primarily in petrochemical and energy-related industries. His major areas of focus have included retrospective event and exposure reconstruction, and event cluster evaluation. In 1979 Miller directed epidemiological field investigations on the relationship between rare brain cancers and chemical workplace exposures, and for this work received an OSHA Special Incentive Award and letter of commendation from the director of NIOSH for "findings of national impact and importance."

Miller has testified as an expert in federal cases involving toxic exposure and has performed onsite risk evaluations of industrial sites in such diverse places as Mexico, Soviet Latvia, Estonia, the Russian Federation (Tataria) and the Arabian Gulf. He has also performed extensive research on the nuclear test program and is the author of the 1986 book *Under The Cloud: the Decades of Nuclear Testing*, chosen by the *Library Journal* as one of the top science-technical books of 1986. Miller is a member of the American Statistical Association and is currently completing a technical book on calculating industrial exposures.

INDEX

INDEX

INDEX

INDEX

INDEX

INDEX

INDEX

INDEX

INDEX

INDEX

INDEX

INDEX

INDEX

INDEX

INDEX

INDEX

INDEX

INDEX

INDEX

INDEX

INDEX

INDEX

INDEX

INDEX

INDEX

INDEX

INDEX

INDEX

INDEX

INDEX

INDEX

INDEX

INDEX

INDEX

INDEX

INDEX

INDEX

INDEX

INDEX

INDEX

INDEX

INDEX

INDEX

INDEX

INDEX

INDEX

INDEX

INDEX

INDEX

INDEX

www.ingramcontent.com/pod-product-compliance
Lightning Source LLC
LaVergne TN
LVHW052207200926
848481LV00001B/49